GALEN
METHOD OF MEDICINE
III

LCL 518

GALEN

METHOD OF MEDICINE

BOOKS 10–14

EDITED AND TRANSLATED BY

IAN JOHNSTON

AND

G. H. R. HORSLEY

HARVARD UNIVERSITY PRESS
CAMBRIDGE, MASSACHUSETTS
LONDON, ENGLAND
2011

First published 2011

LOEB CLASSICAL LIBRARY® is a registered trademark
of the President and Fellows of Harvard College

Library of Congress Control Number 2011921281
CIP data available from the Library of Congress

ISBN 978-0-674-99680-9

*Composed in ZephGreek and ZephText by
Technologies 'N Typography, Merrimac, Massachusetts.
Printed on acid-free paper and bound by
The Maple-Vail Book Manufacturing Group*

CONTENTS

MANUSCRIPTS

Kühn (= K) vol. X has provided the base text for ours printed here.

The Latin text printed in K (abbreviated by us as KLat in this volume) was produced three hundred years before K was printed, and so is not a translation, or a correction of K's Greek text. It clearly draws on other MSS than those used by K, reflecting different readings in the Greek. This may sometimes alert us to a problem in K; but generally we have not privileged the Latin translation over K's Greek text.

For a list of MSS of the *MM,* see Diels, pp. 91–92. The following manuscripts are referred to in our textual notes with the abbreviation listed.

B—British Library MS Add. 6898 (London; 12th cent.)[1]

P1—Parisinus Gr. 2160 (Paris; 14th cent.)

P2—Parisinus Gr. 2171 (Paris; 15th cent.)[2]

[1] We are not attempting to provide here a full collation of B against K. B exhibits many other differences, e.g., in word order and in orthography; and it is interesting for other reasons as well. But we have very rarely noted these.

[2] We have not directly consulted the two Paris MSS. We have made use of some of its readings for Books 1 and 2 as they are reported by R. J. Hankinson (1991), App. 1, pp. 235–37.

Boulogne (2009), pp. 11, 31, draws upon these three MSS, and also upon three others as well as two fragmentary MSS for his translation. We have included no references to these other five in our textual notes.

ABBREVIATIONS

Ce   Celsus. *De Medicina*. Translated by W. G.
      Spencer. Loeb Classical Library. 3 vols. Cam-
      bridge, MA: Harvard University Press, 1935–1938.
CMG   Corpus Medicorum Graecorum
Cu   Nicholas Culpepper. *The English Physician En-
      larged (Culpepper's Herbal)*. London: Folio So-
      ciety, 2007 [1653].
D   Dioscorides. *The Greek Herbal of Dioscorides*,
      translated by John Goodyer [1653]. Edited by
      R. T. Gunther. New York: Hafner, 1968 [1933].
EANS *The Encyclopedia of Ancient Natural Scientists.*
      Edited by P. T. Keyser and G. L. Irby-Massie.
      London: Routledge, 2008.
G   Galen. References to the *MM* are indicated by
      the Kühn page number; references to other Ga-
      lenic works are indicated by the Kühn volume
      and page numbers. His three major pharmaco-
      logical treatises are *De simplicium medicamen-
      torum temperamentis et facultatibus*, XI.379–
      892K and XII.1–377K; *De compositione medica-
      mentorum secundum locos*, XII.378–1007K and
      XIII.1–361K; *De compositione medicamentorum
      per genera*, XIII.362–1058K.

ABBREVIATIONS

Gr   M. D. Grmek. *Diseases in the Ancient Greek World*. Baltimore, MD: Johns Hopkins University Press, 1991.

L&S   C. T. Lewis and C. Short. *A Latin Dictionary*. Oxford: Clarendon, 1993 [1879].

LCL   Loeb Classical Library.

LSJ   H. G. Liddell, R. Scott, and H. Stuart Jones. *A Greek-English Lexicon*. 9th ed. (1940), with revised suppl. by P. G. W. Glare. Oxford: Clarendon, 1996.

M   C. C. Mettler. *The History of Medicine*. Philadelphia: Blakiston, 1947.

OCD   *Oxford Classical Dictionary*. Edited by S. Hornblower and A. Spawforth. 3rd ed. Oxford: Clarendon, 1996.

OED   *Oxford English Dictionary*. 12 vols. Oxford: Oxford University Press, 1978 [1933].

S   *Stedman's Medical Dictionary*. 27th ed. Baltimore: Lippincott, Williams and Wilkins, 2000.

Si   R. E. Siegel. *Galen on the Affected Parts*. Basel: S. Karger, 1976.

T   Theophrastus. *Enquiry into Plants*. Translated by A. Hort. Loeb Classical Library. 2 vols. Cambridge, MA: Harvard University Press, 1916, 1926.

SYNOPSIS OF CHAPTERS

BOOK X

1. Galen offers a general treatment strategy for fevers. There are three primary factors to be considered—the magnitude of the fever, its cause, and the patient's capacity. Which of these is given primary importance depends on the circumstances. If the patient's capacity is strong and the cures of the fever itself and its cause coincide, there is no problem. If, however, there are conflicting indications, judgment must be exercised.

2. Consideration is given to ephemeral fevers and the causes of these becoming longer in duration (oligohemeral or polyhemeral). Much will depend on whether the cause of the fever remains, and this in turn will depend, at least in part, on the skill of the doctor. Certain causes, like heatstroke, are by their nature transient. However, when there is significant blockage of the skin pores adversely affecting transpiration, the cause is likely to remain. The patient's capacity, of little importance as an indicator in the ephemeral fevers, becomes significant in the chronic fevers.

3. Galen presents a very detailed case report about a young man with a prolonged fever. The focus is on the patient's capacity and the timing of nourishment for him. In a vivid description of medical brinkmanship, Galen tri-

umphs over the "diatritarians," escapes the wrath of the relatives, and restores the patient to health.

4. Fasting is recognized as a cause of fever. The generation of shivering and rigors is attributed to "gnawing" vapors and humors which are not properly dispersed because the pores in the skin are blocked. The role of a strong capacity in resisting the effects of the "gnawing" humors is described. The relationship of the severity of the shivering and rigors to the amount of these humors is considered. Brief reference is made to other illustrative cases. The importance of providing nourishment and effecting moistness is stressed.

5. Fasting is seen as very bad for the hot and dry states and as a cause of fever in these in particular. Causes of such states, both innate and acquired, are listed. Fasting is considered harmful for hot, dry bodies that are healthy but deadly in those with fever in a hot, dry summer. Typically these are hectic fevers. A case report is given in which Galen is merely an observer—others are responsible for treatment. There is another case report of a patient for whom Galen is the treating doctor. Details of the treatment are given. The merits or otherwise of nourishment given under different circumstances are discussed. The indications for nourishment according to Hippocrates are listed. The importance of the capacity is again stressed.

6. Galen considers the situation of a hectic fever in a person with a strong capacity and what the "diatritarians" would do (wrongheadedly). He then states what should be done. Next he considers what is to be done if inflammation or putrefaction of humors occurs. When fever arises in a dry body without putrefaction, the mainstays of treatment are seen as moistening and cooling remedies.

7. A short summary of these moistening and cooling remedies is given. They will be the subjects of the subsequent chapters.

8. The first is the surrounding (ambient) air. Basically, you have to make do with what there is, although house placement and design do allow some control over the ambient air. Galen considers how the different organs are differently affected by cool air. There may be conflicting effects on the internal organs and the skin beneficial to the former and detrimental to the latter. Some variations in respiratory pattern are described.

9. Next to be considered are the cooling medications applied externally. Obviously these must be cooling, but they must also be free of astringency as far as possible because of the adverse effects of this on the skin. Various specific medications are described as examples with brief reference to two cases.

10. This is a detailed discussion about baths and bathing including reference to Philippus' opposition to bathing when there is wasting. There follows a digression on the particular difficulties of treating multiple, coexisting conditions. Bathing is identified as having a twofold potency, being effective both primarily and accidentally. There is a further digression, this time on difference in general. Shivering and rigors are considered. The chapter ends with a precise account of the sequence of bathing.

11. Galen discusses moist and cold nutriments for hectic fevers. Mention is made of specific remedies. Details of the method of administration of asses' milk are given. Hectic fevers, being hotter, are seen as requiring colder remedies and greater precision in their use.

BOOK XI

1. Galen introduces the subject of the book—the treatment of continuous fevers caused by the putrefaction of humors. He provides a general review of therapeutic indicators and their basis. Opposition is fundamental to treatment.

2. Treatment based on opposition dictates cooling in the case of ephemeral and hectic fevers. In fevers due to putrefying humors there is also a prophylactic component —eradication of the cause of putrefaction.

3. Galen formulates a general statement of the theory behind the method of arriving at the indicators of the type of treatment to be used.

4. Putrefaction which causes fever may occur in all the vessels, in the large vessels only, or in one particular part. The different forms of putrefaction should be recognized and the causes identified. Prevention or eradication of causative factors is of great importance in treating fevers.

5. To determine the quantity and quality of the nutriments to be provided it is necessary, Galen claims, to evaluate the patient's capacity. He identifies two other determinants: the condition associated with the disease and the bodily *krasis*.

6. Some of the differences between the different kinds of fevers—ephemeral, polyhemeral due to putrefaction of humors, and continuous—are listed. Variations in treatment are seen as depending on these differences.

7. Galen gives a systematic account of the indications (*endeixeis*) on which the treatment of fevers due to putrefaction is based.

8. The causes of putrefaction in general, occurring in both living and nonliving bodies, are identified. Factors that prevent normal transpiration and how to correct them are considered in general terms.

9. The management of continuous fevers due to putrefaction in the whole body, including the problem of conflicting indicators, is detailed.

10. The indicators or aims in curing the fevers due to putrefaction are analyzed and related to the issue of sequence in utilizing the various remedies. If possible, remedies that are heating should be avoided; they will tend to increase both putrefaction and fever.

11. The problem of a significant intervening symptom complicating the sequence of treatment is examined. Various symptoms of this kind are listed and some treatment options considered.

12. Galen considers the Hippocratic dictum that opposites cure opposites, with a digression into terminological issues involving reference not only to Hippocrates but also to Plato and Aristotle.

13. The therapeutic indicators in continuous fevers due to putrefaction of humors are considered in terms of the basic classes of those things contrary to nature—causes, diseases, and symptoms. Indications are also taken from what accords with nature—things such as age, climatic conditions, etc. All must be taken into account in planning the sequence of treatment.

14. The use of phlebotomy in fevers due to putrefaction of humors is examined. A key element is seen to be a strong capacity in the patient. Age and other, external factors must also be taken into account. The question of nourishment is considered.

15. The use of poultices and fomentations applied to the hypochondrium and the dangers of such treatments are described. Galen launches a further attack on the Methodics, this time for their claim to be able to teach the medical art in six months. The dangers of overtreatment, comparing the rich to the poor and slaves, are recognized. The uses of relaxing (loosening) and repulsive remedies are described. Consideration is given to the special features of the treatment of the liver and stomach.

16. There is continuing consideration of the methods of treatment of the liver and stomach when these organs are inflamed. Various agents are discussed. A case report is given. Treatment of other organs and structures, particularly the spleen, is outlined.

17. Galen states that astringents (binding agents) must be used less in the chest because they endanger the heart and lungs. Strength must be preserved here by nourishment, he claims. The use of the cupping glass is described.

18. Treatment of inflammation of the brain and meninges is detailed.

19. Galen discusses the importance of maintaining the patency of the proper channels for the excretion of superfluities, which are detailed for the various parts of the body.

20. The benefits and harms of poultices and baths in the treatment of fevers are examined.

21. The timing of treatment and, in particular, nourishment in both intermittent and continuous fevers is outlined. The key element is identified as close observation of the timing and nature of the paroxysms.

BOOK XII

1. Galen considers the types of symptoms that occur in fevers and the issues surrounding their treatment—in particular, the question of symptomatic versus definitive treatment. The importance of the patient's capacity is stressed, and some paregoric medications are discussed.

2. A brief statement is made on measures which cure a condition also preventing symptoms. There is mention of several specific measures, e.g. phlebotomy.

3. The management of patients who are febrile with an abundance of crude humors is described. Clinical features are outlined, including the risk of syncope. The various therapies applicable under differing circumstances are considered.

4. There is discussion of the relative merits of a number of wines available in either Europe or Asia for the treatment of patients who are adversely affected by yellow bile involving the cardiac orifice of the stomach with possible syncope. The wines are considered in terms of color, age, astringency, flavor, and other qualities. Their tendency, or otherwise, to cause headache is detailed, as is their effect on the distribution of nutriments.

5. Syncope is identified as a sudden collapse of capacity (*dunamis*). The substance of the capacity is said to lie in the *pneuma* and the *krasis* of the solid bodies. These are the two things that must be maintained in health and restored in disease. Galen considers the three forms of *pneuma* (psychical, vital, and physical) and *eukrasia* of solid bodies—what disturbs these and how they might be restored in disease.

6. The treatment of patients in whom the humors are excessively thinned and who are at risk of syncope is detailed. In particular, this depends on the strength or weakness of the capacity. The significance of the Hippocratic *facies* is considered. Some specific aspects of nourishment—its rate, the materials to be used, and its temporal relationship to abatements and paroxysms—are detailed.

7. Galen lists five causes of syncope: severe pain, insomnia, excessive evacuation of the stomach, movement in those with delirium, and *dyskrasia* of the "principles." He details the causes of severe pain: dissolution of continuity and sudden change when it involves the parts capable of pain sensation. Strong heating and cooling may produce pain by either mechanism. The chapter concludes with three relatively detailed case reports.

8. The conditions that bring about pain are outlined and their treatment discussed. Regarding causes, a distinction is made between thick, viscid humors, thin, bitter humors, and vaporous *pneuma*. Different structures are considered—gastrointestinal tract, eyes, ears, teeth—and various measures described. This concludes the discussion of symptoms associated with fevers.

BOOK XIII

1. Galen considers abnormal swellings (tumors), beginning with inflammatory swellings, which are the most common. The general signs and symptoms of these are given.

2. The underlying cause of all inflammations is identified as an influx of blood. The theories of the different schools are summarized. The distinction between prophylaxis and therapy is made in relation to inflammation.

3. The increased flow of blood into a part that becomes inflamed is due either to some other part or parts sending it and the inflamed part receiving it, or to the inflamed part drawing the blood to itself. The role of the separative or excretory capacity in this process is examined.

4. The types of abnormal swelling in general (i.e., the differentiae of the genus *onkos*) are said to depend on the predominant nature of the inflow: blood, *pneuma*, yellow bile, black bile, phlegm, or an unspecified thick, viscid humor.

5. The question of why glandular (lymph node) swellings develop, either with or without wounds or ulcers, is considered. Galen details how to treat glandular swellings, including the use of incisions for the drainage of infected glands, and their management.

6. The treatment of inflammation itself is considered. The key components are identified as evacuation and prevention of the inflow of blood. Recognition of the causes of the latter is deemed important. The methods involved are listed, as is the sequence of treatment in inflammation. The treatments of *dyskrasia* and *plethora* are outlined. The treatment of poisonous bites is described.

7. The distinction is made between indications from *homoiomerous* and from organic parts in inflammation pertaining to evacuation.

8. The roles of heating and cooling agents in the treatment of inflammation are considered.

9. Galen presents a breakdown of the components of the key process of the evacuation of blood from the inflamed part.

10. Certain indications are identified which are common to both *homoiomerous* and organic parts.

4. Edema, the next unnatural swelling considered, is attributed to a flux of phlegmatic humor. Galen identifies other causes of edema in which it is a symptom. The treatment of edema, including the use of oxyrrhodinum, is outlined. The pathogenesis of "stones" in joints and renal calculi is considered.

5. Some treatments of swellings of different parts, including tendons and ligaments that have become scirrhous (indurated, hardened), are described. The uses of emollients and vinegar-based medications are outlined.

6. Galen makes a brief digression on nomenclature in relation to scirrhous and edematous swellings.

7. Unnatural swellings due to flatulent *pneuma* and their differentiation from the edemas, including those occurring in muscles due to a blow and those that are chronic, are described. Details of various medications are given. The treatment of priapism, also a swelling due to inflation by *pneuma*, is detailed and exemplified by two brief case reports.

8. Swelling of the tongue is described and illustrated by a case report.

9. The significance of the color of various swellings is considered, as are issues of nomenclature in swellings. The characteristics of another kind of swelling termed "cancer" are detailed and its causation attributed to black bile.

10. Galen describes an affection that is due to a thick and seething humor producing multiple pustules (*anthrakes*). Treatment with phlebotomy and cooling agents is outlined. Methods of treatment of the lesions themselves are listed.

11. Glandular (scrofulous) swellings are described. Is-

sues of nomenclature are again dealt with. The treatment of these swellings by surgery or medications is outlined.

12. Other inflammatory swellings including abscesses and the variable content of these swellings are described. The nature and treatment of the three abnormal swellings termed atheroma, meliceris, and steatoma are outlined.

13. A general description is given of surgical removal of abnormal swellings, including some of those already mentioned and others—for example, warts, bladder stones, cataracts, fibroids, hydroceles, ascites. The contrasting and complementary roles of surgery and medications are considered.

14. A discussion of what is "proper" in terms of number and magnitude and the terminology relating to this is presented. The examples of supernumerary and congenitally absent digits are considered.

15. The general principles of treatment of excesses and deficiencies in terms of number and magnitude are outlined. The management of obesity is discussed and illustrated with a case report.

16. Galen discusses restoration of wasting or atrophy, both general and in a specific part. The treatment of a deficiency of skin on the penis (*leipodermos*) is considered. The treatment of other skin deficiencies termed *colobomata* is described.

17. Various other diseases in which there is excess or deficiency of what accords with nature, including "spontaneous" ulcers, *phagedaenae*, and herpes (herpetic ulceration), are described. A case report of the last is given. The role of *kakochymia* in causation is identified and its management described. Various (unusual) treatments of warts, both peduculated and sessile, are described.

18. The methods of generation of something that is lacking, including hair in alopecia, *ptilosis*, and baldness, are outlined.

19. The treatments of unnatural swellings in the eyes—pterygium, chalazion, cataract, and *hypopyon*—are detailed. The treatment of gastrointestinal worm infestations is described.

ΓΑΛΗΝΟΥ ΘΕΡΑΠΕΥΤΙΚΗΣ
ΜΕΘΟΔΟΥ

METHOD OF MEDICINE

ΒΙΒΛΙΟΝ Κ

1. Ἐπὶ μὲν δὴ τῶν ἐφημέρων πυρετῶν ἁπλῆ τίς ἐστιν ἡ παρὰ φύσιν ἐν τῷ σώματι διάθεσις, ὡς ἂν τῶν πρῶτον ἐργασαμένων αἰτίων αὐτοὺς οὐκ ἔτ' ὄντων. ἐφ' ὧν δ' ἐστὶν ἔτι τὸ ποιοῦν αἴτιον, οὐκ ἐπὶ τούτων οὔθ' ἡ διάθεσις οὔθ' ἡ ἴασις ἁπλῆ. χρὴ γὰρ τὸ μὲν ἤδη γεγονὸς τοῦ πυρετοῦ λύειν ἐκ τῆς οἰκείας ἐνδείξεως αὐτοῦ, τὸ δὲ γεννησόμενον ἀποτρέπειν τε καὶ κωλύειν γενέσθαι· κωλυθήσεται δὲ τῆς ποιούσης αἰτίας αὐτὸ τελέως ἐκκοπείσης. ὥστ' ἐν μὲν τοῖς ἐφημέροις ἡ τῶν πρακτέων ἔνδειξις ἐκ τοῦ πυρετοῦ μάλισθ' | ἡμῖν ἔσται, συνεπισκοπουμένοις αὐτῷ φύσιν καὶ ἡλικίαν καὶ ἔθος, ὥραν τε καὶ κατάστασιν καὶ χώραν καὶ δύναμιν. ἐν δὲ τοῖς ἄλλοις ὅσοι τὴν ἀνάπτουσαν αὐτοὺς αἰτίαν ἔχουσιν ἔνδον ὁ πρῶτος σκοπὸς τῆς ἐνδείξεως ἀπὸ τῆς αἰτίας ἐστίν.

εἴπερ οὖν ὀρθῶς ἡμῖν ἐδείχθη σῆψις χυμῶν αἰτία τῶν πολυημέρων εἶναι πυρετῶν, ταύτην ἰάσασθαι χρὴ πρώτην, οὐκ ἀμελοῦντας οὐδὲ τοῦ σβεννύναι τὸν ἐξ αὐτῆς ἀνῆφθαι φθάσαντα πυρετόν. εἰ μὲν οὖν εἰς ταὐτὸ συμβαίνοι τά τε τοῦ πυρετοῦ σβεστήρια καὶ τὰ τῆς αἰτίας ἀλεξητήρια, τὸ δηλούμενον ὑπ' ἀμφοῖν ἓν ὂν ἑτοίμως πρακτέον. εἰ δὲ ἡ τῶν ἤδη γεγονότων

2

BOOK X

1. In the ephemeral fevers what is abnormal in the body is 661K the condition pure and simple since the causes initially bringing the fevers about are not still in existence. But in instances where the effecting cause does still exist, neither the condition nor the cure is simple. It is necessary, on the one hand, to resolve what has already occurred of the fever by means of its specific indication and, on the other hand, to turn aside what will be produced and prevent it from occurring. It will be prevented if the cause creating it is completely eradicated. Consequently, in the ephemeral fevers, our indication of what is to be done will be from the fever above all, while we consider in conjunction with this na- 662K ture, age, custom, season, climatic conditions, place and capacity. In the other fevers, which have within themselves the cause that kindles them, the primary object of the indication is from the cause.

Therefore, if we correctly identify putrefaction of humors as the cause of polyhemeral (chronic) fevers, it is this which must be cured first, but not to the neglect of quenching of the fever kindled beforehand by it. If those things that quench the fever and the remedies of the cause come together in the same thing, what you must promptly do is the one thing indicated by both. If, however, the cure of

πυρετῶν ἴασις αὐξάνει τὴν ἀνάπτουσαν αὐτοὺς αἰτίαν, ἢ τὰ τὴν αἰτίαν ἐκκόπτοντα παροξύνει τοὺς
πυρετούς, ἐπισκεπτέον ἀκριβῶς ὑπὲρ τῆς τῶν βοηθημάτων ἰδέας. ἐπίσκεψις δ᾽ ἀκριβὴς ἄν σοι γένηται
διελομένῳ τὴν ὅλην σκέψιν εἰς τρεῖς τούτους σκοπούς,
ἆρά γε τὸν πυρετόν ἐστι μόνον ἰατέον ἀμελοῦντα τῆς
αἰτίας, ἢ τὴν αἰτίαν ἐκκοπτέον οὐδὲν φροντίζοντα τοῦ
πυρετοῦ, ἢ τὸ μὲν πλεῖστον τῶν βοηθημάτων ὡς πρὸς
663K θάτερον αὐτῶν ἐστι ποιητέον, | οὐκ ἀμελητέον δ᾽ οὐδὲ
τοῦ λοιποῦ.

  τὸ μὲν οὖν πρῶτον τῶν ῥηθέντων οὐκ ἐπαινέσομεν·
εἰ γὰρ οἷς ἂν ἰώμεθα βοηθήμασι τὸ γενόμενον[1] ἀεὶ
τοῦ πυρετοῦ, ταῦθ᾽ ἡμῖν αὐξήσει τὴν ποιοῦσαν αὐτὸν[2]
αἰτίαν, οὔτ᾽ ἀναιρήσομέν ποτε τὴν γένεσιν αὐτοῦ[3] οὔτε
τὸ μέγεθος καθαιρήσομεν· ἥ τε[4] γὰρ γένεσις ἀκολου-
θήσει τῇ τῆς ποιούσης αἰτίας ἰδέᾳ τό τε μέγεθος
ἐκείνῃ συναυξηθήσεται. τὸ δὲ δεύτερον τῶν ῥηθέντων
εἰ διορισώμεθα, γένοιτ᾽ ἂν ἡμῖν χρηστόν. εἰ μὲν γὰρ
ἀφόρητον εἴη τῷ κάμνοντι τὸ τοῦ πυρετοῦ μέγεθος, οὐ
χρὴ δι᾽ ὧν τὴν αἰτίαν ἐκκόπτομεν αὐξάνειν τὸν πυρετόν. ἅμα γὰρ, ὡς ἔοικεν, ἀναιρήσομεν ἀμφοτέρους,
τόν τε πυρετὸν καὶ τὸν ἄνθρωπον. εἰ δ᾽ οὕτως εἴη
μέτριος ὁ πυρετὸς ὥστε μὴ προκαταλῦσαι τὴν δύναμιν τοῦ νοσοῦντος ἐν ᾧ χρόνῳ πρὸς τὴν αἰτίαν ἐνιστάμεθα μόνην, οὐκ ἀπόβλητος ὁ τοιοῦτος ἂν εἴη τρόπος.
οὐκ ἀπόβλητος δὲ οὐδὲ ὁ τρίτος, ἀνθίστασθαι μὲν
κελεύων τῷ μείζονα τὴν ἀξίαν ἔχοντι, μὴ μέντοι μηδὲ
θατέρου παντάπασιν ἀμελεῖν. ὡς τὰ πολλὰ μὲν οὖν ἡ

4

the fever that has already occurred increases the cause kindling the fever, or those things that eradicate the cause aggravate the fever, you must give precise consideration to the kind of remedies [to be used]. Your accurate evaluation depends on your dividing the overall consideration into these three objectives: whether you must cure the fever alone neglecting the cause; or whether you must eradicate the cause without giving thought to the fever; or whether you must apply the greatest number of remedies to one of these without neglecting the other.

663K

I shall not recommend the first of the stated options; for if we always cure the fever that is occuring with those remedies that will increase the cause producing it, we shall never remove its genesis nor reduce its magnitude. For the genesis will follow the kind of effecting cause and be increased in magnitude. If I were to define the second of the stated options, it would be useful to us. If the magnitude of the fever is intolerable to the patient, we should not increase the fever through those things by which we eradicate the cause because, at one and the same time, it seems we shall destroy both—the fever and the person. But if the fever is moderate such that it does not break up the capacity of the patient during the time we are opposing the cause alone, such a method is not to be cast aside. Nor is the third [option] to be cast aside when it gives direction to oppose against what has the greater importance without neglecting the other component altogether. Generally speaking, the cause has the greater force. How-

---

αἰτία μείζονα τὴν ἰσχὺν ἔχει. γένοιτο δ᾽ ἄν ποτε
664K τηλικοῦτον⁵ τοῦ πυρετοῦ τὸ | μέγεθος ὡς ὑπόγυιον
ἐπιφέρειν τὸν ὄλεθρον, εἰ μή τις αὐτὸ κωλύσειεν·
ἥνικα γε χρὴ τοῦτ᾽ ἐκκόψαντας πρότερον, ἐπὶ τὴν τῆς
αἰτίας ἀναίρεσιν ἰέναι.

τωυτὶ μὲν οὖν, ὦ Εὐγενιανέ, διωρίσθω σοι κατ᾽
ἀρχὰς εὐθέως ὑπὲρ τῶν ἐναντίων ἐνδείξεων. ἐπεὶ δ᾽ ἐν
τῷ ταύτας ἐξετάζειν οὐ σμικρὰν μοῖραν εἰς τὸ σωθῆναι τὸν ἄνθρωπον ἡ δύναμις ἔχειν ἐφάνη, γίγνοιτ᾽ ἄν
σοι καὶ ἥδε σκοπὸς οὐ σμικρὸς εἰς τὴν τῶν πρακτέων
εὕρεσιν. ὥστε εἶναι τοὺς πρώτους τρεῖς σκοποὺς ἐν-
δεικτικοὺς τῶν ποιητέων ἐν τοῖς ὑποκειμένοις τῇ σκέ-
ψει πυρετοῖς· ἕνα μὲν οὖν αὐτὸν τὸν πυρετόν, ἕτερον δὲ
τὴν γεννῶσαν αὐτὸν αἰτίαν, καὶ τρίτον τὴν δύναμιν.
ἀλλὰ τοὺς μὲν δύο πρώτους σκοποὺς ἀναιρεῖν προσήκει, φυλάττειν δὲ τὸν τρίτον. ἔσται δή σοι κἀνθάδε
πάλιν ἡ αὐτὴ σκέψις ἣν μικρὸν ἔμπροσθεν ἐσκέψω,
παραβάλλοντι τῇ δυνάμει τόν τε πυρετὸν ἐν μέρει καὶ
τὴν αἰτίαν, εἶτα εἰ μὲν τῶν αὐτῶν ἄμφω δέοιντο
βοηθημάτων, ἑτοίμως λαμβάνοντι, μαχομένων δὲ τῶν
ἐνδείξεων, ἤτοι τὴν ἀπὸ τῆς δυνάμεως ἢ τὴν ἀπὸ τῆς
αἰτίας ἢ τοῦ πυρετοῦ προαιρουμένῳ, ἢ τῷ μὲν ἰσχυ-
665K ροτέρῳ | μᾶλλον ἑπομένῳ, μὴ μέντοι γ᾽ ἀμελοῦντι
μηδὲ τῶν ὑπολοίπων. ἐκ γάρ τοι τῶν τοιούτων διορι-
σμῶν διδαχθήσῃ πρώτως μὲν τὴν αἰτίαν ἐκκόπτειν
καὶ τὸν πυρετόν, ὅταν στοχαζομένῳ σοι φαίνηται τοῖς
τοιούτοις ἰάμασιν ἡ δύναμις ἐξαρκοῦσα· πρώτην δ᾽ αὖ
πάλιν ῥωννύναι τὴν δύναμιν, ὅταν ἄρρωστος οὖσα μὴ

6

ever, sometimes the magnitude of the fever is so great that 664K
it carries with it [the threat of] imminent death unless
someone prevents it. When that happens, it is necessary to
eradicate this first before proceeding to the removal of the
cause.

Therefore Eugenianus, let me define for you right from
the start the issue of opposite indications. Since in the
examination of these indications it is clear that the capacity
plays no small part in the salvation of the person, this
would also become for you a significant indicator toward
the discovery of what is to be done. Consequently, the
three primary indicators are indicative of what is to be
done in the fevers that are subject to our inquiry. One is
the fever itself; another is the cause generating it; and a
third is the capacity. What is appropriate is to take away the
first two indicators while preserving the third. Now even
here there will again be for you the same consideration
raised a little earlier: if you compare the fever and the
cause in turn with the capacity, and both have need of
the same remedies, you will readily accept this. If, however, the indications are conflicting, you choose first either
that from the capacity, or that from the cause, or that from
the fever, following more that which is the stronger, with- 665K
out of course neglecting the others. From such distinc-
tions, you will be taught primarily to eradicacte the cause
and the fever, whenever it is obvious to your reckoning
that the capacity is sufficient for such cures. Alternatively,
strengthen the capacity first, whenever it is weak and

---

5 K; τηλικοῦτο B

φέρῃ τά τε τῆς αἰτίας καὶ τὰ τοῦ πυρετοῦ βοηθήματα. καὶ μέντοι καὶ πάντων ἅμα στοχάζεσθαί ποτε δυνατόν, ὡς ὀλίγον ἔμπροσθεν ἐπί τε τοῦ πυρετοῦ καὶ τῆς αἰτίας ἐδείχθη.

2. Τριῶν οὖν ὄντων τούτων οἷς χρὴ προσέχειν τὸν νοῦν ἐν ταῖς προκειμέναις τῶν πυρετῶν ἰάσεσιν, αὐτοῦ τε τοῦ πυρετοῦ καὶ τῆς ποιούσης αὐτὸν αἰτίας καὶ τῆς τοῦ κάμνοντος δυνάμεως, εὑρημένων δὲ καὶ τῶν τῆς ἐνδείξεως τρόπων τῶν γενικῶν, οὓς καὶ καθόλου προσαγορεύομεν, ἐπὶ τὰς διαφορὰς ἤδη τῶν εἰδικῶν τε καὶ κατὰ μέρος ἰέναι χρὴ βοηθημάτων, ἀναμνήσαντας αὖθις τῶν ἐφημέρων πυρετῶν, ἐπειδή τινες ἐξ αὐτῶν μεταπίπτουσιν εἰς τοὺς πολυημέρους. γιγνομένης δὲ τῆς μεταπτώσεως αὐτῶν καθ᾽ ἕνα τρόπον τὸν κοινότατον, δι᾽ ἀμαθίαν τῶν ἰατρῶν, εἰδικωτέρους δὲ δύο, |
666K διότι καὶ αὐτῶν τῶν πυρετῶν ἤτοι γ᾽ ἀπήλλακται τὸ ποιῆσαν αἴτιον ἢ καὶ νῦν ἔτι μένει, περὶ ἀμφοτέρων ἐν μέρει λεκτέον. ἀπήλλακται μὲν οὖν τὸ ποιῆσαν αἴτιον, ὅταν ἐπὶ θυμῷ καὶ λύπῃ καὶ ἀγρυπνίᾳ καὶ φροντίδι καὶ ἡλίῳ θερινῷ καὶ κόπῳ καὶ βουβῶνι παυσαμένῳ γένηται πυρετὸς ἄνευ στεγνώσεως· οὐκ ἀπήλλακται δέ, ὅταν ὑπὸ τοῦ προκατάρξαντος ἢ αὐτοῦ πυρετοῦ γένηταί τις στέγνωσις. ἤ τε γὰρ ψύξις ἀεὶ καὶ ἡ τῶν στυφόντων ὁμιλία διὰ μέσης στεγνώσεως ἐργάζονται πυρετόν, ἥ τ᾽ ἔγκαυσις ἔστιν ὅτε στεγνοῖ τὴν ἕξιν, αὐτή τε πολλάκις ἡ ἀρχὴ τοῦ πυρετοῦ φρικώδης γενομένη στέγνωσιν εἰργάσατο. καὶ μέντοι καὶ γλίσχροι καὶ παχεῖς ἢ καὶ πολλοὶ χυμοὶ μετρίαν ἔμφραξιν

8

might not tolerate the remedies of the cause and the fever. And, of course, it is sometimes also possible to make an estimate of all the factors at the same time, as was shown a little earlier in the case of the fever and the cause.

2. Therefore, since there are these three things to which we must direct our attention in the proposed cures of the fevers—the fever itself, the cause producing the fever, and the capacity of the patient—having discovered the generic modes of the indication, which we also term general, we must now proceed to the differentiae of the remedies which are specific and individual, again making mention of the ephemeral fevers, since some of these change into polyhemeral (chronic) fevers. When change does occur in these, the most common way is through the ignorance of doctors. However, there are two more specific ways, depending on whether the effecting cause of the fever itself has either gone away or still remains. I must speak about both of these in turn. Thus, the effecting cause goes away whenever the fever occurs without stoppage of the pores due to the cessation of anger, grief, insomnia, anxiety, the sun's heat, fatigue or glandular swelling. The effecting cause does not go away whenever some stoppage of the pores occurs due either to the *prokatarktic* [external antecedent] cause or the fever itself. Thus cold and the association of astringents always bring about a fever through the medium of stoppage of the pores, and heatstroke may sometimes obstruct the bodily state; and often the actual commencement of the fever, when there is shivering, brings about stoppage of the pores. Indeed, viscid, thick and abundant humors, when they create a moderate block-

666K

ποιησάμενοι τὸν ἐφήμερον ἐνίοτε γεννῶσι πυρετόν,
ὅταν ἀγαθῶν ἰατρῶν τύχωσιν, ὡς ἔμπροσθεν ἐδείκνυτο. καλοῦνται μὲν οὖν πάντες ἐφήμεροι οἱ τοιοῦτοι
πυρετοί, διότι λύεσθαι πεφύκασιν ὅσον ἐφ᾽ ἑαυτοῖς
ἡμέρᾳ μιᾷ, συναριθμουμένης αὐτῇ δηλονότι καὶ τῆς
νυκτός· ὥσπερ ἐν τῷ λέγειν ἡμερῶν εἶναι τὸν⁶ μῆνα
τριάκοντα καὶ τὸν ἐνιαυτὸν τριακοσίων ἑξήκοντα καὶ
667K   πέντε, καὶ τόδε τι πέπρακται⁷ πρὸ | τριῶν ἡμερῶν, ἢ
πραχθήσεται⁸ μετὰ τέσσαρας.

οὐ μὴν ἡ αὐτή γε διάθεσις ἁπάντων αὐτῶν ἐστιν,
ἀλλ᾽ ἔνιοι μὲν ἀχώριστον ἔχουσι τὴν στέγνωσιν, ἔνιοι
δὲ οὐκ ἀχώριστον μέν, ἤτοι δ᾽ ὡς τὰ πολλὰ συνοῦσαν
ἢ ἀμφιδόξως ἢ σπανίως. ἀχώριστος μὲν οὖν ἡ στέ-
γνωσίς ἐστι τοῖς διὰ τὸ στυπτηριῶδες ἢ χαλκανθῶδες
ὕδωρ ἢ τι τοιοῦτον ἕτερον, ἢ διὰ ψύξιν ἢ διὰ βραχείαν
ἔμφραξιν πυρέξασιν, ὡς τὸ πολὺ δὲ τοῖς δι᾽ ἔγκαυσιν·
ἀμφιδόξως δὲ τοῖς διὰ κόπον. ὅσοι δὲ διὰ θυμὸν ἢ
λύπην, ἢ ἀγρυπνίαν, ἢ σύντονον σκέψιν, ἢ ἀπεψίαν, ἢ
βουβῶνα πυρέττουσι, σπανιάκις αὐτοῖς σύνεστι στέγνωσις. οὗτοι πάντες οἱ πυρετοὶ κατὰ τὸν ἑαυτῶν
λόγον οὐκ ἀναμένουσι δεύτερον παροξυσμόν, ἀλλ᾽
ἐντὸς τῶν εἴκοσι καὶ τεσσάρων ὡρῶν παύονται τοὐπίπαν, ἢ εἴ που κατὰ τὸ σπάνιον ἐνίοτε τινες αὐτῶν ἐπὶ
πλέον ἐκτείνονται, δύο δ᾽ αὐτοῖς αἰτίαι προσγενόμε-
ναι, κωλύουσι λυθῆναι, μία μὲν ἡ ψύξις ἤτοι γ᾽ ἐκ τοῦ
περιέχοντος ἀέρος ἢ ἐξ ἀλείμματός τινος ἔξωθεν
προσενεχθέντος, ἑτέρα δ᾽ ἡ δαιμονία διάτριτος, ἐὰν
ἀναγκασθῶσιν ὑπερβάλλειν αὐτὴν ἰατρῶν ἀμαθίᾳ. εἰ

age, sometimes generate an ephemeral fever, whenever they have good doctors, as was shown before. All such fevers are, then, called ephemeral fevers because their very nature is such that they are resolved in one day, this of course including the night also, just as we say a month is thirty days, a year is three hundred and sixty-five days, and such-and-such a thing has been done "three days ago" or will be done "after four days."

Of course, the condition of all these fevers is not the same—some have an inseparable stoppage of pores, whereas in some it is not inseparable, being present frequently, occasionally, or rarely. Stoppage of the pores is inseparable in those patients who are febrile due to astringent or vitriolic water or some other such water, or due to cooling, or a slight blockage; frequent in those febrile due to heatstroke; and occasional in those who are febrile due to fatigue. Stoppage of the pores is seldom present in those who are febrile due to anger, grief, insomnia, intense concentration, apepsia (failure of digestion) or buboes (glandular swellings). All these fevers, by virtue of their own nature, do not wait for a second paroxysm but cease altogether within twenty-four hours. Or if, somehow, when they happen on a rare occasion to extend beyond this, there are two causes preventing their resolution: one is cold, either from the ambient air or from some unguent applied externally, and the other is the "wondrous three-day period," if they are compelled to go through this by the ignorance of doctors. If the ambient air is warm, and

667K

668K

---

6 B; τὸν om. K

7 K; πεπράχθαι B, recte fort.

8 K; πράχθήσεσθαι B, recte fort.

668K δὲ τὸ περιέχον ἀλεεινὸν | ὑπάρχει καὶ μηδεὶς τῶν τὴν διάτριτον σεβόντων ἰατρῶν παρείη τῷ κάμνοντι, παύσεται μὲν ὁ πυρετὸς ἐπὶ τῷ πρώτῳ παροξυσμῷ μεθ' ἱδρῶτος ἢ νοτίδος ἢ διαπνοῶν ἀτμωδῶν, ὁ κάμνων δ' αἰσθανόμενος ἀπαλλαγῆναι ἑαυτὸν τοῦ πυρετοῦ τὰ συνήθη πράττειν ἐπιχειρήσει λουόμενος ἢ ἀλειφόμενος ἢ καὶ χωρὶς τούτων ἐσθίων, εἴ τις εἴη τῶν ὀρείων ἀγροίκων. οἱ μὲν οὖν ἰδιῶται καθάπερ ἄλλα πολλὰ πρὸς αὐτῆς τῶν πραγμάτων τῆς φύσεως διδασκόμενοι πράττουσιν ἄμεινον τῶν σοφιστῶν, οὕτω κἀν τοῖς ἐφημέροις πυρετοῖς ἱκανοὶ τὸ σύμπαν ἑαυτοῖς ἐκπορίζειν εἰσίν, ὅταν γε χωρὶς ἰσχυροτέρας στεγνώσεως συστῶσιν.

εἰ δὲ μετὰ τοιαύτης εἰσβάλοιεν, ἀτυχήσαντες μὲν ἰατρῶν περιπίπτουσί τινι τῶν σεβόντων τὴν διάτριτον, εὐτυχήσαντες δὲ τῶν λουσόντων τε καὶ θρεψόντων αὐτοὺς καὶ τἆλλα σύμπαντα πραξόντων, ὅπως ἡ στέγνωσις λυθείη. πολλάκις γοῦν, ὡς εἴρηται, δι' ἔμφραξιν βραχεῖαν ἅμα λεληθυίᾳ πυκνώσει τοῦ δέρματος, ἤτοι διὰ λουτρὸν μοχθηρὸν ἢ δι' ἄλλο τι γενομένη, πυρετοῖς ἐφημέροις ἑάλωσαν ἔνιοι· καθ' ὃν

669K καιρὸν ἐὰν μὴ ταχέως τις ἰάσηται τὴν ἔμφραξιν, | εἰς τοὺς πολυημέρους μεταπίπτουσιν. ἔστι γὰρ ὁ τοιοῦτος πυρετὸς ὅσον μὲν ἐφ' ἑαυτῷ κοινὸς τῶν πολυημέρων τε καὶ τῶν ἐφημέρων, ἀλλὰ διὰ τὴν σμικρότητα τῆς αἰτίας μονήμερος γίνεται, καθάπερ γε καὶ διὰ μεγέθους ἐστὶν ὅτε πολυήμερος· ἀλλ' ὅ γε παροξυσμὸς εἷς αὐτοῦ μέχρι τοῦ τέλους ἀπὸ τῆς ἀρχῆς ἐστιν,

12

none of the doctors who revere the three-day period is in attendance on the patient, the fever will cease after the first paroxysm along with sweating, moistness or transpired vapors. The patient, when he feels himself freed from the fever, will attempt to carry out his customary activities, washing or anointing himself and also, apart from these actions, eating if he is one of the rural people from the mountains. Thus, just as in many other things they learn from Nature itself, laymen do better than "experts," in the same way, even in the ephemeral fevers, they are altogether adequate to provide for themselves, at least whenever they are in a state that falls short of a stronger stoppage of the pores.

If they are beset by such a stoppage, they are unlucky if they encounter one of the doctors who reveres the three-day period, whereas they are lucky if it is one of those who bathes and nourishes them and does all the other things so that the stoppage of the pores is relieved. Anyway, as I said, people are often seized by ephemeral fevers due to a slight stoppage along with an occult thickening of the skin, or to faulty bathing, or to the occurrence of something else. Unless someone quickly cures the blockage at the time, [the ephemeral fevers] change to polyhemeral (chronic) fevers. For such a fever, inasmuch as it is a fever of itself, is common to both the polyhemeral and ephemeral, but when it is due to a minor cause it becomes monohemeral, just as when it is due to a major cause it is sometimes polyhemeral. But in fact its paroxysm is one from the beginning

669K

13

ἐὰν μὴ τύχῃ διαδεξάμενος αὐτὸν ὁ ἐπὶ σήψει χυμῶν.
ἔστ᾽ ἂν οὖν μήτ᾽ ἐν τῷ σφυγμῷ μηδέπω τὸ τῆς σήψεως
ὑπάρχει γνώρισμα μήτ᾽ ἐν τῇ θέρμῃ μήτ᾽ ἐν τοῖς
οὔροις, ἐπ᾽ ἐμφράξει τε μόνῃ τὴν γένεσιν ἔχει καὶ
ἴασις αὐτῷ τῆς ἐμφράξεως ἡ λύσις γίγνεται, καὶ ὡς
χρὴ λύειν αὐτὴν ἔμπροσθεν εἴρηται.

πλείους οὖν αἱ κατὰ μέρος διαφοραὶ φαίνονται τῶν
ἐφημέρων πυρετῶν· μία μὲν ὅταν μηδ᾽ ὅλως ᾖ μηκέτι
τὸ ποιῆσαν αὐτοὺς αἴτιον, οἷον ἡ ἔγκαυσις· ἑτέρα δ᾽
ὅταν ἔτι παρείη, καθάπερ ἡ στέγνωσις, ἧς ἐδείχθησαν
οὖσαι διτταὶ διαφοραί, πύκνωσίς τε τῶν μικρῶν πόρων καὶ ἣν νῦν μεταχειριζόμεθα κατὰ τὸν λόγον
ἔμφραξις. ἔστι δὲ δήπου καὶ τῆς πυκνώσεως ἡ μὲν διὰ
670K ξηρότητα, καθάπερ | ἐπ᾽ ἐγκαύσεσιν ἢ κόποις ἢ θυ-
μοῖς, ἡ δὲ διὰ ψύξιν ἤτοι γε ἁπλῆν ἢ μετὰ στύψεως·
ὅπῃ δ᾽ ἀλλήλων αὗται διαφέρουσιν ἐν τοῖς περὶ φαρμάκων διήρηται. καὶ τῆς ἐμφράξεως δὲ ἡ μέν τις διὰ
πλῆθος, ἡ δὲ διὰ ποιότητα γίνεται χυμῶν ἤτοι γλίσχρων ἢ παχέων ὑπαρχόντων. ἐπικρατήσει δὲ κατὰ
τὴν ἴασιν ἐν μὲν τῇ διὰ πλῆθος ἀφαίρεσις αἵματος, ἐν
δὲ τῇ διὰ ποιότητα τῶν λεπτυνόντων ἡ χρῆσις. ὁ δ᾽
ἀπὸ τῆς δυνάμεως σκοπὸς ἐν μὲν τοῖς ἐφημέροις
πυρετοῖς μικρός, ἀξιόλογος δ᾽ ἐν τοῖς πολυημέροις
ἐστὶ καὶ μᾶλλον ὅσῳ περ ἂν ὦσι μακρότεροι. κατὰ
φύσιν μὲν οὖν ἐχούσης τῆς δυνάμεως οὐδὲ τὸ βραχύτατον ἡ τῶν κενωτικῶν βοηθημάτων ἐμποδίζεται
χρῆσις ἐν τοῖς πολυημέροις πυρετοῖς καὶ τοῖς ὀλιγη-
μέροις, οὐ κατὰ φύσιν δὲ ἐχούσης, λέγω δὲ οὐ κατὰ

14

to the end, unless it happens to be succeeded by putrefaction of the humors. Therefore, so long as there are not yet signs of putrefaction in the pulse, or in the heat or urine, it has its genesis in the blockage alone, and the cure for it is the resolution of the blockage, which it is necessary to resolve, as was said before.

The individual differentiae of the ephemeral fevers appear to be multiple; one when the cause producing them no longer exists at all—for example heatstroke; another when it is still present, as in the case of stoppage of the pores of which two differentiae were shown, a thickening of the small pores and blockage, which we are now dealing with in the discussion. Of course, there is also the blockage of thickening due to dryness, as in heatstroke, fatigue or anger, or that due to cooling, either simple or with contraction. How these differ from each other was distinguished in the works on medications.[1] Of blockage, there is that due to abundance or that due to the quality of the humors when they are either viscid or thick. In terms of the cure, removal of blood will prevail in the blockage due to abundance, whereas the use of thinning agents will prevail in that due to the quality [of the humors]. The indicator from the capacity is of little significance in the ephemeral fevers, but is highly significant in the polyhemeral fevers, and especially in those that are longer-lasting. Therefore, when the capacity is normal, the use of the evacuating remedies is not hindered to the slighest degree in the polyhemeral or oligohemeral fevers. However, when the capacity is not in accord with nature—I say "not in accord with na-

670K

---

[1] Possibly a reference to *De simplicium medicamentorum temperamentis et facultatibus*, XII.160K ff.

φύσιν, ὅταν ἀρρωστοτέρα πως ὑπάρχῃ κατὰ τὸ μέγεθος ἀεὶ τῆς ἀρρωστίας, ἐμποδίζεται τὰ κενωτικὰ βοηθήματα. σπανιώτερον μὲν οὖν εὐθύς ἐστιν ἐν τῇ πρώτῃ τῶν ἡμερῶν ἢ τῇ δευτέρᾳ τὴν δύναμιν ὑπάρχειν ἀσθενῆ. γίγνεται δ᾽ ἐνίοτε καὶ διὰ καχεξίαν τοῦ

671K κάμνοντος καὶ διὰ γῆρας. ἐνίοις δὲ καὶ | κόπος ἅμα ἐγκαύσει καὶ λύπῃ καὶ ἀγρυπνίᾳ προσγενομένη καὶ ἀσιτίᾳ δι᾽ ὅλης ἡμέρας ἐπιγιγνομένη, κἄπειτα περὶ τὴν ἑσπέραν εἰσβάλλοντος τοῦ πυρετοῦ, πρὶν τραφῆναι τὸν ἄνθρωπον, ἀγρύπνου τε γενομένου τῆς νυκτὸς ἱκανῶς κατέλυσε τὴν δύναμιν.

3. Καί σοι διηγήσομαι τοιοῦτον ἄρρωστον ἐφ᾽ οὗ πρῶτον ἐτόλμησα τῷ λόγῳ ποδηγούμενος ὑπεριδεῖν μὲν τῆς διατρίτου, στοχάσασθαι δὲ τῆς δυνάμεως. ὕστερον δὲ καὶ ἄλλους ἰδὼν ὁμοίως αὐτῷ θαρρῶν ἤδη τὸν αὐτὸν τρόπον ἰασάμην ὅνπερ κἀκεῖνον. ἡ γὰρ πρώτη πεῖρα μαρτυρήσασα τοῖς ὑπὸ τῆς⁹ ἐνδείξεως εὑρημένοις θαρσαλεωτέρους ἀποτελεῖ πρὸς τὴν ἐκ δευτέρου χρῆσιν. ὁ τοίνυν ἁλοὺς τῷ πυρετῷ νεανίσκος ἦν μὲν ἐτῶν πέντε καὶ εἴκοσι, ἰσχνὸς καὶ μυώδης τὸ σῶμα, καθάπερ κύων ξηρὸς καὶ θερμὸς ἀκριβῶς τὴν κρᾶσιν. ἔχαιρε δέ πως καὶ γυμνασίοις καὶ τἄλλα φροντιστής τε καὶ φιλόπονος ἦν. οὗτος ἀποδημῶν, εἶτα πυθόμενός τι τῶν οὐχ ἡδέων ἠνιάθη τε ἅμα καὶ συντείνας ἑαυτὸν εἰς τὴν πόλιν ἠπείγετο. διὰ μὲν οὖν

672K τῆς προτέρας ἡμέρας ἐκοπώθη τε | μετρίως καὶ λουσάμενος καὶ δειπνήσας ἀνεπαύσατο κατά τι πανδοχεῖον ἀγρυπνήσας τὰ πλείω. κατὰ δὲ τὴν ὑστεραίαν

16

ture" whenever it is in some way weaker—it does temporarily hinder the evacuating remedies in relation to the magnitude of the weakness. It is quite rare for the capacity to be weak immediately on the first or second days. However, sometimes this does also occur due to *kachexia* of the patient and to age. In some also, fatigue greatly dissipates the capacity when it is preceded by heatstroke, grief or insomnia, or followed by fasting for a whole day, and then around evening there is an attack of fever before the patient is nourished, and sleeplessness occurs through the night.

671K

3. I shall set out for you in detail a case of such a weak person in whom, guided by reason, I first dared to disregard the three-day period and evaluate the capacity. Later, when I saw other [patients] like him, I was already confident about curing them in the same way I cured that man. For the first experience, because it confirmed those things discovered by indication, brought a greater confidence regarding their use on the second occasion. Thus, a young man aged twenty-five was seized by fever. His body was lean and muscular, and altogether dry and hot in *krasis* like a dog. He took some pleasure in exercises and was thoughtful and industrious. When he was traveling away from home and learned of something unpleasant, he was distressed, and at once exerted himself and made haste to the city. During the previous day he was moderately fatigued and, having bathed and dined, he rested at an inn where he spent, for the most part, a sleepless night. On the

672K

---

9 B; τοῖς K

17

ἔτι καὶ μᾶλλον ἠπείχθη καὶ διανύσας ὁδὸν παμπόλλην καὶ ψαμμώδη καὶ αὐχμηράν, ἐν ἡλίῳ θερμῷ
σχεδὸν ὥρας ἑβδόμης καὶ ἡμισείας εἰς τὴν πόλιν
ἀφίκετο. πυθόμενος δὲ ἡδίω δι᾽ ἅπερ ἔσπευδεν, εἰς
γυμνάσιον ἐπορεύθη λουσόμενος, εἶτ᾽ ἀλειψάμενος
ἀνετρίψατο σύν τινι τῶν αὐτόθι νεανίσκων. καὶ προ-
τραπεὶς ὑπ᾽ αὐτοῦ κινηθῆναι βραχέα, φιλονεικίας αὐτοῖς ἐγγενομένης, οἵαις πολλάκις εἰώθασι περιπίπτειν
οἱ γυμναστικοί, πλείω τοῦ δέοντος ἐγυμνάσατο· καὶ
ἦν ἤδη ξηρὸς ἀμέτρως.

ἐξελθὼν δὲ τοῦ γυμνασίου καταλαμβάνει μαχο-
μένους τῶν ἑταίρων τινάς· οὓς διαλύων ἔλαθεν αὖθις
ἑαυτὸν ἑτέρῳ περιπεσὼν οὐ μικρῷ γυμνασίῳ, τοὺς μὲν
ἕλκων ἐξ αὐτῶν, τοὺς δὲ ὠθῶν, τοὺς δὲ διαλαμβάνων
μέσους, ἐπιτιμῶν τέ τισιν ἐξ αὐτῶν ὡς ἀδικοῦσι καὶ
θυμούμενος ὑπὲρ τῶν ἀδικουμένων, ὥσθ᾽ ὑποστρέψαι
ξηρὸς ἐσχάτως οἴκαδε, κόπου τε καὶ ἀνωμαλίας
αἰσθανόμενος. ὕδατος οὖν ὡς εἰώθει πιών, ἐπειδὴ μη-
673K δὲν ἐγίγνετο κρεῖττον, ἀλλ᾽ | ἐπετείνετο τὰ τῆς ἀνωμα-
λίας αὐτῷ, τοῦτο μὲν ἤμεσεν. ἄμεινον δ᾽ ὑπολαβὼν
εἶναι μηδέπω τρέφεσθαι, κατέκλινε τότε καὶ ἡσύχαζε
ὥρας σχεδόν τι τῆς ἡμέρας ἑνδεκάτης· τοῦτο πράξας,
ἀγρυπνήσας δὲ μετὰ τοῦ πυρέξαι δι᾽ ὅλης τῆς νυκτός,
ἡσύχαζε κατὰ τὴν ἐπιοῦσαν ἄχρι μεσημβρίας, ἰάσασθαι τὴν ἀγρυπνίαν ἐλπίζων. ἡνίκα δέ τινες αὐτὸν
ἰατροὶ τῆς διατρίτου θεασάμενοι, κατὰ μὲν τὸ παρὸν
ἔφασαν ἀξιόλογον εἶναι πυρετόν, εἰς ἑσπέραν δ᾽ αὖθις
ὄψεσθαι. καὶ τοίνυν καὶ θεασάμενοι πάλιν ἑσπέρας

following day, he urged himself on even more, traversing a very long road which was sandy and rough. After spending almost seven and a half hours in the hot sun, he reached the city. When he learned that the things he was in such a hurry about were rather better, he took himself off to the gymnasium to bathe. Then, after anointing himself, he spent time with one of the young men who was there. And when he was urged by him to stir himself a bit, contention arose between them, of the kind that often happens among those accustomed to exercise, and he exercised more than he should have, although he was already excessively dry.

On leaving the gymnasium, he came upon some of his friends fighting. In parting them, he found himself unexpectedly involved in further, by no means slight, exercise, dragging some of them apart and thrusting others away, grasping them around the waist. He rebuked some of them because they were behaving badly, and was angry on behalf of those who were wronged. So he went home extremely dry and was aware of fatigue and irregularity. Therefore, he took a drink of water, as was his custom, and because there was nothing better, but this intensified the   673K
irregularity in him and he vomited it. Assuming it would be better not to take food yet, he lay down at that time and rested until almost the eleventh hour of the day. Having done this, he was sleepless with fever right through the night. The following day, he rested until noon, hoping to cure his insomnia. On the third day, when several of the doctors of the "diatritarian" persuasion saw him, they said the fever was now significant and they would see him again toward evening. And further, when they saw him again in

παρακμάζοντα τὸν πυρετόν, οὐκ ἠξίωσαν οὐδὲ τότε
θρέψαι, καίτοι γ᾽ ἄλλου τινὸς ἰατροῦ συμβουλεύοντος,
ἀλλὰ ἀντέστησαν ἐκεῖνοι γενναίως, εἰ μὲν γὰρ ἀπύρε-
τος ἐγεγόνει, τάχα ἂν αὐτῷ δοῦναι τροφὴν εἰπόντες,
ἔτι δὲ πυρέττοντι οὐκ ἂν δοῦναι. καὶ τοίνυν καὶ κατὰ
τὴν τρίτην ἡμέραν ἕωθεν ἀφικόμενοι τὴν διάτριτον
ὑπερβάλλειν ἠξίουν. ἦν δ᾽, ὡς εἴρηται πρόσθεν, ἡ
ὕποπτος ὥρα τῆς ἡμέρας ἐκείνης ἑνδεκάτη.

χωρισθέντων οὖν αὐτῶν ἐγὼ παραγενόμενος ἐθεασάμην τοῦ νεανίσκου τὸ πρόσωπον οἷόν περ ὁ Ἱππο-
674K κράτης ἐν Προγνωστικῷ | γράφει διὰ τῆσδε τῆς ῥή-
σεως· Ῥὶς ὀξεῖα, ὀφθαλμοὶ κοῖλοι, καὶ τἄλλα ἅπερ
ἴσμεν ἐφεξῆς αὐτῷ εἰρημένα. πάντως οὖν αὐτὸν ἁλώ-
σεσθαι πυρετῷ ἑκτικῷ τε καὶ μαρασμώδει μὴ τραφέν-
τα πείσας ἐμαυτόν, ὅτι τάχιστα παρασκευάσας ἐκ
χόνδρου ῥόφημα δίδωμι προσενέγκασθαι. ἀλλ᾽ ὅμως
καίτοι τοῦτο προσενεγκάμενος, οὐδὲν ἧττον ἐν τῷ
καιρῷ τοῦ παροξυσμοῦ περὶ τὴν ἑνδεκάτην ὥραν εἰσ-
βάλλοντος, ἀπεψύχθη τε τὰ ἄκρα δυσεκθερμάντως
καὶ ὁ σφυγμὸς αὐτῷ μικρὸς καὶ ἄρρωστος ἐσχάτως
ἐγένετο. διὸ δὴ καὶ κατὰ τὴν τετάρτην ἡμέραν ἕωθέν
τε καὶ εἰς ἑσπέραν ἔδωκα τροφὴν αὐτῷ τήν τε δύναμιν
ἀνακτώμενος καὶ τὸν αὐχμὸν τοῦ σώματος ἐπιτέγγων.
ἦν γὰρ αὐτῷ τὸ δέρμα καρφαλέον ὥσπερ βύρσα.
διαμένοντος δὲ τοῦ πυρετοῦ λεπτοῦ καὶ ὁμοίου, κατὰ

---

2 See Hippocrates, *Prognostic* II. The full description is as fol-

the evening with the fever in abatement, they did not think it was a good idea for him to take nourishment at that time, and indeed, when another doctor advised this, those men vigorously opposed him, saying that, if he became afebrile, perhaps they would give him nourishment whereas, if he were still febrile, they would not. And so, during the third day, when they came early in the morning, they thought it worthwhile for him to go through the three-day period, the anticipated time being the eleventh hour of that day, as was said before.

So when I arrived after their departure and saw the face of the young man, it was as Hippocrates describes in the *Prognostic*, by way of the following statement: "A sharp nose and hollow eyes," and the other things we know he said to follow.[2] All in all, having persuaded myself that, because he would be seized by a hectic fever and marasmus if he were not nourished, I gave him thick gruel to take, which I prepared as quickly as possible. But despite having taken this, no less at the time of the paroxysm, which attacked him around the eleventh hour, he was chilled, his extremities were hard to warm, and his pulse became small and extremely weak. On which account, on the fourth day too, I gave him nourishment early in the morning and toward evening, restoring his capacity and moistening the dryness of his body, as his skin was parched like hide. When a fever that was slight and similar persisted, I

674K

lows: "Nose sharp, eyes hollow, temples sunken, ears cold and contracted with their lobes turned outwards, the skin about the face hard and tense and parched, the color of the face as a whole being yellow or black" (translation after W. H. S. Jones, LCL, *Hippocrates*, vol. 2, p. 9). This is, of course, the "Hippocratic facies."

τὴν πέμπτην αὖθις ἡμέραν ἐδικαίωσα τρέφειν αὐτὸν
οὐχ ἁπλῶς ῥοφήμασιν ὡς ἔμπροσθεν, ἀλλὰ καὶ κόκκους ῥοιᾶς ἐμβαλὼν εἰς χόνδρον ἐξ ὕδατος θερμοῦ
χωρὶς ἀρτύσεως. κάλλιστον γὰρ ἔδεσμα τοῦτο κάμνοντι πικροχόλῳ στομάχῳ. καὶ γὰρ καὶ ῥώννυσιν
675K αὐτὸν ἡ ῥοιά. καὶ χωρὶς τοῦ διαφθαρῆναι μέχρι | πλεί-
στου μένων ὁ χόνδρος ἐν τῇ γαστρὶ πέττεται κατὰ
βραχύ, μήτ᾽ ἀποξυνόμενος μήτ᾽ ἐπιπολάζων, ἅπερ
εἴωθε τοῖς ῥοφήμασιν ἔστιν ὅτε συμβαίνειν. ἐγένετο
δ᾽ οὖν καὶ κατὰ τὴν πέμπτην ἡμέραν ἡ ἀρχὴ τοῦ
παροξυσμοῦ παραπλησία, καὶ πάλιν ἡμῶν κατά τε
τὴν ἕκτην καὶ τὴν ἑβδόμην ἡμέραν ὡσαύτως αὐτὸν
διαιτησάντων ἡ κατὰ τὴν ὀγδόην αὖθις ὁμοία ταῖς
ἔμπροσθεν.

ἔνθα δὴ καὶ μάλιστα τὴν ἄνοιαν ἢ τὴν φιλονεικίαν
ἢ οὐκ οἶδ᾽ ὅ τι φῶ τῶν τὸν διάτριτον αὐτὸν ἐν ἀρχῇ
κελευσάντων ὑπερβάλλειν ἀκριβῶς ἦν καταμαθεῖν.
ἐναργῶς γάρ τοι φαινομένου πᾶσιν, ὡς οὐκ ἂν εἰς τὴν
τετάρτην ἡμέραν ὁ ἄνθρωπος ἀφῖκτο, μὴ τραφεὶς πρὸ
τοῦ κατὰ τὴν τρίτην παροξυσμοῦ, κακῶς ἐκεῖνοι καὶ
τότε καὶ ταῖς ἐφεξῆς ἡμέραις ἔφασαν αὐτὸν τεθράφθαι. ἀλλὰ γὰρ οὐχ οἷόν τ᾽ ἦν, ἵν᾽ ἐξελέγξῃ τις
αὐτούς, προδοῦναι τὸν κάμνοντα μὴ διδόντας[10] τροφὴν ἐν τῇ παροξυντικῇ τῶν ἡμερῶν. ὁμοίως οὖν θρέ-
ψαντες ἐπὶ τῆς ἐννάτης ἡμέρας τὸν ἄνθρωπον καὶ
θεασάμενοι κατὰ τὸν παροξυσμὸν εὐτονώτερον μὲν
ἑαυτοῦ γεγονότα τὸν σφυγμόν, ἔτι μέντοι τὸ ἄρρω-
676K στον ἔχοντα μετὰ τῆς | τῶν ἄκρων ψύξεως, οὐχ ὑπ-

thought it fit to nourish him again on the fifth day, not simply with gruel as before, but also putting pomegranate seeds into the gruel made from hot water without seasoning. For this is the best food for an ailing, bilious (picrocholic) stomach, as the pomegranate seeds also strengthen it. And the gruel, when it remains in the stomach for a long   675K
time without being corrupted, is gradually digested, neither turning sour nor floating to the surface—things which customarily happen to gruel on occasion. Therefore, on the fifth day also, the beginning of the paroxysm was similar, and I fed him during the sixth and seventh days, and again on the eighth day, the same as before.

So then, here also it was particularly possible to recognize precisely the folly or contentiousness (or whatever one might call it) of those ordering him in the beginning to go through the three-day fast. For although it seemed clear to everyone that the man would not reach the fourth day, if he were not nourished before the paroxysm on the third day, those men also said that both at the time and in the days following he was nourished badly. But it was not possible for anyone to convict them of jeopardizing the patient, if they did not giving nutriment on the days of paroxysm. Therefore, when I nourished the man similarly on the ninth day, and saw that during the paroxysm the pulse was more vigorous than it had been, although it was still weak along with the cooling of the extremities, I did not wait to   676K

---

10 *Conj. nos*; διδόντα K; δόντα B

23

ἐμείναμεν ἀνέχεσθαι τὴν γλωσσαλγίαν τῶν ἰατρῶν,
ἀλλὰ κατὰ τὴν ἑνδεκάτην ἡμέραν προειπόντες τοῖς
φίλοις τοῦ κάμνοντος ὡς εἴσονται τήμερον μέχρι τοῦ
δεῦρο δι' ἡμᾶς τὸν ἄνθρωπον σωζόμενον ἐπετρέψαμεν
ὑπερβάλλειν αὐτὸν τὰς παροξυντικὰς ὥρας.

ἀσφυξίας οὖν ἐν αὐταῖς γενομένης πάντως καὶ
καταψύξεως ἰσχυρᾶς ὅλου τοῦ σώματος, ὡς μήτε
φθέγγεσθαι μηκέτι καὶ μόγις τῶν θλιβόντων αἰσθάνεσθαι, κληθέντες ἅμα πάντες οἱ ἐξ ἀρχῆς ὁρῶντες
ἰατροὶ μονονοὺ διασπασθῆναι πρὸς τῶν οἰκείων τοῦ
κάμνοντος ἐκινδυνεύσαμεν, ἐγὼ μὲν ὡς ἑκὼν προδοὺς
τὴν σωτηρίαν αὐτοῦ διὰ φιλονεικίαν, οἱ δ' ἐρασταὶ τῆς
διατρίτης διὰ τὴν ἀμαθίαν, ἅμα δὲ καὶ ἀναισθησίαν.
ἐκεῖνοι μὲν οὖν ὠχρότεροι καὶ ψυχρότεροι τοῦ νοσοῦντος αὐτοῦ γενόμενοι μηχανήν τινα ἐβουλεύσαντο φυ-
γῆς. προνοήσας δ' ἐγὼ τοῦτο τὴν αὔλειον θύραν
ἐκέλευσα κλεισθῆναι, καί τινι τῶν ἑταίρων προσέταξα
λαβόντι τὴν κλεῖν φυλάττειν· εἶτ' ἐν τῷ μέσῳ κατα-
στὰς Ἤδη μὲν οὖν ὑμᾶς, ἔφην, ἀκριβῶς πεπεῖσθαι
τίς ἐστιν ὁ σώσας τὸν ἄνθρωπον ἄχρι τοῦ δεῦρο,
σωθήσεται δὲ καὶ νῦν ὑφ' ἡμῶν. οὐ γὰρ ἂν εἰ πάντως
677K   αὐτὸν ἀπολέσθαι προσεδοκήσαμεν | ἐν τῷδε τῷ παρ-
οξυσμῷ τοῦ τρέφειν ἀπέστημεν ἄν, ἀλλ' ἐπειδὴ τοσοῦτον αὐτοῦ ῥώμης ἐκ τῆς ἔμπροσθεν διαίτης ἔγνωμεν
ὑπάρχειν ὡς δύνασθαι διενεγκεῖν τὸν παροξυσμόν,
ἐπετρέψαμεν ὑπερβάλλειν ἀσίτῳ. κάλλιον μὲν οὖν ἦν
τεθράφθαι πρὸ πολλοῦ τὸν ἄνθρωπον. ἀλλ' ἵνα καὶ
τούτους ἐξελέγξω καὶ πείσω τινὰς τῶν παρεχόντων

24

endure the endless talking of the doctors, but on the eleventh day said in advance to the friends of the patient that they would know today up to what point I allowed the very man I saved to go beyond the paroxysmal hours.

Therefore, when there was complete absence of a pulse and severe cooling of the whole body, so the patient no longer spoke and was scarcely aware of those rubbing him, I and all the doctors called at the same time, who saw him from the beginning, were in danger of all but being torn asunder by the relatives of the patient—I as willingly jeopardizing his safety due to contentiousness and the lovers of the three-day period due to ignorance along with stupidity. The latter then, paler and colder than the patient himself, were devising some means of escape. When I realized this, I ordered the outer door to be shut and assigned one of my companions to watch out for someone removing the bolt. Then, taking my stand in their midst, I said, "You have already been persuaded precisely who it is who has saved the man up to this point, and it is I who will save him now." For if I expected him to perish altogether in this paroxysm, I 677K would not have desisted from nourishing him, but since I knew his strength to be of such a degree from the previous diet as to be able to endure the paroxysm, I relied on him getting through it by fasting. It was better, then, that the man had been nourished long before. But so that I might both refute them and persuade some of those present who

αὐτοῖς τὰ ὦτα, φάσκουσιν ὑφ᾽ ἡμῶν ἐπιτρίβεσθαι τὸν
νεανίσκον, ἀπολέσας ἐκεῖνον τὸν καιρὸν ἐπιδείξω νῦν
αὐτοῖς ὅτι καὶ κατ᾽ αὐτὸν τὸν παροξυσμὸν ἐνίους τῶν
ἀρρωστούντων προσήκει τρέφειν, μήτι γε μὴ πρὸ τοῦ
παροξυσμοῦ. τοῦτ᾽ εἰπὼν καὶ διαστήσας τὰς γνάθους
αὐτοῦ ἐγχέων πτισάνης χυλὸν δι᾽ ἀγγείου στενοστό-
μου κυάθων τὸ πλῆθος τριῶν, εἶτα ὀλίγον ὕστερον
οἴνου λευκοῦ λεπτοῦ κεκραμένου συμμέτρως θερμῷ
δύο κυάθους.

ἐφ᾽ οἷς ἀνέβλεψέ τε καὶ ἀκούειν καὶ φθέγγεσθαι
καὶ γνωρίζειν τοὺς παρόντας ὑπήρξατο, πρότερον οἷόν περ ξύλον ξηρὸν ἐκτεταμένος ἀναίσθητός τε καὶ
ἄφωνος. αὖθις οὖν αὐτῷ δοὺς ἄρτου τι καταπιεῖν ἐξ
οἴνου καθ᾽ ὃν εἴρηται τρόπον κεκραμένου τελέως ἀν-
εκτησάμην. καὶ πάλιν ἐπὶ τὴν ἐξ ἀρχῆς ἀγαγὼν ὁδὸν
678K τῆς διαίτης, εὐφόρως τὸν ἐν τῇ τρισκαιδεκάτῃ | παρ-
οξυσμὸν ὑπομείναντα θεασάμενος, ἐν τῇ τεσσαρεσ-
καιδεκάτῃ πάλιν ἔωθεν θρέψας ἔλουσα μετὰ τοῦτο
περὶ ὥραν ὀγδόην. εἶτ᾽ αὖθις θρέψας γενναιότερον
ἐπιδούς τε πιεῖν οἴνου· τοῦτο δὲ καὶ κατὰ τὴν δωδεκά-
την τε καὶ τρισκαιδεκάτην ἡμέραν ἐπεποιήκειν· ἐν τῇ
πεντεκαιδεκάτῃ πάλιν ἔωθεν ἔθρεψα. μᾶλλον δὲ αὐτοῦ
τότε τὸν παροξυσμὸν ἐνεγκόντος εὐφόρως, αὖθις
ἔλουσα κατὰ τὴν ἑκκαιδεκάτην ἡμέραν, καὶ τἆλλα
ὁμοίως ἔπραξα περὶ τὸν ἄνθρωπον τοῖς ἔμπροσθεν.
ἐπεὶ δὲ κατὰ τὴν ἑπτακαιδεκάτην ἡμέραν ὁ παροξυ-
σμὸς ἄθλιπτος καὶ μικρὸς ἐγένετο, θαρρῶν ἤδη τοῦ
λοιποῦ διῆτων αὐτὸν ἀναληπτικῶς. οὗτος ὁ ἄρρωστος

gave credence to those saying the young man is being destroyed by me, having lost that opportunity, I shall show them now that it is also appropriate to nourish some of those who are sick during the paroxysm itself, but not at all before the paroxysm. Having said this, I prised open his jaws and poured in juice of ptisan through a narrow-necked vessel to an amount of three ladlefuls; then a little later, thin white wine which had been mixed moderately with hot [water] to the amount of two ladlefuls.

After this, he opened his eyes and began to hear and speak, and to know those present, whereas before he had been stretched out like a piece of dry wood, insensible and unable to speak. Therefore, I again gave him some bread to swallow mixed with wine in the manner stated and completely revived [the man]. And once more I led him along the path of diet from the beginning. When I saw him easily   678K tolerate the paroxysm on the thirteenth day, I nourished him on the fourteenth day again, in the early morning, and bathed him after this around the eighth hour. Then, once again having nourished him more liberally, I gave him wine to drink. And I had done this as well on the twelfth and thirteenth days. On the fifteenth day, again early in the morning, I nourished him. Since at that time he bore the paroxysm particularly well, I bathed him again on the sixteenth day and did the other things in regard to the man in the same way as I had before. When, on the seventeenth day, the paroxysm was slight and not severe, I was now confident to manage him restoratively for the remaining time.

GALEN

ἐπαίδευσε πολλοὺς τῶν ἡμιμοχθήρων τε καὶ μὴ παντάπασιν ὄνων ὡς καὶ πρὸ τῶν παροξυσμῶν ἐνίοτε χρὴ
τρέφειν, εἰ καὶ δυοῖν ὡρῶν εἴη τὸ μεταξύ, καὶ πολὺ δὴ
μᾶλλον ἔμπροσθεν τῆς διατρίτου. ἐδίδαξα δέ σε καὶ
κατ' αὐτὴν τὴν εἰσβολὴν τοῦ πρώτου παροξυσμοῦ
τοὺς τοιούτους τρέφεσθαι δεομένους. λέγω δὲ τοὺς
τοιούτους ἐφ' ὧν ἡ δυσκρασία τοῦ σώματος ἐπὶ τὸ
ξηρὸν καὶ τὸ θερμὸν ἐκτρεπομένη πυρετοὺς ἀνάπτει. |

679K
4. Ἔστι μὲν οὖν ἴσως καὶ τοῦτο τοῖς πολλοῖς τῶν
ἰατρῶν ἄπιστον. ἀλλ' εἴτε τῷ λόγῳ χρὴ παρέχειν
αὐτῷ τὴν πίστιν εἴτε τοῖς ἔργοις, ἀμφοτέροις ἡμεῖς
παρεχόμεθα, τοῖς μὲν ἔργοις ἀνθρώπους ἐπιδεικνύντες
αὐτοῖς, οὕς, ἐὰν ἀσιτήσωσιν ἐπὶ δυοῖν ἡμέραιν, ἀνάγκη πυρέξαι· καθάπερ ἐνίους ὀργισθέντας ἢ λυπηθέν-
τας ἢ ἀγρυπνήσαντας. ἐναργέστατα γὰρ ἐπὶ τῶν
ἀναλαμβανομένων ἐκ νόσου μακρᾶς τὰ τοιαῦτα φαίνεται γιγνόμενα καὶ τῶν θερμῶν καὶ τῶν ξηρῶν τὴν
κρᾶσιν. τῷ λόγῳ δ' ἡ πίστις τοῦ γινομένου λέλεκται
μὲν ἤδη καὶ δι' ἄλλων, εἰρήσεται δὲ αὐτοῦ καὶ νῦν τὰ
κεφάλαια.

οἱ δακνώδεις ἀτμοὶ καὶ χυμοὶ διὰ τῶν αἰσθητικῶν
σωμάτων φερόμενοι φρίκας καὶ ῥίγη γεννῶσιν. ἐν οἷς
στεγνουμένου τοῦ δέρματος ἐπισχεθέντα τὰ διαπνεό-
μενα πρόσθεν, ὅταν ᾖ λιγνυώδη, πυρετὸν ἐξάπτει.
τοὺς τοιούτους οὖν ἀτμοὺς καὶ χυμοὺς αἵ τε κινήσεις
αἱ πολλαὶ καὶ σφοδραὶ καὶ ἀγρυπνίαι καὶ οἱ θυμοὶ καὶ
αἱ λῦπαι καὶ αἱ φροντίδες ἐργάζονται πλέονας. ἐφεξῆς
δὲ τούτων αὐτὸ καθ' ἑαυτὸ μόνον ἱκανὸν ἐνίοτε τὸ μὴ

28

This patient taught many of those who were only half bad and not complete asses that it is sometimes necessary to nourish before the paroxysm, if there is a two-hour interval, and much more of course, before the "three-day period." And I taught you that such people need to be nourished at the actual onset of the first paroxysm—I mean people like this, in whom the *dyskrasia* of the body, turned to the dry and hot, kindles fevers.

4. Perhaps, then, this too is not accepted by the majority of doctors. But it is necessary to provide belief either by reason itself or by actions. I provide both since I have shown people, through my very actions, that if they go without food for two days they will inevitably develop a fever, just as some will when they are made angry, or caused to grieve, or made sleepless. For it is very clear in those recovering from a long illness and those who are hot and dry in terms of *krasis* that such things manifestly occur. Belief in their occurrence based on reason has already been stated elsewhere, and it will also be stated here now in summary.[3]

679K

The gnawing vapors and humors, when they are carried through bodies capable of sensation, generate shivering and rigors. In these bodies, when the pores of the skin are blocked, those things previously dispersed are held back, and whenever these are sooty, they kindle a fever. Much vigorous movement, insomnia, anger, grief and anxiety bring about more such vapors and humors. Next to these, and sometimes sufficient by itself alone, is not tak-

[3] Possibly a reference to *De temperamentis*, I.522K.

προσενέγκασθαι τροφήν. ἐκχολοῦται γὰρ ἐπὶ τῶν
680K πικροχόλων | φύσεων ἡ ἕξις ἐπὶ ταῖς μακροτέραις
ἀσιτίαις. ἐὰν οὖν ἅμα τε τοῦτο συμβῇ καὶ κίνησίν
τινα κινηθεὶς ὁ κάμνων ἰσχυροτέραν τῆς δυνάμεως
ἀθροωτέραν ἐργάσηται τῶν δακνωδῶν περιττωμάτων
τὴν εἰς τοὐκτὸς φοράν, ἀναγκαῖόν ἐστιν, ὡς ἐν Τοῖς
τῶν συμπτωμάτων αἰτίοις ἐπεδείκνυμεν, ἤτοι ῥῖγος ἢ
φρίκην ἐπιγίγνεσθαι. ῥῖγος μὲν οὖν ἔσται διά τε τὸ
πλῆθος τῶν φερομένων καὶ τὴν δριμύτητα καὶ τὸ
τάχος τῆς φορᾶς καὶ τὴν τῆς δυνάμεως ἀσθένειαν. τά
τε γὰρ πολλὰ τῶν ὀλίγων καὶ τὰ ταῖς ποιότησιν
ἠκριβωμένα τῶν μὴ τοιούτων δάκνει μᾶλλον, ὥσπέρ
γε καὶ τὰ θᾶττον φερόμενα τῶν βραδυτέρων. ἥ τε
δύναμις ἰσχυρὰ μὲν οὖσα καταφρονεῖ καὶ ἀνέχεται
πάντων, ἀσθενὴς δὲ γινομένη καὶ πρὸς τῶν τυχόντων
ἀνιᾶται. φρίκη δ᾽ ἂν γένοιτο τῆς τε ποιότητος αὐτῶν
καὶ τοῦ τάχους τῆς φορᾶς ἐκλυθέντων ἅμα τῷ πλήθει,
καὶ μάλισθ᾽ ὅταν ἡ δύναμις ἀνθίσταται τοῖς λυποῦ-
σιν. ἐπὶ πλέον δὲ τῶν εἰρημένων ἐκλυθέντων οὐδὲ
φρίκη γένοιτ᾽ ἄν, ἀλλ᾽ ἤτοι τις ἑλκώδης αἴσθησις ἢ
ἀνωμαλία μόνη.

πάντων μὲν οὖν τῶν εἰρημένων συνελθόντων τὸ
681K σφοδρότατον | ἐστι ῥῖγος, ἁπάντων δ᾽ ἐκλυθέντων οὐ-
δεμία τῆς διεξόδου τῶν περιττωμάτων αἴσθησις. εἰ δὲ
τὰ μὲν εἴη, τὰ δὲ οὐκ εἴη, καὶ τὰ μὲν μείζω, τὰ δὲ
σμικρότερα, τὸ ἐν μέσῳ πᾶν ἀπεργασθήσεται πλάτος
ἐν ταῖς τοιαύταις μίξεσι τοῦ τε μεγίστου ῥίγους καὶ

30

ing nourishment. For the state produces bile in bilious   680K
(picrocholic) natures due to overlong fasting. Therefore, if
this should occur, and at the same time the patient, stirred
by some movement stronger than his capacity, creates a
more concentrated passage of the gnawing superfluities to
the outside, it is inevitable, as I showed in [my work] *On
the Causes of Symptoms*, that either rigors or shivering will
supervene.[4] A rigor will be due to the amount of what
is being carried, to the sharpness and swiftness of the onward passage, and to the weakness of the capacity. For
many things [being carried] sting more than few do, and
those that are exactly of these qualities sting more than
those things that are not, just as those things being carried
more quickly sting more than those being carried more
slowly. And the capacity, when it is strong, makes light of
and endures all things, whereas when it becomes weak, it is
distressed by the things it encounters. Shivering may occur
when the quality of these [superfluities] and the swiftness of their outward passage are released along with the
abundance, and particularly whenever the capacity op-
poses those things causing grief. If still more of the things
mentioned are released, shivering may not occur, but either some perception of wounding or irregularity alone.

Therefore, when all the aforementioned things happen
to coincide, the rigor is very strong, whereas when they are   681K
all released, there is no perception of the outward passage
of the superfluities. If some are and some are not, or some
are larger and some smaller, a whole middle range will
be produced in such mixtures between the greatest rigor

[4] *De symptomatum causis*, VII.85–272K; see particularly
VII.182K ff.

31

τῆς παντελοῦς τῶν περιττωμάτων ἀναισθησίας. οὐδὲν
οὖν ἐστι θαυμαστὸν εἰ τῷ μὲν ἀνωμαλίαν, τῷ δὲ
ἑλκώδη τινὰ αἴσθησιν ἢ φρίκην ἢ ῥῖγος ἕκαστον τῶν
αἰτίων αὐτό τε καθ᾽ ἑαυτὸ καὶ σὺν ἄλλοις ὁρᾶται
φέρον. ὑποκείσθωσαν γὰρ ἀσθενεῖς μὲν αἱ δυνάμεις,
αἰσθητικὸν δ᾽ ἱκανῶς τὸ σῶμα· καὶ γὰρ καὶ τοῦτο
ἐδείχθη συντελεῖν οὐκ ὀλίγον εἰς τὴν τῶν τοιούτων
συμπτωμάτων γένεσιν, ἰσχνὴ δὲ ἡ ἕξις καὶ ἡ κρᾶσις
πικρόχολος, ὅ τε στόμαχος ἐκχολούμενος ὁμοίως, ἐπί
τε τούτοις ἅπασιν ἔνδεια γιγνέσθω σιτίων, ἀγρυπνία
τέ τις καὶ λύπη καὶ φροντίς, ἀναγκαία τε πρόοδος
οἴκοθεν ἐπειγούσης τινὸς χρείας. εἶτα τούτων ὑποκειμένων ὅπερ ἐθεάσω ποτὲ γιγνέσθω. προθυμηθήτω διὰ
τὴν χρείαν ὁ τοιοῦτος ἄνθρωπος ὁδὸν μακροτέραν |
682K ἀνύσαι κατὰ τὴν πόλιν ἐπειγόμενος. ἆρ᾽ οὐχ ἑτοίμως
ἐκ τῆς προσγενομένης αὐτῷ κινήσεως τῶν δακνωδῶν
περιττωμάτων ἀνωμαλίαν μέν τινα πρῶτον, αὐτίκα δὲ
καὶ φρίκην γενέσθαι; καὶ εἰ ἐπιμείνειε κινούμενος ἢ μὴ
φθάσειε λαβεῖν σιτίων, αὐτίκα πυρέττειν αὐτόν;
  ἔδειξα δέ σοι πυρέξαντας οὕτω τινὰς καὶ μάλιστα
τῶν ἐκ νόσου μακρᾶς ἀνακομιζομένων· ὧν ἑνὶ κατὰ
τύχην ἀπαντήσας, τῆς φρίκης ἀρχομένης ἄρτι δηλώ-
σαντι τὸ γιγνόμενον, ἄρτον ἐξ οἴνου κεκραμένου δοὺς
προσενέγκασθαι, παραχρῆμα τὴν φρίκην ἔπαυσα.
ἀλλὰ τοῦτον μέν, ὡς οἶσθα, κατὰ τὴν ὁδὸν εἰς ἐργα-
στήριον εἰσαγαγὼν ἐκώλυσα πυρέξαι. ἑτέρους δ᾽ εἰς
τὴν ἑαυτῶν καταγωγὴν ἀφικέσθαι φθάσαντας ἔθρεψα
κατὰ τὴν ἀρχὴν τῆς φρίκης ἢ μικρὸν ὕστερον. ἁπλῶς

and complete nonawareness of the superfluities. It is, therefore, not surprising that each of the causes by itself and with others is seen to bring in one case irregularity, but in another a woundlike sensation, shivering or rigor. Where underlying capacities are weak, the body is excessively sensitive. This was shown to contribute to no small extent to the genesis of such symptoms, while the outward appearance is lean, the *krasis* is picrocholic and the stomach is filled with bile similarly, and in addition to all these things, assume there to be a deficiency of foods, some insomnia, grief and anxiety, and a compulsion to leave one's house driven by some need. And so, assuming those things you once saw, and that a man of this sort is willing to take a longer path when he is impelled to the city by some need, wouldn't some irregularity readily arise in him first from the added movement of the biting superfluities, and shivering immediately also occur? And if the man continues to move or does not take food beforehand, does he immediately become febrile?

682K

I showed you some who were feverish in this way, and particularly among some of those who were recovering from a prolonged illness. One of them I encountered by chance indicated what occurred when the shivering began. I gave him bread to take after mixing it with wine, and immediately stopped the shivering. But, as you know, having taken this man along the road to my workplace, I prevented him becoming febrile. Others, however, before they reached their own homes, I nourished at the start of the shivering or a little later. In brief, in all those in whom

33

δ᾽ εἰπεῖν οἷς ἔτι τὰ τῆς ἀρχῆς τῶν παροξυσμῶν ἐστι
συμπτώματα, τούτοις ἅπασιν ἄρτον ἐξ οἴνου κεκραμένου θερμοῦ διὰ ταχέων προσφέρων ἔπαυσά τε
παραχρῆμα τὴν φρίκην, ἐκώλυσά τε πυρέξαι. ὅσῳ δ᾽
ἂν θᾶττον αὐτοὺς θρέψοις, τοσούτῳ μᾶλλον κωλύσεις
τὸν πυρετόν. εἰ δὲ βραδύνοις ποτὲ βραχύ, πυρετὸς μὲν
683K   οὐδ᾽ οὕτω, | θερμασία δὲ αὐτοῖς ἐπιγίγνεται πολλή,
τρόπον ὁμοιότατον τοῖς ἐν κρύει μὲν ὁδοιπορήσασι
σφοδρῷ, καταχθεῖσι δ᾽ οἴκαδε καὶ τραφεῖσι θερμοῖς,
ὡς εἴ γε βραδύνοις πλέον, ἔτι πλεῖον οὗτοι θερμαν-
θήσονται θερμασίαν τοσαύτην τε καὶ τοιαύτην ὡς
ἀμφιβάλλειν εἰ πυρετὸν ἤδη κλητέον αὐτήν. εἰ δ᾽ ὧραι
δύο μετὰ τὴν ἀρχὴν τῆς ἀνωμαλίας εἴησαν γεγενη-
μέναι, τρέφειν μέντοι καὶ τότε μενούσης γ᾽ ἔτι τῆς
φρίκης, προλέγειν δ᾽ ὅτι πυρέξουσιν οὗτοι πυρετὸν
ἄλυπον, ᾧ νοτίδες ἀκολουθήσουσι. καὶ τούτου γενομένου καὶ λούειν χρὴ καὶ τρέφειν αὖθις. εἰ δὲ ἡ μὲν
φρίκη παύσοιτο, πυρέττοιεν δ᾽ ἤδη σαφῶς, μηκέτι
τρέφειν αὐτοὺς ἐν ἐκείνῳ τῷ καιρῷ, παρακμάζοντος δὲ
τοῦ γενομένου παροξυσμοῦ τρέφειν αὐτίκα μὴ περι-
μένων ἀπυρεξίαν.

ἐν ἁπάσαις γὰρ ταῖς αὐχμώδεσι διαθέσεσιν οὐ
παύονται καλῶς οἱ παροξυσμοὶ πρὶν ἤτοι λουτροῖς ἢ
τροφαῖς ὑγραινούσαις τεγχθῆναι. ταῦτα ἐγὼ πάντα
δι᾽ ἔργων ἐπιδειξάμενος ἐφ᾽ οἷς, ὡς οἶσθα, παραδοξο-
ποιὸς ὑπὸ τῶν ἐπαινούντων ἢ φθονούντων ἐκλήθην.
ὁπότε δὲ λόγῳ διηρχόμην ὑπὲρ τῶν αὐτῶν ἐπιδεικνὺς
684K   ὡς χρὴ τινὰς μὲν πρὸ[11] τοῦ | παροξυσμοῦ τρέφειν,

there are still the symptoms of the start of the paroxysm, when I administered bread after promptly mixing it with warm wine, I immediately stopped the shivering and prevented the person developing a fever. The quicker you nourish them the more you will prevent the fever. But if you delay a little sometimes, even when there is no fever, a 683K marked heat supervenes in them in a way very similar to that in which those who have walked in a severe frost are afflicted when they come home and are nourished with hot foods. So if you actually delay further, these people will be heated still more, and to such a degree of heat that there is doubt as to whether you must now call it a fever. If, however, two hours have passed since the beginning of the irregularity, nevertheless nourish at that time while the shivering still persists, prefering that they will become febrile with a painless fever which moistness will follow. And when this occurs, you should also bathe and nourish them again. If the shivering stops but they are now clearly febrile, no longer nourish them at that time, but when the paroxysm that has occurred abates, nourish immediately without waiting for them to become afebrile.

In all parched conditions the paroxysms do not properly cease before the patients are moistened, either with baths or with moistening nutriments. I have shown all these things through my actions, due to which, as you know, I was called a miracle worker, both by those who praised me and those who resented me. When I went over these things in theory and established that it is necessary to 684K

---

11 B; πρό om. K

ἐνίους δὲ καὶ κατ᾽ αὐτὸν ἤτοι παρακμῆς ἀρξαμένης ἢ
καὶ νὴ Δία τῆς ἀκμῆς ἐνεστηκυίας, οἶσθ᾽ ὡς τηνικαῦτα παραδοξολόγον ἐκάλουν με. τί ἂν οὖν τις πάθοι
πρὸς ἀνθρώπους μήτε ἔργῳ μήτε λόγῳ πεισθῆναι
δυναμένους; Ἀλλ᾽ ἀδύνατόν, φασι, διαγνῶναι τὰς τοιαύτας διαθέσεις. πῶς οὖν ἡμεῖς τὰς τοιαύτας διαγινώσκομεν; ἀδύνατον εἶναί σοι δοκεῖ διαγνῶναι ξηρὰν
καὶ θερμὴν κρᾶσιν, ἢ ἐξ ἀρχῆς γεγενημένην, ἢ ἐν τῷ
νῦν χρόνῳ τοιαύτην ἀποτελεσθεῖσαν; ἐμοὶ μὲν γὰρ
οὕτω ῥᾷστον, ὡς καὶ ταυτὶ γνωρίζειν τὰ γράμματα. τί
δ᾽ ὅτι τούτῳ μὲν οὐδέπω σηπεδὼν οὐδεμία χυμῶν ἐστι,
τούτῳ δ᾽ ἐστὶν ἀδύνατον εἶναι διαγνῶναι νομίζεις;
ἀλλ᾽ οὐχ ἡμεῖς γε. καὶ γὰρ διὰ τῶν σφυγμῶν καὶ διὰ
τῶν οὔρων καὶ δι᾽ αὐτῆς ἔστιν ὅτε τῆς κατὰ τὴν
θέρμην ποιότητος ἐναργῶς διαγινώσκομεν. εἰ δ᾽ ἐκ
τοῦ μηδὲν αὐτοὶ μήτε μαθεῖν ἐθελῆσαι μήτ᾽ ἀσκῆσαι
τοῖς ἐπιστήμοσιν ἀπιστοῦσι, τί κωλύει καὶ τοῖς γεω-
μέτραις αὐτοῖς καὶ τοῖς ἀριθμητικοῖς καὶ τοῖς ἄλλοις
ἅπασι τεχνίταις ἀπιστεῖν; |

5. Ἐάσαντες οὖν ἤδη τοὺς τοιούτους ἀνθρώπους
αὖθις ἀνέλθωμεν ἐπὶ τὸ συνεχὲς τοῦ λόγου. ταῖς
θερμαῖς καὶ ξηραῖς ἕξεσιν ἐναντιώτατόν ἐστι καὶ
πυρετωδέστατον αἴτιον ἀσιτία. λέγω δ᾽ ἕξεις θερμὰς
καὶ ξηρὰς οὐ μόνον ὅταν ἐξ ἀρχῆς ὦσι τοιαῦται κατὰ
τὴν οἰκείαν κρᾶσιν, ἀλλὰ κἂν ἐξ ὑστέρου γεννηθῶσιν
ἐκ διαίτης θερμῆς καὶ ξηρᾶς καὶ κινήσεων πλειόνων,
ἀγρυπνίας τε καὶ φροντίδος καὶ λύπης καὶ χωρίου
θερμοῦ καὶ ξηροῦ καὶ ὥρας θερινῆς καὶ καταστάσεως

nourish some patients before the paroxysm, some also either during it or when the abatement begins, or also, by Zeus, when the abatement is established, at that time people called me, as you know, a narrator of marvels. Why, then, would someone be tolerant toward men who cannot be persuaded, either by action or reason? "But," they say, "it is impossible to diagnose such conditions." How, then, do we diagnose such conditions? Does it seem impossible to you to diagnose a dry and hot *krasis* that has either occurred from the beginning or is such that is being brought to completion at the present time? It is very easy for me to do this since I also know these particular books. Do you think it is impossible to recognize that in one case there is not yet any putrefaction of humors but in another there is? I, at least, don't think like this. I clearly recognize [such putrefaction] through the pulse, the urine and sometimes through the actual quality in the heat. However, if there are those who, because they wish neither to learn nor to practice, do not believe those who do know, what is to stop them disbelieving geometricians and mathematicians and all the other specialists?

5. Now, letting such people be, let me return once more 685K to the thread of the argument. Fasting is the most inimical and fever-inducing cause for the hot and dry states. I say "hot and dry states" not only when they are such from the beginning, in terms of the specific *krasis*, but also if they are generated later from a hot and dry diet, too much movement, insomnia, anxiety, grief, a place that is hot and dry, the summer season, and hot, dry climatic conditions.

θερμῆς καὶ ξηρᾶς. ἔξεστι δὲ καὶ χωρὶς τοῦ τῆς ὥρας
ὀνόματος ἐν τοῖς τοιούτοις λόγοις μόνης μεμνῆσθαι
τῆς καταστάσεως. ἄνωθεν γοῦν ἀναληφθέντος τοῦ
λόγου μᾶλλον πεισθήσῃ τῷ λεγομένῳ. τὰς θερμὰς καὶ
ξηρὰς κράσεις, εἴτ᾽ ἐκ τῆς ἀρχαίας φύσεως εἴτ᾽ ἐκ τῆς
διαίτης καὶ πόνων ἀμέτρων καὶ παθῶν ψυχικῶν καὶ
χωρίου καὶ καταστάσεως εἰς τοῦτο εἶεν ἠγμέναι,
βλάπτουσιν ἱκανῶς αἱ ἀσιτίαι. ἐν τούτῳ τῷ λόγῳ
παραλέλειπται μὲν ἡ ὥρα κατά γε τὴν λέξιν, οὐ
παραλέλειπται δὲ τῇ δυνάμει· περιέχεται γὰρ ἐν τῇ
καταστάσει τῆς ἄκρως θερμῆς καὶ ξηρᾶς καταστά-
686K σεως τὴν εἰρημένην κρᾶσιν ἐργαζομένης, | ἣν οὐχ οἷόν
τε γενέσθαι κατ᾽ ἄλλην ὥραν οὐδεμίαν ὅ τι μὴ τοῦ
θέρους· οὔτε γὰρ ἐν φθινοπώρῳ δυνατὸν οὔτ᾽ ἐν ἦρι,
πολὺ δὲ μᾶλλον οὐδ᾽ ἐν χειμῶνι, θερμοτάτην ἅμα καὶ
ξηροτάτην γενέσθαι κατάστασιν· ὥσπερ οὐδὲ ψυχρο-
τάτην οὐδὲ ὑγροτάτην ἐν ἄλλῃ τινὶ πλὴν ἐν χειμῶνι.
διὰ ταῦτα μὲν δὴ κἀγὼ παραλιπὼν ἐνίοτε τὴν ὥραν
ἀρκοῦμαι τῇ καταστάσει. ἐξέστω μὲν ἑκάστῳ χρῆσθαι τοῖς ὀνόμασιν ὡς ἂν ἐθέλοι, φυλάττοντι τῶν
πραγμάτων τὴν φύσιν. ἔστι δὲ ἡ φύσις τῶν πραγμάτων οἵαν ἤδη τε πολλάκις ἔφην, οὐδὲν δὲ ἧττον ἐρῶ
καὶ νῦν· ὅσα γὰρ ἀναγκαιότατα μὲν εἰς τὰς ἰάσεις
ἐστίν, ἀγνοεῖται δὲ μάλιστα, ταῦτ᾽ οὐδ᾽ Ἱπποκράτης
ὀκνεῖ διδάσκειν πολλάκις.

ἀσιτία τοίνυν τοῖς ξηροῖς καὶ θερμοῖς σώμασιν
ὑγιαίνουσι μὲν βλαβερά, πυρέττουσι δ᾽ ἐν θέρει θερμῷ καὶ ξηρῷ τὸν ἐπ᾽ ἐγκαύσει πυρετόν, ἢ πόνοις

It is also possible in such discussions to mention climatic conditions alone, apart from the name of the season. Anyway, if I take up the argument from the beginning, you will be the more persuaded by what I say: [this is] that fasting is extremely harmful to hot and dry *krasias*, whether brought to this by the original nature, or by regimen, immoderate labors, cooling affections, place, and climatic conditions. In this discussion the season (at least in terms of the word) is left out, although it has not been left out in potency; for it is encompassed in the weather of the utterly hot and dry climatic conditions creating the aforementioned *krasis* which cannot possibly occur in any other season but summer. It is impossible for very hot and dry weather to occur in autumn, or spring, or, above all, in winter, just as it is impossible for weather that is simultaneously very cold and very moist to occur in any other season apart from winter. So then, because of this, if I sometimes leave out the season, I am satisfied with the climatic conditions. Let each person be allowed to use these terms according to preference, while preserving the nature of the matters. The nature of the matters is of a kind I have often stated already, and will state no less now too; for whatever is absolutely necessary to cures is nevertheless particularly unknown. These are things Hippocrates did not shrink from teaching often.[5]

686K

Thus, fasting is harmful for healthy bodies that are dry and hot, whereas for bodies susceptible to fever in a hot, dry summer, fasting [brings on] a fever due to heatstroke,

[5] This is taken as a general reference to Hippocrates' *Airs, Waters, Places*.

39

ἀμέτροις ἢ ἀγρυπνίαις ἢ τοῖς εἰρημένοις πάθεσι τῆς
ψυχῆς οὐχ ἁπλῶς βλαβερόν, ἀλλ᾽ εἴπερ τι καὶ ἄλλο
τῶν πάντων ὀλέθριον· ἤτοι γὰρ εἰς καυσώδεις ἐμ-
πίπτουσι πυρετοὺς ἐξ αὐτῆς, ἐξ ὧν ἐὰν μὴ φθάσωσιν
687K ἀποθανεῖν, εἰς τοὺς ἑκτικοὺς | μεταπίπτουσιν, αὖθις δ᾽
ἐκ τούτων εἰς μαρασμόν· ἢ ἐξ ἀρχῆς εὐθέως ὁ ἑκτικὸς
αὐτοῖς συμπίπτει πυρετός, ὑπερβὰς τὸν καυσώδη.
μάλιστα δ᾽, ὡς εἴρηται πολλάκις, αἱ ξηραὶ καὶ θερμαὶ
κράσεις ἁλίσκονται τοῖς ἑκτικοῖς πυρετοῖς εὐθὺς ἐξ
ἀρχῆς, ὥσπερ καὶ ἐπὶ τῆς ἀνθρώπου, τῆς φθινοπώρου
μὲν ἀρξαμένης πυρέττειν δι᾽ ἀγρυπνίαν καὶ λύπην, ἐπὶ
πλεῖστον δὲ παρατεινάσης τοῦ χειμῶνος· ἣν τεθεά-
μεθα μέν, ὡς οἶσθα, τεταρταίαν, ἐγνωρίσαμεν δ᾽ εὐθέως ἑκτικὸν εἶναι πυρετὸν ἐπιπεπλεγμένον ἑτέρῳ τινὶ
τῶν ἐπὶ χυμοῖς. ἴασις δ᾽ ἦν μόνη διδόναι ψυχρὸν ἐν
καιρῷ μέτριον εἰθισμένη πίνειν αὐτὸ καὶ παρὰ τὸν
ὑγείας χρόνον· ἀλλ᾽ ἕτεροι μὲν ἐπὶ ταύτης ἐνεπιστεύ-
οντο τὴν θεραπείαν, ὅθεν ἡμεῖς ἔγνωμεν σιωπᾶν. ἐπ᾽
ἄλλου δὲ κατὰ τὸν αὐτὸν χρόνον, ὃς ἡμῖν ἑαυτὸν
ἐπέτρεψε, καθ᾽ ἑκάστην τροφὴν ἐδίδομεν, ἀκραιφνοῦς
πηγαίου ψυχροῦ ποτε μὲν δύο κυάθους, ἔστι δ᾽ ὅτε
τρεῖς· ἀθρόον γὰρ οἱ τοιοῦτοι ψυχρὸν οὐ φέρουσιν
ἄνευ τοῦ βλαβῆναι. διὸ κάλλιστόν ἐστιν εὐθὺς ἐν τῷ
πρώτῳ παροξυσμῷ διαγνόντα τοῦ πυρετοῦ τὴν ἰδέαν
688K ἀκινδυνότερον χρήσασθαι ψυχρῷ πλείονι | μηδέπω
ξηρῶν ἱκανῶς τῶν σωμάτων γεγονότων.

ὁ γοῦν ἐκ θυμοῦ πυρέξας ἐν τοῖς ὑπὸ κύνα καύμασι,
θερμὸς καὶ ξηρὸς νεανίσκος ἐν τῷ πρώτῳ παροξυσμῷ

immoderate labors, insomnia, or the aforementioned affections of cold, which is not simply harmful but, above all, deadly. For people either fall into a burning fever from this, and if they don't die from it first, they change to a hectic fever, and again, from this to marasmus, or a hectic fever immediately befalls them from the start, bypassing the burning [fever]. Most of all, as I have often stated, the dry and hot *krasias* are seized by the hectic fevers right from the start, as happened in the case of the woman who, during autumn, began to be febrile due to sleeplessness and grief, extending for most of the winter. I observed her on the fourth day, as you know, and straightaway recognised it to be a hectic fever mixed with another of those fevers due to humors. The only cure was to give her cold water at an appropriate time and in moderation, accustomed as she was to drink this throughout the time of health. But others were entrusted with the treatment of this woman, so I knew to remain silent. In the case of another person at the same time, who did entrust himself to me, I gave at each nourishment pure water from a cold spring, sometimes two ladlefuls and sometimes three, for such people do not bear concentrated cold water without being harmed. Wherefore, it is best to use more cold water immediately in the first paroxysm when you have diagnosed the kind of fever as less dangerous, since bodies have not yet become excessively dry.

At all events, someone febrile from anger in the burning heat of the "dog days,"[6] a hot and dry young man in the

687K

688K

[6] This is the period when the Dog Star rises and sets with sun (generally reckoned as July 3 to August 11).

πιὼν ὕδατος ψυχροῦ δύο κοτύλας, αὐτίκα μὲν ἤμεσε χολὴν ξανθοτάτην, ἐξέκρινε δ᾽ ὀλίγον ὕστερον καὶ κάτω. κἄπειτ᾽ αὖθις ἐπὶ τῇ τροφῇ λαβὼν ὁμοίως ὕδατος ὅσον κοτύλην οὐκέτι ἐπύρεξεν. ἀλλ᾽ ἔνιοι τῶν ἰατρῶν ἐν ἀρχῇ μὲν οὐ γνωρίζοντες οὐδεμίαν ἰδέαν πυρετοῦ, γνόντες δ᾽ ὕστερον, ὅτε οὐδὲν ὄφελος, εἰς ἑκτικὸν ἢ καὶ νὴ Δία ἤδη μαρασμώδη πυρετὸν ἐμπίπτοντα τὸν ἄρρωστον, ἔδοσαν αὐτῷ ψυχρὸν πιεῖν, ἡμᾶς μιμησάμενοι. καὶ πάντες ἀπέτυχον τοῦ σκοποῦ· πολλῆς γὰρ ἀκριβείας δεῖται κατὰ τὸ μέτρον ἐπὶ τῶν οὕτως ἐχόντων ἡ δόσις τοῦ ψυχροῦ καὶ λεχθήσεται περὶ αὐτῆς ὀλίγον ὕστερον, ἐπειδὰν πρότερον ἐπιθῶ τελευτὴν τοῖς ἐνεστῶσιν. ὁ γάρ τοι προγεγραμμένος ἄρρωστος ἁπάντων παράδειγμά ἐστι τῶν εὐθὺς ἐξ ἀρχῆς τρέφεσθαι δεομένων διὰ ξηρότητα, κἂν ἤδη πυρέττειν ἄρχωνται. γινομένου γὰρ αὐτοῖς τοῦ πυρετοῦ διὰ τὸ δακνῶδες τῶν ἐν ταῖς ἡλιώσεσι | καὶ ἀσιτίαις καὶ πόνοις ἐκθερμανθέντων καὶ λεπτυνθέντων χυμῶν, ἡ ὑγραίνουσα τροφὴ μέγιστον ἴαμά ἐστιν.

ὥσθ᾽ ὅπου γε διὰ πλῆθος, ἢ ἔμφραξιν, ἢ φλεγμονήν, ἢ ἁπλῶς εἰπεῖν σηπεδόνα τινῶν χυμῶν ὁ πυρετὸς γεννᾶται, μέγιστον κακόν ἐστιν ἡ τροφή. τούτους μὲν οὖν οὐδ᾽ ἐν ταῖς παρακμαῖς ἀβλαβῶς ἂν τρέφοις, μήτι γε δὴ κατὰ τὴν εἰσβολὴν τῶν παροξυσμῶν· τοὺς προειρημένους δὲ ἅπαντι μὲν καιρῷ, μάλιστα δ᾽ ἐν τῇ παρακμῇ. καί με πολλάκις ἐθεάσω τοὺς ἐπὶ πλήθει καὶ φλεγμονῇ νοσοῦντας ἐν ἀσιτίαις διαφυλάξαντα μακραῖς οὕτως ὡς μηδ᾽ ὅλως θρέψαι πρὸ τῆς ἑβδόμης

first paroxysm, when he drank two cupfuls of cold water, immediately vomited very yellow bile and excreted a little later and downward. And then again, when he took water in like manner with the nourishment, as much as a cupful, he was no longer febrile. But some of the doctors, although they did not in the beginning recognize any kind of fever, did so later when it was of no help. The sick man had fallen into a hectic and, by Zeus, already marasmic fever, and they gave him cold water to drink, imitating me. And they all failed to hit the mark, for the administration of cold water requires great precision in terms of amount in those so affected, a topic on which something will be said a little later once I put an end to the present matters. For, certainly, the sick man previously written about is an example of all those who need to be nourished immediately from the start due to dryness, even if they are already beginning to be febrile. When the fever has occurred in them due to the biting nature of the humors heated and thinned by the 689K sun, and by fasting and labor, moistening nourishment is the greatest cure.

As a result, where the fever is generated by abundance, blockage, inflammation or, in a word, putrefaction of some humors, nourishment is the greatest evil. You would not then nourish these [patients] in the abatements without harm, much less during the attack of the paroxysms. You could, however, nourish those previously mentioned at any time, but particularly in the abatement. And you often saw me maintain those sick due to abundance and inflammation by means of a prolonged fast in this way, such that I did not nourish them at all before the seventh day, but was sat-

ἡμέρας, ἀλλ᾽ ἀρκεσθῆναι μόνῳ μελικράτῳ τῆς δυνάμεως ἐρρωμένης δηλονότι, τινὰς δ᾽ αὐτῶν οἷς ἡ δύναμις οὐκ ἦν εὔρωστος, ἢ χολώδης ὁ στόμαχος, ἢ
ἀσθενής, ἢ περιττῶς αἰσθητικός, ἐπὶ χυλῷ πτισάνης
μόνῳ διεφύλαξα μέχρι τῆς ἑβδόμης ἡμέρας μέλλον-
τός γε δηλονότι κατὰ ταύτην ἤτοι παρακμάζειν τοῦ
νοσήματος ἢ καὶ παντάπασι λυθήσεσθαι διὰ κρίσεως.

ἔστωσαν δή σοι καθ᾽ ἕκαστον ἄρρωστον εὐθέως
ἀπὸ πρώτης ἡμέρας οἱ σκοποὶ τοῦ τρέφειν, ὡς Ἱππο-
690K κράτης | ἐκέλευσεν, ἥ τε τοῦ νοσήματος ἀκμὴ καὶ ἡ
τοῦ κάμνοντος δύναμις. εἶτ᾽ εἰ μὲν ἐγχωρεῖ μηδ᾽ ὅλως
τρεφόμενον ὑπερβάλλειν τὴν ἀκμήν, ἐν ἀσιτίαις φύ-
λαττε τὸν ἄνθρωπον· εἰ δὲ βραχείας τινὸς βοηθείας
δέοιτο, τὸ μελίκρατον ἀρκείτω μόνον· εἰ δὲ ἔτι μείζο-
νος ἢ κατὰ μελίκρατον, ὁ χυμὸς τῆς πτισάνης. εἰ δ᾽
οὐκ ἐγχωρεῖ τὴν ἀκμὴν τοῦ νοσήματος ὑπερβάλλειν,
ἤτοι λεπτῶς διαιτώμενον ἢ ἀσιτοῦντα, πειρᾶσθαι τρέφειν τοῦτον εὐθὺς ἐν τῇ τοῦ πρώτου παροξυσμοῦ
παρακμῇ. τινὰς δ᾽ αὐτῶν, ὡς εἴρηται, καὶ μέλλοντος
εἰσβάλλειν καὶ ἀρχομένου. ξηροὶ δ᾽ εἰσὶν οὗτοι καὶ
θερμοὶ τὴν κρᾶσιν, ἤτοι γ᾽ ἐξ ἀρχῆς ἢ κατ᾽ ἐκεῖνον τὸν
χρόνον, ἀσθενεῖς τε τὴν δύναμιν ἐξ ἀνάγκης· οὐ γὰρ
ἐνδέχεται ξηρὸν καὶ θερμὸν ἱκανῶς σῶμα πυρέξαν
ἐρρῶσθαι ταῖς δυνάμεσιν. εἰ μέντοι ποτὲ συνέλθοι
φλεγμονὴ τοῖς τοιούτοις πυρετοῖς τε καὶ σώμασιν
ἐπικαίρου μορίου, τεθνήξονται πάντως· ἐθεάσω γὰρ
ἡμᾶς καὶ τοῦτο προειπόντας μὲν ἀεί, ψευσαμένους δ᾽

isfied with melikraton alone when the capacity was obviously strong. On the other hand, some of those in whom the capacity was not robust, or the stomach bilious, weak or unduly sensitive, I maintained with juice of ptisan alone until the seventh day, since the disease was clearly either going to abate or be completely resolved through a crisis.

Make your indicators of nourishment in each sick person immediately from the first day both the peak of the disease and the capacity of the patient, as Hippocrates directed. Then, if it is possible to go beyond the peak without nourishing him at all, keep the person fasting. If, however, he requires some small degree of medical aid, let melikraton alone suffice. If he needs more than melikraton, the juice of ptisan [will suffice]. However, if it is not possible to go beyond the peak of the disease, either feeding him little or fasting him, attempt to nourish him immediately in the abatement of the first paroxysm. [Attempt to nourish] some other patients, as was stated, both when the paroxysm is about to attack and as it is beginning. Those who are hot and dry in terms of *krasis*, either from the beginning or at that time, are necessarily weak in respect of the capacity, for it is not possible for a body that is excessively dry and hot to have strength in the capacities after it has become febrile. If, however, at some time inflammation combines with such fevers, and [involves] a vital part of the body, patients will invariably die. You saw me always state this beforehand and I have never been wrong. This is

690K

οὐδέποτε. καὶ μᾶλλον, εἰ πνεύμονος, ἢ τοῦ τὰς πλευ-
691K ρὰς ὑπεζωκότος χιτῶνος, ἢ | γαστρὸς ἢ ἥπατος· ἀεὶ
γὰρ ἐν τούτοις τοῖς μέρεσι φλεγμοναὶ μὴ ὅτι τοῖς
οὕτως ἀσθενέσι τὴν δύναμιν, ἀλλὰ καὶ τοῖς ἰσχυροτέροις αὐτῶν ὀλέθριαι.

καὶ μία σωτηρία πλευριτικοῖς τε καὶ περιπνευμονι-
κοῖς, προσκείσθω δὲ καὶ συναγχικοῖς, ἡ ῥώμη τῆς
δυνάμεως· ἔτι τε τούτων οὐδὲν ἧττον οἷς ἧπαρ ἢ
γαστὴρ ἐφλέγμηνεν. ὅσον μὲν γὰρ ἐπὶ ταῖς φλεγμοναῖς αὐτῶν ἥκιστα προσήκει τρέφειν, ὅσον δ᾽ ἐπὶ τῇ
τῆς δυνάμεως ἀρρωστίᾳ πολλάκις. ὥστ᾽ ἀναγκαῖον ἢ
καταλῦσαι τὴν δύναμιν ἢ τὰς φλεγμονὰς αὐξῆσαι.
ταυτὶ μὲν οὖν ἐπὶ πλέον ἴσως ἢ τοῖς ἐνεστῶσιν ἁρμόττει λέλεκταί μοι διεγνωκότι γε καὶ τοῦτο τὸ γράμμα
περὶ τῶν ἄνευ φλεγμονῆς διαλεχθῆναι πυρετῶν. ἀλλὰ
διὰ τὴν τοῦ λόγου κοινωνίαν ἐξέβην ἐφ᾽ ὅσον ἦν
ἀναγκαῖον.

6. Αὖθις οὖν ἐπανέλθωμεν ἐπὶ τούσδε τοῖς προειρημένοις πυρετοῖς συγγενεῖς, οὓς ὀνομάζειν εἰώθαμεν
ἑκτικούς. ὁ γάρ τοι[12] νεανίσκος, ὃν προέγραψα, τὸν
ἑκτικὸν ἂν ἐπύρεξε πυρετόν, ὑπερβάλλειν ἀναγκα-
σθεὶς τὴν θαυμαστὴν διάτριτον, εἴπερ γε μὴ φθάσας
692K ἀπέθανεν ὑπὸ τῶν | εἰθισμένων ἄνευ μεθόδου διαιτᾶν.

12 K; τι B

46

particularly so if [the inflammation] involves the lung, or the membrane underlying the ribs, or the stomach, or the 691K
liver. For always in these parts it is not only that inflammations are fatal for those who are weak like this in capacity, but also for those who are stronger than them.

One salvation in the pleurisies and peripneumonias, and, it must be added, in cynanche,[7] is the strength of the capacity, and besides these, no less in those in whom the liver or stomach has become inflamed. For to the extent that it is least appropriate to nourish due to inflammations of these [organs], so much is it appropriate to nourish often due to weakness of the capacity, so that inevitably either the capacity is dispersed or the inflammation is increased. Perhaps in making a diagnosis, I have stated these particular things more than is fitting for the present circumstances; and in fact they have been discussed in the book on fevers without inflammation.[8] But because of the common features of the argument, I overstepped to the extent that was necessary.

6. Therefore, let me return once more to those fevers that are congeners of the previously mentioned fevers—those we are accustomed to call hectic. For the young man whom I wrote about earlier was febrile with a hectic fever when he was forced to get through the wondrous "three-day period," at least if he did not die beforehand at the hands of those doctors accustomed to feed without 692K

[7] A somewhat obsolete term for sore throat, possibly to be equated with quinsy.   [8] Galen considers this matter in *De morborum causis*, chapter 2 (VII.2K ff.), and *De totius morbi temporibus*, VII.445K ff. The various kinds of fever are covered in detail in *De differentiis febrium*, VII.273–405K.

ὑποκείσθω τοίνυν ἄλλος τις ἅπαντα μὲν ἔχων τὰ αὐτὰ
τῷ προειρημένῳ, τὴν δύναμιν δὲ ἰσχυρότερος εἰς τοσοῦτον ὡς καὶ τὴν διάτριτον ὑπερβάλλων ἀντέχειν.
ἀνάγκη τὸν τοιοῦτον οὕτω διαιτώμενον ὡς διαιτῶσιν
οἱ τὴν διάτριτον ὑμνοῦντες, εἰς τὸν ἑκτικὸν πυρετὸν
ἐμπεσεῖν. ἔθρεψαν γὰρ ἂν αὐτὸν ἐν τῇ τετάρτῃ τῶν
ἡμερῶν, εἶτ᾽ αὖθις ἐν τῇ ἕκτῃ, κἄπειτ᾽ ὀγδόῃ τε καὶ
δεκάτῃ· τοιοῦτος γοῦν τις ὁ τύπος τῆς διαίτης αὐτῶν
ἐστιν, ὡς ἐθεάσω πολλάκις· ἡ μὲν πρώτη τροφὴ μετὰ
τὴν πρώτην διάτριτον, αἱ δ᾽ ἄλλαι παρὰ μίαν. εὐθὺς δ᾽
ἄν, οἶμαι, καὶ δι᾽ ἀρτομέλιτος ἐξήραναν αὐτοῦ τὰ
ὑποχόνδρια· καὶ γὰρ καὶ τοῦτο ἐκ τοῦ νόμου τῶν
τοιούτων ἰατρῶν ἐστιν, εἴτε φλεγμαίνοι σπλάγχνον,
εἴτε καὶ μή.

κατ᾽ ἀρχὰς μὲν οὖν ὅπως χρὴ διαιτᾶν τοὺς οὕτω
κάμνοντας εἴρηται μὲν καὶ πρόσθεν, οὐδὲν δὲ ἧττον ἐν
κεφαλαίοις εἰρήσεται καὶ νῦν. ὅταν ἤτοι γε ἐκ κόπων,
ἢ θυμῶν, ἢ φροντίδων πλεόνων, ἢ ἀγρυπνίας, ἢ λύπης, ἢ ἐνδείας μακρᾶς, ἢ καὶ πάντων ἅμα συνελθόντων, ὥσπερ ἐπὶ τοῦ προγεγραμμένου νεανίσκου, ξη-
693K ρανθέντος τοῦ | σώματος εἰσβάλῃ πυρετός, αὐχμηρὸν
μὲν ποιῶν τὸ δέρμα, πυρώδη δ᾽ ἔχων τὴν θέρμην,
ἐλπὶς μὲν δήπου τοῦτον τὸν ἄνθρωπον ἑκτικῷ ληφθή-
σεσθαι πυρετῷ καὶ μᾶλλον ἐν θέρει καὶ ἐν θερμῇ καὶ
ξηρᾷ καταστάσει. χρὴ δὲ ὅτι τάχιστα διά τε πτισάνης
χυλοῦ καὶ τῶν διὰ χόνδρου ῥοφημάτων θρέψαντα καὶ
ἄρτου τι προσεπιδιδόντα θεραπεῦσαι πόσει ψυχρᾷ
τὸν κάμνοντα. σβέννυται γὰρ αὐτίκα καὶ παύεται

48

method. Let us suppose, then, someone else having all things the same as the man previously described, but a capacity stronger to the extent that he is also able to endure the course of the three-day period. Of necessity, when such a person is fed in the way that those who praise the three-day period feed, he would fall into a hectic fever. They would nourish him on the fourth day, and again on the sixth, and then on the eighth and the tenth days. Anyway, this is the pattern of their feeding, as you saw often; the first nourishment after the first three-day period, the others on alternate days. They would, I believe, also immediately dry out his hypochondrium with a poultice of bread and honey because this is also part of the custom of such doctors whether a viscus is inflamed or not.

First, how you must feed those suffering in this way was stated earlier, but will also be stated in no less detail now under the chief points. Whenever a fever attacks after the body is dried out due to fatigue, anger, extreme anxiety, insomnia, grief, prolonged starvation, or all of these coming together at the same time, as in the case of the young man previously described, after this makes the skin dry and it has a fiery heat, the expectation is, of course, that this sickly person will be taken by a hectic fever, especially in summer and in hot and dry weather. It is necessary, after nourishing with the juice of ptisan and porridge made of wheat, and adding some bread, to treat the patient as quickly as possible with a cold drink. For the fever is immediately

693K

τελέως ὁ πυρετὸς ὅταν γε, ὡς εἴρηται, μήτε φλεγμονή
τις αὐτῷ συνῇ μήτε σῆψις χυμῶν. εἰ δέ γε τούτων τι
συνεπιφαίνοιτο, προσέχειν ἀκριβῶς χρὴ καὶ διακρί-
νειν ἐν τίνι καιρῷ πρώτως τὸ ψυχρὸν ὕδωρ αὐτῷ
δοῦναι τολμήσομεν. ἐγὼ γὰρ ἔδωκα πολλοῖς καὶ τῶν
τοιούτων, ἄμεινον εἶναι νομίσας αὐξῆσαι τὰς φλεγμονὰς τό γε παραυτίκα τοῦ περιπεσεῖν ἀνέχεσθαι τὸν
ἄνθρωπον ἑκτικῷ πυρετῷ. παντελῶς τοίνυν, ὡς ἀεὶ
λέγομεν, ἡ θεραπεία τῇ διαγνώσει τε καὶ προγνώσει
συνέπεται.

διαγνῶναι μὲν γὰρ χρὴ τὰ παρόντα, προγνῶναι δ᾽
ἐξ αὐτῶν τὰ γενησόμενα τὸν μέλλοντα χρήσεσθαι
βοηθήματι δραστηρίῳ. εἰ μὲν τοίνυν μεγάλην ἔσε-
σθαι στοχάζοιο τὴν βλάβην ἤτοι | τῆς σήψεως τῶν
χυμῶν ἢ τῆς φλεγμονῆς, ἐπισχεῖν μὲν τὴν δόσιν τοῦ
ψυχροῦ, τοῖς δ᾽ ἔξωθεν ἐπιτιθεμένοις ψυκτηρίοις ἰάμα-
σι χρῆσθαι, ποτὲ μὲν ἐπὶ στόματι τῆς γαστρὸς ἢ καθ᾽
ὅλων τῶν ὑποχονδρίων, ἔστι δ᾽ ὅτε καὶ κατὰ τοῦ
θώρακος ἐπιτιθέντα, καθ᾽ ἅπερ ἂν ἡγησώμεθα μάλιστα τὸ πλεῖστον εἶναι τῆς θέρμης. εἴρηται δ᾽ ὅτι καὶ
βαλανεῖα τοῖς οὕτω κάμνουσιν ἐπιτήδεια, τουτέστι
τοῖς ἐξηρασμένοις μὲν ἱκανῶς τὸ σῶμα, πυρέττουσι δ᾽
ἄνευ χυμῶν σηπεδόνος ἐπί τινι τῶν προειρημένων
αἰτίων. ἐπιδέδεικται δὲ ἡμῖν ἐν τοῖς περὶ τῶν πυρετῶν
λογισμοῖς ὅτι καὶ ταῖς φλεγμοναῖς διὰ τὴν σηπεδόνα
τῶν ἐν αὐταῖς χυμῶν ἕπονται πυρετοί. διὰ τοῦτο οὖν
ὅταν εἴπω ποτὲ τοῖς ἐπὶ σήψει πυρετοῖς ἤτοι συμφέρειν ἢ μὴ συμφέρειν τόδε τι, καὶ τοὺς ἐπὶ φλεγμο-

694K

quenched and completely ceases, at least whenever neither inflammation nor putrefaction of humors is present with it, as I said. If, however, one of these [complications] does appear at the same time, we must give it close attention and decide at what particular time, primarily, we shall dare to give the patient cold water. I gave such things to many patients, thinking it better to increase the inflammation, at least in the short term, so as to prevent the person from falling into a hectic fever. Thus, as I always say, treatment invariably follows diagnosis and prognosis.

It is necessary for anyone who intends to use effective remedies to diagnose those things that are present and to prognosticate from them what will occur. Therefore, if you suspect the harm of either putrefaction of the humors or of 694K inflammation will be great, hold back the administration of cold water and use cooling cures applied externally, sometimes over the mouth of the stomach or the whole hyochondrium, and sometimes also to the chest at whatever place you might particularly expect the major part of the heat to be. It is said also that bathhouses are useful for those suffering in this way; that is to say, for those who are dried excessively in the body but are febrile without putrefaction of humors due to one of the previously mentioned causes. In the deliberations about fevers,[9] I have shown that they also follow inflammation due to the putrefaction of the humors in them. Because of this, whenever I say at any time there is either this particular connection or there is not in fevers due to putrefaction, it is also necessary for

[9] Presumably in the *De differentiis febrium*, VII.273–405K, referred to in the previous note.

ναῖς ἀκούειν σε χρὴ περιλαμβανομένους ἐν τῷ κοινῷ
γένει τῆς σηπεδόνος. ὅταν δ' ἐξῆς ἀλλήλοις ἀμφοτέρους ὀνομάζοντος ἀκούῃς μου, τοὺς ἐπὶ φλεγμοναῖς
ἴσθι τῶν ἐπὶ μόνῃ σήψει τηνικαῦτα διοριζομένους.
ὁποία γάρ τίς ἐστιν ἡ τῆς φλεγμονῆς διάθεσις εἴρηται
μὲν ἔν τε τῷ Περὶ τῆς ἀνωμάλου δυσκρασίας κἀν τῷ |
695K   Περὶ τῶν παρὰ φύσιν ὄγκων.

ἀναμνῆσαι δέ σε καὶ νῦν ἀναγκαῖόν ἐστιν ὅτι
κατασκήψαντος αἵματος θερμοῦ πλείονος εἰς τὸ τοῦ
ζῴου μόριον ἐξαίρεται μὲν αὐτίκα τὰ μείζω τῶν ἀγγείων, μὴ στέγοντα τὸ πλῆθος, ἐξῆς δ' αὐτοῖς τὰ
σμικρότερα. κἄπειθ' οὕτως, ἐπειδὰν μηδ' ἐν τούτοις
στέγηται, διαδροῦται πρὸς τοὐκτὸς εἰς τὰς μεταξὺ τῶν
ἀγγείων εὐρυχωρίας, ὡς καὶ τὰς ἐν τῇ συνθέτῳ σαρκὶ
χώρας ἁπάσας καταλαμβάνειν. αὕτη μὲν ἡ τῆς φλεγμονῆς διάθεσις. ἐγχωρεῖ δὲ καὶ χωρὶς αὐτῆς ἐν
αὐτοῖς τοῖς ἀγγείοις σήπεσθαί τινας χυμούς, ἤτοι
καθ' ἕν τι τοῦ ζῴου μόριον ἐν τοῖς πέρασιν αὐτῶν
σφηνωθέντας, ἢ καὶ καθ' ὅλας αὐτῶν τὰς εὐρύτητας,
ὁπόσαι μεταξὺ βουβώνων τέ εἰσι καὶ μασχαλῶν. ὅταν
οὖν, ὡς ἐλέγομεν, ἄνευ σηπεδόνος αὐχμώδει σώματι
πυρέττειν συμβῇ, τὸ κεφάλαιον τῆς ἰάσεως ἐν τοῖς
ὑγραίνουσι καὶ ψύχουσι βοηθήμασι τίθεσθαι χρή.

7. Δύο δ' εἰσὶν αἱ τούτων ὕλαι κατὰ γένος αἱ
πρῶται· μία μὲν ἐν τοῖς προσπίπτουσιν ἔξωθεν, ἑτέρα
δὲ ἐν τοῖς εἴσω τοῦ σώματος λαμβανομένοις, ἃ δὴ καὶ
696K   προσφερόμενα | καλοῦσιν. ἔξωθεν μὲν οὖν προσπίπτει

you to understand this with regard to inflammation, this being encompassed in the common class of putrefaction. But whenever you hear me naming both, one after the other, you must, under these circumstances, know that I am always making a distinction between the fevers with inflammation and those with putrefaction alone. The kind of condition inflammation is, was stated in [the works] *On Irregular Dyskrasia* and *On Abnormal Swellings*.[10]

It is now necessary to remind you that, when a rather large amount of hot blood rushes down to a part of the organism, it immediately distends most of the vessels, although the abundance does not obstruct them, whereas it does obstruct the smaller vessels that come after them. And then, in this way, when it is not retained in these vessels, it transudes toward the outside to the open spaces between the vessels, so as also to occupy all the spaces in the compound flesh. This is the condition of inflammation. And it is possible, apart from this, for certain humors to putrefy in the vessels themselves, either in relation to one part of the organism where they are plugged up in the ends of these, or in relation to the whole of their spaces which are between the groins and the axillae. Therefore, when fever occurs in a dry body without putrefaction, you must, as I said, place the chief point of the cure in the moistening and cooling remedies.

7. The primary materials of these remedies are two in terms of class. One comprises those remedies that are applied externally; the other comprises those taken into the body, which [doctors] also call "exhibited." Externally,

695K

696K

---

10 *De inaequali intemperie*, VII.733–52K, and *De tumoribus praeter naturam*, VII.705–32K.

τό τε περιέχον ἡμᾶς αὐτὸ καὶ ὅσα δυνάμει φαρμάκων
ὑγραίνειν τε καὶ ψύχειν πεφύκασι, καὶ τρίτον πρὸς
τούτοις ὑδάτων γλυκέων λουτρά. τὰ δ᾽ εἴσω τοῦ σώματος λαμβανόμενα τά τε ἐσθιόμενα καὶ τὰ πινόμενα καὶ
ὁ εἰσπνεόμενος ἀήρ ἐστιν. ἐκ τούτων χρὴ πειρᾶσθαι
τήν θ᾽ ὑγρότητα καὶ τὴν ψύξιν ἐκπορίζειν τοῖς ἐν τῷδε
τῷ λόγῳ προκειμένοις πυρετοῖς. ἀλλ᾽ ἡ μὲν ψύξις ὡς
ἂν ὑπὸ δραστηρίου ποιότητος γινομένη ὀλιγοχρόνιός
τ᾽ ἐστὶ καὶ σφαλερά, πολυχρόνιος δὲ ἡ τῶν ἐξηρασμέ-
νων ἐπανόρθωσις ὑπάρχει δι᾽ ὑγρότητος, ἧττόν τε
σφαλερὰ τῆς διὰ ψύξεως. ἐπέλθωμεν οὖν ἐφεξῆς τὰς
εἰρημένας ἐξ διαφορὰς τῆς ὕλης, ἀπὸ τοῦ περιέχοντος
ἀρξάμενοι.

8. Τούτῳ τοίνυν εὐκράτῳ μὲν ὄντι χρηστέον ὡς
ἔχει, μηδὲν περιεργαζομένους. εἰ δ᾽ ἤτοι θερμὸν ἱκανῶς ἢ ψυχρὸν ὑπάρχοι, τῷ μὲν θερμῷ τοὐναντίον
ἐπιτεχνᾶσθαι καταγείους οἴκους, ψυχροτάτους τε καὶ
εὐπνουστάτους ἐξευρίσκοντας, πρὸς ἄρκτον ἐστραμ-
697K μένους αὔρας τέ τινας ἡδείας | αὐτοῖς μηχανωμένους,
ἐνίοτε μὲν ἐξ εὐρίπων εἰς οὓς καταρράσσουσιν ὕδατος
κρουνοὶ πλείους, ἐνίοτε δ᾽ ἐξ ἀγγείων εἰς ἀγγεῖα μετ-
αρρέοντος ὕδατος ψυχροῦ. εὐθὺς δὲ τοῦτο καὶ ὑπνω-
τικὸν ὑπάρχει, ῥαίνοντάς τε συχνῶς τὸν οἶκον ἀκραι-
φνεῖ ψυχρῷ καὶ ῥόδα πολλὰ κατὰ τῆς γῆς ἐκχέοντας,
ἢ ἀμπέλων ἕλικας, ἢ βάτων ἀκρέμονας, ἢ σχίνων
κλῶνας, ἤ τι τῶν ἄλλων φυτῶν ὅσα ψύχει· λέλεκται δὲ
ὑπὲρ αὐτῶν ἤδη καὶ πρόσθεν. εἴργοντας δὲ δηλονότι
καὶ πλῆθος ἀνθρώπων εἰσιέναι· καὶ γὰρ καὶ τοῦτο

there are the ambient air falling upon us, those medications that moisten in potency and cool by nature, and third, besides these, baths of sweet waters. Things taken into the body are foods, drinks and the inspired air. It is from these things that we must attempt to provide moisture and cooling for the fevers before us in this discussion. But cooling, as it may occur over a short time due to an efficacious quality, is also dangerous, whereas the correction of those who are dried by what is moistening occurs over a long time, and is less dangerous than correction by cooling. Therefore, let me go through successively the six different materials mentioned, starting with the ambient air.

8. You must make use of this as it is when it is *eukratic*, and not overdo things. If it is excessively hot or dry, build houses underground against the heat, searching out those that are very cool and very well ventilated, and are turned toward the north wind, contriving certain sweet breezes in them, sometimes from channels into which many founts of water cascade down and sometimes from cold water flowing back and forth between vessels. This also is immediately sleep-inducing. Sprinkle the house continuously with pure cold water and strew on the ground many roses, or tendrils of vines, or branches of brambles, or twigs of mastich, or one of the other plants that cool. I have already spoken about these things earlier. Obviously you must also prevent a large number of people entering the house, as

697K

θερμαίνει τὸν οἶκον. οὕτω μὲν τὸν θερμότατον ἀέρα πειρᾶσθαι ψύχειν, ἐν θέρει δηλονότι θερμῷ καὶ ξηρῷ συνιστάμενον· οὐ γὰρ ἂν χειμῶνί ποτε γένοιτο θερμότατος, ὥσπερ οὐδὲ ψυχρότατος ἐν θέρει. τὸν ψυχρὸν δὲ μετρίως μὲν ὑπάρχοντα τοιοῦτον ἀγαπᾶν καὶ δέχεσθαι, μηδὲν αὐτὸν ἐπιτεχνώμενον ἢ ἐξαλλάττοντα τὴν κρᾶσιν αὐτοῦ· ψυχρότατον δ' ὄντα, καθ' ὅσον μὲν εἰσπνεῖται προσίεσθαι· ψύχει γὰρ ἱκανῶς τὴν ἐν τῇ καρδίᾳ θέρμην· οὐ μὴν καθ' ὅσον ἔξωθέν γε ἅπτεται τοῦ κάμνοντος, πυκνοῖ γὰρ καὶ συνάγει τὸ δέρμα καὶ κωλύει τὰς ἀπορροὰς διαπνεῖσθαι τῶν σηπεδονωδῶν |

698K χυμῶν· διὰ μὲν γὰρ τοῦ πνεύμονος ἀραιοῦ σπλάγχνου χαλεπὸν οὐδὲν ἐπὶ τὴν καρδίαν ἀφικνεῖσθαι ῥᾳδίως οὐ μόνην τὴν ποιότητα, ἀλλὰ καὶ τὴν οὐσίαν αὐτοῦ· διὰ μέντοι τοῦ δέρματος οὔτε τὴν ποιότητα δυνατὸν οὔτε τὴν οὐσίαν ἐπὶ τὴν καρδίαν ἀφικνεῖσθαι, φθάνοντος πυκνοῦσθαι.

τὸ γὰρ ἐν τοῖς Περὶ τῆς τῶν ἁπλῶν φαρμάκων δυνάμεως ῥηθὲν ἐπὶ τῶν ἰσχυρῶς στυφόντων ἀναμνῆσαι χρὴ κἀνταῦθα καὶ γνῶναι τὸν ἄγαν ψυχρὸν ἀέρα τὰς ὁδοὺς ἀποκλείειν ἑαυτῷ, τὸ δέρμα στεγνοῦντ' ἀεί. μείζονα μέντοι γνωστέον ἐσομένην ἐκ τῆς ψυχρᾶς εἰσπνοῆς τὴν ὠφέλειαν ἤπερ ἐκ τῆς τοῦ δέρματος πυκνώσεως τὴν βλάβην, καὶ μάλισθ' ὅταν δι' ἐπιβλημάτων οἷόν τε θάλπειν αὐτό. χρῆσθαι δ' ἀέρι τοιούτῳ κατ' ἐκείνους μάλιστα τῶν ἑκτικῶν πυρετῶν ἐν οἷς ἡ καρδία πάσχει πρώτως, καθάπερ γε καὶ εἰ ἡ γαστὴρ πρώτη ἐπεπόνθει, διὰ τῶν ἐσθιομένων καὶ

this also warms it. In this way, attempt to cool the very hot air which clearly exists in a hot, dry summer; very hot air would not occur at any time in winter just as very cold air would not in summer. Be content with air that is moderately cold and accept this, doing nothing to it and not changing its *krasis*. When the air is very cold, allow it to be breathed in to the extent that is sufficient to cool the heat in the heart, but not to the extent that it affects the patient externally, for it thickens and contracts the skin and prevents the outflow of the putrefied humors from being transpired. It is not difficult for both the quality of the air and 698K also its substance to come to the heart easily through the lung which is an organ of loose texture. However, it is not possible for either the quality or the substance of the air to come to the heart through the skin when it has been thickened beforehand.

It is necessary to call to mind here what was said in the writings *On the Potencies of Simple Medications* in the case of things that are strongly astringent, and to be aware that very cold air closes off the channels in the skin, always obstructing its pores.[11] You must realize, of course, that the benefit from inhaling cold air will be greater than the harm from the thickening of the skin, particularly when it is possible to warm the skin with coverings. Use such air especially in those hectic fevers in which the heart is affected primarily just as also, if the stomach is the first to have been affected, help it with foods and drinks, because

[11] See *De simplicium medicamentorum temperamentis et facultatibus*, XII.160K ff.

πινομένων ἀρήγειν αὐτῇ· ταύτῃ μὲν γὰρ ἐκεῖνα τὴν δύναμιν εἰλικρινῆ φυλάττοντα προσπίπτει, τῇ καρδίᾳ δὲ ὁ εἰσπνεόμενος ἀήρ. ἥπατι δὲ ἐκ μὲν τῆς εἰσπνοῆς οὐδὲν ὄφελος, ὠφελεῖται δ᾽ ἧττον μὲν τῆς κοιλίας, οὐ

699K μὴν ἀμυδρῶς γε, διὰ τῶν ἐσθιομένων τε καὶ πινομένων.

κοινὴ δ᾽ ἁπάντων ἴασις ἡ διὰ τῶν ἔξωθεν ἐπιτιθεμένων ὑγραινόντων τε καὶ ψυχόντων. ἐπὶ τούτοις μὲν οὖν τοῖς σπλάγχνοις ὡς τὰ πολλὰ τοὺς ἑκτικοὺς καὶ μαρασμώδεις ἐθεασάμεθα πυρετούς. ἐπιγίγνονται μὴν ἐνίοτε καὶ ταῖς τοῦ πνεύμονος ξηραῖς καὶ θερμαῖς δυσκρασίαις· ἀλλ᾽ οὐκ ἐπιτήδειον εἰς τοὺς τοιούτους πυρετοὺς τὸ σπλάγχνον, ὑγρὸν καὶ χαῦνον ὑπάρχον. ἐπιγίγνονται δὲ καὶ θώρακι καὶ μεσαραίῳ καὶ κύστει καὶ νήστει καὶ κώλῳ καί ποτε καὶ μήτρᾳ καὶ νεφροῖς. ἐπὶ δὲ τῷ διαφράγματι μαρασμὸν μὲν οὐκ εἶδον, ἑκτικὸν δ᾽ ἐθεασάμην πυρετόν, ἅπαξ μὲν ἀκριβῶς ἀπεργασθέντα, πολλάκις δ᾽ ἀποκτείναντα πρὶν ἀκριβῶς συμπληρωθῆναι. δύσπνοιά τε γὰρ ἕπεται καὶ παραφροσύνη ταῖς τοιαύταις διαθέσεσιν ἐφ᾽ αἷς ἀποθνήσκουσιν ὡς τὰ πολλὰ πρὶν ἑκτικὸν ἀκριβῶς γενέσθαι τὸν πυρετόν. ῥᾴστη δ᾽ αὐτῶν ἡ διάγνωσις τῇ τε σκληρότητι τοῦ σφυγμοῦ καὶ τῷ τὸ ὑποχόνδριον ἐπὶ πλεῖστον ἀνασπᾶσθαι καὶ τῷ δυσπνοεῖν ἀνώμαλόν τέ τινα καὶ πολυειδῆ δύσπνοιαν. ἐπὶ πλεῖστον μὲν γὰρ

700K ἐνίοτε σμικρὸν καὶ πυκνὸν ἀναπνέουσιν, αὖθις δ᾽ ἔστιν ὅτε βραδύνουσι σαφῶς. εἶτ᾽ ἐξαίφνης ὥσπερ στενάζοντες ἀνέπνευσαν, ἢ διπλῆν τὴν εἴσω φορὰν

58

when foods and drinks that are pure encounter it, they preserve the capacity, as the inspired air does for the heart. There is no benefit to the liver from the inspired air, and it is benefited less than the stomach by what is eaten and drunk, which is hardly surprising.   699K

The general cure for all these is the external application of moistening and cooling agents. In the case of these viscera, I saw for the most part hectic and marasmic fevers. Sometimes these also supervene in the dry and hot *dyskrasias* of the lung, but that organ is not favorable to such fevers, being moist and spongy. They also supervene in the chest wall, mesentery, bladder, jejunum and colon, and sometimes also in the uterus and kidneys. I have not seen a marasmic fever in the diaphragm, although I have seen a hectic fever run its complete course once only, as they often cause death before they are entirely completed. For dyspnea and delirium follow such conditions in those who die, and in most instances before the fever becomes strictly hectic. The diagnosis of these is very easy through the hardness of the pulse, and still more through palpation of the hypochondrium, and by the irregular dyspnea and a certain variable dyspnea. For the most part the patient's breathing is either shallow and rapid or   700K
clearly slow. Then, all of a sudden, just as though sighing deeply, they breathe with either a double inspiration

τοῦ πνεύματος οἷον ἐπεισπνέοντες ἢ διπλῆν τὴν ἔξω ποιοῦνται, καθάπερ ἐπεκπνέοντες. ἀναπνέουσι δὲ καὶ τῷ ἄλλῳ θώρακι παντὶ πολλάκις αἰσθητὸν καὶ μέγα, τὰς ὠμοπλάτας ἐξαίροντες· ἐνίοτε δ᾽ ἱκανῶς ἀραιὰν καὶ μεγάλην ποιοῦνται ἀναπνοήν, ὅταν ἐπικρατῇ τὰ τῆς παραφροσύνης.

ἀλλ᾽ οὐ πρόκειται νῦν λέγειν οὔτε τὰ γνωρίσματα τῶν πεπονθότων τόπων οὔτε τὰς αἰτίας αὐτῶν. αὖθις οὖν ἐπὶ τὴν θεραπείαν ἴωμεν, ἐπισκοπούμενοι τῶν προκειμένων ὑλῶν τὰς δυνάμεις ὁποίας τινὰς ἔχουσιν ὡς πρὸς τοὺς ἑκτικοὺς πυρετούς. εἴρηται δὲ ἤδη περὶ ἀέρος οὐκ ὀλίγα θερμοῦ καὶ ψυχροῦ καὶ μάλιστα μὲν ὠφελοῦντος, ὅταν, ὡς εἰώθασιν ὀνομάζειν, ἡ καρδία πρωτοπαθεῖ· βοηθοῦντος δ᾽ οὐκ ὀλίγον οὐδ᾽ ἐν τοῖς ἄλλοις πυρετοῖς, ἐπειδὴ κἂν τούτοις ἐξ ἀνάγκης ἡ καρδία παραπλησίαν ἀναδέχεται δυσκρασίαν. ἧς χωρὶς οὐχ οἷόν τε μὴ ὅτι μαρασμὸν ἢ ἑκτικὸν πυρετόν, ἀλλὰ μηδὲ τῶν ἄλλων μηδένα γενέσθαι. δῆλον οὖν ὅτι

701K κἂν | ὁ πνεύμων ποτὲ πάθῃ πρώτως, ἐμψύχοντος αὐτὸν ἀέρος δεήσεται τῶν ἄλλων ἁπάντων μᾶλλον.

9. Ἐφεξῆς δ᾽ ἂν εἴη σκέψασθαι περὶ τῶν ἔξωθεν ἐπιβαλλομένων φαρμάκων ὑγραινόντων τε καὶ ψυχόντων. ὅτι μὲν δὴ καὶ τούτων ἐκλεκτέα τὰ ψύχοντα χωρὶς τοῦ στύφειν ἰσχυρῶς πρόδηλον παντί. πρὸς γὰρ τῷ μηδ᾽ ὑγραίνειν τὰ τοιαῦτα καὶ δικνεῖσθαι πρὸς τὸ βάθος ἡ ψύξις αὐτῶν ἀδυνατεῖ, συναγόντων τε καὶ σφιγγόντων τὸ δέρμα. κάλλιστα δ᾽ οὐ μόνον ὅσα ψύχει χωρὶς τοῦ στύφειν, ἀλλ᾽ εἰ καὶ τῇ συστάσει τοῦ

like breathing in again, or a double expiration like breathing out. They also breathe with all the rest of the chest wall, often perceptibly and markedly raising their shoulder blades. Sometimes they make an excessively infrequent and large inspiration whenever the features of delirium prevail.

But I don't propose to state now the signs of the affected places or their causes. Therefore, let me go back to the treatment, giving consideration to what kinds of potencies the materials before us have as regards the hectic fevers. Quite a lot has already been said about air that is hot or cold, and particularly that it is beneficial whenever the heart is primarily affected, as they are accustomed to term it, and [the air] provides no little help, even in the other fevers, since in these also the heart necessarily receives a similar *dyskrasia*. Apart from this *dyskrasia* it is not possible for marasmus or a hectic fever to occur, but neither is it possible for any of the others. It is clear, then, that even if the lung is sometimes affected primarily, it will require air 701K that cools it more than all the other organs.

9. The next matter to consider is the medications applied externally that are moistening and cooling. It is clear to everyone that of these, the ones to be chosen are those that are cooling without being strongly astringent. Apart from the fact that such things do not moisten, their cooling also cannot penetrate deeply because they contract and close off the skin. The best are those that not only cool without being astringent but are also fine-particled in

σώματος εἴη λεπτομερῆ. τελεωτάτην μὲν οὖν εὑρεῖν
οὐσίαν λεπτομερῆ καὶ ψυχρὰν ἀκριβῶς ἴσως ἀδύνατον. ἀκριβῶς δ᾽ εἶπον, ἐπειδὴ τὸ πάντων ὧν ἴσμεν
ψυχρῶν τῇ δυνάμει λεπτομερέστατον, τὸ ὄξος, ἔχει
τινὰ μεμιγμένην ἑαυτῷ θερμότητα. καὶ μέντοι καὶ
ξηραίνει τὰ πλησιάζοντα σώματα κἂν ὑγρὸν εἴη κατὰ
τὴν φαντασίαν. ὅθεν οὐδὲ μόνῳ αὐτῷ ποτε χρώμεθα
πρὸς τὰς ὑγραίνεσθαί τε ἅμα καὶ ψύχεσθαι δεομένας
702K διαθέσεις, ἀλλ᾽ ὕδατι ψυχρῷ | τοσούτῳ μιγνύντες ὡς
δύνασθαι πιεῖν. ἀλλ᾽ εἰ καὶ τὸ τελέως λεπτομερὲς ἅμα
καὶ ψυχρὸν σῶμα μὴ δυνατὸν εὑρεῖν, ὅμως ἐκλέγεσθαι χρὴ τὰς ἐπιτηδειοτάτας ὕλας εἰς τὴν τοῦ τοιούτου σύνθεσιν φαρμάκου. λέλεκται μὲν οὖν ἐπὶ πλέον
ὑπὲρ αὐτῶν ἐν ταῖς περὶ τῶν φαρμάκων πραγματείαις.

εἰρήσεται δὲ καὶ νῦν ἐπὶ παραδειγμάτων ὀλίγον
ἕνεκα τοῦ γυμνάσασθαί σε καὶ κατὰ τοῦτο τὸ γένος ἐν
τοῖς κατὰ μέρος. ἄρξομαι δ᾽ ἀπὸ τοῦ πάντων ἁπλου-
στάτου φαρμάκου τῶν ὑγραινόντων τε καὶ ψυχόντων,
ᾧ καὶ κατὰ τῶν ἐρυσιπελάτων χρῶμαι καὶ κατὰ τῶν ἐν
αἰδοίοις φλεγμονῶν ἐν ἀρχῇ, πρὶν ὑποφαίνεσθαί τινα
νομώδη σηπεδόνα. χρὴ δ᾽ εἰς αὐτὸ παρεσκευάσθαι
κηρὸν ὡς κάλλιστον πεπλυμένον. εἴη δ᾽ ἂν κάλλιστος
ὅ τε Ποντικὸς ὁ λευκός, ὅ τε ἐξ Ἀττικῶν κηρίων. ἔστω
δὲ καὶ ῥόδινον ἐξ ἐλαίου τοῦ καλουμένου πρός τινῶν
μὲν ὀμφακίνου, πρὸς ἄλλων δὲ ὠμοτριβοῦς, ἐσκευασμένου χωρὶς ἁλῶν. ἄριστον δὲ εἶναι καὶ τοῦτο τὸ
ἔλαιον ἀκριβῶς λεπτομερές, ὥσπερ τὸ Σαβῖνον. ἐπὶ δὲ

the consistency of their bodies. It is, perhaps, impossible to discover a substance that is completely fine-particled and entirely cooling. I said "entirely" since we know that of all the agents cooling in potency, the most fine-particled, which is vinegar, has some heat mixed with it. Besides, it also dries the bodies that are adjacent, even if we imagine it to be moistening. On this account, we never use it alone for the conditions that need to be moistened and cooled at the same time. Instead, we mix it with as much cold water 702K as to be drinkable. But if it is not possible to discover a body that is completely fine-particled at the same time as being cold, it is nevertheless necessary to choose the most useful materials for the compounding of such a medication. More has been said about them in the treatises on medications.[12]

I shall also say a little more now by way of example for the sake of your becoming practiced in this class [of medications] individually. I shall begin with the simplest medications of all—those that are moistening and cooling, which I also use against erysipelas and inflammation of the genitals in the beginning, before the appearance of a spreading putrefaction. It is necessary to prepare a wax, washed as well as possible, for this. What would be best would be white Pontic and also the one derived from the Attic honeycomb. There should also be oil of roses made from the oil that some call "omphacinum" and others call "omotribes,"[13] prepared without salt. It is best for this oil to be absolutely fine-particled, as the Sabine is. As regards

[12] See, for example, *De simplicius medicamentorum temperamentis et facultatibus*, XI.628K.

[13] The oil of unripe olives; see Dioscorides, I.29.

703K τῆς χρείας ἄμφω μιχθέντα τηκέσθω δι' ἀγγείου | διπλοῦ καὶ γενομένης ὑγρᾶς κηρωτῆς. ἔσται δὲ τοιαύτη τριπλάσιον ἢ τετραπλάσιον ἔχουσα τοῦ κηροῦ τὸ ῥόδινον. ἐπειδὰν δὲ αὐτὴ ψυχθῇ, μιγνύσθω ψυχθείσῃ κατὰ βραχὺ τοσοῦτον ὕδατος ὅσον ἂν ἐν θυείᾳ μαλαττομένῃ σὺν αὐτῷ δέξασθαι δύναιτο. χρὴ δὲ καὶ τὴν κηρωτὴν αὐτὴν ἱκανῶς ἐψῦχθαι καὶ τὸ μιγνύμενον ὕδωρ αὐτῇ ψυχρότατον ὑπάρχειν. ἱκανῶς δὲ ψύξεις τὴν κηρωτήν, ἐπειδὰν μετρίως παγῇ, καθιεὶς εἰς ὕδωρ ψυχρότατον ὅλον τὸ ἀγγεῖον ἐν ᾧ περιέχεται. μίξας δὲ εἰ βούλοιο καὶ ὄξους ὀλίγον ἱκανῶς λεπτοῦ καὶ διαυγοῦς, ἔτι δὲ καὶ μᾶλλον ὑγραῖνόν τε ἅμα καὶ ψῦχον ἐργάσῃ φάρμακον. οὐ χρὴ δ' ἀναμένειν ἐπιχεῖσθαι κατὰ τοῦ σώματος αὐτὸ τοῦ κάμνοντος εἰς τοσοῦτον ὡς θερμανθῆναι σαφῶς, ἀλλ' ὑπαλλάττεσθαι συνεχῶς.

ἄλλο φάρμακον. ὀξαλίδος ἢ ὀξυλαπάθου χυλός, ἀλφίτων λεπτῶν ὀλίγων μιχθέντων ἀναλαμβανέσθω διπτύχῳ ῥάκει τριβακῷ, ψυχρὸν δ' ἱκανῶς ἐπιτιθέσθω καὶ τοῦτο. μὴ παρόντος δὲ τοιούτου ῥάκους, ὀθόνιον δίπτυχον ἀναδεύσας, ἐπιτίθει τῷ ψύχεσθαι δεομένῳ

704K μορίῳ, | καὶ μὲν δὴ καὶ ὁ τῆς ἀνδράχνης καὶ ὁ τοῦ ἀειζώου χυλὸς ἅμα τῷ τῆς ὄμφακος ὁμοίως ἀλφίτοις μίγνυται. πρὸς τούτοις δ' ἔτι φακὸς ὁ ἀπὸ τῶν τελμάτων καὶ τριβόλου χλωροῦ καὶ πολυγόνου καὶ θριδακίνης καὶ σέρεως χυλός, ὅσα τ' ἄλλα ψύχειν ἐλέχθη σὺν ἀλφίτοις λεπτοῖς πάντα. καὶ χωρὶς δ' ἀλφίτων ὅλας τὰς πόας ἔξεστι λειοῦντα χρῆσθαι. καὶ μὲν δὴ

its use, after both have been mixed, let them be dissolved in a double vessel and made into a moist salve. Such a salve will be one that has three or four times as much of the oil of roses as the beeswax. When it cools, mix in gradually as much water with the cooled salve as it can receive, as it is being softened in a mortar with it. And it is necessary for the salve itself to be cooled sufficiently and the water mixed with it to be very cold. You will cool the salve sufficiently, when it has become moderately stiff, if you let down the whole vessel in which it is contained into water that is very cold. After you have mixed it, if you wish to also add a little vinegar which is sufficiently thin and clear, you will make an even more moistening and, at the same time, cooling medication. It is necessary not to let it remain poured over the body of the patient for so long that it clearly becomes heated but to change it continuously. 703K

Another medication: the juice of sorrel or curled dock. When it is mixed with a little thin barley groats, take this up on a double thickness lint. Apply it extremely cold. If such a lint is not to hand, soak double thickness linen and place it on the part that needs to be cooled. And as the next step, similarly mix the juice of purslane, sengreen and unripe olives with the barley groats. Besides these as well, there are lentils from the marshes, the juice of water chestnut, knotgrass, wild lettuce and endive, and all the other things said to cool, mixed with the thin barley groats. Apart from barley groats, it is possible to use whole grasses that have been 704K

καὶ τὸ διὰ τῶν φοινίκων ἐπίθεμα τῶν λιπαρῶν, ὀνομάζουσι δ' αὐτοὺς πατητούς, ἀγαθὸν φάρμακον. ἕψειν
δὲ χρὴ καὶ τούτων τὴν σάρκα, τὸ ὑμενῶδες ἅπαν
ἐξαίροντα ποτὲ μὲν δι' ὄξους μόνου διαυγοῦς, ἔστι δ'
ὅτε καὶ ὕδατος αὐτῷ μιγνύντα· τακερᾶς δ' ἱκανῶς
γενηθείσης, λειοῦντα χρῆσθαι. σκληρὸν δ' εἰ φαί-
νοιτό σοι κατὰ τὴν σύστασιν, ἄμεινον μιγνύναι τῆς
προγεγραμμένης κηρωτῆς. εἰ δὲ καὶ ῥοδίνου ποτὲ τῶν
εἰρημένων ἑκάστῳ μῖξαι βουληθείης, οὐδὲν ἔσται σοι
χεῖρον τὸ φάρμακον· ἔστω δὲ ὀλίγιστόν τε τὸ μιγνύμενον καὶ ἁπλοῦν καὶ ψυχρὸν ἱκανῶς, κἀξ ἐλαίου
γεγονὸς ἄλας οὐκ ἔχοντος. ἁπλοῦν δ' ὅταν εἴπω ῥόδινον ἢ ἄλλο τι τῶν τοιούτων, ἀκούειν σε χρὴ τὸ χωρὶς

705K   τῶν ἀρωμάτων | ἐσκευασμένον. ἀγαθὸν φάρμακον εἰς
τὰ τοιαῦτα καὶ ὁ τῶν ῥόδων χυλὸς ἅμα τοῖς ἀλφίτοις
ψυχρός. ἀγαθὸν δὲ καὶ αὐτὰ λειωθέντα· καθάπερ οὖν
καὶ αἱ λεῖαι βλάσται τῶν μετρίως στυφόντων φυτῶν,
οὕτω δ' ἔστιν ὅτε καὶ βοτάναις χρήσασθαι μετρίως
στυφούσαις δυνατόν, ὥσπερ τῷ στρύχνῳ.

ἀλλ' ἡ μὲν τοιούτων φαρμάκων ὕλη δι' ἑτέρας
πραγματείας εἴρηται πᾶσα· νῦν δ' ἀρκεῖ τὰ λελεγμένα
παραδείγματος ἕνεκα. χρῆσθαι δ' οὐχ ὡς οἱ πολλοὶ
καθ' ὅλου τοῦ θώρακος ἢ συμπάσης τῆς γαστρός,
ἀλλὰ κατ' ἐκείνου μάλιστα τοῦ πρώτως πεπονθότος.
οὐδὲ γὰρ ἀναγκαῖον ἢ σὺν τῷ ψύχεσθαι σφοδρῶς

---

14 The Greek term *strychnon* covers a number of different

brayed, and particularly, also, the application made from the oily dates (they call them "trodden"), which is a good medication. It is, however, also necessary to boil the flesh of these, taking away everything membranous, sometimes mixing it with clear vinegar alone and sometimes also mixing water with it. When this is made sufficiently tender, use it brayed. If it seems to you hard in consistency, it is better mixed with the previously described salve. If you also wish to mix oil of roses on occasion with each of the aforementioned [things] your medication will be none the worse. Make sure that what is mixed is very little, simple, sufficiently cold and made from oil which is without salt. As for "simple," whenever I speak of oil of roses or anything else of this kind, you must understand that this is something prepared without aromatic herbs. The juice of roses which is cold, together with barley groats, is a good medication for such purposes. This is also good when brayed, just as the soft shoots of the moderately astringent plants also are, it being sometimes possible in this way to use moderately astringent plants like the nightshade.[14]

705K

But the material of such medications is spoken of fully in another treatise. What has been said now is enough for the sake of exemplification. Do not, as many do, use these on the whole chest or abdomen, but on that which has been particularly affected in the first place. It is essential that you do not cool one of the parts that does not need strong cooling along with cooling what does need to be

plants; see Dioscorides, IV.71–74. Both Linacre and Peter English take the reference here to be to nightshade. For solanum (nightshade), see *De simplicium medicamentorum temperamentis et facultatibus*, XI.588, 740, and 767K.

δεομένῳ ψυχθῆναί τι τῶν οὐχ ὁμοίως δεομένων, ἢ θᾶττον τοῦ δέοντος παύσασθαι, δεδιότα βλάψαι τι τῶν γειτνιώντων. ἐγὼ γοῦν οἶδα κατὰ τῶν ὑποχονδρίων ἐπιτεθέντος ποτὲ ψύχοντος φαρμάκου παραχρῆμα δυσπνοήσαντα τὸν ἄνθρωπον, ἄλλον δ᾽ αὐτίκα βήξαντα καὶ μικρὸν ὕστερον ἑκατέρῳ παυσάμενον τὸ σύμπτωμα, τοῦ ψύχοντος ἀρθέντος. μὴ βουλομένῳ δέ

706K σοι γενναίως ψύχειν οὐκ ὀλίγον | ὑπάρχει πλῆθος ἐμπλαστῶν φαρμάκων κηρωτοειδῶν μετρίως ψυχόντων, ὧν οὐ μόνης τῆς ὕλης τὰς δυνάμεις, ἀλλὰ καὶ τῆς συνθέσεως ἔμαθες τὴν μέθοδον.

10. Ἐγὼ δ᾽ αἰσθανόμενος ὕστερον τὸ μέτρον ἤδη τῆς προκειμένης πραγματείας ἐκπίπτων ἐπὶ τὰ συνεχῆ τοῦ λόγου μεταβήσομαι. συνεχὴς δ᾽ ἐπὶ τοῖς εἰρημένοις ἡ περὶ βαλανείου σκέψις αὐτοῦ τε τοῦ πράγματος ἕνεκα ἔτι τε μάλιστα ἐπειδὴ Φίλιππος ἡγεῖται βλάπτειν αὐτὸ τοὺς μαραινομένους. ἐγὼ τοίνυν ἣν ἔχω γνώμην ὑπέρ γε τῆς δυνάμεως ἁπάσης τῶν βαλανείων καὶ προσέτι τῆς[13] εἰς τοὺς ἑκτικοὺς καὶ μαρασμώδεις ἢ ἁπλῶς εἰπεῖν ἅπαντας, ἐν τῷδε τῷ λόγῳ μάλιστ᾽ ἂν διέλθοιμι. μέμνημαι γὰρ ὅτι καὶ πρόσθεν ἐπὶ τῶν ἐφημέρων πυρετῶν ἐπήνουν αὐτά· καὶ κατὰ τὸν ἕβδομον λόγον ἐπὶ τῶν κατὰ ξηρότητα δυσκρασιῶν τῆς γαστρὸς ἀπεπτούντων τε καὶ λεπτυνομένων ἐκέλευον χρῆσθαι πολλάκις. οὐ μὴν περί γε τῆς συμπάσης δυνάμεως ἐν ἑτέρῳ τινὶ λόγῳ διῆλθον,

707K ἀλλ᾽ εἰς τοῦτον ἐφύλαξα καθ᾽ ὃν | ἀγωνιστικωτέρα[14]

strongly cooled, and that you don't too quickly stop cooling what does need cooling, fearing to harm one of the adjacent parts. Anyway, I know a man who, on the occasion of a cold medication being applied to the hypochondrium, immediately became dyspneic, and another who was immediately seized with coughing. A little while later the symptom ceased in each case when what was cooling was removed. But if you do not wish to cool strongly, there is no 706K small number of ceratelike, emplastic medications that are moderately cooling, of which you learned not just the potencies of the material but also the method of composition.

10. Realizing belatedly that I have already digressed somewhat from the subject proposed, I shall pass to those things connected with the discussion. Linked with what has been stated is the consideration of bathing and, apropos the matter itself, even more so since Philippus believes it harms those who are wasting.[15] Therefore, in this discussion particularly, I would like to go over the knowledge I have about the potency of baths, and besides this, their potency for the hectic and marasmic fevers, or put simply, all of them. I mentioned before that I approved of baths in the ephemeral fevers, and in the seventh book, I urged their frequent use often in the *dyskrasias* involving dryness of the stomach when there was failure of digestion and thinning. I did not go over the whole potency in any other book, but reserved [the subject] for this one,

---

15 This is Philippus of Rome (ca. 45–95), a Pneumatist mentioned quite frequently by Galen; see *EANS*, pp. 648–49.

---

13 B; τὸν K
14 K; ἀγωνιστικωτάτη B

μάλιστά ἐστιν ἡ χρῆσις αὐτοῦ καὶ τὸν Φίλιππον ἔχει κωλύοντα.

δέδεικται μὲν οὖν ἤδη καὶ πρόσθεν ἐπὶ τῆς τῶν ἑλκῶν ἰάσεως ἡ αἰτία τῆς ἀγνοίας τῶν καθ᾽ ἕκαστον πάθος οἰκείων σκοπῶν· εἰρήσεται δ᾽ οὐδὲν ἧττον καὶ νῦν, εἴς τε τὰ παρόντα καὶ τὰ μέλλοντα χρήσιμος ὑπάρχουσα. μιᾶς μὲν γὰρ ἐν τῷ σώματι διαθέσεως οὔσης, εἰ καὶ μὴ τῷ λόγῳ, τῇ πείρᾳ γοῦν εὑρεῖν οὐ χαλεπὸν αὐτῆς ἐστι τὴν ἴασιν· ἐπιπλεκομένων δὲ δυοῖν ἢ τριῶν, καὶ μάλισθ᾽ ὅταν ἐναντιωτάτων ἀλλήλοις δέωνται βοηθημάτων, ἀδύνατον μὲν εὑρεῖν τῇ πείρᾳ τὸ ποιητέον, οὐ ῥᾴδιον δ᾽ οὐδὲ τῷ λόγῳ. καὶ γὰρ καὶ τὴν οὐσίαν ἑκάστης τῶν διαθέσεων ἀκριβῶς χρὴ γνῶναι καὶ τὴν οἰκείαν τῆς θεραπείας ἐφ᾽ ἑκάστης αὐτῶν ἔνδειξιν λαβεῖν, ἥντινά τε χρὴ πρώτην τῶν ἄλλων ἢ μᾶλλον, ἥντινα δ᾽ ἧττον ἢ δευτέραν ἢ τρίτην ἰάσασθαι. διὰ ταύτην οὖν τὴν αἰτίαν οὔθ᾽ ὅτι πυρετῶν ἢ πυρετοὶ συμπάντων ἐστὶν ἴδιον ἴαμα τὸ ψυχρόν, εἴτε οὖν κατ᾽ ἐνέργειαν εἴτε κατὰ δύναμιν εἴη τοιοῦτον, ἐγνώσθη τῷ πλήθει τῶν ἰατρῶν, οὔθ᾽ ὅτι κατὰ συμβεβηκὸς ἕτερα πολλὰ πέφυκε ψύχοντα πυρετῶν γίνεσθαι | βοηθήματα. περὶ ὧν εἴρηται μὲν ἤδη τι κἂν τοῖς ἔμπροσθεν εἰρήσεται δὲ κἂν τοῖς αὖθις. ἔνια γὰρ τῶν βοηθημάτων ἐπιπεπλεγμένην τέ πως ἔχει καὶ διττὴν δύναμιν, ὡς καὶ πρώτως ὀνινάναι καὶ κατὰ συμβεβηκός, οἷόν πέρ ἐστι καὶ τὸ λουτρόν.

ἡ μὲν γὰρ πόσις τοῦ ψυχροῦ καθ᾽ ἑαυτὴν μὲν ὀνίνησιν, ἐνεργείᾳ ψύχουσα· χυλὸς δ᾽ ὄμφακος τῇ

in which its use is especially rather contentious and has
Philippus forbidding it.

I have already shown previously, in relation to the cure of wounds and ulcers, the cause of the ignorance of the particular indicators in respect of each affection, and I shall say no less now, that it is useful both for those things [being considered] and those things to come presently. For when there is one condition in the body, it is not difficult to discover what the cure of this is, if not by reason, at least by experience. When two or three conditions are intermingled, and particularly when they require entirely opposite remedies to each other, it is impossible to discover by experience what one must do; nor is it easy by reason. It is also necessary to know the substance of each of the conditions precisely, and to take the specific indication of treatment in the case of each of them, and which one it is necessary to cure first or more than the others, and which second or less, and which third. This is the reason why the majority of doctors don't know that cooling is a specific cure of all fevers, insofar as they are fevers, whether it is this by virtue of function or by virtue of potency, or that contingently there are many other things, cooling by nature, that are remedies of fevers. Something was already
said about these matters in what has gone before and will be said again in what follows. For some remedies have an intermixed and, somehow or other, twofold potency as they bring benefit both primarily and contingently—bathing is an example.

A drink of cold water brings benefit of itself, being cooling in its action. The juice of unripe grapes is a cooling

δυνάμει ψυκτικὸν φάρμακον. ταυτὶ μὲν οὖν ἀμφότερα
ψύχει πρώτως, ὅπερ ἐστὶ καθ᾽ ἑαυτά, διὰ μέσου μηδε-
νός· αἵματος δ᾽ ἀφαίρεσις, ὡς ἔμπροσθεν ἐδείκνυμεν,
ἐπὶ τῶν συνόχων καλουμένων πυρετῶν οὐκέτι πρώτως,
ἀλλὰ κατὰ συμβεβηκὸς ἰαταί ποτε τὰς θερμὰς δυσκρασίας. ὡσαύτως δὲ καὶ κλυστὴρ καὶ κάθαρσις ἐπι-
βροχή τε καὶ κατάπλασμα διαφορητικὸν ἢ πεπτικόν.
ἑκατέρας δὲ τὰς δυνάμεις ἔχει συλλαβὸν ἐν ἑαυτῷ τὸ
προκείμενον ἐν τῷ λόγῳ λουτρόν, ὡς ἂν οἶμαι σύνθετον ὑπάρχον ἐκ διαφερόντων ταῖς δυνάμεσι τῶν
ἑαυτοῦ μερῶν. εἰσελθόντες μὲν γὰρ ὁμιλοῦσιν ἀέρι
θερμῷ, μετὰ δὲ ταῦτα εἰς ὕδωρ εἰσίασι θερμόν, εἶτ᾽
709K ἐξελθόντες εἰς ψυχρόν, εἶτ᾽ ἀπομάττονται τὸν | ἱδρῶτα.

δύναται δὲ τὸ μὲν πρῶτον αὐτοῦ μέρος θερμῆναί τε
δι᾽ ὅλου τοῦ σώματος καὶ χέαι τὰς ὕλας, ὁμαλῦναί τε
τὰς ἀνωμαλίας, ἀραιῶσαί τε τὸ δέρμα καὶ κενῶσαι
πολλὰ τῶν ἔμπροσθεν ὑπ᾽ αὐτοῦ κατεχομένων. τὸ δὲ
δεύτερον, ὅταν ἐπὶ ξηρᾷ διαθέσει τοῦ σώματος αὐτῷ
τις χρῆται, νοτίδα χρηστὴν ἐνθεῖναι τοῖς στερεοῖς τοῦ
ζῴου μορίοις. ἡ δὲ τρίτη μοῖρα τῶν λουτρῶν, ἐπειδὰν
τῷ ψυχρῷ τύχωμεν χρώμενοι, ψῦξαί τε σύμπαν τὸ
σῶμα καὶ πυκνῶσαι τὸ δέρμα καὶ ῥῶσαι τὰς δυνάμεις.
τὸ δὲ τέταρτον ἐκκενῶσαι δι᾽ ἱδρώτων τὸ σῶμα χωρὶς
τῆς ἐκ τοῦ ψύχεσθαι βλάβης. ταῦτ᾽ οὖν δυναμένου
περὶ τὸν ἄνθρωπον ἐργάζεσθαι πάντα τῶν ποτίμων
ὑδάτων εὐκράτου λουτροῦ καὶ πρὸς τούτοις ἔτι φρίκας
τε καὶ πυκνώσεις τοῦ δέρματος, ὅταν ἀκαίρως λού-
σωνται, πειρατέον ἐφεξῆς διορίσαι πάντα, περὶ πρώ-

medication in its potency. Thus, both these cool primarily, which is [to say] by virtue of themselves, there being no intermediary. However, removal of blood, as I showed before, does on occasion cure hot *dyskrasias* in the so-called continuous fevers, no longer primarily but contingently. In like manner also, clysters, purging, fomentations and poultices promote dispersion or digestion. Bathing, as proposed in the discussion, possesses each of the two potencies which it brings together in itself, just as, I believe, a compound derived from the components of itself that differ in their potencies [would do]. People come into contact with hot air when they go into [the bathhouse], and after that, when they enter the hot water, and then when they go out into the cold, they wipe off the sweat.

709K

The first component of this [sequence] is able to heat and dissolve the materials through the whole body, regulate irregularities, reduce the density of the skin, and empty out many of those things previously retained by it. The second [component], when someone uses it due to a dry condition of the body, puts useful moisture into the solid parts of the organism. The third component of bathing, when we happen to use cold water, cools the whole body, thickens the skin, and strengthens the capacities. The fourth component evacuates the body through sweats without the harm of cooling. Since a *eukratic* bath of fresh water is able to effect all these things for the person, and further, in addition to these, to bring about shivering and thickening of the skin whenever people bathe at an inappropriate time, we must attempt to distinguish everything in order,

τῶν ἐκείνων τὸν λόγον ποιησαμένους ὅσοι φρικώδεις
γίγνονται λουόμενοι. μία μὲν γὰρ αὐτοῖς ἐστιν ἡ ὡς
ἂν εἴποι τις αἰτία συνεκτικὴ τοῦ γινομένου συμπτώματος, ἄλλαι δ' ἐκείνης πλείους προηγοῦνται.

710K   τὴν μὲν δὴ τὸ φρίττειν ἐργαζομένην αἰτίαν | εἶναί
φημι κίνησιν ἀθρωοτέραν τῶν δακνωδῶν περιττωμάτων. ὅθεν οἷς ταῦτα μὲν πολλά, πυκναὶ δὲ αἱ σάρκες
ἢ τὸ δέρμα φρίττουσιν, ἄν τε εἰς βαλανεῖον εἰσέλθω-
σιν, ἄν τε ἐν ἡλίῳ στῶσιν, ἄν τε κινηθῶσιν ὁπωσοῦν
σφοδρότερον ἤτοι δι' αἰωρήσεων ἢ τρίψεων ἢ γυμνα-
σίων· οἷς δ' ὀλίγα ταῦτ' ἐστὶ καὶ ἡ τοῦ σώματος ἕξις
ἑτοίμη χαλασθῆναι καὶ ἀραιωθῆναι πρὸς τῆς ἀμφ'
αὐτοῦ θερμασίας, οὐ μόνον οὐ φρίττουσιν, ἀλλὰ καὶ
βελτίους γίγνονται κενουμένων τῶν περιττωμάτων.
εἰκότως τοίνυν οὔτ' ἐν τῇ τῆς ἐπισημασίας ἀρχῇ τις
ἔλουσεν ἄρρωστον οὔτ' ἐν ἐπιδόσει· πεπύκνωται γὰρ
ἐν ἐκείνῳ τῷ χρόνῳ καὶ πεπίληται τὸ δέρμα καὶ ἡ
ὑποκειμένη μετ' αὐτὸ σαρκώδης οὐσία. κατὰ δὲ τὰς
παρακμὰς ἤτοι γε ἀρχομένας ἢ προιούσας ἢ προελ-
θούσας ἐπιπλέον ἡμεῖς τε πολλάκις ἄλλοι τέ τινες
ἰατροὶ πολλοὺς τῶν καμνόντων λούσαντες ὠφέλησαν
οὐ σμικρά. πότε μὲν οὖν χρὴ τῆς παρακμῆς ἀρχομένης ἢ προελθούσης ἢ καὶ κατὰ τὴν ἀκμὴν ἐνίοτε,
καὶ γὰρ καὶ τοῦτο συμβαίνει ποτὲ σπανίως, ἐπὶ τὸ
βαλανεῖον ἀπάγειν ἐν τοῖς ἑξῆς διοριῶ.

711K   νυνὶ δὲ ὑπὲρ τῆς καθόλου δυνάμεως | ἑκάστου τῶν
τοῦ λουτροῦ διελθεῖν προθέμενος ἓν τὸ πρῶτον πάν-

74

beginning the discussion with those who develop shivering when they bathe. For, in them, one cause is what one might call the "synektic"[16] cause of the symptom when it occurs, although there are other causes—and more of them—which precede this.

So then, I say the effective cause of the shivering is the   710K
concentrated movement of the gnawing superfluities. This is why those who have many of these superfluities, but in whom the flesh or skin is thick, shiver if they go to the bathhouse, or stand in the sun, or move too vigorously in any way whatsoever, either through passive exercises, rubbing, or active exercises. On the other hand, in those who have few [of these superfluities], the state of the body is readily relaxed and made less dense by the heat around it. Not only do they not shiver, but they also become better when the superfluities are evacuated. Therefore, it is reasonable for someone to bathe the patient neither in the accession of the manifestation nor in the progression, because the skin is thickened at that time and condensed, and the underlying fleshy substance with it. Often both I myself and certain other doctors helped many patients quite considerably by bathing in the abatement, accession, progression or exacerbation [of the fever]. I shall define in what follows when you should lead [the patient] to the bathhouse, either while the abatement is beginning or after it has advanced, or sometimes even at the peak, for this also happens (albeit infrequently).

Now, since I have undertaken to go over the general po-   711K
tency of each of the components of bathing, I shall dis-

16 This is the only use of this Stoic causal term in the *MM*; see I. Johnston (2006), pp. 34–35, and the excerpt from *De causis pulsuum* translated on p. 111.

των δίειμι καί φημι πολλοὺς μὲν καὶ τῶν ἀπεπτη-
σάντων, ἔτι δὲ πλείους οἷς πλῆθός ἐστι δακνωδῶν
περιττωμάτων, ἅπαντάς τε τοὺς ἐν ἐπισημασίαις ἢ
ἀναβάσεσι, καὶ τοὺς ἐν ἀκμῇ δὲ πλὴν ὀλίγων δή
τινων, εἰκότως φρίττειν, ἄν τ' εἰς βαλανεῖον εἰσέλθω-
σιν, ἄν τ' ἐν ἡλίῳ θερμῷ στῶσιν, ἄν τε γυμνασίοις ἢ
τρίψεσιν ἢ αἰωρήσεσιν ἐπιχειρήσωσιν· ἕκαστον γὰρ
τῶν εἰρημένων ἀθρόαν ὁρμὴν ἐργάζεται τῶν περιττωμάτων. ὅταν δὲ ἔτι μὲν ὑπάρχῃ πυκνὸν οὑτωσὶ τὸ
δέρμα τῶν λουομένων, ὡς κατὰ τὴν πρώτην τοῦ περιέχοντος ἀέρος προσβολήν, μὴ δύνασθαι τὴν κατὰ
φύσιν ἀπολαβεῖν ἑαυτοῦ διάθεσιν, ἀλλὰ χρόνου πλέονος εἴη εἰς τοῦτο δεόμενον, ἀθρόα δὲ ὁρμήσῃ πρὸς
τοὔκτὸς φέρεσθαι τὰ περιττά, κατέχεσθαί τε τοὐντεῦ-
θεν ἀναγκαῖόν ἐστιν αὐτοῖς καὶ δάκνειν, ἀθροιζομέ-
νοις ὑπὸ τὸ δέρμα καὶ διὰ τῶν σαρκῶν φερομένοις.
δέδεικται δ' ἐν Ταῖς τῶν συμπτωμάτων αἰτίαις ὡς τὰ
δακνώδη περιττώματα διὰ τῶν αἰσθητικῶν σωμάτων
φερόμενα ῥίγη τε καὶ φρίκας ἐργάζεται. τοῦ μὲν δὴ
712K φρίττειν ἐν τοῖς λουτροῖς ἅπαντας | τοὺς εἰρημένους
ἐξευρήκαμεν ἤδη τὰς αἰτίας, ὥσπερ γε καὶ τοῦ μὴ
φρίττειν οἷς ἤτοι μηδ' ὅλως ἐστὶ δακνῶδες μηδὲν ἐν
τῷ σώματι περιττὸν ἢ τοσοῦτον ὡς ῥᾳδίως ἐκκενοῦσθαι.

περὶ τούτων οὖν αὖθις αὐτῶν ὁ λόγος ἡμῖν γιγνέσθω, δι' οὓς καὶ τῶν φριττόντων ἐμνημονεύσαμεν. εἰ
γάρ τις εἰσελθὼν εἰς βαλανεῖον οὐκ ἔφριξεν, ἀλλ'
ἐχαλάσθη τε καὶ ἠραιώθη τὸ δέρμα, τούτῳ πάντως

cuss the first one of all and say that many of those who suffer failure of digestion (apepsia) in whom there is a large quantity of gnawing superfluities, and all those in the accessions or progressions [of the fever], and those at the peak, apart from a few, are likely to shiver should they go into the bathhouse, or stand in hot sun, or attempt gymnastic exercises, rubbings or passive exercises, because each of the things mentioned brings about a concentrated movement of the superfluities. And besides, whenever the skin of those bathing is thick such that, at the first impact of the surrounding air, it is unable to recover its own normal condition but requires a longer time for this purpose, it will set in motion a rush to carry the collected superfluities to the outside. And it is inevitable for them to be held back from here and to gnaw, since they are collecting together under the skin and being carried through the flesh. I have shown in the work *On the Causes of Symptoms* that the gnawing superfluities which are carried through bodies endowed with sensation bring about rigors and shivering.[17] So then, we have now sought out all the stated causes of shivering in baths, just as we also have of not shivering in baths where there is altogether no gnawing superfluity in the body, or it is such that is easily evacuated.

712K

Therefore, let my discussion once more be about those very things due to which I made mention of those who shiver. If someone does not shiver on entering the bath but has skin that is relaxed and of loose texture, what will undoubtedly and inevitably follow are, as previ-

[17] *De symptomatum causis*, Book 2, chapters 5–6, VII.180–84K, and I. Johnston (2006), pp. 254–55.

ἀναγκαῖον ἕπεσθαι τὰ πρόσθεν λεχθέντα, κένωσιν
τῶν περιττῶν, ὁμαλὴν θερμότητα δι' ὅλου τοῦ σώματος, ἀραίωσιν τῶν πόρων, χάλασιν τῶν συντεταμέ-
νων, χύσιν τῶν πεπιλημένων. ἡ μὲν οὖν κένωσις τῶν
περιττωμάτων χρησιμωτάτη πᾶσι πυρετοῖς ἐστιν·
ὡσαύτως δ' ἀραίωσίς τε καὶ χάλασις· οὔτε δὲ ἡ χύσις
οὔθ' ἡ θερμότης. ἀλλ' ἡ μὲν θερμότης ἅπασιν ἐναντία·
ψύχεσθαι μὲν γὰρ αὐτῶν ἡ διάθεσις, οὐ θερμαίνεσθαι
δεῖται. τὸ διαχεῖσθαι δ' ὁμαλῶς τοῖς μὲν στερεοῖς τοῦ
ζῴου μορίοις οὐκ ἀνεπιτήδειον, τοῖς χυμοῖς δ' ἀεὶ
λυσιτελές, ἀλλ' ὅταν ἤτοι φλεγμονή τις ἢ ὁμοία φλεγμονῇ διάθεσις ὑπάρχῃ κατὰ τὸ ζῷον, ἢ πλῆθος
713K ὁποτερονοῦν, εἴτε τὸ πρὸς τὴν δύναμιν εἴτε τὸ | πρὸς
τὴν εὐρυχωρίαν τῶν ἀγγείων, ἁπάντων ἐστὶ βλαβερώτατον. αὐξάνονται μὲν οὖν αἱ φλεγμοναὶ τῶν θερμανθέντων καὶ χυθέντων χυμῶν ἐπιρρεόντων αὐταῖς.
διατείνονται δ' οἱ χιτῶνες τῶν ἀγγείων, μὴ στέγοντες
τὸ πλῆθος ἐν τῇ χύσει πνευματωθέν· ἡ δύναμις δ'
αὐτῷ τούτῳ κακοπαθεῖ. ταῖς τοίνυν θερμαῖς καὶ
ξηραῖς διαθέσεσι τοῦ σώματος, ὑπὲρ ὧν ὁ λόγος ἦν,
ἀποχεῖσθαι μὲν τὰ δακνώδη χρηστόν, ἀκίνδυνος γὰρ
ἡ χύσις, οὐκ ἀβλαβὴς δ' ἡ θέρμανσις. ἐκ μὲν δὴ τοῦ
πρώτου μέρους τῶν λουτρῶν μεμνῆσθαι χρὴ ταῦθ'
ὑπάρξαντα τοῖς οὕτω κάμνουσιν, ὅταν ἐν καιρῷ λού-
ωνται.

μετίωμεν δήπου πρὸς τὸ δεύτερον αὐτῶν μέρος,
ὅπερ ἦν αὐτὸ τὸ κυριώτατον προσαγορευόμενον λουτρόν. ἐν τούτῳ τοίνυν ὑγραινόμενοι μὲν ὀνίνανται,

ously stated: evacuation of superfluities, an even heat through the whole body, rarefaction of the pores, relaxation of those things under tension, and dissolution of those things that have been condensed. The evacuation of the superfluities is very useful for all fevers. The same applies to rarefaction and relaxation but not to dissolution or heating. Heating is inimical to all [fevers] because their condition needs to be cooled, not heated. The regular dispersal to the solid parts of the organism is not without benefit, although it is not always useful to the humors. But whenever either inflammation or a condition akin to inflammation exists in the organism, or an abundance of either, it is most harmful of all to either the capacity or the lumen of vessels. Thus, inflammations are exacerbated when humors that are heated and dissolved flow into them. The walls of the vessels become distended if they do not contain the *pneuma*-infused abundance in the flow, while the capacity suffers harm due to this very thing. Thus, for the hot and dry conditions of the body, which are what the discussion is about, it is useful for what is gnawing to pour out because the dissolution is without danger, whereas heating is not without harm. So then, it is necessary to call to mind the things which befall patients in this way from the first component of bathing whenever they bathe at the proper time.

713K

Let me proceed now to the second component of baths which was called the most important. In this, when patients are moistened, they are benefited, whereas when

θερμαινόμενοι δ' οὐδὲν ὀνίνανται. μετέλθωμεν οὖν
αὖθις ἐπὶ τὸ τρίτον, ἐν ᾧ ψύχεται μὲν ἀλύπως τὰ
τεθερμασμένα, ῥώννυται δ' ἡ δύναμις. ὅσα δ' ἠραιώθη
τε καὶ περαιτέρω τοῦ προσήκοντος ἐχαλάσθη, ταῦτα
εἰς τὴν κατὰ φύσιν ἐπανέρχεται συμμετρίαν, ὡς δηλοῖ
καὶ τὸ τέταρτον τοῦ λουτροῦ μέρος. ἐφ' ὧν γὰρ
714K ἐπράχθη καλῶς | πάντα, καὶ μετὰ τὴν τοῦ ψυχροῦ
χρῆσιν ἱδροῦσιν ἔτι καὶ πάντ' αὐτῶν ἐκκενοῦται τὰ
περιττά.

καὶ τοίνυν ἐκ πάντων ὧν εἴπομεν περὶ βαλανείων
ἀθρόον κεφάλαιον συμβαίνει, κεκενῶσθαι μὲν ὅσον
ἦν ἐπὶ τῷ σώματι λιγνυῶδές τε καὶ καπνῶδες· ἐν δὲ τῇ
κατὰ φύσιν ὑπάρχειν συμμετρίᾳ τὰς σάρκας καὶ τὸ
δέρμα, τὸν δὲ τῶν στερεῶν μορίων αὐχμὸν πεπαῦ-
σθαι, καὶ τὴν θερμασίαν οὐ μόνον τὴν ἐκ τοῦ λουτροῦ
προσγινομένην, ἀλλὰ καὶ τὴν ἔμπροσθεν οὖσαν ὑπὸ
τῆς τοῦ ψυχροῦ χρήσεως ἀναιρεῖσθαι. μέγιστον δὲ
τοῦ λόγου τεκμήριον ἡ συμβαίνουσα κατάστασις ἐπὶ
τοῖς τοιούτοις λουτροῖς, ὅταν ὁδοιπορήσωμεν ἐν ἡλίῳ
θερμῷ. παραγινόμεθα μὲν ἐπ' αὐτὰ μηδὲ φθέγξασθαι
δυνάμενοι διὰ τὴν ξηρότητα τῆς γλώττης καὶ φάρυγ-
γος, ἅπαν τε τὸ σῶμα καρφαλέον ἔχοντες. ἐξελθόντες
δὲ τοῦ ψυχροῦ παραχρῆμα πάντ' ἀνακτώμεθα τὰ κατὰ
φύσιν, οὔτε τῇ πυρώδει θερμασίᾳ κάμνοντες οὔτε τῇ
ξηρότητι δυσφοροῦντες ἑτοίμως τε φθεγγόμενοι καὶ
τῆς δίψης τὸ πλεῖστον ἰαθέντες.

ἆρ' οὖν ἐναργέστερον ἔτι δύναμιν λουτρῶν γνῶναι
ποθεῖς ἐπὶ ξηραῖς καὶ θερμαῖς σωμάτων διαθέσεσιν;

80

they are heated they are not. Let me move on once more to the third [component of bathing] in which those things that have been heated are painlessly cooled and the capacity is strengthened. Those things that are rarefied and relaxed beyond what is appropriate return to a normal balance, as the fourth component of bathing makes clear. For in those patients in whom everything was done properly, after the use of cold [water] they both sweat further 714K and evacuate all their superfluities.

Therefore, if all those things I said about bathing are gathered together under one heading, there is evacuation of whatever in the body is sooty and smoky, both flesh and skin exist in a normal balance, the dryness of the solid parts ceases, and the heat, not only that which was added by the bath but also that which previously existed, is taken away by the use of cold [water]. The state which occurs due to such baths, whenever we travel in the hot sun, provides the greatest proof of the argument. We come to the baths unable to speak due to the dryness of the tongue and throat, and with our whole body parched. However, when we come out of the cold [water], we immediately recover our accord with nature in every way, we are not distressed by the burning heat, nor are we adversely affected by the dryness. We speak with ease and are cured of thirst to a great extent.

Do you, then, need to know more clearly still the potency of baths in dry and hot conditions of bodies? I don't 715K

715K ἐγὼ μὲν οὐκ | οἶμαι. πάρεστι δὲ τοῖς βουλομένοις πει-
ραθῆναι μετὰ τὴν τοιαύτην ὁδοιπορίαν ἀλουτήσασι
τῆς ἑπομένης βλάβης. ἢ γὰρ εὐθέως πυρέξουσιν ἢ
πολλῆς ἄσης ἀνάπλεοι διατελέσουσι βαρυνόμενοι
τὴν κεφαλὴν καὶ μάλιστα ἐὰν μηδέπω τῷ ψυχρῷ
σβέσωσι τὸ καῦμα. πολλοὶ μέντοι νεανίσκοι μετὰ τὰς
τοιαύτας ὁδοιπορίας εὐθέως εἰς ὕδωρ ψυχρὸν ἑαυτοὺς
ἐπιρρίψαντες ὠνίναντο, καὶ μάλισθ᾽ ὅσοι περ ἂν ὦσιν
ἰσχυροὶ καὶ ψυχρῶν λουτρῶν ἐθάδες. οὕτω δὲ κἂν τοῖς
ἀγροῖς, ἐν οἷς οὐκ ἔστι βαλανεῖα, πράττουσιν, εἰς
λίμνας ἢ ποταμοὺς ἑαυτοὺς ἐμβάλλοντες οὐδενὸς ἰατροῦ συμβουλεύσαντος αὐτοῖς, ἀλλ᾽ ὑπὸ τῆς διοι-
κούσης τὸ σῶμα φύσεως ἀγόμενοι πρὸς τὸ δέον, ἥτις
καὶ τοῖς ἀλόγοις ζῴοις τὰς ἐπὶ τἀναντία τῶν λυπούν-
των ὁρμὰς ἐντίθησι· λούεται γὰρ κἀκεῖνα ψυχρῷ τῷ
θάλπει καταπονούμενα, καθάπερ γε καὶ θερμὰς εὐνὰς
ἐξευρίσκει τῷ κρύει κάμνοντα. κατὰ δὲ τὴν τοιαύτην
ἐναντίωσιν ἐσθίει μὲν πεινῶντα, πίνει δὲ διψῶντα, καὶ
τἆλλα πάντα πράττει φύσει. καὶ εἴπερ γε διάγνωσιν
ἀκριβῆ τῆς τῶν πυρεττόντων εἴχομεν φύσεως, ἐτολ-
716K μῶμεν ἄν, οἶμαι, συνεχῶς ἐξ αὐτῶν | λούειν ἐν ὕδασιν
ἐνίους ψυχροῖς βαλανείου χωρίς. ὅτι μὲν γάρ εἰσί
τινες οἱ δεόμενοι τούτου δῆλον ἐκ τῶν ὠφεληθέντων,
οὓς οὐκ ἂν ὤνησε τὸ ψυχρὸν λουτρόν, εἴπερ μὴ δι-
έκειντο κατ᾽ ἐκεῖνον τὸν καιρὸν ἐπιτηδείως πρὸς αὐτό.

τῷ δ᾽ ἀγνοεῖν ἡμᾶς ἀκριβῶς τὰς διαθέσεις μεγίστην τε τὴν ἐκ τῆς ἀποτυχίας ὑπάρχειν βλάβην
ἀφιστάμεθα τῶν τοιούτων βοηθημάτων καὶ μάλιστα

think so! It is possible for those who wish to take the chance of subsequent harm not to bathe after such a journey. They will either become febrile immediately or will carry on, full of great distress and with a heavy head, especially if the heat is not yet quenched by cold [water]. However, many young men are immediately benefited after such journeys when they throw themselves into cold water. This particularly applies to those who are strong and accustomed to cold baths. And even in the countryside where there are no bathhouses they act in this way, throwing themselves into lakes or rivers without any doctor advising them, but being led to what is needed by the nature which governs the workings of the body, and which also puts into irrational animals the impulses to do what is opposite to those things that are distressing. They too bathe in cold water to counter the effects of heat, just as those suffering from freezing cold will find hot beds. In respect of this kind of opposition, those who are hungry, eat and those who are thirsty, drink, and they do all other things naturally. And if, indeed, we do have a precise diagnosis of the nature of those who are febrile, we will be confident, I believe, to bathe some of them continuously in cold waters quite 716K apart from the bathhouse. For it is clear from those who have derived benefit that there are some who do need this, but whom cold bathing would not have benefited if they were not suitably disposed toward it at that time.

But because we don't know the conditions precisely—the greatest harm being from failure—we avoid such remedies, and particularly in those with hectic fevers, as the

ἐπὶ τῶν ἑκτικῶν πυρετῶν, ὡς ἂν ὀλιγαίμου καὶ ὀλιγο
σάρκου κἀπιμέλου τοῦ σώματος αὐτοῖς γεγονότος.
οὐδὲν γὰρ ἔχουσι τῶν στερεῶν μορίων πρόβλημα τὴν
ἐκ τοῦ ψυχροῦ προσβολὴν ἀλύπως ἐκδεχόμενον, ἀλλὰ
αὐτοῖς εὐθέως προσπίπτει τοῖς ὁμοιομερέσιν ὀνομαζο
μένοις, ὑφ' ὧν ἅπασι τοῖς ζῴοις αἱ ἐνέργειαι γίνονται.
νέος δ' ἄν τις εὔσαρκος ὥρᾳ θέρους ἐν ἀκμῇ πυρε
τώδους νοσήματος, ἄνευ σπλάγχνου φλεγμονῆς, εἰς
ψυχρὸν ἑαυτὸν ἐμβαλὼν ὕδωρ ἱδρῶτα κινήσειεν· εἰ δὲ
καὶ ψυχρολουσίας συνήθης εἴη, πάνυ θαρρῶν χρή
σοιτο τῷ βοηθήματι. ἀλλὰ περὶ μὲν τῶν τοιούτων
σωμάτων αὖθις εἰρήσεται.

τοῖς δὲ τὸν ἑκτικὸν πυρέττουσι πυρετὸν ἐπὶ θερμῇ
καὶ ξηρᾷ κράσει καὶ μάλιστα τοῖς ἤδη μαραινομένοις |
717K οὐκ ἀσφαλὲς εἰς ὅλον ἀναρρίπτειν, ἐπὶ τοιοῦτον ἰοῦσι
βοήθημα· καθάπερ οὐδ' ὅσοι θέρους ὥρᾳ θερμοῦ καὶ
ξηροῦ μακροτέραν ὁδὸν ἀνύσαντες ἰσχνοὶ καὶ ἀσθε
νεῖς ὄντες ἐμψυχθῆναι δέονται· οὐδὲ γὰρ οὐδὲ τούτοις
ἀκίνδυνος ἡ τοῦ ψυχροῦ χρῆσις ἄνευ τοῦ κατὰ βαλα
νεῖον ὁμαλῶς προθερμανθῆναι. τοιοῦτον γάρ τοι συμ
βαίνειν ἔοικεν ἡμῖν εἰς τὴν ψυχρὰν δεξαμενὴν εἰσιοῦ
σιν ἐπὶ τοῖς βαλανείοις, οἷόν τι καὶ τῇ τοῦ σιδήρου
βαφῇ· καὶ γὰρ ψυχόμεθα καὶ τονούμεθα, καθάπερ
ἐκεῖνος, ἐπειδὰν διάπυρος γενόμενος ἐμβάπτηται τῷ
ψυχρῷ. καὶ τούτου χάριν ἐπὶ τῶν ἀσθενεστέρων σω
μάτων εὕρηται τὰ βαλανεῖα, προθερμαίνοντα καὶ προ
παρασκευάζοντα τῷ ψυχρῷ λουτρῷ. τοιοῦτον δὲ δή τι
καὶ οἱ χωρὶς τοῦ βαλανείου χρώμενοι τῷ ψυχρῷ πράτ

body is relatively bloodless in them, and lacking in flesh and fat. They have no barrier of the solid parts to receive the impact of the cold [water] painlessly, but this immediately falls upon the so-called *homoiomeres* themselves through which the functions arise in all living creatures. Should some corpulent young man, in the season of summer, at the peak of a febrile disease, without inflammation of an internal organ, throw himself into cold water, he would bring on sweating. If, however, he is also accustomed to bathing in cold water, one can use the remedy with great confidence. But I shall speak further about such bodies again.

In those who are febrile with a hectic fever due to a hot and dry *krasis*, and especially in those already marasmic, it   717K
is not safe to run the risk of them proceeding to such a remedy, just as it is not for those who need to be cooled when, in the heat and dryness of summer, they have made an overly long journey and are thin and weak. Nor is the use of cooling without danger to them, if they have not been previously heated evenly in the bathhouse. Such a thing seems to happen to us when we go into the cold tank at the bathhouses, like the tempering of iron, for we too become cooled and braced, just as the iron does when, having become red hot, it is dipped into cold [water]. This is why bathhouses have been devised for weaker bodies—as prior heating and preparation for the cold bath. However, people also act in this way who use cold water apart from the

τουσιν, ὅταν προγυμνασθέντες εἰς αὐτὸ καθάλλωνται. οἷον γάρ τι τὸ βαλανεῖον ἡμῖν ἐστι, τοιοῦτον ἐκείνοις τὸ γυμνάσιον οὐ μόνον ἐκθερμαῖνον, ἀλλὰ καὶ τὴν ἐκ τοῦ βάθους κίνησιν τῆς ἐμφύτου θερμασίας πρὸς τοὐκτὸς ἐργαζόμενον, ὥστ᾽ ἀπαντῆσαί τε τῷ προσπίπτοντι ψυχρῷ καὶ ἀπομαχέσασθαι καὶ κωλῦσαι

718K βιαίως ἐμπεσεῖν τῷ βάθει | καὶ πλῆξαί τι τῶν σπλάγχνων. οὔτε γὰρ τὴν οὐσίαν αὐτὴν τοῦ ψυχροῦ μέχρι τῶν σπλάγχνων ἐξικέσθαι καλὸν οὔτε ἀκραιφνῆ τὴν ποιότητα· βέλτιον δὲ καὶ ἀσφαλέστερόν ἐστι τῷ τῆς διαδόσεως λόγῳ πρὸς τὸ βάθος ὁδοιπορῆσαι μόνην τὴν ποιότητα χωρὶς τῆς οὐσίας.

καὶ τοίνυν καὶ ὅσοι τῶν πυρεττόντων ἑκτικῶς ἐπὶ τὸ βαλανεῖον ἥκουσιν, εἰ μὴ τῷ ψυχρῷ βαφεῖεν, οὐδὲν ὀνίνανται. τῶν μὲν γὰρ ἄλλως πυρεττόντων οὐδέπω τὰ στερεὰ μόρια δύσλυτον ἔχει τὴν θέρμην. ὥσθ᾽ ἱκανὴ βοήθεια τούτοις ἐστὶν ἡ ἐκ τῶν ἱδρώτων ἅμα ταῖς ἀδήλοις αἰσθήσει διαπνοαῖς. καὶ τῶν γ᾽ ἐφημέρων πυρετῶν αὐτὴν σχεδόν γε τὴν οὐσίαν ἐκκενοῦσθαι συμβαίνει δι᾽ αὐτῶν. ἐπὶ δὲ τῶν διὰ ξηρότητα καὶ θερμασίαν ἑκτικοῖς ἁλόντων πυρετοῖς ὁμοία τοῖς διαπύροις σιδήροις ἐστὶν ἡ τῶν στερεῶν μορίων διάθεσις. ὥστ᾽ οὐκ ἀρκεῖ θερμῆναι καὶ τέγξαι μόνον ἐν τοῖς βαλανείοις αὐτούς, τροφὴν γὰρ ἂν οὕτω γε καὶ αὔξησιν δοίημεν τῷ πυρετῷ· βέλτιον δ᾽ ἐμβάπτοντας τῷ ψυχρῷ σβέσαι τὴν θέρμην.

719K ἀλλ᾽ ἐν τούτῳ κίνδυνος εἰς τοὐναντίον | ἀγαγεῖν καὶ ψῦξαι τὸ σῶμα. τίς δ᾽ οὔ φησιν; οὐ μὴν διὰ τοῦτό γε

bathhouse, when they leap down into it after exercising beforehand. For just as bathing is to us, so exercise is to them,
not only heating but also bringing about the movement of
the innate heat from the depths toward the exterior, so that
it encounters the cold which falls upon them, and counter-
acts and strongly prevents it intruding deeply and striking   718K
one of the internal organs. It is neither good for the actual
substance of the cold nor for the quality in its pure form to
reach as far as the internal organs. It is better and safer, in
theory, for the distribution of the quality only to come to
the depths without the substance.

Therefore, when those who have a hectic fever come to
the bath, if they are not dipped in cold water, they don't
benefit. Among those who are febrile for other reasons, the
solid parts do not yet have heat that is difficult to dissipate.
Consequently, an adequate remedy for them is that from
sweating along with the invisible but perceptible transpirations. And in the ephemeral fevers, what happens is that
the substance itself is almost evacuated through these
transpirations. In those who are seized by hectic fevers due
to dryness and heat, the condition of the solid parts is like
red-hot iron. As a result, it is not enough for them to be
heated and moistened only in the baths because, in this
way, we would be giving them nourishment and increasing
the fever. Better to dip them in cold water to quench the
heat.

But in this there is a danger of leading to the opposite
and cooling the body. Who denies this? For it is not, in fact,   719K

τὸ κατορθούμενον ἐν τῷ βοηθήματι προσήκει ψέγειν, οὐδ' ὅτι δύσληπτον τὸ μέτρον ἀφίστασθαι παντάπασιν. εἰ μὲν γὰρ οἷόν τ' ἦν ἑτέρως ἰᾶσθαι τούς τ' ἄλλους ἑκτικοὺς πυρετοὺς καὶ τοὺς μαρασμώδεις, ἄμεινον ἂν ἦν τὴν ἀσφαλεστέραν ἰέναι. ἐπεὶ δ' ἐν μὲν τῷ ψῦξαι καὶ ὑγρᾶναι τὸ κῦρος τῆς θεραπείας ἐστίν, ἅπασι δὲ τοῖς ψυκτικοῖς βοηθήμασιν ἐφεδρεύει βλάβη διὰ τὴν ἰσχνότητα τοῦ σώματος, ἀναγκαῖον οἶμαι γίγνεσθαι τὸ τοῦ Θουκυδίδου δράσαντάς τι καὶ κινδυνεῦσαι. οἷς μὲν γὰρ ἑτέρα μὲν οὐχ ὑπάρχει τῆς σωτηρίας ὁδός, ἡ δὲ οὖσα μόνη σφαλερὰ καθέστηκεν, ἀναγκαῖον, οἶμαι, τούτοις ἐστὶν ὁμόσε τοῖς δεινοῖς ἰέναι. οὐ μὴν οὐδὲ τὸ σφάλμα θανατῶδες οὐδ' ἀβοήθητον. οὓς γὰρ ἂν ἐν τῷ σβεννύναι τὸν πυρετὸν εἰς τὴν ἐναντίαν διάθεσιν ἀγάγῃ τὸ ψυχρόν, ἔνεστιν ἰάσασθαι θερμαίνοντας ἐν τῷ μετὰ ταῦτα χρόνῳ παντί. ἀλλ' οὐδ' ὁ κίνδυνος ἴσος οὔθ' ὁ τῆς διαθέσεως τῆς ψυχρᾶς οὔθ' ὁ τῶν ἰαμάτων αὐτῆς. ἐδείχθη γὰρ 720K ἤδη καὶ πρόσθεν ἐν τῷ τῆσδε τῆς πραγματείας | ἑβδόμῳ γράμματι, δέδεικται δὲ κἂν τῷ περὶ μαρασμοῦ βιβλίῳ καὶ πρὸ τούτων ἔτι κατὰ τὴν ὑγιεινὴν πραγματείαν, ὡς οὐχ οἷόν τε τὴν τῶν στερεῶν μορίων ξηρότητα τελέως ἰάσασθαι, καὶ ὡς, εἴπερ ἦν τοῦτο πρᾶξαι δυνατόν, ἀγήρως ἄν τις ἐγένετο τοιαύτῃ διαίτῃ χρώμενος.

εἴπερ οὖν ἀνίατος μέν ἐστιν ἡ ξηρότης τῶν στερεῶν σωμάτων, ὠκυτάτη δ' ἐπ' αὐτὴν ὁδὸς τέτμηται ἡ διὰ τῶν ἑκτικῶν πυρετῶν, ἄμεινον εἰς ψυχρὰν δυσκρα-

appropriate to find fault with what is successfully accomplished by the remedy, nor is it appropriate to avoid it altogether because the measure is hard to understand. If it were possible to cure both the hectic and marasmic fevers in another way, it would be better to travel the safer path. Since, however, the principle of treatment lies in cooling and moistening, and injury is associated with all the cooling remedies due to the thinness of the body, I think the statement of Thucydides, "doing something and facing the danger," necessarily arises.[18] For them there is no other path to safety, while the one path that does exist is perilous, making it necessary, I believe, to meet these terrible things head on. However, it is neither a fatal mistake nor is it irredeemable. For those whom the cooling brings to the opposite condition by quenching the fever can be cured by heating during the whole time after this. But the danger of the condition that is cold is not equal to the danger of its cures. It was already shown before, in the seventh book of this work, and it has been demonstrated in the book on 720K marasmus, and before these also in the work on health, that it is not possible to cure dryness of the solid parts completely, and that, if it were possible to do this, someone could defy aging using such a regimen.[19]

Therefore, if dryness of the solid bodies is incurable, then the swiftest path to have cut to it is that through the hectic fevers. It is better that the person, after changing to

[18] See Thucydides, 1.20.2.

[19] See Galen's *De marcore*, VII.666–704K, and *De sanitate tuenda*, particularly Book 5, chapters 3, 4, and 9.

σίαν μεταστήσαντα τὸν ἄνθρωπον ἔχειν ᾧ θεραπεύ-
σομεν. ὁ μὲν γὰρ τοῦτο πράξας ἀναμαχέσαιτ᾽ ἂν ἐξ
ὑστέρου τὴν βλάβην, ὁ δ᾽ ἐπιτρέψας ἰέναι τὴν ἐπὶ
θάνατον, ἀνέλπιστον τῷ κάμνοντι τὴν σωτηρίαν εἰργάσατο. ὅσῳ τοίνυν ἄμεινόν ἐστι τοῦ χωρὶς ἐλπίδος
ἀπολέσθαι βεβαίως τὸ σὺν ἐλπίδι χρηστῇ δράσαντάς
τι καὶ κινδυνεῦσαι, τοσούτῳ τὸ μετὰ μεγάλων βοηθημάτων ἀγωνίσασθαι τοῦ μηδὲν πράξαι βέλτιον,
ἐμψύχωμεν οὖν ἅπαντι τρόπῳ τοὺς ἑκτικῶς πυρές-
σοντας αὐτίκα, πρὶν προσελθόντας εἰς μαρασμὸν
τελευτῆσαι. τὸν δ᾽ ἀκριβῶς μαρανθέντα μηδ᾽ ἐπι-
χειρῶμεν ἰᾶσθαι. καὶ γὰρ εἰ τὴν θέρμην αὐτοῦ σβέ-
721K σαιμεν, | ἀλλ᾽ ἥ γε ξηρότης ὑπολειφθεῖσα γήρως
τρόπῳ τὸν ἄνθρωπον ἀπολεῖ, τοσοῦτον ἐπιβιώσαντα
χρόνον ὅσον ἀντισχεῖν τὰ στερεὰ μόρια πρὸς τὴν
ἐσχάτην ξηρότητα. τάχα δ᾽ ἄν τις ἕλοιτο τῶν οἰκείων
τοῦ νοσοῦντος, ἔτι τε μᾶλλον ὁ κάμνων αὐτὸς ἐπι-
βιῶναί τινα χρόνον ἐν γέροντος σχέσει μᾶλλον ἢ
τεθνάναι παραχρῆμα. θεραπεύειν οὖν καὶ τούτους
μετὰ προρρήσεως, εἰς ἕτερον μεθιστάντας μαρασμόν,
ὃν ἤδη καλεῖν ἔθος ἡμῖν ἐστιν ἐκ νόσου γῆρας·
ἐγχωρεῖ γὰρ αὐτοὺς οὐ μόνον ἡμέρας πλείους, ἀλλὰ
καὶ μῆνας ζῆσαι.

νεανίσκον γοῦν ἐγώ τινα τῶν οὕτως ἐχόντων ἰασά-
μενος, εἶθ᾽ ἑξῆς ἀνακομίζων καὶ οἷον γηροτροφῶν
αὐτοῦ τὴν πρεσβυτικὴν ἀσθένειαν, οὐ μόνον εἰς μῆνας, ἀλλὰ καὶ εἰς ἔτη διεφύλαξα ζῶντα. χρὴ δ᾽ οὐ νῦν
ἀκούειν ποθεῖν τοῦ τοιούτου μαρασμοῦ τὴν ἐπιμέλει-

a cold *dyskrasia*, has that by which we shall treat this. For when someone has done this, he may rejoin the battle against the harm later, whereas when he has allowed the harm to proceed toward death, he brings no hope of salvation for the patient. Therefore, by as much as it is better to do something with some hope, albeit with danger, than to face certain death bravely without hope, it is better to take up the fight with strong remedies rather than do nothing. So let us cool by every means those with a hectic fever straightaway before, progressing to marasmus, they die. Let us not, however, attempt to cure someone who is obviously marasmic. For even if we do quench his heat, the   721K dryness that remains will, in the manner of old age, destroy the man after he has lived for as long a time as the solid parts can hold out against the extreme dryness. Perhaps one of the relatives of the sick person, or even more, the sufferer himself might choose to survive for a time in an aged state rather than die immediately. Therefore, also treat these people with a prognostic statement that, after they have changed to another marasmus, which it is now our custom to call "aging through disease," it is possible for them to live not only for more days but even for months.

Anyway, when I cured a young man who was someone affected in this way, then in due course restored his "geriatric" weakness (like caring for the aged), not only did I keep him alive for months but even for years. You should not need to hear now the care of such a marasmus. It was

αν· ἔν τε γὰρ τῷ τῆσδε τῆς πραγματείας ἑβδόμῳ
προείρηται δυνάμει, τὰς ξηρὰς καὶ ψυχρὰς δυσκρασίας ἰωμένων ἡμῶν, ἔν τε τῷ γηροκομικῷ μέρει τῆς
Ὑγιεινῆς πραγματείας. ὁ δ' ἐνεστὼς λόγος οὐ μαρα-
σμῶν ἐπηγγείλατο θεραπείαν διδάξαι, ἀλλὰ πυρετῶν
722K ἑκτικῶν, ἐξ οὗ γένους εἰσὶ καὶ οἱ | μαρασμώδεις· εἴρη-
ται δή μοι περὶ τούτων ἄλλα τε πολλὰ λόγῳ καὶ πείρᾳ
κεκριμένα καὶ ὡς οὐ χρὴ δεδιέναι βαλανεῖον, ὥσπερ ὁ
Φίλιππος εὐλαβέστερον ἴσως αὐτῷ χρώμενος,

ὃν τρόπον ἐπὶ τῶν ἄλλων εἰθίσμεθα πυρετῶν ὅσοι
διὰ φλεγμονὰς καὶ σήψεις γίνονται χυμῶν. ἐπ' ἐκείνων
μὲν γὰρ οὐδὲν βλάψεις, εἰ καὶ μὴ βάπτοις εἰς τὸ
ψυχρὸν ὕδωρ αὐτούς, ἀλλ' ὡς εἴθισται περιχέοις χλιαρὸν
μὲν πρῶτον, εἶθ' οἷον ἐξ ἡλίου θερινοῦ, κἄπειτα τούτου
βραχύ τι ψυχρότερον. ἐπὶ δὲ τῶν ἑκτικῶν πυρετῶν οὐ
τὸ θερμὸν λουτρόν ἐστι τὸ τὴν ὠφέλειαν παρέχον,
ἀλλὰ τὸ ψυχρόν, ᾧ παρασκευάζει τὰ βαλανεῖα τὸ τοῦ
κάμνοντος σῶμα, καθάπερ ἐπὶ τῶν ἄλλων ἁπάντων
τῶν ἐν ὑγίᾳ λουομένων. οὐ μόνον δὲ διὰ τοῦτο δοκεῖ
μοι Φίλιππος ἀγνοεῖν ἐπὶ τῶν μαραινομένων ὁποῖόν τι
χρῆμά ἐστι βαλανεῖον, ὅτι τοὺς οὕτως ἰσχνοὺς οὐδεὶς
τολμᾷ βάπτειν εἰς τὴν ψυχρὰν δεξαμενὴν ἢ τὸ γοῦν
ὕστατον ὕδωρ καταχεῖν ψυχρόν, ἀλλὰ καὶ διότι πολλάκις ἔζευκταί τις ἄλλος αὐτοῖς πυρετὸς ἐπὶ χυμοῖς
723K σηπομένοις ἢ σπλάχνου φλεγμονῇ· | καὶ μὲν δὴ καὶ
ὅτι πάθος ὀλέθριον τῶν μαραινομένων αὐτὸ ὑπάρχει,
καὶ διὰ τοῦτο οὐκ ὀρθῶς ἔνιοι πάντα μέμφονται τὰ

previously spoken of in the seventh book of this treatise in relation to capacity, when I cured the dry and cold *dyskrasias*, and in the part on the care of the aged in the treatise *On the Preservation of Health*.[20] The present discussion made no promise to teach a treatment of marasmus—only of the hectic fevers from which class the marasmic disorders also are. Now I have said many other 722K things about these matters, determined by reason and experience, and that you must not be afraid of bathing, like Philippus, who perhaps used it too cautiously.

It is a method I am accustomed to use in the case of the other fevers such as arise due to inflammation and putrefaction of humors. In those fevers you will do no harm at all, even if you don't immerse the [patients] in cold water, but as is the custom, you first pour lukewarm [water] over them, next for example, water warmed by the sun's heat, and then water that is a little colder than this. In the hectic fevers, it is not the hot but the cold bathing that provides the benefit for which the baths prepare the patient's body, just as in the case of all the other people who bathe when healthy. Because of this, it seems to me that Philippus did not know what use bathing is for those who are wasting, not only because nobody dares to dip those who are thin in this way into the cold tank, or at least to pour extremely cold water on them, but also because some other fever is frequently joined with them due to putrefying humors or inflammation of an internal organ. Furthermore, this is a 723K fatal affection of those with wasting, and because of this, some quite wrongly blame all the things applied. There is

[20] See the reference to Galen's *De sanitate tuenda* in the previous note.

προσφερόμενα, δέον οὐκ ἐκείνοις, ἀλλὰ τῇ διαθέσει τὴν αἰτίαν ἀναφέρειν.

ἀναλαβόντες οὖν αὖθις ἅπερ ἐλέγομεν ἐπιθῶμεν ἤδη τῷ λόγῳ κεφαλήν. ἅπαντας τοὺς ἑκτικὸν νοσοῦντας πυρετὸν καὶ μᾶλλον ἐξ αὐτῶν ὅσοι περ ἂν ἤδη μαραίνωνται, χωρὶς τοῦ τινα ἕτερον ἐπιπεπλέχθαι πυρετὸν αὐτοῖς, ἤτοι γ' ἐπὶ σήψει μόνῃ χυμῶν ἢ μετὰ φλεγμονῆς, λοῦε θαρρῶν ἄνευ καμάτου παντός, ὡς μὴ καταλῦσαι τὴν δύναμιν· ὅπερ οὐχ ἥκιστα καὶ αὐτὸ βλάπτον ἰσχυρῶς αἴτιον γίγνεται τοῦ ψέγεσθαι τὸ λουτρόν. τὸ δ' ἄνευ καμάτου τοιόνδε τι λέγω. τὸν ἀρρωστοῦντα βούλομαι κομίζεσθαι μὲν ἐπὶ τοῦ σκίμποδος εἰς τὸ βαλανεῖον, ἑτοίμως δ' αὐτῷ παρεσκευασμένης σινδόνος θερμῆς, ἐπ' ἐκείνῃ ἐν τῷ πρώτῳ τῶν τριῶν οἴκων τοῦ βαλανείου μεταφέρεσθαι γυμνωθέντα. τέσσαρες δ' ἔστωσαν οἱ κρατοῦντες αὐτόν, εἷς καθ' ἕκαστον πέρας. εἰ μὲν οὖν αὐτάρκως εἴη θερμὸς ὁ πρῶτος οἶκος, ἐν αὐτῷ γυμνωθεὶς ὁ κάμνων εἰς τὸν δεύτερον εἰσκομιζέσθω· εἰ δ' ἔτι σοι φαίνοιτο ψυχρότερον εἶναι, μὴ γυμνός, ἀλλ' ἐπιβεβλημένος ἤτοι γ' 724K ἑτέραν σινδόνα | μὴ ψυχρὰν ἤ τι τοιοῦτον ἐπίβλημα. δῆλον δ' ὡς χρὴ τὴν μέλλουσαν ὀχεῖν αὐτὸν σινδόνα ἰσχυροτέραν ὑπάρχειν· εἰ δ' ἀσθενὴς εἴη, καθ' ἑαυτὴν ἐπιπτύσσοντα διπλῆν ἐργάζεσθαι βέλτιον. ἔστω δὲ ὁ μέσος οἶκος οὐ μόνον τῇ θέσει, ἀλλὰ καὶ τῇ κράσει τοσούτῳ τοῦ πρώτου θερμότερος ὅσῳ τοῦ τρίτου ψυχρότερος. ἐν τούτῳ τῷ οἴκῳ τὸ ἔλαιον ἔστω χλιαρόν, ὡς εὐθέως περιχυθῆναι τῷ κάμνοντι κατὰ τῆς σιν-

no need to attribute the cause to those things. Attribute it, rather, to the condition.

Therefore, taking up again what I was saying, let me now bring completion to the discussion. Confidently bathe without any trouble all those who are ill with a hectic fever, and particularly those who are already wasting, except when there is some other fever involved for them, due either to putrefaction of the humors alone, or with inflammation, as long as you don't break down their capacity. Not least, when this (i.e. bathing) causes severe harm, it becomes a reason to blame bathing. This is what I mean when I say "without any trouble." I like to convey the sick person to the bathhouse on a bed, having prepared for him a warm linen cloth ready to hand by which to transfer him in the first of the three rooms of the bathhouse after he is undressed. There should be four [people] to take hold of him, one to each corner [of the cloth]. If the first room is sufficiently hot, when the patient is undressed in it, let him be carried to the second room. However, if this still seems to you to be too cold, don't strip him but lay on either another linen cloth that is not cold or some such covering. It is clear   724K
that the linen cloth that is going to bear him must be quite strong. If it is weak, it is better to make it double, folding it on itself. The middle room must not only be middle in position but also in terms of *krasis*, the first room being hotter than it to the same extent that the third is colder. In this room the oil should be lukewarm so that it can be immediately poured over the patient after he has been carried in

δόνος εἰσκομισθέντι. καὶ τοῦτο πράξαντες, εἰς τὸν
τρίτον εἰσίτωσαν, εἰς τὴν δεξαμενὴν ἀποκομίζοντες
αὐτόν, ὥστε διόδῳ χρήσασθαι μόνῃ τοῖς τρισὶν οἴκοις
τοῦ βαλανείου, μὴ ταχέως βαδιζόντων τῶν εἰσκο-
μιζόντων τὸν κάμνοντα, ἀλλὰ τοσοῦτον ἐπιμεινάντων
κατὰ τοὺς πρώτους οἴκους ὅσον ἐν μὲν θατέρῳ μετα-
τεθῆναι μόνον ἀπὸ τοῦ σκίμποδος, ἐν δὲ τῷ μέσῳ
περιχυθῆναι τὸ ἔλαιον.

ἔστω δὲ καὶ ὁ ἀὴρ ὁ τῶν οἴκων ἁπάντων μήτε
θερμὸς ἄκρως μήτε ψυχρός, ἀλλ' εὔκρατος ἱκανῶς καὶ
μετρίως ὑγρός· ἔσται δὲ τοῦτο προεκχυθέντος ὕδατος
εὐκράτου δαψιλῶς ἐκ τῆς δεξαμενῆς, ὡς διαρρυῆναι
διὰ πάντων τῶν οἴκων. ἐν αὐτῷ δὲ τῷ τῆς κολυμ-
725K βήθρας ὕδατι χρονιζέτω μετρίως, | ὀχούμενος ἐπὶ τῆς
σινδόνος, οὐδεμιᾶς ἐπαντλήσεως αὐτῷ γενομένης, οἵ-
ας ἐπ' ἄλλων εἰθίσμεθα πράττειν. ἀλλ' οὐδὲ κατὰ τῆς
κεφαλῆς αὐτοῦ καταχεῖσθαι βουλοίμην ἄν· ἱκανὸν
γὰρ καὶ ταύτῃ δῦναι δὶς ἢ καὶ τρὶς ἅμα τῷ παντὶ
σώματι κατὰ τοῦ ὕδατος, ὑφιέντων μετρίως εἰς τὸ
κάτω τὴν σινδόνα, καὶ αὖθις ἀνακομιζόντων αὐτὸν
ὁμοίως νεανίσκων τεττάρων. ἐκκομισθεὶς δ' ἐντεῦθεν
εἰς ὕδωρ ψυχρὸν ἀθρόως βαπτέσθω[15] μηδ' ἐπὶ βραχὺ
χρονίζων ἐν αὐτῷ. δεδιδάχθαι δὲ χρὴ καὶ τἆλλα μὲν
ἀκριβῶς ἅπαντα τοὺς βαστάζοντας αὐτόν· ἐξαίρετον
δ' ἔτι τοῦτ' ἔστω δίδαγμά τε ἅμα καὶ παράγγελμα, τὸ
τάχος τῆς βαφῆς. εὐθὺς δ' ἕτερος ἑτοίμην ἔχων σινδόνα κατ' αὐτοῦ βαλέτω, κἄπειθ' ἑξῆς ἐπιβληθεὶς τῷ
σκίμποδι διὰ σπόγγων μὲν ἀποματτέσθω τὰ πρῶτα,

on the linen cloth. When [the attendants] have done this, let them enter the third room, carrying him to the bathing receptacle so as to make use of a single passageway between the three rooms of the bathhouse. Nor should those carrying the patient walk quickly; they should remain in the first room only for as long as it takes for him to be transferred onto something else from the bed, and in the middle room while the oil is poured over him.

Also, let the air of all the rooms be neither very hot nor very cold, but adequately *eukratic* and moderately moist. This will be so if *eukratic* water is abundantly poured forth from the receptacle so as to flow through all the rooms. Let the patient spend a moderate amount of time in the actual water of the bathing receptacle, supported by the linen cloth, without there being any pouring of water over him of the kind I am accustomed to do in other cases. But I would not wish water to be poured over his head, since it is enough for this to plunge two or three times into the water along with the whole body by letting the linen cloth downward in a measured manner, and then having the four young men lift him up again similarly. When he is carried out from there to the cold water, he must be immersed completely but he must not stay in it for long. And it is necessary for those bearing him to have been instructed in all the other tasks precisely, and there must be specific teaching and instruction on the speed of the immersion. Someone else, who has a linen cloth ready, must throw it over him, and then next, after he is laid on the bed, he must be wiped down, first with sponges and then with soft linen

725K

---

15 B (cf. mergatur KLat); βλαπτέσθω K

μαλακοῖς δ' ὕστερον ὀθονίοις, μηδ' αὐτῶν τῶν ἀπο-
ματτόντων αὐτὸν βιαίως ψαυόντων, ἀλλ' ὡς ἔνι μάλιστα πραότατα. μετὰ δὲ ταῦτα ἐπαλείψαντάς τε καὶ
περιθέντας ἱμάτια κομίζειν αὖθις ἐπὶ τοῦ σκίμποδος
εἰς τὸν οἶκον ἐν ᾧ διαιτᾶται, δώσοντας τροφήν. εἴρηται δὲ κἀν τοῖς πρόσθεν, ἡνίκα ἰώμεθα τὰς ξηρὰς
δυσκρασίας, ἐν τῷ τῶνδε τῶν ὑπομνημάτων ἑβδόμῳ, |
726K   τὸν οἶκον ἐζεῦχθαι χρῆναι τῷ βαλανείῳ. περὶ μὲν οὖν
λουτρῶν αὐτάρκης ὁ λόγος εἰς τὰ παρόντα.

11. Συνεχῆ δ' ἐπὶ τοῖς εἰρημένοις τὰ περὶ τῆς
δυνάμεως τῶν τροφῶν. ὅτι μὲν οὖν ὑγρὰς καὶ ψυχρὰς
εἶναι προσήκει τὰς μελλούσας ὀνήσειν τοὺς ἑκτικοὺς
πυρετοὺς εὔδηλον παντί. τοιαῦται δ' εἰσὶν ὅ τε χυλὸς ὁ
τῆς πτισάνης ψυχρὸς λαμβανόμενος, ὅ τε χόνδρος
ὁμοίως πτισάνης χυλῷ σκευασθείς, ὅπερ ἐστὶ δι'
ὕδατος καὶ πράσου βραχέος ἀνήθου τε καὶ ἁλῶν
ἐλαίου τε καὶ ὄξους. ἐπιτήδειος δὲ τοῖς ὧδε κάμνουσι
καὶ ὁ βραχὺς ἄρτος ἐν ὕδατι ψυχρῷ. μηδενὸς δ' αὐτῶν
ἡ ψύξις εἰλικρινὴς ἔστω, καθάπερ ἡ τοῦ ποτοῦ. ἐκεῖνο
μὲν γὰρ καὶ θερμαίνεται ταχέως καὶ διεξέρχεται,
ταῦτα δ' ἐπὶ πλέον ἐν τῇ γαστρὶ διαμένοντα τάχ' ἄν
που καὶ διαψύξαιεν αὐτὴν ἀμέτρως ὄντα ψυχρά. χρο-
νίου δ' ὑπάρχοντος ἤδη τοῦ πυρετοῦ καὶ τὸ τῆς ὄνου
γάλα δοτέον, ἀκριβῶς προσέχοντα τὸν νοῦν, μὴ τυρω-
θῇ ποτε κατὰ τὴν γαστέρα.

τοῦτο δὲ ἐνίοτε μὲν αὐτὸς ὁ κάμνων ἡμᾶς διδάξει
τῆς ἑαυτοῦ φύσεως ἐμπείρως ἔχων, ἐνίοτε δὲ ἡμᾶς
727K   αὐτοὺς ἐξευρίσκειν δεήσει, κατὰ | βραχὺ τῇ χρήσει

98

cloths, those who wipe him not touching him vigorously but, as far as possible, very gently. After this, when they have smeared him with oil and covered him with a cloak, they should convey him again on the bed to the house in which he is living and give him nourishment. I said earlier, in the seventh book of this treatise, when we were curing the dry *dyskrasias*, that the house must be joined to the 726K bathhouse. This is enough discussion about baths for the present.

11. Connected with what has been said are matters pertaining to the potency of the nutriments. It is clear to everyone that it is appropriate for moist and cold nutriments to be those which will help the hectic fevers. Such nutriments are the juice of ptisan when taken cold, and gruel prepared similarly with juice of ptisan made with water, a little piece of leek, dill, salt, oil and vinegar. A little bread with cold water is also useful for those suffering in this way. However, none of these things should be pure cold, like the coldness of the drink is, because the latter heats up quickly and passes through, whereas the former, if they remain in the stomach longer and are cold, perhaps in some way may also cool it, if they are very cold. When the fever has already existed for a long time, you must also give asses' milk, paying careful attention that it does not, at any time, become curdled in the stomach.

Sometimes the patient himself will teach us this, having an experience of his own nature; at other times, we ourselves need to discover it, approaching it gradually by use. 727K

Figure 10. Two plans of ancient bathhouses. From Andrew Farrington, *The Roman Baths of Lycia: An Architectural Study* (The British Institute of Architecture at Ankara, 1995). Descriptions are given on p. 150 of that work; with permission.

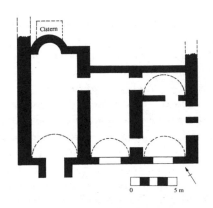

Cistern

0   5 m

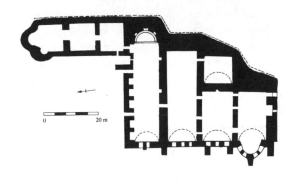

0   20 m

προσερχομένους. ἕτοιμον μὲν οὖν ἤτοι μέλιτος αὐτῷ
μιγνύντας ἢ καὶ βραχύ τι τῶν ἐδωδίμων ἁλῶν κωλῦσαι τὸ σύμπτωμα. φευκτέον δ᾽ εἰς ὅσον οἷόν τε τὸ μέλι
κατὰ τοὺς ἑκτικοὺς πυρετοὺς καὶ διὰ τοῦτο καὶ τὸ
μελίκρατον, ἐκχολοῦται γὰρ αὐτίκα καὶ ξηροτέρους
ἅμα θέρμῃ δριμείᾳ τοὺς πυρετοὺς ἀπεργάζεται. δοτέον οὖν τὴν πρώτην τὸ γάλα τῆς ὄνου κυάθου μὴ
πλέον, ὡς ἐθεάσω διδόντας ἡμᾶς εἰσαχθείσης μὲν εἰς
τὸν κοιτῶνα τῆς ὄνου, πρὸς τὸ μηδένα χρόνον ἐν τῷ
μεταξὺ γενέσθαι, καταπιόντος δ᾽ αὐτὸ τοῦ κάμνοντος
αὐτίκα, θεωρούντων δ᾽ ἡμῶν ἀκριβῶς καὶ παραβαλ-
λόντων τὴν κίνησιν τῶν ἀρτηριῶν, τὴν ἐπὶ τῇ προσ-
φορᾷ γιγνομένην, τῇ πρὸ τοῦ ληφθῆναι τὸ γάλα. εἰ
μὲν γὰρ διαφθείροιτο, τοὺς σφυγμοὺς σμικροὺς καὶ
ἀνωμάλους εὑρήσεις οἵους περ ἐπὶ θλίψει στομάχου
συμβαίνοντας οἶσθα· εἰ δὲ μὴ διαφθείροιτο, τοὺς
σφυγμοὺς μείζους τε καὶ ἰσχυροτέρους γιγνομένους.
ὥστ᾽ ἐξέσται σοι μετὰ ταῦτα προσφέροντι τὸ πλῆθος
αὐξῆσαι τοῦ γάλακτος ἥμισυ κυάθου, κἄπειθ᾽ ἥμισυ
προσθεῖναι πάλιν, εἶθ᾽ οὕτω κατὰ βραχὺ μέχρι τοῦ
728K συμμέτρου προσαυξῆσαι. τοῦ μέτρου | δὲ οὐ μόνον
τοῦ γάλακτος, ἀλλὰ καὶ τῶν ἄλλων ἁπάντων σιτίων
ἔμαθες τοὺς σκοπούς, ἔνθα τὰς τῆς γαστρὸς ἰώμεθα
δυσκρασίας, ἐν τῷ τῶνδε τῶν ὑπομνημάτων ἑβδόμῳ.
πρώτην γὰρ ἐν ἐκείνῳ τῷ λόγῳ τὴν ξηρὰν δυσκρασίαν αὐτῆς ὑποθέμενοι κεχρονισμένην εἰς τοσοῦτον
ὡς σύμπαν ἤδη τὸ σῶμα λεπτὸν ὑπάρχειν ἐπεδεί-
ξαμεν, ὅπως ἄν τις ἅμα μὲν ἄμφω καλῶς ἰῷτο, καὶ τὴν

Thus, when you mix either honey or a little of one of the edible salts with it, you readily prevent the symptom. However, you must avoid honey as far as possible in the hectic fevers and, because of this also, melikraton, since it immediately turns into bile and makes the fevers drier along with a biting heat. At first you must give not more than a cupful of asses' milk, as you saw me give it, leading the ass into the bedroom, so that no time interval occurs and the patient drinks it immediately, while we make precise observations and compare the movement of the arteries occurring after the administration with that before the milk was taken. If the milk is corrupted, you will find a small and irregular pulse, such as you know to occur due to compression of the stomach. If it is not corrupted, the pulse is larger and stronger. As a result, it will be possible for you, by adding an amount after this, to increase the milk by half a cupful, and then to add half again, and so gradually increase it again in this way up to an appropriate amount. You learned the indicators of the limit not 728K only of milk but also of all other foods in the seventh book of this treatise where I was curing the *dyskrasias* of the stomach. For, primarily, in that discussion, having assumed a dry *dyskrasia* of the stomach of such chronicity that the whole body was already thin, I showed how someone might properly cure *dyskrasia* of the stomach and

δυσκρασίαν τῆς γαστρὸς καὶ τὴν ἰσχνότητα τοῦ σώματος. καὶ τοίνυν ἴσως ἄμεινον ὑπὲρ τοῦ μὴ γράφειν πολλὰ πρὸς ἐκεῖνον ἀναπέμψαι τὸν λόγον ὑπὲρ τῶν ἄλλων ἁπάντων ἐδεσμάτων· ὅμοιαι γὰρ αἱ διαθέσεις οὖσαι παραπλησίων χρήζουσι καὶ τῶν ἰαμάτων.

ὅσον δ' ἐστὶν ἡ τῶν ἐκτικῶν πυρετῶν θερμοτέρα, τοσοῦτον καὶ τῶν βοηθημάτων δεῖται ψυχροτέρων, ἀκριβείας τε πλείονος, ὅπως μή τι βλάψαιμεν ἰσχυροῖς βοηθήμασι χρησάμενοι. καὶ τοὺς μὲν ἤδη μαραινομένους, ὡς εἴρηται, λουστέον, ὅσοι δ' ἐν τοῖς ἐκτικοῖς εἰσι πυρετοῖς οὐδέπω μαραινόμενοι, τούτους οἷόν τ' ἐστὶ καὶ χωρὶς σινδόνος ὑποβεβλημένης λούειν, ὡς ἂν ἰσχυροὺς ἔτι τὴν δύναμιν ὄντας. εἰσφέρειν μέντοι 729K καὶ τούτους | εἰς τὸν οἶκον, ἤτοι γε ἐπὶ σκίμποδος ἢ φορείου τινός, εἶθ' ἑαυτοῖς βαδίζοντας ἐπὶ τὸ θερμὸν ὕδωρ ἀκτέον, ὅπως ἐν ἐκείνῳ χρονίζοιεν. ὅσοι δ' ἀκρατέστεροι τῶν τοιούτων εἰσί, σφαλερὸν ἐπιτρέπειν αὐτοῖς ἐμβαίνειν ταῖς ψυχραῖς κολυμβήθραις, ἀλλ' ἔξω μᾶλλον αὐτῶν καταχεῖν ἐπὶ χλιαρῷ τῷ προτέρῳ τὸ δεύτερον ψυχρόν. ἔστω δ' οὕτω ψυχρὸν τότε κατὰ τὴν δεξαμενὴν ὕδωρ αὐτοῖς παρεσκευασμένον, ὅσον ἐξ ἐκείνης ἄν τις ἀρυσάμενος, αὔταρκες ἀθρόως εἰς ἅπαξ ἐπιχέαι τοῖς λελουμένοις, ὡς μηδεμίαν ἐμφαίνειν σαφῆ ποιότητα μήτε τῶν ἀκριβῶς ψυχρῶν μήτε τῶν χλιαρῶν ὑδάτων, ἀλλ' οἷον τὸ κρηναῖον καλούμενον ἐν μέσῳ ψύξεως πληκτικῆς καὶ χλιαρότητος ἐκλυτικῆς. εἰ μὲν οὖν ἀκριβῶς ἅπαντα γίγνοιτο, κατά τε τὰ λουτρὰ καὶ τὰς τροφὰς ἐλπὶς σωθήσεσθαι τοὺς μα-

thinness of the body both at the same time. Perhaps, therefore, it would be better to refer back to that discussion about all the other foods so I don't write too much because, as the conditions are alike, they also need similar cures.

By as much as the hectic fevers are hotter, so they require colder remedies and greater precision, lest we shall cause harm when we use strong remedies. And you must bathe those who are already marasmic, as I stated, whereas those with hectic fevers who are not yet marasmic, you can also bathe without the linen cloth being placed under them because they are still strong in capacity. Nevertheless, also bring them into the room, either on a bed or a litter, and then you must lead them, walking unaided, to the warm water so that they spend time in that. It is dangerous to allow those of them who are more helpless than such patients to go into the cold bathing pool. Pour water over them externally instead, first lukewarm and second cold. At that time have prepared for them cold water in the tank, as much as someone, having drawn off from this a sufficient amount, can suddenly pour all at once over those who have bathed—water that displays no clear quality of either those waters that are thoroughly cold or those that are lukewarm, but is like a so-called spring, lying between overpoweringly cold and debilitatingly warm. If all these things happen exactly as they should, there is hope that, with the baths and nutriments, those with marasmus will

729K

ραινομένους· εἰ δ᾽ ἁμαρτηθείη τι κἂν ἕν, ἀνατρέπει
τἄλλα σύμπαντα· τὸ γάρ τοι τῆς διαθέσεως αὐτῶν
ὀλέθριον οὐδὲ βραχύτατον ἁμάρτημα φέρει. καὶ εἴ τις
ἐσώθη τῶν τοιούτων, τὸ σαρκῶδες αὐτοῦ γένος ὁ
μαρασμὸς κατειλήφει· τῶν στερεῶν δ᾽ αὐτῶν ἐξηρα-
730K σμένων | ἀδύνατόν ἐστι τὸν ἄνθρωπον ὑγιασθῆναι τε-
λέως· ἀλλ᾽ ἀναγκαῖον ἤτοι γ᾽ ἐξ αὐτοῦ τοῦ πυρετοῦ
τὸν θάνατον ἥκειν εὐθέως ἢ μεταπεσόντος εἰς τὸ
καλούμενον ἐκ νόσου γῆρας. πρῶτον μὲν γὰρ ὁ τοιοῦτος πυρετὸς ἐκβόσκεται τὴν οἰκείαν ἰκμάδα τῶν
μορίων, ὑφ᾽ ἧς τρέφεται· μεταβαίνει δ᾽ ἐντεῦθεν ἐπὶ τὸ
σαρκῶδες γένος, ὃ ταῖς ἰσὶ καὶ τοῖς ὑμενώδεσι μορίοις
τῶν στερεῶν σωμάτων περιπέφυκεν· εἶθ᾽ οὕτως ἅπτεται καὶ αὐτῶν τῶν στερεῶν μορίων.

καθ᾽ ἕκαστον γάρ τοι μόριον τῶν ἁπλῶν καὶ πρώτων ὀνομαζομένων, ἐθεάσω κατὰ τὰς ἀνατομὰς τὸ μὲν
οἷον ἰνῶδές τί ἐστι τῆς οὐσίας αὐτῶν, τὸ δ᾽ ὑμενῶδες,
τὸ δὲ σαρκῶδες. οἷον εἰ τύχοι τῆς φλεβὸς ἕνα χιτῶνα
κεκτημένης λεπτόν, ἔστιν εὑρεῖν κατ᾽ αὐτὴν ἶνας
πολλὰς ἀραχνοειδέσι διατάσεσι συναπτομένας· αἷς
ἀμφοτέραις ἡ οἰκεία τῆς φλεβὸς οὐσία περιπέφυκεν,
ἄλλη κατ᾽ ἄλλο μόριον ὑπάρχουσα, κοινὸν οὐδὲν ὄνομα κεκτημένη· καλεῖν γοῦν αὐτὴν οὐδὲν κωλύει διδασκαλίας ἕνεκα σαρκοειδῆ φύσιν ἢ καὶ νὴ Δία τὴν
τοῦδέ τινος τοῦ μορίου σάρκα, κατὰ γαστέρα μὲν
731K ἑτέραν | ὑπάρχουσαν, καθ᾽ ἧπαρ δ᾽ ἑτέραν, ὥσπερ γε
καὶ κατὰ ἀρτηρίαν καὶ μῦν. ὠνόμασται δὲ μόνη σὰρξ
ἡ ἐν μυσί, τῶν ἄλλων οὐδεμίαν ὀνομάζουσι σάρκα,

be saved. However, if a mistake is made in even one of them, this upsets all the others. For the fatal nature of their condition does not allow the slightest mistake. And if any one of these patients is saved, the marasmus has seized his flesh as a class, whereas when the solid parts themselves have become dried, it is impossible for the person to be completely restored to health. But, inevitably, either death comes immediately from the fever itself or the patient undergoes a change to the so-called "age from disease." In the first place, such a fever consumes the proper moisture of the parts from which they are nourished. From here, it passes to the fleshy class, which grows around the fibrous and membranous parts of the solid bodies. Then, in this way, it also involves the solid parts themselves.

730K

In each of the parts termed simple and primary, you saw from dissections that there is, as it were, a fibrous, membranous or fleshy component of their substance—for example, in the single tunic acquired by a vein it is possible to discover many fibers bound together with weblike extensions, the specific substance of the vein being surrounded by both of these.[21] Other membranes exist in other parts but have not acquired a common name. At any rate, nothing prevents you from calling this, for didactic purposes, a "natural flesh" or also, by Zeus, the "flesh of this particular part," there being one for the stomach, and another for the liver, just as there is also one for an artery and one for a muscle. It is only called "flesh" in muscles; with very few exceptions no one calls any of the others

731K

21 See, for example, *De facultatibus naturalibus*, II.181K, *De anatomicis administrationibus*, II.601K, and *De usu partium*, III.457K.

πλὴν ὀλίγοι δή τινες· ἀλλὰ τὰ μὲν ἐν τοῖς σπλάγχνοις, οἷον ἥπατι καὶ νεφροῖς καὶ σπληνὶ καὶ πνεύμονι, παρεγχύματα καλοῦσι· τὰς δ' ἐν ἐντέροις καὶ
γαστρὶ καὶ στομάχῳ καὶ μήτρᾳ ἀνωνύμους ἀπολεί-
πουσιν. ἀλλὰ σὺ τῶν μὲν ὀνομάτων μηδὲν φρόντιζε,
γίγνωσκε δὲ ἑκάστου τῶν μορίων τὸν ὄγκον τῆς οὐσίας ὑπὸ τῆς τοιαύτης μάλιστα συμπληροῦσθαι φύσεως, φθορὰν καὶ γένεσιν ἐπιδεχομένης, ὡς ἔνεστι
θεάσασθαι σαφῶς ἐπὶ τῶν κοίλων ἑλκῶν. οὐ μὴν οὔθ'
ἡ τῶν ἰνῶν φύσις οὔθ' ἡ νευρώδης ἢ ὑμενώδης ὁρᾶται
τὴν αὐτὴν ἔχουσα γένεσιν· οὐδ' ἐγχωρεῖ τὰς ἐν ἑκάστῳ μορίῳ τοιαύτας ἶνας ὑπὸ τῶν κακοηθῶν πυρετῶν
ἐκτακῆναι, καθάπερ τὰς σάρκας. αἵ γε πρὶν ξηραν-
θῆναι τελέως διαφθείρουσι τὸ ζῷον.

εἰσὶ μὲν οὖν καὶ οἱ μαρασμώδεις πυρετοὶ πάντες ἐκ
τοῦ τῶν συντηκόντων γένους, ἀλλὰ διαφέρουσι τῷ τὸ
συντηκόμενον ἑκάστοτε τῆς σαρκὸς ἐπὶ μὲν τούτων
732K διαφορεῖσθαι, καταρρεῖν δ' ἐπ' | ἐκείνων εἰς γαστέρα.
φαίνεται δὲ τοῦτο κἀπὶ τῶν κρεῶν γιγνόμενον ἐναργῶς, ὀπτωμένων ἐπὶ τῆς καλουμένης ἐσχάρας· ἐνίων
μὲν γὰρ ἀπορρεῖ πάμπολυ τὸ σύντηγμα καὶ φέρεται
κατὰ τῶν ἀνθράκων ἀθρόως· ἐνίων δ' ὅλως οὐδὲν
αἰσθητὸν ἀπορρέον φαίνεται, κἂν ἴσον ᾖ τὸ πλῆθος
τῶν ἀνθράκων. οἱ γοῦν ἄγριοι σύες ἢ οὐδ' ὅλως ἢ
ὀλίγιστον ἔχουσι τὸ ἀπορρέον, οἱ δ' ἥμεροι πάμπολυ·
καὶ ὅσον περ ἂν ᾖ πιμελωδέστερον καὶ σαρκωδέστε-
ρον τὸ ζῷον, τοσοῦτον τὸ ἀπορρέον αὐτῶν τῶν σαρκῶν ἐπὶ τῆς ἐσχάρας ὀπτωμένων πλέον φαίνεται. τὸν
αὐτὸν οὖν τρόπον ἐπὶ τῶν θερμοτάτων πυρετῶν, εἰ μὲν

"flesh." But the [fleshes] in the viscera, such as the liver, kidneys, spleen and lung, they call parenchymas. That in the intestines, stomach, esophagus and uterus, they leave without a name. But give no thought to the names. Know, however, that the bulk of the substance of each of the parts is mostly filled up by such a basic substance which allows of destruction and generation, as can be seen clearly in hollow wounds and ulcers. But in fact the nature of fibers, or nerves (sinews), or membranes is seen not to have the same generation. Nor is it possible for such fibers in each part to be dissolved by the *kakoethical* fevers, like flesh is. In fact, before these fibers are dried out completely, the fevers destroy the organism.

All the marasmic fevers are, then, also from the class of fevers that are colliquative, but they differ on each occasion in the dissolving away of flesh; in some instances it is dissipated and in other instances it flows down to the stomach. This quite obviously happens with meats when they are roasted on a so-called gridiron as the liquefied material of some meats flows away in large part and is carried collectively down to the coals. Of other meats, however, nothing at all perceptible obviously flows away, even if there is an equal quantity of coals. At all events, wild pigs have either nothing at all or very little that flows away, whereas those that are tame have a large amount. And to the degree that the animal is more fatty and fleshy, so what flows away when their flesh is roasted on a gridiron is discernibly more. In the same way, in the very hot fevers, if the flesh is

732K

οὖν εἶεν σάρκες ὑγραὶ καὶ μαλακαὶ μετὰ πλήθους
πιμελῆς, τὰς συντήξεις αἰσθητὰς ἴσχουσιν· εἰ δὲ ἄνευ
πιμελῆς καὶ ξηραί, τοῖς μαρασμώδεσιν ἁλίσκονται
πυρετοῖς. ἐναργέστατον δὲ τούτου τεκμήριον οἱ τὰς
συντήξεις ἐκκρίνοντες ἄρρωστοι διὰ γαστρός, ἢν μὴ
φθάσωσιν ἀποθανεῖν, ἤτοι γ' ὑπὸ βίας τοῦ νοσήματος
ἢ διὰ τὴν τῶν ἰατρῶν ἀμαθίαν εἰς τοὺς μαρασμώδεις
ἐμπίπτοντες πυρετούς. ἄχρι μὲν γὰρ ἂν ἔχωσι τὴν
ὕλην δαψιλῆ, ταύτην συντήκουσιν· ὅταν δὲ ἐπιλεί-
πωσιν αὐτοὺς αἵ τε πιμελαὶ καὶ αἱ σάρκες αἱ μαλακαί,
733K   ξηραίνουσι | τὰς σκληρὰς καὶ δυστήκτους σάρκας, ὧν
οὐδὲν ἀπορρεῖ διὰ ξηρότητα, καθάπερ οὐδὲ τῶν ταρι-
χηρῶν κρεῶν, εἰ καὶ ταῦτ' ἐθελήσειας ὀπτᾶν, ἐπιβάλ-
λων ἴσῳ τοῖς ἄλλοις πυρί· ξηρότεραι μὲν γὰρ αἱ
σάρκες αὐτῶν ἔσονται καὶ σκληρυνθήσονταί γε βύρ-
σης τρόπον· οὐ μὴν ἐκρυήσεταί γε αὐτῶν οὐδεμία
τηκεδὼν αἰσθητή. ταῦτ' οὖν γινώσκοντί σοι καὶ ἡ τῶν
συντηκτικῶν πυρετῶν ἐπιμέλεια κατὰ τὰς αὐτὰς γινέ-
σθω μεθόδους ἐπί τε ψυχροῦ δόσιν ὅτι τάχιστα παρα-
γιγνομένῳ καὶ διὰ τῶν ὑγραινόντων καὶ ψυχόντων
τρέφοντι. μελίκρατον δ' οἰομένῳ πολεμιώτατον εἶναι,
ᾧ μάλισθ' ὁρᾷς χρωμένους ἐπὶ τῶν τοιούτων πυρετῶν
τοὺς πλείστους τῶν ἰατρῶν.

ἀλλὰ περὶ μὲν τῶν συντηκόντων πυρετῶν αὖθις
ἐξέσται διελθεῖν, ὥσπερ γε καὶ περὶ τῶν ἐν λοιμοῖς
πυρετῶν ἑκτικῶν, οἷος ὁ νῦν ἐπιδημῶν ἐστιν, εἰρήσεται γὰρ περὶ τοῦδε κατὰ τοὺς λοιμώδεις πυρετούς.
τῶν δ' ἄλλων ἑκτικῶν καὶ μαρασμωδῶν αὐτάρκως
εἰρῆσθαι νομίζω τὴν μέθοδον τῆς ἰάσεως.

moist and soft with an abundance of fat, they have a perceptible colliquescence. On the other hand, if they are without fat and are dry, they are seized by marasmic fevers. The clearest evidence of this comes from sick people who excrete what is liquefied through the stomach. If they don't die beforehand, either from the force of the disease or because of the ignorance of doctors, they fall into marasmic fevers. For as long as they have abundant material, they melt this away. However, when fat and soft flesh are deficient in them, they dry out the hard and difficult to 733K dissolve flesh, nothing of which flows away due to the dryness, just as nothing of pickled meats does. And if you wished to roast this, you would throw it with other things onto an adequate fire, for the flesh of these [meats] will be drier and hardened like hide. In fact, no perceptible liquefied material will flow away from those meats. Therefore you, who know these things, must carry out the care of the colliquative fevers according to the same methods, administering a dose of cold [water] as quickly as possible and nourishing with foods that are moistening and cooling. However, you must regard melikraton as highly inimical, although you do see the vast majority of doctors using it in such fevers.

But it will be possible to go through the colliquative fevers again, as it will be to go through the hectic fevers in plagues, such as the one now visiting us, for I will speak about this in relation to the pestilential fevers. But I think enough has been said on the method of cure of the other hectic and marasmic fevers.

ΒΙΒΛΙΟΝ Λ

1. Τῶν δ' ἐπὶ χυμοῖς σηπομένοις ἀναπτομένων πυρετῶν οἱ πρῶτοι μὲν εἴρηνται σκοποὶ πρόσθεν ἡνίκα ἐδείκνυον ὅπως ἐγχωρεῖ καὶ δύο καὶ τρεῖς αὐτοὺς ποιῆσαι, κἂν εἰ βουληθείημεν ἕνα. νυνὶ δ' ὥσπερ ἐπὶ τῶν ἄλλων πυρετῶν ἐκ τῆς τῶν πρώτων γενῶν τομῆς ἄχρι τῶν ἐσχάτων εἰδῶν ἀφικόμεθα τέμνοντες, οὕτω πρᾶξαι πειραθῶμεν κἀπὶ τῆς προκειμένης διαφορᾶς, ἐκεῖνο πρότερον ἀναμνήσαντες ὃ κἀν τοῖς περὶ διαφορᾶς τῶν σφυγμῶν ὑπομνήμασιν ἐδείξαμεν, ὡς εἴτε πρώτας διαφοράς, εἴτε γένη πρῶτα καλεῖν ἐθέλοι τις, |

εἴτε γενικωτάτας ἰδέας, εἴθ' ὁπωσοῦν ἄλλως φυλάττων ἀκριβῆ τοῦ πράγματος τὴν ἔννοιαν, οὐ διοίσει. τῶν μὲν δὴ πυρετῶν αὐτῶν τοὺς ἐφημέρους ὅπως χρὴ θεραπεύειν ἐν τῷ τῶνδε τῶν ὑπομνημάτων ὀγδόῳ διῆλθον· ὥσπερ γε καὶ ὅπως τοὺς συνόχους ἐν τῷ μετ' αὐτό· καὶ τρίτους γε τοὺς ἑκτικοὺς ἐν τῷ μετ' ἐκεῖνο τῆς ὅλης πραγματείας ὄντι δεκάτῳ, μετὰ τῶν ἰδίων ἑκάστῳ πλεῖστα καὶ τῶν κοινῶν ἐπελθών. νυνὶ δ' ὅπως ἄν τις ἰῷτο μεθόδῳ τοὺς ἐπὶ σήψει χυμῶν συνισταμένους πρόκειται διελθεῖν.

BOOK XI

1. The primary indicators of the fevers kindled by putrefy-   734K
ing humors have been stated before when I showed how it
is possible to make them two or three, or even one, if we
wish. Now, when I make divisions as in the case of the
other fevers, I proceed from the division of the primary
classes as far as the ultimate kinds. So let me attempt to do
this in the case of the differentiae before us, first calling to
mind what I demonstrated in the treatise *On the Differentiae of the Pulses:*[1] that whether someone might wish to call
them primary differentiae, or primary classes, or the most   735K
generic kinds, or name them in whatever other way, it will
make no difference as long as he accurately preserves the
concept of the matter. I went through in detail how we
should treat the ephemeral fevers in the eighth book of the
present treatise, just as I did with the continuous fevers in
the book after that, and the hectic fevers in the one after
that, which was the tenth of the whole work, when I covered the majority of those things specific to each of them
and those that are common [to them all]. I now propose to
go through in detail how someone might cure by method
the continuous fevers which arise due to putrefaction of
humors.

[1] One of Galen's four major tracts on the arterial pulse: *De
differentiis pulsuum*, VIII.493–765K.

οἱ σκοποὶ δ᾽ οἱ θεραπευτικοὶ κατὰ μὲν τὴν εἰς δύο
τομὴν ἥ τε διάθεσις ἣν θεραπεύομεν ὑπάρχει καὶ ἡ
τῶν τοῦ κάμνοντος μορίων κρᾶσις. ἐγχωρεῖ δ᾽, ὡς
ἐλέχθη, καὶ δι᾽ ἑνὸς ἑρμηνεῦσαι κεφαλαίου, θεραπευτικὸν εἰπόντων ἡμῶν εἶναι σκοπὸν ἕνα κοινὸν ἁπάντων νοσημάτων, τὴν ἐναντίωσιν· οὗ τεμνομένου γεν-
νᾶσθαι τοὺς δύο, τοῦ μὲν νοσήματος ἐνδεικνυμένου
τὴν τῶν ἐναντίων ἑαυτῷ βοηθημάτων χρῆσιν, τῆς δὲ
τοῦ θεραπευσομένου κράσεως ὁριζούσης ἅμα τῷ νοσήματι τὸ μέτρον τῆς ἐναντιώσεως. ἐρρέθη δ᾽ ἐν
736K ἐκείνοις ὡς καὶ τὸ περιέχον ἡμᾶς ἕνα ποιήσασθαι | τὸν
σκοπὸν ἐγχωρεῖ. καὶ ὡς ὁπόταν εἰς δύο μόνους τὴν
πρώτην ποιησώμεθα τομήν, ἤτοι γ᾽ ἐν τοῖς νοσεροῖς
αἰτίοις ἢ ἐν τοῖς ὑγιεινοῖς, τίθεσθαι χρὴ τὸ περιέχον.
ὄντων δ᾽ αἰτίων ὑγιεινῶν ἁπάντων τῶν ποιούντων
ὑγείαν, ἓν ἐξ αὐτῶν ἐστι τὸ καλούμενον βοήθημα,
κατὰ διαφόρους ἐννοίας ἔσχατον καὶ πρῶτον ὑγιεινὸν
αἴτιον ὀρθῶς λεγόμενον. ἔσχατον μὲν γὰρ ἔσται τῷ
χρόνῳ, διότι φύσις καὶ τέχνη καὶ τύχη τῷ χρόνῳ
πρότερα τῶν βοηθημάτων ἐστὶν ὑγείας αἴτια· πρῶτον
δὲ καθ᾽ ὅσον αὐτὸ ψαύει τοῦ νοσοῦντος, ἀλλοιοῦν τὴν
διάθεσιν αὐτοῦ· καὶ διὰ τὴν ἐκ τούτου γινομένην
ἀλλοίωσιν τῶν νοσούντων σωμάτων εἰς ὑγείαν ἀγο-
μένων ἥ τέχνη καὶ ὁ τεχνίτης καὶ ἡ τύχη αἴτια τῆς
ὑγείας γίνεται, οὐ τῷ πρώτως αὐτὰ τὰς νόσους ἐκκόπτειν, ἀλλὰ τῷ δι᾽ ἑτέρων ὑλῶν ἐπιτηδείων, αἵτινες
ὀνομάζονται βοηθήματα.

φλεβοτομία γὰρ αὐτὴ μὲν καθ᾽ ἑαυτὴν αὐτὸ δὴ

The therapeutic indicators, based on a twofold division, are the condition which we are treating and the *krasis* of the parts of the patient. However, it is also possible, as I said, to describe them under one heading, if we say there is one common therapeutic indicator of all diseases—opposition. When this is divided, it generates the two, since the disease indicates the need of remedies opposite to itself while the *krasis* of the person who will be treated, along with the disease, determines the measure of the opposition. It was said in that treatise that it is also possible to make the indicator a single one: what surrounds us (the ambient air). It was also said that, whenever we make the primary division into two only, we must place the ambient air either in the causes of disease or the causes of health. Since the causes of health are all those things that bring about health, one of these is what we call a remedy which, depending on different concepts, is rightly termed both a last and a first cause of health. It will be last in time because nature, craft and chance are prior in time to remedies as causes of health. It is first to the extent that it affects the one who is diseased and changes his condition and, because of the change occurring from this when diseased bodies are brought to health, the craft, the craftsman and chance become causes of health, not because they eradicate the diseases primarily, but because [they do so] through other suitable materials which are termed remedies.

Thus phlebotomy, in and of itself, has actually acquired

736K

τοῦτο τὸ νῦν εἰρημένον ὄνομα κέκτηται· βλάψασα δὲ
τὸν ἄνθρωπον ἐκ τῶν αἰτίων γίνεται τῶν νοσωδῶν,
ὥσπερ γε καὶ ὠφελήσασα τῶν ὑγιεινῶν. καὶ διὰ τοῦτο

737K καλεῖται βοήθημα, καθ᾽ ὃν ἂν ὠφελῇ καιρόν· ὡς | καὶ
πάσης ὕλης τῆς ὁπωσοῦν ἀλλοιούσης τὸ σῶμα καθ᾽
ὃν ἂν ὠφελῇ χρόνον ὀνομαζομένης βοηθήματος. οὕτως οὖν καὶ περιέχοντος ἡμᾶς ἴδιον μὲν ὄνομα κατὰ
τὴν ἑαυτοῦ φύσιν ὁ ἀήρ· ἐν δὲ τῇ πρὸς ἡμᾶς σχέσει
ποτὲ μὲν βοήθημα καὶ τῶν αἰτίων ἕν τι τῶν ὑγιεινῶν
γίγνεται, ποτὲ δὲ τῶν ἐναντίων αὐτοῖς τῶν νοσωδῶν.
ἐπεὶ δ᾽ ἔκ τε τῆς ἐνεστώσης καταστάσεως καὶ τῆς
ὥρας τοῦ ἔτους καὶ τῆς τοῦ χωρίου φύσεως ὁ ἀὴρ
ἴσχει τὰς καθ᾽ ὑγρότητα καὶ ξηρότητα καὶ θερμότητα
καὶ ψυχρότητα μεταβολάς, ἐξ ἐκείνων πάλιν ἑκάστου
γίγνονταί τινες ἐνδείξεις μερικώτεραι.

τύχη μὲν οὖν καὶ τέχνη καὶ τεχνίτης διὰ μέσων τῶν
ὑλῶν ἔχουσι τὸ δρᾶν· ἡ φύσις δ᾽ αὐτὴ δι᾽ ἑαυτῆς, ἐκ
τριῶν τούτων συμπληρουμένη μορίων τῆς τε τοῦ πνεύματος οὐσίας καὶ τῆς τῶν ὄντως στερεῶν, ἅπερ ἐκ
σπέρματος ἐδείχθη γιγνόμενα, καὶ τρίτης ἐπ᾽ αὐτοῖς
τῆς σαρκοειδοῦς οὐσίας, ἐν ἑκάστῳ τῶν μορίων ἰδίας
ὑπαρχούσης. ἕκαστον δὲ τῶν εἰρημένων τριῶν ἰδίαν
τινὰ ἔχει ποσότητα καὶ ποιότητα κατὰ φύσιν. ἡ μὲν
οὖν ποιότης αὐτῶν ἐκ τῆς πρεπούσης κράσεως ὑγρᾶς

738K καὶ ξηρᾶς καὶ ψυχρᾶς καὶ θερμῆς | συνίσταται. τὸ
ποσὸν δὲ τῶν μὲν στερεῶν σωμάτων ἴσον ἀεὶ διαμένει,
τοῖς μὲν αὐξανομένοις ἔτι κατὰ τὸν ἀριθμὸν μόνον,
τοῖς δ᾽ ἤδη τετελειωμένοις καὶ κατὰ τὸ μέγεθος. τῆς

116

this very name just mentioned. When it harms a person, it becomes one of the causes of disease just as, when it brings benefit, it is one of the causes of health. Because of this, at the time it brings benefit it is called a remedy, like every 737K material which changes the body in any way whatsoever, should it be of benefit at the time, is also called a remedy. Similarly, the specific name for what surrounds us, based on its own nature, is the air. In its state in respect to ourselves, it sometimes becomes a remedy and one of the causes of health, while at other times it becomes one of the oppositions to these; that is, a cause of diseases. Since the air is susceptible to changes in moistness, dryness, heat and cold due to the existing climatic conditions, the season of the year and the nature of the place, more specific indications arise from each of these in turn.

Chance, craft and craftsman act through the medium of the materials. The nature acts through itself, being made up of the following three components: the substance of the *pneuma*, the substance of the truly solid parts (which, it was shown, arises from a seed), and third, in addition to these, the fleshy substance which is specific in each of the parts. Each of these three components has some specific quantity and quality in accord with nature. Thus, their quality consists of a suitable *krasis* of moist, dry, cold and hot. The quantity of the solid bodies always remains the 738K same: in those that are still increasing [in magnitude], in number alone; in those already brought to completion, also in terms of magnitude. However, the magnitude of

σαρκοειδοῦς δ᾽ οὐσίας ἐξαλλάττεται τὸ μέγεθος, εἰ καὶ
μηκέτι αὔξοιτο τὸ ζῷον. οὕτως δὲ καὶ ἡ τοῦ πνεύματος
οὐσία μονονοὺ καθ᾽ ἑκάστην ῥοπὴν ἐλάττων τε καὶ
πλείων γίνεται. καὶ τοίνυν καὶ ἡ δύναμις ἐφ᾽ οἷς
ἔμπροσθεν εἶπον σκοποῖς ἤτοι μόνη προστίθεται
κατὰ τὴν τῶν εἰρημένων τριῶν οὐσιῶν συναρίθμησιν,
ἢ τῆς κράσεως ἀφορισθείσης ἰδίᾳ καθ᾽ ἑαυτήν. ἥτις
αὖ πάλιν καὶ αὐτὴ ποτὲ μέν, ὡς ἐδείκνυτο, διχῇ τέμνε-
ται, εἴς τε τὴν ἐξ ἀρχῆς φύσιν καὶ τὴν ἐπίκτητον
διάθεσιν· ὥστε ἀπὸ τούτων καὶ τῶν ἄλλων τῶν προειρημένων γίγνεσθαί τινα ἔνδειξιν ἰαμάτων ἰδίαν
ἑκάστου. πολλάκις δὲ ἡ ἐπίκτητος κρᾶσις εἰς τὰς
ποιητικὰς αἰτίας ἀναχθεῖσα τὴν ἔνδειξιν ἐκείναις ἐχα-
ρίσατο. κατὰ τοῦτον οὖν ἔφαμεν τὸν τρόπον ἐξ ἡλικίας ἐπιτηδευμάτων τε καὶ ἐθῶν ἔνδειξιν βοηθημάτων
γίγνεσθαι, συμφωνεῖν δ᾽ ἀλλήλαις ἁπάσας τὰς εἰρημένας ἐνδείξεις· ἐν γὰρ ταῖς ἰδικωτέραις ἀεὶ περιέχε-
739K   σθαι τὰς | γενικωτέρας. ἄγεσθαι δ᾽ εἰς χρῆσιν τὰς
ἰδικωτέρας, εὑρισκομένας ἐκ τῆς τῶν γενικωτέρων
τομῆς.

2. Ἡ γάρ τοι πρώτη πασῶν ἔνδειξις θεραπευτικὴ ἐξ
αὐτοῦ τοῦ θεραπευομένου νοσήματος γιγνομένη τὴν
ἐναντίωσιν ἐδίδαξεν. ἐπεὶ δ᾽ ἔστι πλείονα γένη τῶν
νόσων, ἕκαστον αὐτῶν ἰδίαν ἕξει τὴν ἐναντίωσιν.
ἀλλὰ τὰ μὲν ἄλλα παρείσθω τό γε νῦν εἶναι. προκεχω-
ρήσθω δὲ μόνον ἡμῖν εἰς τὰ παρόντα τῆς θεραπευομένης νόσου τὸ γένος ἐν δυσκρασίᾳ κείμενον. ἰαθήσεται τοίνυν τοῦτο διὰ τῆς ἐναντίας αὐτῷ δυσκρασίας.

118

the fleshy substance changes if the organism is no longer increasing. In the same way too, the substance of the *pneuma* in effect becomes less or more every moment. And accordingly, also, the capacity is added to the indicators I spoke of before, either on its own in relation to the enumeration of the aforementioned three substances, or when the *krasis* is distinguished individually by itself. And this itself is again sometimes divided in two ways into the original nature and the acquired condition, as was shown. As a result, from these things and from others previously mentioned, there arises a certain indication of cures specific for each. Often the acquired *krasis*, when it is reckoned in regard to the effecting causes, yields the indication for those. In this way, then, I said that the indication of the remedies arises from age, habits and customs, and that all the aforementioned indications agree with one another, for always those indications that are more general are included in those that are more particular. And [I said that] 739K we are led to the use of the more particular indications which are discovered from the division of the more general ones.

2. Certainly, the foremost therapeutic indication of all arising from the actual disease being treated directs us to opposition. Since there are more classes of diseases, each of these will have a specific opposition. But let the other classes be disregarded, at least for the moment. Let us, instead, bring forward for present consideration only the class of disease being treated—that which lies in a *dyskrasia*. Accordingly, this will be cured through the *dyskrasia* opposite to it. But since it is a hot *dyskrasia* that

ἀλλ' ἐπεὶ τοῦ γένους ὅλου τῶν πυρετῶν ἡ δυσκρασία
θερμότης ἐστί, διὰ ψυχρότητος ἰαθήσεται, ὥστε ἀλλήλαις ὁμολογεῖν ἁπάσας τὰς ἐνδείξεις. ἐάν τε γὰρ
εἴπωμεν ἁπλῶς οὕτως ὡς χρὴ τὸν πυρετὸν ἰᾶσθαι διὰ
τῶν ἐναντίων ἐάν τε προσθέντες ὡς διὰ τῶν ἐναντίων
τῇ κράσει μάχην οὐδεμίαν ἕξουσιν οἱ λόγοι· κατὰ δὲ
τὸν αὐτὸν τρόπον οὐδ' εἰ φαίημεν ἰᾶσθαι χρῆναι διὰ
τῶν ψυχόντων τὸν πυρετόν, ὥσπερ γε οὐδ' εἰ τέμνοντες τοῦτο φαίημεν ἐνίοτε μὲν ἰᾶσθαι τὸν πυρετὸν ἡμᾶς
740K διὰ τῶν ἐνεργείᾳ ψυχόντων, ἐνίοτε | δὲ διὰ τῶν δυνά-
μει, καί ποτε αὖθις διὰ τῶν κατὰ συμβεβηκός, ἢ καὶ
συντιθέντες ταῦτα μάχην οὐδεμίαν ἕξει τοῖς προειρημένοις ὁ λόγος, ὥσπερ γε οὐδ' εἰ φαίημεν ὅσα μὲν
κατ' ἐνέργειαν ἢ δύναμιν ἢ κατὰ συμβεβηκὸς θερμαίνει χείρονας ἑαυτῶν ἀποτελεῖν τοὺς πυρετούς, ὅσα
δ' αὖ κατ' ἐνέργειαν ἢ δύναμιν ἢ κατὰ συμβεβηκὸς
ψύχει θεραπεύειν αὐτούς. ἡ γὰρ ἀπὸ τῆς διαθέσεως
ἔνδειξις εἰς τοσούτους ἐτμήθη τοὺς κατὰ μέρος σκοπούς. αὐτοὶ δὲ πάλιν οὗτοι οἱ σκοποὶ τὴν ὕλην τῶν
βοηθημάτων εὑρίσκουσι κατὰ τὰς ἐν τοῖς περὶ φαρμάκων εἰρημένας μεθόδους· ἀρχὴ γὰρ ἐκείνων ἐστὶν
εἰς ὅπερ ἐτελεύτησεν ἡ ἐκ τῆς θεραπευομένης διαθέσεως τομή.

κυριωτάτη μὲν οὖν ἐν τῷ λυθῆναι τὴν νόσον θερα-

---

[2] The three works, often referred to, on simple and compound
medications: *De simplicium medicamentorum temperamentis et*

characterizes the whole class of fevers, it will be cured through cooling, so that all the indications are in agreement with one another. And if we were to say simply that it was necessary to cure the fever in this way through the opposites, and if we add that it is through the opposites in terms of *krasis*, the arguments will have no point of contention. Similarly, there would be no point of contention, if we were to say the fever needs to be cured through cooling agents. Likewise, when we divide this, if we were to say that we sometimes need to cure the fever through agents that are cooling by action and sometimes through those that are cooling by potency and, at other times again, through those agents that cool contingently, or also when we add these together, the argument will have no point of contention in regard to the previously mentioned matters. The same would apply, if we were to say that those things which heat, in relation to action, capacity or contingency, make the fevers worse than they were, whereas those things which cool, in relation to action, capacity or contingency, cure them. For the indication from the condition was divided into such indicators individually. Again, these same indicators reveal the material of the remedies in respect to the methods spoken of in the works on medications.[2] For the principle of those [methods] is division, which reaches completion from the condition being treated.

740K

Therefore, the treatment considered most effective in

facultatibus (XI.379–892K and XII.1–377K), *De compositione medicamentorum secundum locos* (XII.378–1007K and XIII.1–361K), and *De compositione medicamentorum per genera* (XIII.362–1038K).

πεία νοεῖται, καταχρωμένων δὲ καὶ τὸ μικτὸν γένος ἔκ τε τῆς ὄντως θεραπείας καὶ ἣν ὀνομάζουσι προφυλακήν. τῶν μὲν οὖν ἐφημέρων πυρετῶν ἡ ὄντως ἐστὶ θεραπεία, καθάπερ γε καὶ ἡ τῶν ἑκτικῶν. ὅσοι δ' ἐπὶ χυμοῖς συνίστανται σηπομένοις ἡ λεγομένη θεραπεία τὴν προφυλακὴν προσείληφεν, ἥτις ἐστὶ τὸ τὴν νόσον ἐργαζόμενον αἴτιον ἐκκόπτειν. ὀνομάζεται δὲ καὶ αὐτὸ |
741K τοῦτο τὸ μέρος τῆς τέχνης θεραπευτικόν, ἀποκεχωρισμένον τοῦ καθαρῶς τε καὶ εἰλικρινῶς προφυλακτικοῦ τῷ γενέσθαι τὴν νόσον μὲν αἰσθητῶς ὑπὸ τῆς αἰτίας ἤδη, μέλλειν δὲ ἔσεσθαι κατὰ τὰς προφυλακτικάς. τοῖς μὲν οὖν ὀνόμασιν ὡς ἂν ἐθέλοι τις χρήσθω, τοῦτο γὰρ ἀναμιμνήσκειν ἀεὶ χρή, φυλαττέτω δὲ τὴν ἐφ' ἑκάστου τῶν πραγμάτων ἔνδειξιν τῶν ἰαμάτων, ὡς εἴ γε παραλείποι τι κἂν ἓν ἐξ αὐτῶν, ἀνάλογον ἐκείνου τῷ μεγέθει τε καὶ τῇ δυνάμει βλάψει τὴν ἴασιν.

3. Ὅνπερ δὲ τρόπον ἀπὸ τῆς διαθέσεως ἐπὶ τὰς ὕλας τῶν βοηθημάτων ἀφικόμεθα, τὸν αὐτὸν τρόπον ἀφ' ἑκάστου τῶν ἄλλων γενῶν τῶν πρώτων κατέρχεσθαι χρὴ πρὸς αὐτάς· οἷον ἀπὸ τῆς δυνάμεως· ἀντιτέτακται γὰρ αὕτη τῇ νόσῳ καθάπερ τις ἀνταγωνιστής· καὶ πρόκειταί γε τῷ ἰατρῷ βοηθεῖν ταύτῃ καὶ συναγωνίζεσθαι τρόπῳ παντί. τίς οὖν πρῶτος σκοπὸς ἀπὸ τῆς δυνάμεως ἢ τίς ἔνδειξις προτέρα; καὶ γὰρ κἀνταῦθα λεκτέον ὡς ἂν ἐθέλοι τις, αὐτὴν μὲν τὴν
742K δύναμιν ἕνα τὸν πρῶτον σκοπὸν τιθέμενος, τὸ δ' | ὑπ' αὐτῆς ἐνδεικνύμενον ἕτερον σκοπὸν δεύτερον. ἐπεὶ

resolving the disease is when we make full use of the class compounded from what is truly treatment and what people term prophylaxis. This is actually the treatment for the ephemeral fevers, just as it is for the hectic fevers. In those fevers which exist due to humors that are putrefying, the treatment spoken of adds a prophylactic component as well, which is to eradicate the cause bringing about the disease. This itself is also called the therapeutic part of the 741K craft, and is separated from what is purely and absolutely prophylactic by virtue of the fact that the disease already perceptibly exists due to the cause, whereas prophylaxis pertains to what is going to exist in the future. Therefore, however someone might wish to make use of these terms, it is always necessary to remember this: he must preserve the indication of the cures in each of the matters because, if he were to leave out even one of these, he would harm the cure in proportion to both the magnitude and the power of that [which is overlooked].

3. We arrive at the materials of the remedies from the condition in the very same way by which we must come down to them from each of the other classes that are primary; for example, from the capacity. The capacity is set in opposition to the disease like some opponent; and what confronts the doctor is to aid this [capacity] and fight alongside it in every way. What, then, is the primary indicator from the capacity or what the prior indication? And even here one must speak as one wishes, putting the capacity itself as one primary indicator while there is another, 742K second indicator displayed by this capacity. Therefore,

GALEN

τοίνυν ἐδείχθη τὴν ἑαυτῆς ἐνδεικνυμένη φυλακήν,
ὥσπερ ἡ νόσος τὴν ἀναίρεσιν, ἔσται πάλιν ὁ δεύτερος
ἐπὶ τῇ δυνάμει σκοπὸς ἡ φυλακή, καθάπερ ἐπὶ τῶν
νοσημάτων ἡ ἀναίρεσις. ἀλλ' ἡ μὲν φυλακὴ τὴν τῶν
ὁμοίων ἐνδείξεται προσαγωγήν, ἡ δ' ἀναίρεσις τὴν
τῶν ἐναντίων. ἐπειδὴ δὲ τῆς δυνάμεως οὐσία συνεπλη-
ροῦτο διά τε τοῦ πνεύματος καὶ τῆς σαρκώδους ἰδέας
καὶ τῶν στερεῶν, ἐκ τῶν ὁμοίων ἑκάστῳ πορίζεσθαι
χρὴ τὴν διαμονήν· τῷ μὲν πνεύματι διὰ τῆς ἀναπνοῆς
τε καὶ διαπνοῆς καὶ τῆς ἐκ τοῦ αἵματος ἀναθυμιάσεως,
ἐξέστω δὲ κἀνταῦθα τῷ βουλομένῳ καλεῖν ἀέρωσιν, ἢ
λεπτοποίησιν αἵματος, ἢ εἰς ἀτμοὺς λύσιν, ἢ χύσιν, ἢ
ὅπως ἂν αὐτὸς ἐθέλοι· τῷ δὲ τῶν στερεῶν γένει διὰ τῆς
στερεᾶς τροφῆς, ὥσπερ γε καὶ τῷ τῶν σαρκωδῶν διὰ
τῆς ἐν μέσῳ φύσεως ὑγρῶν τε καὶ στερεῶν σωμάτων.

ἑκάστου δὲ τούτων εὑρήσεις τὸ μὲν ποσὸν ἐκ τῆς
κατὰ τὴν φυλαττομένην οὐσίαν ποσότητος, τὸ δὲ
ποιὸν ἐκ τῆς κράσεως. αὐτῆς δὲ τῆς κράσεως καθ'
743K ἕκαστον αὐτῶν ἤτοι | γε ἀρίστης ὑπαρχούσης ἢ
μεμπτῆς κατά τι, τὴν μὲν ἀρίστην ἀεὶ φυλάττειν χρή,
τὴν μεμπτὴν δὲ ἐπὶ μὲν τῶν ὑγιαινόντων ἤτοι φυλάττειν ἢ ἀλλοιοῦν· ἐπὶ δὲ τῶν νοσούντων φυλάττειν, ἐξ
οὗ λογισμοῦ πρόσθεν ἐδείκνυτο μεγίστην ἔνδειξιν
ἔχον τὸ ἔθος, ἢ γὰρ τὴν ἐξ ἀρχῆς αὐτοῦ φυλάττειν
κρᾶσιν ἢ τὴν ἐπίκτητον ἐργάζεσθαι. γίγνεσθαι δ', ὡς
ἐλέγετο, τὴν ἔνδειξιν τοῦ μέτρου τῶν βοηθημάτων ἐκ
τῆς νῦν κράσεως, ὥσπερ γε κἀκ τῆς ἡλικίας, οὐχ ἧς

since the capacity was shown to indicate its own preservation, just as the disease was shown to indicate its own removal, preservation in the case of the capacity will in turn
be the second indicator, just as in the case of the diseases
removal will be. But preservation will indicate the administration of like things while removal will indicate the administration of opposite things. Since the substance of the
capacity is jointly made up by the *pneuma*, [bodies] of the
fleshy kind and those that are solid, it is necessary to bring
about the continuation of each [of these] from like things;
for the *pneuma* through respiration, transpiration and the
vapor from the blood (and here let it be acceptable for
someone who should so wish, to call this rarefaction or
making the blood fine, or dissolution to a vapor, or dispersion, or whatever he may prefer); for the class of the solid
[bodies] through solid nutriment; and for what is fleshy by
that which is intermediate in nature between liquid and
solid bodies.

You will discover the quantity of each of these from the
quantity of substance being preserved, and the quality
from the *krasis*. Since, however, the actual *krasis* of each of
these is either optimum or at fault in some way, it is always   743K
necessary to preserve what is optimum, while in respect of
what is at fault in the case of those who are healthy, it is
necessary either to preserve it or change it. In the case of
those who are diseased, it is necessary to preserve it, and
from the previous reckoning, custom was shown to be the
most important indication for either preserving its own
original *krasis* or bringing about the acquired *krasis*. It was
shown, on the other hand, that the indication of the measure of the remedies arises, as I said, from the current
krasis, just as it does from the age, and not from the *krasis*

ἔμπροσθεν εἶχεν ὁ κάμνων, ἀλλ' ἧς νῦν ἔχει. διαφορᾶς δ' οὐκ ὀλίγης ὑπαρχούσης ἐν τῇ κράσει τῶν
τοῦ ζῴου μορίων, ἴδιον ἑκάστῳ τὸ μέτρον ἔσεσθαι τῶν
ὁμοίων ἑαυτῷ, τῷ μὲν γεωδεστέρῳ τῶν γεωδεστέρων,
τῷ δὲ ὑγροτέρῳ τῶν ὑγροτέρων· οὕτω δὲ καὶ τῷ μὲν
ἀερωδεστέρῳ τῶν ἀερωδῶν, τῷ δὲ θερμοτέρῳ τῶν
πυρωδῶν.

ἐντεῦθεν οὖν ἤδη τὰς ὕλας ἐξευρίσκειν ἁπάντων
τῶν κατὰ μέρος, ἐν τροφαῖς καὶ πόμασι καὶ ἀέρι καὶ
τοῖς ἐπιτηδευομένοις ἅπασιν. ἔνθα δ' ἀλλήλαις ἐναν-
744K τιοῦνταί τινες ἐνδείξεις, τῇ πρόσθεν εἰρημένῃ | χρῆ-
σθαι μεθόδῳ, πρὸς τὴν τῶν πρακτέων εὕρεσιν ἐπιβλέποντα μέγεθός τε καὶ ἀξίωμα τῶν ἐνδεικνυμένων
σκοπῶν. αἱρεῖσθαι γὰρ ἐδείκνυμεν χρῆναι τὰς ἀπὸ
τῶν ἀξιολογωτέρων τε καὶ μειζόντων ἐνδείξεις σκοπῶν. ἀξίωμα μὲν οὖν αὐτῶν εἰς ζωὴν ἢ ὑγείαν ἀποβλε-
πόντων κρίνεσθαι, μέγεθος δ' εἶναι διττόν, ἤτοι κατὰ
τὴν οἰκείαν οὐσίαν ἢ τὴν τοῦ κατὰ φύσιν ἐκτροπήν.
ἐπὶ ταύταις οὖν ἤδη ταῖς μεθόδοις γυμνασθῶμεν ἐν τῷ
γένει τῶν ἐπὶ σηπεδόνι πυρετῶν.

4. Ἐπεὶ δὲ ἡ σηπεδὼν ἤτοι γε ὁμοτίμως ἐν ἅπασι
συνίσταται τοῖς ἀγγείοις ἢ ἐν τοῖς μεγίστοις καὶ
κυριωτάτοις, ἃ δὴ μεταξὺ βουβώνων τ' ἐστὶ καὶ μασχαλῶν, ἢ καθ' ἕν τι μόριον ἤτοι φλεγμαῖνον, ἢ καὶ
χωρὶς φλεγμονῆς, ἐν ἑαυτῷ περιέχον οἷόν περ ἐν
ἑστίᾳ τινὶ τὴν ἀρχὴν τοῦ πυρετοῦ, καθ' ἑκάστην τῶν
εἰρημένων διαφορῶν ἰδίᾳ χρὴ γυμνάσασθαι. καὶ πρώτην γε τὴν ἤτοι κατὰ τὰ μέγιστα τῶν ἀγγείων ἢ

which the patient previously had but from that which he has now. Since there is no small difference in the *krasis* of the parts of the organism, the measure of the things similar to itself will be specific to each part—of those things more earthy to what is more earthy, of those things more moist to what is more moist and, in the same way, of those things that are airy to what is more airy, and of those things that are fiery to what is hotter.

Thus, from here you now discover the materials of all those things individually in the foods, drinks, air, and all the customary activities. Then, when some indications are in opposition to each other, use the previously mentioned 744K method for the discovery of what has to be done, paying close attention to the magnitude and worth of the demonstrated indicators. I showed that you need to choose the indications from the more important and significant indicators. Their importance is to be judged by giving consideration to life and health, while their magnitude is twofold, pertaining to either the specific substance or to the deviation from an accord with nature. Therefore, let us now become practiced in these methods in the class of the fevers due to putrefaction.

4. Since the putrefaction arises either uniformly in all the vessels or in the largest and most important ones, which are those between the groins and the armpits, or in one particular part, either inflamed or without inflammation, containing in itself, very much as if in a hearth, the origin of the fever, it is necessary to be practiced in each of the stated differentiae individually. And let us choose, as the first differentia for discussion, that involving the larg-

σύμπανθ' ἅμα γινομένην τῷ λόγῳ προχειρησόμεθα.
τὸν γὰρ ἐπὶ τῇ τοιαύτῃ διαθέσει πυρετὸν εἶναι μὲν
δήπου σύνοχον ἀναγκαῖον, ἰαθῆναι δ' ἀδύνατον ἄνευ
745K   τοῦ | παύσασθαι τὴν σηπεδόνα. παύσασθαι δ' οὐ
δύναται μενούσης τῆς αἰτίας. ἐκκόπτειν οὖν χρὴ τὴν
αἰτίαν αὐτῆς, εἰ μέλλει προτέρα μὲν ἡ σῆψις, ἐπ' αὐτῇ
δ' ὁ πυρετὸς ἰαθήσεσθαι. τίς οὖν ἡ αἰτία τῆς ὁμοτίμου
σηπεδόνος ἐν ἅπασι τοῖς ἀγγείοις ἀναμνησθῆναι χρὴ
πρῶτον, ἢ τίς ποθ' ἡμῖν ἐδείκνυτο σηπεδόνος αἰτία,
κἄπειθ' ἑξῆς ζητῆσαι τὴν τῆς ὁμαλῶς γινομένης ἐν
ἅπαντι μέρει τοῦ ζῴου.

δέδεικται τοίνυν ἐν ταῖς τῶν νόσων αἰτίαις ὡς
ἐκεῖνα μόνα σήπεται τῶν σωμάτων ὅσα φύσιν ἔχοντα
θερμὴν καὶ ὑγρὰν ἀδιάπνευστα καὶ ἀρρίπιστα μένει
ἐν θερμῷ καὶ ὑγρῷ χωρίῳ. εἰ δ' ἐν κινήσει τε εἴη καὶ
ῥιπίζοιτο καὶ διαπνέοιτο, δύναιτ' ἂν ἄσηπτα φυλάττεσθαι, καὶ μάλισθ' ὅσα σώματα διοικοῦσιν αἱ φύσεις,
ὥσπερ τὰ τῶν ζῴων τε καὶ τῶν φυτῶν· αὗται γὰρ ἐξ
ἀρχῆς ἑαυταῖς ἔχουσι συμφύτους κινήσεις, ὡς ἐν τοῖς
Τῶν φυσικῶν δυνάμεων ὑπομνήμασιν ἐδείκνυμεν, αἷς
ἀποκρίνουσι τὰ περιττά. δέδεικται δὲ καὶ δι' ἄλλων
δυοῖν βιβλίων πηλίκη τίς ἡ χρεία τῆς τ' ἀναπνοῆς
746K   ἐστι καὶ τῶν σφυγμῶν, ὑπὲρ τοῦ | ῥιπίζεσθαι καὶ
διαπνεῖσθαι τὸ σῶμα καὶ τὴν κατὰ φύσιν ἑαυτοῦ
φυλάττειν θερμασίαν.

εἴπερ οὖν τι μέλλει σαφῶς σήπεσθαι τῶν κατ'

---

3 See De morborum causis, chapter 2, VII.2–10K.
4 See De naturalibus facultatibus, II.1–2K.

est of the vessels or occurring in all the vessels at the same time; for surely the fever due to such a condition is inevitably continuous, and cannot be cured without the putrefaction being brought to an end. However, it is not possible for the putrefaction to be brought to an end if the cause remains. We must, then, eradicate the cause of the putrefaction, if the intention is to cure the putrefaction first, followed by the fever. Therefore, it is necessary to call to mind initially what the cause of the putrefaction that is of equal significance in all the vessels is, or what we showed once to be the cause of putrefaction, and then to inquire next into the cause of the putrefaction occurring equally in every part of the organism.

745K

Accordingly, it was shown in relation to the causes of diseases that only those bodies putrefy which have a hot and moist nature and remain unventilated and uncooled in a hot and dry place.[3] If, however, they are in motion, and are fanned and ventilated, it would be possible for them to be maintained without putrefaction, and particularly those bodies whose natures govern them, as is the case with animals and plants. For these have in themselves from the beginning innate movements, as I showed in the treatise *On the Natural Faculties*,[4] with which they separate the superfluities. And it has been shown through two other books how great the use of respiration is and of the pulse for fanning and bringing air to the body, and preserving its own heat in accord with nature.[5]

746K

Therefore, if one of the things in the body is clearly go-

[5] The two works referred to are *De utilitate respirationis*, IV.470–511K, and *De usu pulsuum*, V.149–80K. Both are translated into English by D. J. Furley and J. S. Wilkie (1984).

αὐτό, τὰς εἰρημένας χρὴ διαπνοὰς ἐπισχεθῆναι. πῶς
δ᾽ ἂν καὶ δύναιντο αἱ καθ᾽ ὅλον τὸ ζῷον εἰς τοῦτ᾽
ἐλθεῖν ἄνευ στεγνώσεως, ἤτοι γ᾽ ἐν τοῖς πέρασι τῶν
ἀγγείων γινομένης ἢ κατὰ σύμπαν τὸ δέρμα; κατὰ
μὲν οὖν τὰ πέρατα τῶν ἀγγείων ἡ στέγνωσις γίγνοιτ᾽
ἄν, ἢ διὰ σφοδράν τινα ἔξωθεν ψῦξιν, ἢ διὰ πάχος, ἢ
πλῆθος, ἢ γλισχρότητα τῶν περιεχομένων ἐν αὐτοῖς
χυμῶν, ἀθροώτερον ῥευσάντων ἐπὶ τὴν ἔξω φοράν.
γίγνεται δ᾽ αὐτοῖς τοῦτο καὶ διὰ γυμνάσια τὰ κατὰ
παλαίστραν ἢ ἄλλως ἐπιτηδευθέντα καὶ δι᾽ ὁδοιπορίας συντόνους ἢ διὰ τὸ περιέχον ἐξαίφνης ἐκ κρύους
εἰς θάλπος μεταβαλόν. ἐνίοτε δὲ καὶ διὰ θυμὸν οἱ
χυμοὶ ζέσαντες ἀθρόα καὶ χρησάμενοι τῇ πρὸς
τοὐκτὸς φύσει τὰς εἰρημένας στεγνώσεις εἰργάσαντο.

χρὴ τοίνυν ὅστις ἰᾶσθαι μέλλει τὸν προκείμενον ἐν
τῷ λόγῳ πυρετόν, ἅμα μὲν ἐκκόπτειν αὐτοῦ τὴν αἰτίαν,
ἅμα δὲ καὶ τὴν ἤδη γεγενημένην ὑπ᾽ αὐτῆς ἐν τῷ ζῴῳ
θερμασίαν ἐμψύχειν. ὅπως μὲν οὖν ἐκείνην χρὴ ψύχειν
ἔμπροσθεν εἴρηται· τὴν δ᾽ αἰτίαν ἐκκόπτειν προσήκει, |
747K   κατὰ τὴν ἰδίαν ἑκάστης φύσιν ἐξευρίσκοντα τὴν ἐναν-
τίωσιν· εἰ μὲν ὑπὸ ψυχρᾶς αἰτίας ἐπιλήθη τε καὶ
πυκνὸν εἴργασται τὸ σῶμα, χαλῶντα καὶ ἀραιοῦντα
παντοίως αὐτό· δι᾽ ἔμφραξιν δὲ τοῦτο παθὸν ἐκφράτ-
τοντα. τὸ δὲ ἐκφράττον εὑρήσεις κἀνταῦθα, τῷ τῆς
ἐναντιώσεως προσέχων σκοπῷ· τοὺς μὲν γὰρ πολλοὺς
χυμοὺς κενώσεις, τοὺς παχεῖς δὲ καὶ γλίσχρους ἐργάσῃ λεπτούς τε καὶ ῥυτούς. εἰ δὲ καὶ πλείω συνέλθοι
ποτὲ αἴτια, πρὸς ἅπαντα ἐνιστάμενος ἐξ ὑπεναντίου·

ing to putrefy, it is necessary for the aforementioned transpirations to be held back. But how would it be possible, in
the whole organism, for these transpirations to come to
this without obstruction occurring either in the ends of the
vessels or in the whole skin? The obstruction in the ends of
the vessels could occur due to severe cooling externally, or
to a thickening, or an abundance, or the viscidity of the humors contained in them when they flow in a more concentrated fashion in their outward passage. This happens to
them due to exercises, either in the wrestling school or carried out otherwise, or due to arduous journeys, or due to a
sudden change in the ambient air from very cold to very
hot. Sometimes, also, the humors seethe due to anger and,
by needing to flow outward all at once, bring about the
aforementioned obstructions.

It is necessary, therefore, for someone who intends to
cure the fever proposed in the discussion, to eradicate the
cause of it and, at the same time, to cool the heat that has
already occurred in the organism due to this cause. How
you must cool that heat was stated earlier; it is appropriate
to eradicate the cause by finding out what is opposite to the   747K
specific nature of each cause. If the body is condensed due
to a cold cause and this brings about thickening, relax and
rarefy it in various ways. If the body is affected by blockage, remove the blockage. And even here you will find
what removes the blockage when you direct your attention
to the indicator of contrariety; for you will evacuate the
abundant humors and make those that are thick and viscous, thin and fluid. And if, at some time, more causes
come together, counteract all of them with their opposites,

δυνατὸν γὰρ δήπου καὶ πεπιλῆσθαι τὸ σῶμα καὶ
πεπυκνῶσθαι τοὺς πόρους καὶ πολλοὺς και γλίσχρους
εἶναι τοὺς χυμούς. ἐν δὲ ταῖς τοιαύταις ἐπιπλοκαῖς, εἴ
τι μεμνήμεθα τῶν ἐν τοῖς ἔμπροσθεν εἰρημένων, ἄρχεσθαι προσήκει ἀπὸ τῆς φλεβοτομίας, ἐκκενοῦντα τὸ
πλῆθος, ἀφικνεῖσθαι δ' ἐπὶ τῷ λεπτύνειν τοὺς χυμούς,
εἶθ' ἑξῆς ἐπὶ τὸ χαλᾶν τὰ πεπιλημένα καὶ ἀραιοῦν τὰ
πεπυκνωμένα. ταυτὶ μὲν οὖν ἐκ τῆς νοσώδους διαθέσεως ἐνδεικτικῶς ληπτέον.

5. Ἀπὸ δὲ τῆς δυνάμεως, ἐπειδὴ σιτία καὶ ποτὰ καὶ
748K πνεύματα ταύτην ἐφύλαττον,[1] ἐν μέτρῳ τέ τινι | καὶ
ποιότητι δεούσῃ προσφερόμενα, καὶ ἦν αὐτῶν τὸ μὲν
τοῦ ποσοῦ μέτρον ἀπὸ τῆς οὐσίας τῆς δυνάμεως
εὑρισκόμενον, τὸ δὲ τῆς ποιότητος ἀπὸ τῆς κράσεως,
ἐπισκεπτέον ὅπως ἔχουσιν αἱ διοικοῦσαι τὸ σῶμα
δυνάμεις. εὐρώστων μὲν γὰρ οὐσῶν θαρρῶν χρῶ τοῖς
κενωτικοῖς βοηθήμασιν, ἅπερ ἐκ τῆς κατὰ τὴν νόσον
ἐλήφθη διαθέσεως· ἀρρωστοτέρων δὲ γεγενημένων
εὐλαβῶς μεταχειρίζου τὰ κενωτικὰ βοηθήματα. καὶ
μὲν δὴ καὶ τὸ τῶν τροφῶν ποσὸν ἐντεῦθέν σοι ληπτέον. ἐρρωμένων γὰρ τῶν δυνάμεων κατὰ τὸν ἑαυτῶν
λόγον καὶ τῆς ἀκμῆς τοῦ νοσήματος ἐν τάχει προσ-
δοκωμένης ἔξεστι λεπτότατα διαιτᾶν, ἀρρωστοτέρων
δ' οὐσῶν οὐκ ἔξεστιν ἄνευ ζημίας μεγάλης· ἀλλὰ χρὴ
προστιθέναι τοῖς τρέφουσι τοσοῦτον ὅσον ἀφῄρηνται
τῆς εὐτονίας αἱ δυνάμεις.

[1] K; ἐφύλαττεν B

132

insofar as it is possible, of course, in a body that has been condensed, for the pores to be closed up, and for the humors to be both many and viscous. In such complicated circumstances, if we call to mind some of those things previously spoken of, it is appropriate to begin with phlebotomy, and when the abundance has been evacuated, to come to the thinning of the humors, and then in turn to the relaxation of what has been condensed and the rarefaction of what has been thickened. These things must be chosen indicatively from the diseased condition.

5. From the capacities (since foods, drinks and *pneumas* preserve these when provided in a certain moderation and required quality, the measure of their quantity being discovered from the substance of the capacity and that of the quality from the *krasis*), it behooves us to consider how the capacities which govern the body are.[6] When these are strong, use with confidence the evacuating remedies, which are found from the condition in the disease. However, when the capacities have become weaker, employ the evacuating remedies cautiously. But here too you must choose the amount of the nutriments. For when the capacities are strong in their own right and you expect the peak of the disease within a short time, it is possible to use a very thin diet, whereas when they are weaker, this is not possible without very considerable damage. But it is necessary to add to the nutriments an amount commensurate with the extent to which the capacities are deprived of their vigor.

748K

6 In this complicated sentence Linacre's use of parentheses is followed, as is his use of the plural for the first *dunamis*; see his p. 552.

ἐκ τούτων μὲν τῶν σκοπῶν τὸ ποσὸν τῆς τροφῆς
ληπτέον, ἐκ δὲ τῆς διαθέσεως τῆς κατὰ τὴν νόσον ἅμα
τῇ κατὰ φύσιν κράσει ἐν τῷ τότε χρόνῳ τῶν στερεῶν
σωμάτων τὸ ποιόν, ἀπὸ μὲν τῆς διαθέσεως, εἰ ἔμφραξις εἴη, τὰς λεπτυνούσας, ἀπὸ δὲ τῆς κατὰ φύσιν
αὐτῶν κράσεως ἅμα τοῖς ἔθεσι, καθὼς καὶ περὶ |
749K   τούτων διώρισται πρόσθεν. ὁ δὲ καιρὸς τῆς τροφῆς
ἐπὶ μὲν τῶν ἄλλων ἁπάντων πυρετῶν, ὅσοι σήψεσιν
ἕπονται, προϊόντος εἰρήσεται τοῦ λόγου· τοῖς δ' ἕνα
μόνον ἀπ' ἀρχῆς ἄχρι τέλους ἔχουσι παροξυσμόν,
ὑπὲρ τούτων γὰρ ἡμῖν ὁ ἐνεστὼς λόγος, ἥ τ' εὐφορία
καὶ τὸ ἔθος οἱ σκοποί· τηνικαῦτα γὰρ ἂν αὐτοῖς
δοτέον ὅταν εὐφορώτατοι σφῶν αὐτῶν ὑπάρχωσι, καὶ
μάλιστα κατ' ἐκεῖνον τὸν καιρὸν τῆς ἡμέρας ἐν ᾧ καὶ
πρόσθεν ὑγιαίνοντες ἔθος εἶχον σιτεῖσθαι· μάλιστα
γὰρ ἂν εὐφόρως ἐνέγκαιεν τὰ σιτία κατὰ τούτους τοὺς
σκοποὺς λαβόντες. ὅτι δὲ καὶ τὸ ψυχρὸν τοῖς οὕτω
νοσοῦσι δοτέον, ὅταν ἤδη πέττωνται μὲν οἱ χυμοί, τὸ
δὲ πάχος αὐτῶν ᾖ προλελεπτυσμένον, ἐκ τῶν ἐν τοῖς
ἔμπροσθεν εἰρημένων εὔδηλον.

6. Ἐπαμφοτερίζοντος γὰρ τοῦ τοιούτου πυρετοῦ
τοῖς τ' ἐπὶ σήψει χυμῶν ἐν περιόδοις τισὶ παροξυνο-
μένοις καὶ τοῖς ἐφημέροις ὀνομαζομένοις, ἐν ἑκατέρῳ
τῷ γένει προσηκόντως ἐμνημονεύσαμεν τούτων, ἕνα
μὲν ἐχόντων παροξυσμὸν ὡς ἐφημέρων, ἐπὶ σήψει δὲ
750K   γιγνομένων | καὶ ἤδη πολυημέρων πως ὄντων. ὥσθ' ἓν
μὲν ἕξουσι κοινὸν πρὸς τοὺς ἐφημέρους, δύο δὲ πρὸς

From these indicators, you must choose the amount of nutriment; from the condition, which relates to the disease together with the natural *krasis* of the solid bodies at the actual time, you must choose the quality of the nutriment. From the condition, if this is an obstruction, choose those things that are thinning; from the natural *krasis* of the bodies along with the customs, [choose] according to what was previously defined regarding these. The time for nourishment in all the other fevers that follow putrefactions will be spoken of as the discussion proceeds. In those that have only one paroxysm from beginning to end (for our present discussion is about those), the indicators are the sense of well-being and the custom. Under these circumstances, you should give patients [nourishment] whenever they have the greatest sense of well-being in themselves, and especially at that time of day when it was customary for them to eat when they were previously healthy, because they would be particularly able to tolerate foods easily when taken on the basis of such indicators. It is clear from what was said in the previous [books] that you must also give cold [water] to those who are sick in this way whenever the humors are already concocted and their thickness has previously been made thin.

6. Since such a fever can be of both kinds, that is, those due to putrefaction of humors in which [patients] suffer certain periodic paroxysms, and those termed ephemeral, it was appropriate that I made mention of those in each class, because those that are ephemeral have the one paroxysm while those occurring due to putrefaction are already in some way of many days' duration (polyhemeral). As a consequence, they will have one thing in

749K

750K

τοὺς ἐπὶ σήψει πολυημέρους. τὸ δ' ἕτερον γένος τῶν συνόχων δύο μὲν ἔχει κοινὰ πρὸς τοὺς ἐφημέρους, τό τε χωρὶς σήψεως γίγνεσθαι καὶ χωρὶς περιόδου· τὸ τρίτον δ' οὐ κοινόν, εἰς γὰρ τὴν πέμπτην ἡμέραν ἐκτείνεται πολλάκις. ἡ μὲν οὖν τῶνδε τῶν πυρετῶν θεραπεία γέγραπται πρόσθεν, ἡ δὲ τῶν ἐπὶ σήψει συνόχων ἑκατέρῳ τῷ γένει κοινωνοῦσα καὶ ἡμᾶς ἠνάγκασεν ἐν ἀμφοτέροις τοῖς λόγοις ὑπὲρ αὐτῆς διελθεῖν. ἀλλ' ἡ μὲν οὐσία καὶ ἡ σύμπασα φύσις αὐτῶν ἐκ τοῦ γένους ἐστὶ τοῦ νῦν ἡμῖν προκειμένου. καθ' ἓν δέ τι τῶν συμβεβηκότων αὐτοῖς ἐκοινώνησαν τοῖς ἐφημέροις, ὅπερ ἦν εἷς ὁ σύμπας παροξυσμὸς ὅλης τῆς νόσου. χρὴ τοίνυν καὶ τὰ τῆς ἰάσεως αὐτοῖς καθ' ἓν μὲν τοῦτο μόνον εἰς κοινωνίαν ἥκειν τοῖς ἐφημέροις, κατὰ δὲ τἆλλα σύμπαντα διχῇ τμηθῆναι, καὶ τοὺς μέν τινας σκοποὺς θεραπευτικοὺς ἐκ τοῦ γένους τῶν ἐπὶ σήψει λαβεῖν, τοὺς δέ τινας ἐκ τῆς 751K οἰκείας αὐτῶν οὐσίας. οὕτω δὲ καὶ τὰς ἄλλας | διαφορὰς τῶν ἐπὶ σήψει χυμῶν ἰᾶσθαι προσήκει, καθάπερ ἤδη πολλάκις ἔμπροσθεν εἴρηται, τὴν ἔνδειξιν ἀπό τε τοῦ κοινοῦ γένους αὐτῶν λαμβάνοντας καὶ τῆς καθ' ἕκαστον ἰδίας διαφορᾶς.

7. Ἀναλαβόντες οὖν αὖ εἴπωμεν ὁπόσαι μέν εἰσιν αἱ πᾶσαι διαφοραὶ τῶν ἐπὶ σηπεδόνι χυμῶν ἀναπτομένων πυρετῶν, τίς δ' ἡ κοινὴ τοῦ γένους αὐτῶν ἔνδειξις τῶν ἰασομένων βοηθημάτων, εἶθ' ἑξῆς ὁποία τις ἡ καθ' ἕκαστον ἰδία. χρὴ γὰρ δήπου τό τε κοινὸν

common with the ephemeral, but two with the polyhemeral due to putrefaction. The other class (that is, the continuous) has two things in common with the ephemeral: they occur without putrefaction and without periodicity. The third factor, which is not common, [is that] they often extend to the fifth day. I have previously written about the treatment of these fevers. However, because the treatment of the continuous fevers due to putrefaction has features common to each class, I was compelled to go over it in both discussions. But the substance and the whole nature of these fevers is from the class of what now lies before us. In one of those things that happen to them they have common ground with the ephemeral—there is, in all, one paroxysm throughout the whole disease. Accordingly, it is necessary that those things pertaining to the cure of these fevers relate to this one thing alone which brings them to a common ground with the ephemeral [fevers]. However, in all other respects it is necessary to make a twofold division, and to take some therapeutic indicators from the class of those fevers due to putrefaction, and other [indicators] from their specific essence. In the same way too, it is appropriate to cure the other differentiae due   751K
to putrefaction of the humors, as was previously stated, and often, when we took the indication from their common class and from the specific differentia relating to each.

7. Therefore, taking up the matter again, let me state of what kind all the differentiae of the fevers kindled by putrefaction are, what the common indication of the class of those remedies that will cure them is, and then in turn, what kind of indication is specific in each case. It is, I presume, necessary that what is common to them all also has a

αὐτοῖς ἅπασι κοινὴν καὶ τὴν ἔνδειξιν ἔχειν, τό τ᾽ ἐν
ἑκάστῳ διάφορον εἰς ὅσον ἀποκεχώρηκε τῶν ὁμο-
γενῶν, εἰς τοσοῦτον καὶ τὴν ἔνδειξιν ἀφωρισμένην
ἐκείνων ἔχειν. εἰσὶ δὲ τῶν ἐπὶ σήψει χυμῶν αἱ διαφοραὶ αἵδε.² μία μὲν ἐκ τοῦ παρεῖναι τὰς αἰτίας αὐτοῖς
τῆς σηπεδόνος, ἢ μὴ παρεῖναι, δευτέρα δὲ ἐκ τοῦ καθ᾽
ὅλον τὸ ζῷον ἢ κατὰ μέρος ὑπάρχειν τὴν σηπεδόνα,
752K καὶ τρίτη καθ᾽ ὅσον ἤτοι μετὰ φλεγμονῆς | ἢ χωρὶς
ταύτης εἴη.

ἐπισκεπτέον οὖν ἡμῖν ἐστι τί μὲν ἔχει κοινὸν ἡ ἀπὸ
τῆς αἰτίας ἔνδειξις ὑπαρχούσης ἔτι, τί δ᾽ ἡ τῆς σηπεδόνος αὐτῆς· κἄπειθ᾽ ἑξῆς τί μὲν ἡ καθ᾽ ὅλον τὸ ζῷον ἢ
τά γε κυριώτατα τῶν ἀγγείων αὐτοῦ· τί δὲ ἡ καθ᾽ ἕν τι
μόριον, ἔπειθ᾽ ἑξῆς τῶν καθ᾽ ἕν τι μορίων τὰς διαφοράς. ἄλλην μὲν γὰρ εἰκὸς ἔνδειξιν εἶναι τὴν μετὰ
φλεγμονῆς, ἑτέραν δὲ τὴν χωρὶς ταύτης. ἐν αὐταῖς δὲ
ταύταις πάλιν ἑτέραν μὲν ἔτι παρούσης τῆς ποιούσης
αἰτίας, ἑτέραν δ᾽ οἰχομένης αὐτῆς. εἶθ᾽ ἑξῆς τὰς καθ᾽
ἑκάστην αἰτίαν ἔτι παροῦσαν διαφοράς, ἐπὶ πάντων
τῶν εἰρημένων κατὰ τὴν τομὴν εἰδῶν· ἐνταῦθα γὰρ
ἡμῖν ἀναγκαῖόν ἐστιν εἰς τὰς ἀπὸ τῶν θεραπευομένων
διαθέσεων ἐνδείξεις τελευτῆσαι. μετὰ δὲ ταύτας δηλονότι
τὰς ἀπὸ τῶν θεραπευσομένων μορίων ἐνδείξεις ληψόμ-
εθα κατὰ τὴν προειρημένην μέθοδον· αἷς προσθέντες
τὰς ἀπὸ τῆς δυνάμεως τελείαν ἕξομεν ἤδη τὴν μέθ-
οδον τῆς ἁπάντων τῶν ἐπὶ σήψει πυρετῶν ἰάσεως. |

² Κ; αἵδε om. Β

138

common indication and, to the extent that what is different in each case has departed from its congeners, the indication is also distinct from those to the same degree. The differentiae due to putrefaction of the humors are as follows: one is from the presence or otherwise of the causes of putrefaction in them; a second is whether the putrefaction involves the whole organism or a part only; and a third is the extent to which it is accompanied by inflammation or not. 752K

We must consider, then, what the indication from the cause that still exists has that is common and what the indication from the putrefaction itself has that is common. Next, we must consider what the indication is in relation to the whole organism, or at least in relation to the most important of its vessels, and what the indication is in relation to some one part; and then, sequentially, the differentiae in relation to one of the parts. It is probable that there is one indication in connection with inflammation and another when this is not present. Again, among these indications themselves, there is one if the effecting cause is still present and another if this has gone. Then, in turn, there are the differentiae in relation to each cause that is still present according to the division of all the kinds mentioned. Then it is necessary for us to bring finality to the indications from the conditions being treated. After these, obviously, we shall take the indications from the parts that will be treated according to the previously mentioned method. When we have added to these the indications from the capacity, we shall now have the complete method of cure of all fevers due to putrefaction.

753K 8. Ἀρξώμεθα οὖν ἀπὸ τοῦ πρώτου σκοποῦ κατὰ τὴν
διαίρεσιν, ὅπερ ἦν ἡ σηπεδὼν τῶν χυμῶν, καὶ σκεψώ-
μεθα τίσιν ἄν τις προσέχων σκοποῖς καὶ τίσιν ἐνδεί-
ξεσι χρώμενος ἐξεύροι τὴν θεραπείαν αὐτῆς. ἀρχὴ δ᾽
εἰς τὴν εὕρεσιν ἡ φύσις ἔσται τοῦ πράγματος ὑπὲρ οὗ
σκοπούμεθα· καὶ γὰρ καὶ τοῦτο ἐμάθομεν ἐν ταῖς
ἀποδεικτικαῖς μεθόδοις. τίς οὖν ἡ φύσις τῆς σηπεδόνος ἐστίν; ἡ μεταβολὴ τῆς ὅλης τοῦ σώματος σηπομέ-
νης οὐσίας ἐπὶ φθορὰν ὑπὸ τῆς ἔξωθεν θερμασίας. οὐ
γὰρ δὴ ὑπό γε τῆς οἰκείας τι φθείρεται, τοὐναντίον δ᾽
ἅπαν αὐξάνεταί τε καὶ ῥώννυται καὶ ὑγιαίνει καὶ ζῇ
τῶν ὄντων ἕκαστον ὑπὸ τῆς οἰκείας θερμασίας διοι-
κούμενον· ὥσπερ ἀμέλει καὶ αὐτὰ τὰ τῶν ζῴων σώματα καίτοι τὸ πλεῖστον τῆς ἑαυτῶν οὐσίας ὑγρὸν
καὶ θερμὸν ἔχοντα, διαρκεῖ πάμπολυν ἐτῶν ἀριθμὸν
ἄσηπτά τε καὶ ὑγιῆ καὶ ζῶντα φυλαττομένης αὐτῶν,
ὡς ἐπιδέδεικται, τῆς οἰκείας θερμασίας, ἐν μὲν τῷ τῆς
καρδίας σώματι διὰ τῆς ἀναπνοῆς, ἐν ἅπασι δὲ τοῖς
ἄλλοις μέρεσι διά τε τῆς πρὸς τὴν καρδίαν κοινωνίας
754K καὶ δι᾽ ἑτέρου τινὸς | εἴδους ἀναπνοῆς, ὃ καθ᾽ ὅλον τὸ
δέρμα γιγνόμενον ὠνόμασται διαπνοή.

τῇ δὲ τούτων βλάβῃ συγκακοῦται μὲν τὸ κατὰ
φύσιν θερμόν· ὀθνεῖον δέ τι καὶ παρὰ φύσιν ἕτερον
ἐγγινόμενον τοῖς σώμασι τοὺς μὲν χυμοὺς πρώτους
δι᾽ ὑγρότητα σήπει τε καὶ διαφθείρει, τῷ χρόνῳ δὲ καὶ
τῆς πιμελῆς ἅπτεται καὶ τῆς σαρκός. ὥσπερ οὖν ὅσα

8. Let me begin from the first indicator according to the   753K
division, which is the putrefaction of the humors, and let
me consider to what indicators someone might direct his
attention, and by using what indications he might discover
the treatment of this putrefaction. The nature of the matter will be the starting point for discovery of what we are
considering. This is also what we learned in [my work] on
the demonstrative methods.[7] What, then, is the nature of
putrefaction? It is the change of the whole putrefying substance of the body toward corruption due to external heat.
For it is surely not due to its own heat that something is
corrupted; on the contrary, each and every living thing
is increased, strengthened, made healthy and lives when
governed by its own heat. Of course, even though the actual bodies of animals do indeed have the greatest part
of their own substance moist and hot, they nevertheless
last a great number of years without putrefaction, and are
healthy and alive when, as has been shown, their own heat
is preserved in the body of the heart by respiration, and in
all the other parts through their connection to the heart
and through some other form of respiration which, when it   754K
occurs in the whole skin, is called transpiration.

The natural heat is, however, jointly harmed by injury
to these processes, while something else alien and contrary
to nature, when it supervenes in bodies, putrefies and cor-
rupts the primary humors due to moisture, and in time also
involves the fat and the flesh. Therefore, just as in the case

---

[7] Presumably the lost work *Logical Demonstration* or *On
Demonstration* in fifteen books; see Galen's *De libris propriis*,
XIX.39K ff. for reference to this and other works on similar mat-
ters.

τῶν μὴ ζώντων σωμάτων σήπεται, πρῶτον μὲν ἀπο-
φράττουσιν αὐτῶν ὅσον ἤδη σέσηπται, εἶθ' ἑξῆς τὸ
λοιπὸν αἰωρήσαντες ἐν ἀέρι ψυχρῷ διαπνοὰς εὐψυχεῖς
μηχανῶνται, κατὰ τὸν αὐτὸν τρόπον ἰασόμεθα τὴν ἐν
τοῖς ζῴοις σηπεδόνα, τὸ μὲν ἤδη διεφθαρμένον ἐκκε-
νοῦντες ἅπαντι τρόπῳ, τὸ δὲ ὑπόλοιπον αἰωρήσεσι
μετρίαις καὶ διαπνοαῖς εὐψυχέσιν εἰς τὴν ἀκριβῆ
συμμετρίαν ἐπαναγαγόντες. ἡ μὲν δὴ κένωσις αὐτῶν
δι' οὔρων τε καὶ διαχωρημάτων ἐμέτων τε καὶ ἱδρώτων
ἔσται· μετὰ δὲ ταῦτα κινήσεις μετρίας ἅμα τῇ τοῦ
περιέχοντος εὐκρασίᾳ προσάξομεν. ὅπως δ' ἄν τις
ἕκαστον τούτων ὀρθῶς μεταχειρίζοιτο, μικρὸν ὕστερον ἐπισκεψόμεθα.

νυνὶ δ' ἐπὶ τὸ δεύτερον ἐκ τῆς διαιρέσεως ἤδη
755K μεταβῶμεν, | ὡς ἄν τις κάλλιστα τῆς ποιούσης αἰτίας
τὴν σῆψιν ἔτι μενούσης ἐξεύροι τὴν ἴασιν. ἔστι δ' οὐδ'
ἐνταῦθα χαλεπὸν οὐδὲν ἐξευρεῖν, ὅτι διὰ τῶν ἐναντίων
ἀναιρουμένης αὐτῆς. ἐπεὶ τοίνυν ἡ κώλυσις τῆς διαπνοῆς εἰργάζετο τὴν σῆψιν, ἀναιρετέον αὐτήν. ἀλλ'
ἐπεὶ πολυειδῶς ἐγίγνετο καὶ πυκνουμένων τῶν πόρων
καὶ πιλουμένων τῶν σωμάτων καὶ διὰ πλῆθος καὶ
πάχος καὶ γλισχρότητα τῶν χυμῶν ἐμφραττομένων,
ἑκάστην τῶν εἰρημένων διαθέσεων ἰᾶσθαι χρὴ διὰ
τῶν ἐναντίων· τὴν μὲν πύκνωσιν τοῖς ἀραιοῦσι, τὴν δὲ
πίλησιν τοῖς χέουσι, τὸ δὲ πλῆθος τοῖς κενοῦσι, καὶ
τὸ πάχος τοῖς τέμνουσι, καὶ τὴν γλισχρότητα τοῖς
ῥύπτουσιν. καὶ εἰ μὲν καθ' ὅλον εἴη τὸ σῶμα τῶν
εἰρημένων ἕκαστον, ὅλῳ τῷ σώματι διὰ τῶν ἐναντίων

142

of those nonliving bodies that putrefy, people first get rid of however much of them has already putrefied, then next, after lifting up what remains into cold air, they contrive an agreeably cool ventilation, in the same way we shall cure the putrefaction in living creatures, when we have evacuated in every way what has already been corrupted and have stirred up what remains with moderate passive exercises and agreeably cooling transpiration to achieve a precise balance. Their evacuation will be through the urine, feces, vomit and sweat. After this, we shall introduce moderate movements along with *eukrasia* of the ambient air. I shall consider a little later how someone might handle each of these things correctly.

But for the present, let me pass now to the second of the divisions, so that someone might best discover the cure 755K when the cause bringing about putrefaction still remains. In this case the cure is not difficult to discover because the cause is removed through its opposites. Accordingly, when the prevention of transpiration brings about putrefaction, you must remove this. But when it has occurred in various ways—that is, when the pores are compressed and the bodies are condensed and blocked up due to the abundance, thickness and viscidity of the humors, it is necessary to cure each of the conditions mentioned through their opposites: condensation by those things that rarefy, contraction by those things that cause flow, abundance by those things that evacuate, thickness by those things that cut, and viscidity by those things that cleanse. And if there is each of the aforementioned things in the whole body, then we give succor to the whole body through the oppo-

τιμωροῦντες· καθ᾽ ἓν δέ τι τῶν μορίων ὑπαρχούσης
τῆς σηπεδόνος, ἐν ἐκείνῳ τὰ εἰρημένα μηχανώμεθα. οὐ
τῶν αὐτῶν δὲ δηλονότι δεήσεται πάντα κενώσεως
φαρμάκων, ὡς ἂν καὶ ταῖς κράσεσι καὶ ταῖς θέσεσι
καὶ ταῖς διαπλάσεσι καὶ τῇ συμπάσῃ κατασκευῇ καὶ
φύσει διαφέροντα ἀλλήλων. |

756K   9. Ὅπως οὖν χρὴ βοηθεῖν ἑκάστῳ μετὰ ταῦτα
σκεψόμεθα κατὰ τὴν οἰκείαν τοῦ λόγου τάξιν· ἧς τὴν
ἀρχὴν ἀπὸ τῆς ἐν ὅλῳ τῷ σώματι σηπεδόνος ποιησόμεθα, γεννώσης καὶ αὐτῆς τοὺς συνόχους ὀνομα-
ζομένους πυρετούς. ἐκκαθαίρειν οὖν χρὴ τὰ σεσηπότα
τῶν οὕτω καμνόντων σωμάτων δι᾽ οὔρων τε καὶ γαστρὸς καὶ ἱδρώτων· εἰ δ᾽ αὐτόματόν ποτε τὴν ὁρμὴν
ἐπὶ τὸ στόμα τῆς γαστρὸς ποιήσαιτο, καὶ δι᾽ ἐμέτων·
ἄλλως δὲ οὐ χρὴ παρὰ φύσιν ἐρεθίζειν αὐτά. ἔστι μὲν
οὐκ ὀλίγη τις ὕλη τῶν τὰς εἰρημένας ἐκκρίσεις προ-
τρεπόντων, ἀλλ᾽ εἰ πλείους αὐτῶν θερμαὶ καὶ ξηραὶ
ταῖς δυνάμεσιν οὖσαι τὸν πυρετὸν ἐπαύξονται. ἐκλεκτέον οὖν τῷ ἰατρῷ ὅσα χωρὶς τοῦ θερμαίνειν καὶ
ξηραίνειν ἱκανῶς τὰς εἰρημένας κενώσεις ἐργάζονται,
καθάπερ ὅ τε τῆς πτισάνης χυλὸς καὶ τὸ μελίκρατον
ὀξύμελί τε καὶ ἀπόμελι καὶ ἡ τοῦ σελίνου ῥίζα κατὰ τὸ
πινόμενον ὕδωρ ἑψομένη.

αὐτὰ δὲ ταῦτα καὶ κατὰ κοιλίαν ἀγαθά· μὴ δια-
χωρούσης δὲ τῆς κοιλίας, κλύζειν μελικράτῳ μετ᾽
757K   ἐλαίου. τὸ δ᾽ ὅλον | σῶμα πρὶν μὲν κενῶσαι, μανοῦν οὐ
χρή· κενώσαντα δ᾽ ἐγχωρεῖ μανοῦν φαρμάκῳ χλιαρὰν
ἔχοντι τὴν θερμασίαν, ὁποῖόν ἐστι τὸ διὰ τοῦ χαμαι-

144

sites. When putrefaction exists in one of the parts, we contrive the things stated in that part. Obviously, not all of these will require evacuating medications, as they differ from each other in terms of *krasis*, position, conformation, and their whole constitution and nature.

9. After these things, according to the proper order of 756K the discussion, I shall consider how it is necessary to provide help in each case. I shall make a start from the putrefaction in the whole body, since this is also what generates the fevers termed continuous. It is necessary to purge those things that have putrefied in patients' bodies in this way through the urine, stomach and sweat, while if at some time they produce a spontaneous impulse toward the opening of the stomach, also through vomiting. Otherwise, if there is not, you should stir up those things that are contrary to nature. There is no little material of the things that impel the aforementioned excretions onward, but if the majority of them are hot and dry in their capacities, they increase the fever. Therefore, what the doctor must select are those things that bring about the evacuations referred to without being excessively heating and drying—for example, the juice of ptisan, melikraton, oxymel, apomel, and the root of wild celery (smallage) boiled in water and drunk.

These same things are also good for the stomach. If, however, the stomach is not excreting, wash it out using melikraton with oil. You should not rarefy the whole body 757K prior to evacuation. However, when it has been evacuated, it is permissible to rarefy it with a medication that is luke-

μήλου. ἐν τούτῳ τῷ καιρῷ καὶ οἶνος ὑδατώδης πινό-
μενος ἁπάσας κινεῖ τὰς ἐκκρίσεις καὶ λουτρὸν εὔκρατον ἐκ γλυκέος ὕδατος. εἰ δὲ καὶ μὴ προσκέοιτό
ποτε κατὰ τὴν λέξιν τὸ γλυκύ, προσυπακούειν χρὴ
γινώσκοντας ὅτι περὶ τοῦ τοιούτου λουτροῦ διὰ παντὸς ὁ λόγος ἡμῖν ἐπὶ τῶν πυρεττόντων ἐστίν, οὐκ
ἀσφαλτώδους ἢ θειώδους, ἢ στυπτηριώδους, ἢ χαλ-
κανθώδους, ἢ ἁλμώδους ἢ θαλάττης αὐτῆς. ἀλλ᾽ ὅταν
ἀνθίσταται τὸ μέγεθος τοῦ πυρετοῦ, καθάπερ ἐπὶ τῶν
συνόχων, οὔτ᾽ οἴνῳ χρηστέον οὔτε λουτρῷ οὔτε ἀλείμ-
μασιν ἀραιωτικοῖς· ἀλλὰ τὸ ψυχρὸν ὕδωρ ἐν τούτοις
τοῖς πυρετοῖς πινόμενον, ὡς ἔμπροσθεν εἴπομεν, ἐπιτηδειότατον, εἰ μὴ καὶ τοῦτο κωλύει τι τῶν εἰρημένων.
αὐτοῦ μὲν γὰρ τοῦ πυρετοῦ διαπαντὸς ἴαμα τὸ ψυχρὸν
ποτόν, οὐ μὴν τοῦ γ᾽ ἐκκενωθῆναι τὰ σηπόμενα διὰ
γαστρός, οὔρων ἢ ἱδρώτων.

διὸ χρὴ γεγυμνασμένον ἐν ταῖς νῦν διδασκομέναις
758K μεθόδαις, ἐπὶ τῶν καμνόντων | σκοπεῖσθαι, τίνα μέν
ἐστιν αὐτοῦ τοῦ πυρετοῦ τὰ ἰάματα, τίνα δὲ τῆς
σήψεως αὐτῆς καθ᾽ ἑαυτὴν ἐγκειμένης μόνης, τίνα δὲ
τῆς ποιούσης αὐτὴν αἰτίας, καὶ ταύτης αὐτῆς καθ᾽
ἑαυτὴν ἐγκειμένης. οὐ γὰρ ὁμολογεῖ πάντα ἀλλήλοις
διὰ παντός, ἀλλ᾽ ἐναντιοῦται πλειστάκις. ἔνθα χρὴ
μεμνημένον ὧν ἐμάθομεν ἐπὶ τὸ μέγιστον ἁπάντων
ἔρχεσθαι καὶ τοῦτ᾽ ἤτοι πρῶτον ἢ μᾶλλον τῶν ἄλλων
ἰᾶσθαι. χαλεπὸν μὲν οὖν εὖ ἴσθι καὶ γνώμης ὀξείας
δεόμενον ἀλλήλοις παραβάλλειν τὸν πυρετὸν καὶ τὴν
σηπεδόνα καὶ τὴν ποιοῦσαν αἰτίαν, ὅταν ἔτι καὶ αὐτὴ

warm in terms of heat, such as that made from chamomile. At this time also, if watery wine is drunk, it stimulates all the excretions, as does a bath made *eukratic* by sweet water. And if ever the term "sweet" is not added to the statement, you should understand it, knowing that our discussion throughout is about such a bath in the case of febrile patients, and not one that is full of asphalt, or sulfurous, or astringent, or vitriolic, or salty, or of seawater itself. But whenever the magnitude of the fever stands in opposition, as in the case of the continuous fevers, you must not use wine, bathing, or rarefying unguents. Rather, as I said before, cold water when drunk during these fevers is very suitable unless any of the things previously mentioned also prevents this. Cold water is invariably a cure of the fever itself, but definitely not of the putrefied materials being evacuated through the stomach, urine or sweat.

On this account, it is necessary to become practiced in those methods now being taught, and to consider in patients what the cures of the fever itself are, and of the actual putrefaction that exists in and of itself, and of the cause bringing this putrefaction about, which also exists in and of itself. For these do not all invariably agree with each other but are very often opposite. Then, having called to mind those things I taught, it is necessary to come to the most important of all, and to cure this either first or more than the others. Be well aware that it is difficult and requires keen judgment to compare the fever, the putrefaction, and the effecting cause with each other, whenever

758K

παρῇ· χαλεπώτερον δὲ εἰ καὶ τὴν δύναμιν αὐτοῖς παραβάλλοις, ὅπερ ἀναγκαῖον μὲν ἀεὶ διὰ τὸν τῆς ζωῆς σκοπόν, ὡς ἔμπροσθεν ἐδείκνυμεν, ἐνίοτε δὲ καὶ κατὰ τὸ συμβεβηκὸς ὡς πρὸς τὴν ἴασιν τῶν νοσημάτων, ὅπερ ὑπεσχόμεθα δείξειν. εἰ μὲν γὰρ ἱμάτιον εἴη τὸ σηπόμενον, ἤ τι τῶν ἀψύχων ἄλλων σωμάτων, ἐκ τῶν εἰρημένων ἂν ὑπῆρχεν ἡμῖν μόνον ἡ ἔνδειξις· ἐπεὶ δ' ἐν ζῶντι σώματι γίνεται σῆψις, ἀλλοιοῦν δυναμένῳ καὶ πέττειν, ἐπανάγειν τε πρὸς τὸ χρηστὸν ὅσα

759K τῶν σηπομένων οἷον ἡμισαπῆ τ' ἐστὶ καὶ ἡμιμόχθηρα, τὴν δύναμιν αὐτοῦ ταύτην ᾗ πέττειν πέφυκεν ἐπεγείρων καὶ ῥωννύων ἰάσῃ τὰ σηπόμενα.

διὸ καὶ τότε μάλιστα τολμῶμεν ἤτοι λούειν ἐν τοῖς βαλανείοις αὐτοὺς ἢ τοῖς ἀραιωτικοῖς χρῆσθαι φαρμάκοις, ἢ ὕδωρ ψυχρὸν ἢ οἶνον διδόναι πίνειν, ὅταν ἴδωμεν τὰ γνωρίσματα τῶν πεττομένων χυμῶν. οὐ γὰρ δὴ τῆς γε σφυγμικῆς δυνάμεως ἐρρωμένης μόνης εἰς τὰ τοιαῦτα χρῄζομεν, ὥσπερ οὐδὲ τῆς καθ' ὁρμὴν ἡμᾶς κινούσης, ἀλλ' ὡς εἴρηται τῆς πεπτικῆς μᾶλλον. εἰ μὲν οὖν αἵ τε δυνάμεις ἰσχυραὶ πᾶσαι τυγχάνοιεν οὖσαι καὶ ὁ πυρετὸς διακαέστατος καὶ τὰ τῆς πέψεως ἐναργῆ σημεῖα, τὸ ψυχρὸν αὐτῷ διδόναι θαρροῦντα· δῆλον γὰρ ὅτι μηδὲ πρεσβύτης ἐστὶν ὁ τοιοῦτος ἐξ ὧν προειρήκαμεν ἁπάσας ἰσχυρὰς ὑπάρχειν αὐτῷ τὰς δυνάμεις. εἰ δὲ καὶ εὔσαρκος εἴη καὶ ἡ κατάστασις θερμὴ καὶ ξηρά, κἂν εἰς κολυμβήθραν αὐτὸν ἐμβάλῃς ψυχράν, οὐ βλαβήσεται. κατὰ τὸν τοιοῦτον γοῦν καιρὸν οἱ ῥίψαντες σφᾶς αὐτοὺς εἰς ὕδωρ ψυχρὸν ἵδρω-

this [cause] is still present. It is more difficult if you compare the capacity with these, which is always necessary because it is the indicator of life, as I showed before. Sometimes, also, it pertains contingently to the cure of diseases which I undertook to show. For if what is putrefied is a cloak or one of the other inanimate bodies, the indication would arise for us only from those things spoken of, but since putrefaction arises in a living body that is able to change, concoct and restore to usefulness those things that are putrefied and are, as it were, half putrid and half bad, when you stir up and strengthen its capacity, which is of a nature to concoct, you will cure those things that are putrefied. 759K

On which account also, at that time particularly, we may undertake either to wash the patient in the bathhouse, or use rarefying medications, or give cold water or wine to drink, whenever we see the signs of the humors being concocted. For we do not, surely, need the pulse capacity alone to be strengthened in regard to such things, just as we do not need that of voluntary movement, but as I said, that of concoction particularly. Therefore, if all the capacities happen to be strong, and the fever very hot, and the signs of concoction clear, you should be confident to give cold water to the patient, for it is clear from those things I said before that such a person is not old, and all the capacities are strong in him. If he is well fleshed and the climatic conditions are warm and dry, even if you plunge him into a cold swimming bath, he will not be harmed. At any rate, those who throw themselves into cold water at such a time

760K σάν τε | πάντως αὐτίκα καὶ ἡ γαστὴρ οἷστισιν αὐτῶν
κατέρρηξε χολώδη.

μετρίου δ᾽ ὑπάρχοντος τοῦ πυρετοῦ καὶ τῶν δυνάμεων οὐσῶν οὐκ ἰσχυρῶν ἅμα τοῖς τῆς πέψεως σημείοις ὠφελεῖ τοὺς τοιούτους τά τε βαλανεῖα καὶ ἡ τοῦ
οἴνου πόσις, ὅσα τε τῶν ἀλειμμάτων ἐστὶ μανωτικά,
καὶ πολὺ μᾶλλον ὅταν ᾖ τὸ περιέχον ψυχρόν. οὐδὲ
γὰρ ἐπιθυμοῦσιν ὡς τὰ πολλὰ ψυχρῶν λουτρῶν ἢ
πόσεων ἐν ταῖς τοιαύταις καταστάσεσιν, ἀναψυχό-
μενοι διὰ παντὸς ταῖς εἰσπνοαῖς· ὡς εἴ γε συμβαίη
ποτὲ ψυχροῦ τοῦ περιέχοντος ἀέρος ὄντος ἰσχυρῶς
καυσοῦσθαι τὸν κάμνοντα, βραχείας ἐπ᾽ αὐτοῦ σωτηρίας ἐλπίδας ἔχειν. εἰ δὲ μηδὲ τὰ τῆς πέψεως εἴη
σημεῖα, μηδὲ τὰς δυνάμεις ἰσχυρὰς ἔχειν φαίνοιτο,
σωθῆναι τοῦτον ἀδύνατον· ὥστε οὔτε λούσεις αὐτὸν
οὔτ᾽ ἀλείψεις τοῖς μανωτικοῖς φαρμάκοις, οὔτ᾽ οἶνον
δώσεις οὔτε ψυχρόν· ἐφ᾽ οὗ γὰρ ἀνέλπιστος ἡ σωτηρία μάταιον ἂν εἴη διαβάλλειν τοῖς ἰδιώταις τὰ πολλοὺς σῴζοντα βοηθήματα. ἐγὼ γοῦν οἶδά τινας
ἰατροὺς ἀμεθόδους μιμησαμένους τὰ ὑφ᾽ ἡμῶν πρατ-
τόμενα καὶ τοῖς αὐτοῖς χρησαμένους βοηθήμασιν ἐπὶ
761K τῶν πάντως τεθνηξομένων, ἀλλ᾽ | οὔτ᾽ ἀνύσαντάς τι
καὶ τὴν εὔκαιρον αὐτῶν χρῆσιν ὕποπτόν τε καὶ φοβε-
ρὰν ἐργασαμένους. ἀλλ᾽ ἡμεῖς γε καιροὺς δηλονότι
καὶ βοηθήματα γράφομεν ἐπὶ τῶν σωθῆναι δυναμένων, ὡς ὅσοι γε ἀνίατοι τῶν νοσούντων εἰσίν, οὔτε
καιρὸς ἐπιτήδειος οὔτε βοήθημα τούτοις ἐστί. καὶ χρὴ
μεμνῆσθαι τοῦδε παρ᾽ ὅλον τὸν λόγον, ὑπὲρ τοῦ μὴ

sweat all over immediately and in some of them the stom-   760K
ach breaks down what is bilious.

When the fever is moderate and the capacities are not
strong, along with the signs of concoction being present,
baths benefit such people, as does a drink of wine and
those unguents that are rarefying, and much more so when
the ambient air is cold. They do not, for the most part, desire cold baths or drinks in such climatic conditions, since
they are being cooled continuously by inspirations. So if it
should happen at any time that the patient burns up in-
tensely although the ambient air is cold, there is little hope
of salvation for him. If there are neither the signs of concoction nor capacities that appear to be strong, it is impossible for him to be saved. As a result, you will neither bathe
him nor will you smear him with rarefying medications,
and you will not give him wine or cold water. When re-
covery is not anticipated, it would be rash to discredit, in
the eyes of laymen, those remedies that bring salvation to
many. I, at any rate, know certain amethodical doctors who
have mimicked the things I have done and have used these
same remedies in those who will undoubtedly die. When   761K
they don't accomplish anything, they make the timely use
of these [remedies] a matter of suspicion and fear. But I, at
least, write quite clearly about times and remedies in those
who can be saved; for those who cannot be cured of their
diseases, there is neither a suitable time nor a suitable
remedy. And it is necessary to remember this throughout

δόξαι ποτὲ τὸν κυριώτατον καὶ πρῶτον σκοπὸν ἁπάντων διαβάλλεσθαι.

λέγοντος γοῦν Ἱπποκράτους τὰ ἐναντία τῶν ἐναντίων ἰάματα τί κωλύει πᾶσιν ἑξῆς τοῖς πυρέττουσιν ὕδωρ ψυχρὸν διδόναι; ἀλλ᾽ οὔτε δοτέον οὔτ᾽ ἀνατρέπεται κατ᾽ ἐκεῖνον τὸν χρόνον ὁ καθόλου σκοπὸς ἁπάσης τῆς θεραπείας. ἐν μὲν γάρ τι τῶν ψυχόντων ἐνεργείᾳ τὸ ψυχρὸν ὕδωρ ἐστί, πλείω δ᾽ ἄλλα τὰ μὲν κατὰ δύναμιν, ὡς εἴρηται, ψύχει, τὰ δὲ κατὰ συμβεβηκός· ὧν ἄλλοτ᾽ ἄλλῳ χρώμεθα διὰ τὴν ἐπιπλοκὴν τῶν ἐναντιουμένων ἀλλήλοις σκοπῶν. εἴρηται δ᾽ ἡ μέθοδος τῆς τοῦ συμφέροντος ἑκάστοτε πρὸ τῶν ἄλλων αἱρέσεως. ἐδείχθη δὲ καὶ ὡς ὅσον μὲν ἐπ᾽ αὐτῷ τῷ πυρετῷ τὸ ψυχρὸν ὕδωρ ἀεὶ βοήθημά ἐστιν, ὅσον δ᾽ 762K ἐπὶ τοῖς ἄλλοις | οὐκ ἀεί· καθάπερ οὐδ᾽ ὅταν ἔμφραξις γλίσχρων καὶ παχέων χυμῶν ἐργάσηται τὴν σηπεδόνα τῶν χυμῶν. ὑποκείσθω δὲ σὺν αὐτοῖς εἶναι καὶ πλῆθος ἐν ὅλῳ τῷ σώματι καὶ ἡ δύναμις οὐκ ἀσθενής. ἐν γὰρ ταῖς τοιαύταις ἐπιπλοκαῖς πολλῶν ἐν τῷ σώματι παρὰ φύσιν ὑπαρχόντων ἔνια μὲν αἴτια προηγούμενα γενήσεται, καθάπερ τὸ πλῆθος ἅμα τοῖς γλίσχροις καὶ παχέσι χυμοῖς· ὑπὸ τούτων γὰρ ἡ ἔμφραξις ἀπειργάσθη, νόσημα οὖσα τῶν ἐμφραττομένων σωμάτων ὀργανικόν. ἐπὶ ταύτῃ δὲ σύμπτωμα ἐγένετο τῆς διαπνοῆς ἡ ἐπίσχεσις, ἐφ᾽ ᾗ πάλιν ἡ σηπεδὼν τῶν χυμῶν αἴτιόν τι προηγούμενον τοῦ πυρετοῦ.

---

8 See Hippocrates, *Aphorisms*, II.22.

the whole discussion so the most important and primary indicator of all will not seem to be discredited at any time.

Anyway, since Hippocrates said that opposites are cures of opposites, what prevents you giving cold water to everyone with a fever as a routine?[8] But you must not give it, nor is the general indicator of every treatment overturned at that time. Cold water is just one of the things that are cooling in action. There are, however, many other things that cool, as I said—some in terms of potency and some contingently. Of these, we use different ones at different times due to the combination of indicators which oppose each other. The method of choosing which of these is appropriate over the others on each occasion has been discussed. And it was shown that as much as cold water is always a remedy for the fever itself, it is not always also a 762K remedy for the other things, just as it is not whenever blockage of the viscous and thick humors brings about putrefaction of the humors. Let us assume that together with these, there is also abundance in the whole body and the capacity is not weak. In such combinations, when many things contrary to nature exist in the body, some of them will be *proegoumenic* causes[9] like the abundance along with the viscous and thick humors are. For it is by these things that the blockage is brought about, being an organic disease of obstructed bodies. Due to this, a symptom occurs—stoppage of transpiration—and due to this again the putrefaction of the humors is a *proegoumenic* cause of the fever.

[9] The two causal terms, *proegoumenic* (as used here) and *prokatarktic* (used later in this book), may be understood as "internal antecedent" and "external antecedent" respectively; see Introduction, section 6, on terminology.

10. Τοῖς οὖν ἐπιχειροῦσιν ἰᾶσθαι τὸν πυρετὸν ἀναγκαῖόν ἐστι παύειν τὴν σῆψιν, ὥστε δύο γενέσθαι σκοπούς, τὸν μὲν ἀπὸ τοῦ πυρετοῦ, τὸν δὲ ἀπὸ τῆς σήψεως. ἄλλοι δ' αὖ πάλιν ἔσονται δύο σκοποί, ἵνα τὸ μὲν γεγονὸς ἤδη τοῦ πυρετοῦ θεραπεύηται, τὸ δὲ γιγνόμενον κωλύηται. ἄλλοι δ' αὖ πάλιν ἔσονται ἀπὸ τῆς σήψεως δύο σκοποί, τὸ μὲν γεγονὸς ἤδη τῆς

763K σήψεως ἰώμενοι, τὸ δὲ γινόμενον | κωλύοντες. ἢν δὲ τὸ γιγνόμενον ἐκ τῆς ἀδιαπνευστίας· οὐδὲν γὰρ χεῖρον οὕτως αὐτὴν ὀνομάσαι σαφοῦς ἕνεκα διδασκαλίας. ὥστε καὶ περὶ ταύτης ἕτεραι δύο ἐνδείξεις ἔσονται, τὸ μὲν ἐπεσχημένον ἐκκενούντων, τὸ δ' ἐπέχεσθαι μέλλον κωλυόντων. κωλυθήσεται δὲ διὰ τῶν τὴν ἔμφραξιν ἰωμένων. ἧς πάλιν καὶ αὐτῆς τὸ μὲν ἤδη γεγονὸς ἰατέον ἐστί, τὸ δὲ ἐσόμενον κωλυτέον. ἰαθήσεται μὲν οὖν τὸ γεγονὸς ὑπὸ τῶν ἐκφραττόντων, κωλυθήσεται δὲ τὴν ἐπιρροὴν τῶν ἐμφραττόντων χυμῶν ἀνειργόντων. ὅπερ οὖν ἔσχατον εὑρέθη κατὰ τὴν ἀναλυτικὴν μέθοδον, ἐν τῇ θεραπείᾳ πρῶτον ἁπάντων χρὴ πραχθῆναι.

δέδεικται γὰρ ἐν τοῖς ἔμπροσθεν λόγοις ὡς, ἐὰν ἐθελήσωμεν τοῖς ἐκφρακτικοῖς βοηθήμασι χρῆσθαι, ἄνευ τοῦ προκενῶσαι τὸ πλῆθος οὐ μόνον οὐδὲν ἀνύσομεν, ἀλλὰ μείζονα τὴν διάθεσιν ἐργασόμεθα, μαχομένων ἀλλήλαις τῶν ἐνδείξεων. ἀλλ' ἐὰν κενώσαντες πρότερον ἐπὶ τὸ θεραπεύειν ἴωμεν τὴν ἔμφραξιν, ἐκφραττόντων ἡμῖν δεήσει φαρμάκων. ἀλλ' ἐπεὶ τούτων |

764K ἐστὶ τὰ πλεῖστα θερμά, κίνδυνος αὐξῆσαι καὶ τὴν

154

10. Therefore, for those who are attempting to cure the fever, it is necessary to stop the putrefaction, so that two indicators arise: one from the fever and one from the putrefaction. In turn, there will be two other indicators, so that the fever which has already occurred is treated and the fever which is in the process of occurring is prevented. Once again, there will be two other indicators from the putrefaction: that we cure the putrefaction that has already occurred and that we prevent the putrefaction which is in the process of occurring. That which is in the process of occurring is from the suppression of transpiration—we would do well to term it thus for the sake of clear teaching. As a result, there will also be two other indications pertaining to this: the evacuation of what is retained and the prevention of future retention. The latter will be prevented by those things that cure the blockage. Of this [blockage] in turn, we must also cure what has already occurred and we must prevent what will occur. What has occurred will be cured by those things that remove the blockage, whereas [what will occur] will be prevented if we hold back the flow of the obstructing humors. Therefore, what is discovered last in the analytic method is what must be done first of all in the treatment.

For it has been shown in prior discussions that, if we wish to use the remedies that clear away blockages without first evacuating the abundance, not only will we accomplish nothing, but we will make the condition more severe, since the indications are at odds with each other. But if, after we have first evacuated, we proceed to the treatment of the blockage, we shall require medications for removing obstructions. But since the majority of these are hot, there is a danger of increasing both the putrefaction and the fe-

763K

764K

155

σηπεδόνα καὶ τὸν πυρετόν. ὅσα τοίνυν ἄνευ τοῦ θερμαίνειν ἐκφράττειν δύναται, τούτοις χρησόμεθα· μετὰ
δὲ τὸ διαρρύψαι καὶ τεμεῖν τὰ ἐμφράττοντα, κενοῦν
αὐτοὺς πειρασόμεθα διά τε γαστρὸς καὶ οὔρων καὶ
ἱδρώτων. οὐσῶν δὲ καὶ τῶν ταῦτα ἐργαζομένων ὑλῶν
θερμῶν ἀναγκαῖον μὲν ἐν τούτῳ τήν τε σῆψιν αὐξάνειν καὶ τοὺς πυρετούς. ὥστε καθ᾽ ὅσον ἐνδέχεται
πειρατέον ἐκλέγεσθαι τὰς ἧττον θερμαινούσας ὕλας·
ἢ εἴ τις εὑρίσκοιτο μὴ θερμαίνουσα, καθάπερ ἐν τῷ
καιρῷ τῷδε τὸ βαλανεῖον, ἐπ᾽ αὐτὴν ἔρχεσθαι. πειρᾶσθαι δ᾽ ἐν τούτῳ τῷ χρόνῳ καὶ τὴν δύναμιν ῥωννύειν, ὅπως πέττῃ τοὺς χυμούς, ἐκνικῶσα τὴν σῆψιν.
αὐτοῦ δὲ τοῦ πυρετοῦ τὸ γεγονὸς ἤδη ψυκτέον· ἔσχατος γὰρ ἐν τῇ τῶν βοηθημάτων τάξει σκοπὸς ὁ ἀπὸ
τοῦδε, καίτοι καθ᾽ ἕτερον τρόπον ὑπάρχων πρῶτος·
ἀπ᾽ αὐτοῦ γοῦν ἤρξατο καὶ ἡ τῶν βοηθημάτων εὑρε-
τικὴ μέθοδος.

11. Εἰ μὲν οὖν μηδὲν ἰσχυρὸν ἐν τῷ μεταξὺ σύμπτωμα προσγενόμενον ἐάσειεν οὕτω προελθεῖν ἀπὸ
τῆς | ἀρχῆς τοῦ νοσήματος ἄχρι τῆς τελευτῆς, τὰ
βοηθήματα ταχίστη λύσις ἐπακολουθήσει τοῦ πυρε-
τοῦ· παρεμπεσόντος δὲ τοιούτου τινὸς ὡς ἐφ᾽ ἑαυτοῦ
τὴν ὅλην ἐπιστρέψαι θεραπείαν, ἀναγκαῖόν ἐστι βρα-
δῦναι τὸ τέλος τῆς ἰάσεως. τὸ γὰρ παρεμπῖπτον τοῦτο
τοὐπίπαν ἤτοι γ᾽ ἐναντίον ἐστὶ τοῖς λύουσι βοηθήμασι τὸν προειρημένον στοῖχον, τῶν τ᾽ αἰτίων καὶ
διαθέσεων καὶ συμπτωμάτων, ἢ πάντως γ᾽ οὐδὲν ὀνίνησιν. εἰ μὲν οὖν ἡ τοῦ παρεμπίπτοντος ἴασις ἐναντία

ver. Accordingly, if there are those things which are able to remove blockages without heating, we shall use them. After thoroughly cleaning out and cutting those things that are causing the blockage, we shall attempt to evacuate them through the stomach, urine and sweat. Since the materials which bring about these evacuations are hot, in [doing] this it is unavoidable that the putrefaction and the fever are increased. Consequently, as far as possible, we must attempt to choose the materials that heat least. Or if something that was nonheating were to be found, as bathing is at the right time, proceed to this. Also attempt at this time to strengthen the capacity so that, as it prevails over the putrefaction, it may concoct the humors. We must cool what has already occurred of the fever itself. In the sequence of remedies, the last indicator is from this—and yet in another way, it is the first. At any rate, the method of discovery of the remedies begins from it.

11. If no significant symptom supervenes in the meantime which prevents it, so that the remedies are allowed to go forward from the beginning of the disease to the end, resolution of the fever will very quickly follow. If, however, some such symptom does intervene, so as to turn about the whole treatment by itself, the completion of the cure is, of necessity, delayed. For what does intervene in this way is either opposite to the remedies that resolve the previously mentioned sequence—that of causes, conditions and symptoms—or is, at least, not helpful in any way at all. Therefore, if the cure of what intervenes is in opposition to

765K

ἢ τῇ προειρημένῃ τάξει τῶν βοηθημάτων, ἀναγκαῖον
ἐν ἐκείνῳ τῷ χρόνῳ πάντα χείρω γενέσθαι· εἰ δὲ μήτ᾽
ὠφελοῖ τι πάντῃ μήτε βλάπτοι τὴν ἐξ ἀρχῆς διάθεσιν
ἡ τοῦ μεταξὺ γενομένου παθήματος ἐπανόρθωσις, εἰς
τὸ τάχος τῆς ὅλης θεραπείας ἐμποδισθησόμεθα· βλα-
πτούσης δ᾽ ἀναγκαῖον ἤτοι γ᾽ εἰς ἔτι πλέονα χρόνον
ἐκταθῆναι τὴν σύμπασαν ἴασιν ἢ κίνδυνον ἀκολου-
θῆσαί τινα ἢ συναμφότερον γενέσθαι. φέρε γὰρ ἐν τῷ
μεταξὺ κακωθέντος στομάχου συγκοπὴν ἐμπεσεῖν, ἣν
ἀναγκαζομένων ἰᾶσθαι τροφὰς ἀκαίρους δοθῆναι χρὴ
σὺν οἴνῳ καὶ ψυχρῷ. τῇ γὰρ τούτων χρήσει τὰς
ἐμφράξεις καὶ τὴν στέγνωσιν καὶ τὴν σῆψιν ἀναγ-
766K   καῖον αὐξηθῆναι, καὶ διὰ τὴν τούτων | αὔξησιν καὶ
αὐτὸν τὸν πυρετόν.

ἀλλ᾽ ὅταν ἥ τε δύναμις ἐξαρκῇ καὶ μηδέν, ὡς
εἴρηται, παρεμπέσῃ καθάπερ ἄρτι μὲν ἐρρέθη συγκοπῇ κακωθέντος στομάχου, οὐ γένοιτο δ᾽ ἄν ποτε καὶ
αἱμορραγία καὶ ἀγρυπνία καὶ ἄλγημα καί τι τῶν
ἁμαρτανομένων ὑπὸ τοῦ κάμνοντος ἤ τινος τῶν ἀμφ᾽
αὐτόν, ἡ εἰρημένη μικρὸν ἔμπροσθεν ἅπασα τῶν βοηθημάτων τάξις ἰάσεται τὸν ἄνθρωπον. οὔτ᾽ αὐτὴ καθ᾽
ἕκαστον βοήθημα τὰ παρὰ φύσιν ἐκθεραπεύουσα
πάντα. τὸ γοῦν μελίκρατον, ὅσον μὲν ἐπὶ τὸ τέμνειν
τοὺς παχεῖς χυμοὺς καὶ ῥύπτειν τοὺς γλίσχρους καὶ
τὰς ἐκκρίσεις προτρέπειν, ἄριστον ἂν εἴη βοήθημα
κατὰ τοὺς ἐπὶ στεγνώσει καὶ σήψει συνόχους· ὅσον δ᾽
ἐπὶ τὸ τὴν θέρμην αὐξάνειν τοῦ πυρετοῦ βλαβερόν·
ὅθεν εἴπερ ἄμετρος ἥδε εἴη, φείδεσθαι μὲν χρὴ τοῦ

the previously mentioned sequence of remedies, inevitably, in that time, everything becomes worse. If, on the other hand, the correction of the affection occurring in the meantime neither benefits nor harms the original condition in any way, we shall hold back the speed of the whole treatment. If the correction does cause harm, inevitably either the whole cure is extended over a still greater time, or some danger will follow, or both will occur. Suppose that meanwhile, because the opening of the stomach (cardiac orifice) is adversely affected, syncope occurs, which we are compelled to cure, making it necessary for foods to be given at inappropriate times with wine and cold water. By the use of these things, blockages, stoppage of the pores and putrefaction are inevitably exacerbated, and by their increase, the fever itself is also exacerbated.

766K

But whenever the capacity is sufficient and, as I said, nothing intervenes, like syncope when the opening of the stomach is adversely affected, as was mentioned just now, and if hemorrhage, insomnia, pains and an error by the patient or one of his attendants does not occur, the whole sequence of remedies spoken of a little earlier will cure the person. This sequence, in relation to each remedy, does not act as a complete cure of all the things contrary to nature. At any rate, melikraton, to the extent that it cuts the thick humors, cleans out those that are viscous and impels the excretions onward, would be the best remedy in the continuous fevers due to stoppage of the pores and putrefaction. However, to the extent that it increases the heat of the fever, it is harmful. As a result, if this heat is severe, it

μελικράτου, χρῆσθαι δὲ τῷ χυλῷ τῆς πτισάνης· εἰ δὲ
τὰς ἐμφράξεις δυσλύτους εἶναι λογιζόμεθα, καὶ ὀξυ-
μέλιτι. καὶ τοῦτο δ᾽ αὖ πάλιν αὐτῷ χρησαμένων ἀμέτρως, ἔντερόν τε ξύει καὶ βῆχα κινεῖ καὶ τὰ νευρώδη
βλάπτει. ὥστ᾽ εἶναι τῶν δυσχερεστάτων εὑρεῖν τι
767K βοήθημα | τοιοῦτον ὃ μηδὲν βλάπτον ὀνίνησι μεγά-
λως. καὶ τοῦθ᾽, ὡς ἐδείχθη, γίνεται διὰ τὴν ἐπιπλοκὴν
τῶν παρὰ φύσιν ἐν τῷ σώματι συνισταμένων διαθέσεων ἅμα τοῖς προηγουμένοις αὐτῶν αἰτίοις καὶ
τοῖς ἑπομένοις συμπτώμασιν.

12. Οὐ μὴν διά γε τοῦτο ποτὲ μὲν ἀληθὲς γίνεται τὸ
τὰ ἐναντία τῶν ἐναντίων ὑπάρχειν ἰάματα, ποτὲ δὲ
ψεῦδος· ἀλλὰ διὰ παντὸς μένει ἀληθές· οὐδὲ γὰρ οἷόν
τε τὰ παρὰ φύσιν ἰαθῆναι δι᾽ ἄλλης τινὸς ἰαμάτων
ἰδέας. οὐ μὴν ἀεί γε βέλτιον ἰᾶσθαι πάντα τὰ παρὰ
φύσιν, ἀλλ᾽ ὅταν μόνα τύχῃ καθ᾽ ἑαυτὰ συνιστάντα·
μετ᾽ ἄλλων δ᾽ εἰ γένοιτο, πολλάκις ἕτερα χρὴ πρὸ
ἐκείνων ἰᾶσθαι, καθάπερ ἐπὶ τῶν κοίλων ἑλκῶν, ὅταν
ἅμα φλεγμονῇ συστῇ. μένει γὰρ καὶ νῦν ἀληθὲς τὸ
ὑπὸ τῶν σαρκούντων αὐτὰ θεραπεύεσθαι· χρῆσθαι δὲ
οὐχ οἷόν τε τοιούτοις πρὶν ἰάσασθαι τὴν φλεγμονήν.
ἀλλ᾽ ὁπότ᾽ ἄν γε ταύτην ἰώμεθα, τὸ ἕλκος τὸ κοῖλον
οὐχ ὅπως ἑαυτοῦ κρεῖττον, ἀλλὰ καὶ κοιλότερον γίγνεται. κἂν εἰ βουληθείης δὲ χρήσασθαι τοῖς σαρκω-
768K τικοῖς | φαρμάκοις, οὔτε σαρκώσεις αὐτὸ καὶ τὴν
φλεγμονὴν αὐξήσεις. τὰ γάρ τοι σαρκωτικὰ τῶν
ἀφλεγμάντων σωμάτων, οὐ τῶν ἔτι φλεγμαινόντων,
ἐστὶ σαρκωτικά.

is necessary to steer clear of melikraton and use the juice of ptisan instead. If, however, we reckon the blockages to be difficult to resolve, we also use oxymel. And this again, if we use it immoderately, abrades the intestines, provokes coughing and harms the neural parts. As a result, it is a matter of the greatest difficulty to discover a remedy of a 767K kind that helps greatly but does no harm. And this, as was shown, occurs due to the combination of the conditions contrary to nature existing in the body at the same time with their *proegoumenic* causes and their consequential symptoms.

12. It is not the case that because of this it is sometimes true that opposites are cures of opposites and sometimes false. This remains true at all times. It is impossible for those things contrary to nature to be cured by any other kind of cures. Nor is it always better to cure all things contrary to nature, but only when they happen to exist independently. If they occur with other things, it is often necessary to cure the other things before curing them, as in the case of hollow wounds and ulcers when they coexist with inflammation. For it also remains true that the treatment of these is by the enfleshing agents. It is, however, not possible to use such medications before curing the inflammation. But whenever we cure the inflammation, the hollow wound is not thereby made better than it was; it actually becomes more hollow. And even if you wish to use the 768K enfleshing medications, you will not enflesh the wound, but will increase the inflammation. For truly, those things that are enfleshing for bodies that are not inflamed are not enfleshing for bodies that are still inflamed.

τοῦτο οὖν τὸ οὕτω μικρὸν ἐν τῇ λέξει, τὸ τὰ ἐναντία
τῶν ἐναντίων ὑπάρχειν ἰάματα, μέγιστον εὑρίσκεται
τῇ δυνάμει. συγχεῖται γὰρ ἡ θεραπευτικὴ μέθοδος
ἅπασα τῷ περιδόντι τοῦτον τὸν σκοπόν, ἁπάντων τῶν
παρὰ φύσιν ὑπὸ τῶν ἑαυτοῖς ἐναντίων θεραπευομένων.
διττῆς δὲ οὔσης αὐτῶν τῆς πρώτης διαφορᾶς, τὰ μὲν
γὰρ ὑπάρχει καθ' ἑαυτά, τὰ δ' ἐν τῷ γίνεσθαι τὸ εἶναι
λαμβάνει ταῖς διαθέσεσιν ἑπόμενα, δίκην σκιῶν, ἡ
μὲν ἴασις ἀναίρεσίς ἐστι τῶν διαθέσεων, ἀκολουθεῖ δ'
αὐτῇ ἡ τῶν συμπτωμάτων ἀναίρεσις. οἷς οὐδὲν αὐτοῖς
προσάγεται πρώτως, ἀλλὰ κἀπὶ τούτων γε μένει τὸ
καθόλου, τὸ τὰ ἐναντία τῶν ἐναντίων ὑπάρχειν ἰάμα-
τα· τὰς γὰρ διαθέσεις αἷς ἕπεται τὰ συμπτώματα
θεραπεύοντες εἰς ἐναντίαν κατάστασιν ἄγομεν. ὥσθ'
ὅσα συμπτώματα διὰ πύκνωσιν ἐγένετο πόρων, ὑπὸ
769K   τῆς ἀραιώσεως ἀναιρεθήσεται. | εἰ δὲ δὴ καὶ αὐτῷ
τῷ συμπτώματι τῷ διὰ τὴν πύκνωσιν τῶν πόρων
γιγνομένῳ, τουτέστι ταῖς ἐλάττοσι τῶν συμμέτρων
ἀπορροαῖς, ὅπερ ἐκαλέσαμεν ὀλίγον ἔμπροσθεν ἀδια-
πνευστίαν, ἕτερον ἐθέλοις φάναι συμπτώματος εἶδος
ἀντεισάγειν ἐναντίον, οὐδ' οὕτως ἄπορος ὁ λόγος· εἰσὶ
γὰρ αἱ πλείους τῶν συμμέτρων ἀπόρροιαι ταῖς ἐλάτ-
τοσιν ἐναντίαι.

ὥστε κἂν ἐν τοῖς συμπτώμασιν αὐτοῖς ἐθέλῃ τις
λέγειν τὰ ἐναντία τῶν ἐναντίων ἰάματα, φυλάττεσθαι
κἀπὶ τούτων τὸν πρῶτον ἁπάντων τῶν παρὰ φύσιν
σκοπὸν τῆς ἰάσεως. ἄμεινον δ', ὡς ἐλέγετο, τῶν ὄντως
τε καὶ πρώτως θεραπευομένων ὑπολαμβάνειν τὰ ἐναν-

In this way, then, the brief statement that opposites are the cures of opposites is found to be very powerful. For the therapeutic method as a whole is confounded if deprived of this indicator: that all things contrary to nature are treated by those things opposite to themselves. And since there is a twofold primary differentiation of these (for there are those that exist of themselves and those that derive their existence from being concomitants of the conditions, like shadows), the cure is the removal of the conditions, while the removal of the symptoms follows it. Nothing is applied primarily to the symptoms themselves, but even in their case the general principle at least remains, which is that opposites are the cures of opposites. When we treat the conditions, we bring the symptoms that follow them to the opposite state. As a result, those symptoms that arise due to a condensation of pores will be abolished by rarefaction. Certainly, if you wish to say you are 769K restoring another and opposite kind of symptom to the actual symptom arising from the condensation of the pores, that is to say to the reduced outflows compared to the norm, which we called a little earlier *adiapneusis*, the argument is not in this way invalidated; for outflows greater than the norm are the opposites to those that are less.

So even in respect of the symptoms themselves, someone might wish to say that opposites are the cures of opposites, and to preserve in their case the primary indicator of the cure of all those things contrary to nature. However, it is better, as I said, to suppose that opposites are remedies for those things actually and primarily being treated—that

τία εἶναι βοηθήματα, τουτέστι τῶν ἐν τῷ σώματι
διαθέσεων, αἵπερ ὄντως εἰσὶν ὄντα. τὰς γὰρ ἐνεργείας
αὐτῶν, ὥσπερ οὖν καὶ τὰς βλάβας, γινόμενα μὲν εἶναι
φατέον, ὄντα δὲ ἁπλῶς οὐ ῥητέον. οὔτε γὰρ τὸ ἐκτείνειν οὔτε τὸ κάμπτειν τὴν χεῖρα τῶν ὄντων ἁπλῶς
ἐστιν εἰπεῖν, ἀλλὰ τῶν γιγνομένων, οὔτε πολὺ μᾶλλον
τὴν ἀκινησίαν αὐτῆς, ὅπερ καὶ αὐτὸ σύμπτωμά ἐστι
770K τῆς ἐν τοῖς | μυσὶν ἢ τοῖς νεύροις διαθέσεως, ἀλλὰ τὴν
μὲν διάθεσιν, ἔν τι τῶν ὄντων ὑπάρχουσαν, ἐναντίων
δεῖσθαι βοηθημάτων, ἕπεσθαι δ' αὐτῇ λυομένῃ τὴν
ἐναντίαν κατάστασιν, ἥπερ ἐστὶν ἡ ὑγεία· καὶ ταύτῃ
πάλιν ἐνέργειάν τινα φύσεως, ἐναντίαν τῷ παυσαμένῳ
συμπτώματι, μὴ μέντοι μηδ' ὑπὸ τῶν ὀνομάτων ἐξ-
απατᾶσθαι, πολλάκις μὲν ἐν ἑτέρῳ σχήματι λεγομέ-
νων ἢ ὡς τοῖς ἐναντίοις προσήκει, πολλάκις δ' ὅλως
οὐκ ὄντων ὀνομάτων, ἀλλ' ἐξ αὐτῆς τῶν πραγμάτων
τῆς φύσεως εὑρίσκειν τὸ ἐναντίον, ἔχοντά γε δὴ καὶ
τούτου σκοπὸν ὁμολογούμενον, ὡς ἔστιν ἐναντία τὰ
πλεῖστον ἀλλήλων διεστῶτα³ καθ' ἕν τι γένος. ἐννοήσας οὖν τὸ σύμμετρον ἐν ἐκείνῳ τῷ γένει, τοῦτο δ',
ὡς πολλάκις ἐδείχθη, μέσον τῶν ἄκρων ἐστίν, εὑρήσεις ἐντεῦθεν ἄπειρόν τι πλῆθος ἐναντίων πραγμάτων,
ἐν τῷ μᾶλλόν τε καὶ ἧττον ἀλλήλων διαφερόντων. ἐν
μὲν γὰρ ταῖς διαθέσεσιν, εἰ οὕτως ἔτυχε, τὴν συμμετρίαν τῶν πόρων, ἧς ἐφ' ἑκάτερα διττὰς ἀμετρίας
ἀλλήλαις ἐναντίας· ἐν δὲ ταῖς ἐνεργείαις τὴν συμ-
771K μετρίαν τῶν κενουμένων, ἧς καὶ αὐτῆς ἐστιν | ἑκατέ-
ρωθεν ἀμετρία. ἀλλὰ κατὰ μὲν τοὺς πόρους ὀνόματα

is to say, for the conditions in the body which actually exist. One must speak of their functions, and also of the damages [to function], as being things that arise and not say they are things that simply exist. For example, it is not possible to speak of extending or flexing the hand as things that simply exist; they are things that occur; much more is it not possible with failure of movement of the hand, which is a symptom in its own right pertaining to the condition in the muscles and nerves. Rather, [one speaks of] the condition, which is one of those things that exists, and requires opposite remedies, while the opposite state, which is health, follows its resolution. And following this in turn is a certain function of nature opposite to the symptom when it has ceased. Do not, however, be deceived by the names, since these are often stated in a different form than that which is appropriate for opposites. Often there are no names at all, but it is appropriate for someone to discover the opposite from the actual nature of the matters, when he has, in fact, the agreed indicator of this also—that opposites are those things that stand apart from each other to the greatest extent in a single class. If you give thought to the mean in that class, which, as was frequently shown, is the midpoint of the extremes, you will discover here a countless number of opposite matters differing from each other in terms of more or less. In the conditions, for example, there is the balance of the pores of which each of the two imbalances is opposite to the other, whereas in the functions, in respect to the balance of the evacuations, there are imbalances on either side of this balance. But in relation to the pores, 771K

770K

---

3 K; ἀφεστῶτα B

κεῖται ταῖς ἀμετρίαις, πύκνωσίς τε καὶ μάνωσις, ἐν δὲ τοῖς κενουμένοις οὐ κεῖται.

διόπερ ἀναγκαζόμεθα λέγειν ἐλάττους τε καὶ πλείους ἀπορροίας, ὥσπερ εἰ καὶ τῶν πόρων ἀναγκαζοίμεθα τοὺς μὲν ἐλάττους τῶν κατὰ φύσιν, τοὺς δὲ μείζους λέγειν, οὐκ ἔχοντες οὔτε τὸ τῆς πυκνώσεως, οὔτε τὸ τῆς μανώσεως ὄνομα. τοῦτο οὖν ὡς ἔφην, ἐν τοῖς μάλιστα φυλακτέον ἡμῖν ἐστιν, ἅμα δὲ καὶ γυμναστέον ἀμφ' αὐτὸ πρὸς τὸ ταχέως εὑρίσκειν δύνασθαι παντὸς τοῦ λεχθέντος οὕτω τὸ ἐναντίον. οὐ γὰρ ἐν τοῖς ποιοῖς σώμασιν μόνον, ἀλλὰ κἂν τοῖς ποσοῖς ἐστιν εὑρεῖν τὴν τοιαύτην ἐναντίωσιν, ἣν οἱ περὶ τὸν Ἀριστοτέλη καλοῦσιν ἀντίθεσιν, οὐκ ἐναντίωσιν. οὔτε γὰρ τὸ μέγα τῷ μικρῷ φασιν ὑπάρχειν ἐναντίον, ἀλλ' ἀντικείμενον ἐν τῷ πρός τι, οὔτε τὸ πολὺ τῷ ὀλίγῳ· κατὰ δὲ τὸν αὐτὸν τρόπον οὐδ' ἀραιὸν τῷ πυκνῷ, οὐδὲ τῷ συντεταμένῳ τὸ κεχαλασμένον, οὐδὲ τῷ κατὰ φύσιν αὖ τὸ παρὰ φύσιν. ἀλλὰ κατά γε τὸν Ἱπποκράτη πάντα τὰ τοιαῦτα τὴν τῶν ἐναντίων
772K ἔχει προσηγορίαν, ὥσπερ γε καὶ κατὰ Πλάτωνα | τὰς γενέσεις ἐκ τῶν ἐναντίων εἶναι φάσκοντα. δέδεικται δέ μοι καὶ δι' ἑτέρων ὅτι καὶ Ἀριστοτέλης αὐτὸς οὐκ ἐφύλαξε τὴν ἑαυτοῦ νομοθεσίαν ἐν τοῖς ὀνόμασιν, ἡνίκα τὰς ἀρχὰς τῶν ὑπὸ φύσεως διοικουμένων ὕλην ἔθετο καὶ εἶδος καὶ στέρησιν. ἀλλ' ὅπερ ἀεὶ λέγομεν ἑπόμενοι τῷ θείῳ Πλάτωνι, καταφρονεῖν μὲν χρὴ τῶν ὀνομάτων, μὴ καταφρονεῖν δὲ τῆς τῶν πραγμάτων

names are established for the imbalances (condensation and rarefaction), whereas in the evacuations they are not.

For this reason, we are forced to speak of lesser and greater outflows, just as we would be forced to speak of the pores as being less or more than accords with nature, if we did not have the terms condensation and rarefaction. This, as I said, is among the things we must be particularly on our guard against. At the same time, too, we must become practiced in that by which we are quickly able to discover the opposite of anything said in this way. For not only in the qualities of bodies, but also in the quantities as well, there is the kind of opposition to discover, which the followers of Aristotle call an "opposition" and not a "contrariety."[10] They do not say the large is opposite to the small but that it is set in opposition to it in respect of opposing something. Nor do they say the many is opposite to the few, nor in the same way is the rarefied [opposite] to the dense, nor the tense to the relaxed, nor again is what is contrary to nature [opposite] to what is in accord with nature. But as far as Hippocrates is concerned, all such things have the designation of opposites, just as they also do for Plato, who says   772K
that the generations are from their opposites. I have also shown elsewhere that Aristotle himself did not preserve his own ruling on names when he set out the principles of those things governed by Nature in respect to material, kind and negation.[11] But what I always say, following the divine Plato, is that we should think little of names but not think little of knowledge of the matters. It is this that

[10] The Greek terms are *antithesis* and *enantiōsis*; see Aristotle's *Metaphysics*, 986b.   [11] See Aristotle, *Categories*, 12a96, and *Metaphysics*, 1004b27.

ἐπιστήμης. αὕτη μὲν γὰρ εἰς σωτηρίαν ἀνθρώπων
διαφέρει καὶ τὸ σφάλμα αὐτῆς εἰς ὄλεθρον τελευτᾷ.

τοῖς δ' ὀνόμασιν ἄν τε κυρίως ἄν τε ἀκύρως χρη-
σώμεθα, τοῖς κάμνουσιν οὐδὲν οὔτε πλέον οὔθ' ἧττον
ἐκ τοῦ τοιούτου. μάθοις δ' ἂν ἐναργέστερον, εἰ μὴ τὰ
ἐναντία τῶν ἐναντίων, ἀλλὰ τὰ ἀντικείμενα λέγοις,
ᾗπερ ἐκείνοις φίλον, ἀλλήλων ὑπάρχειν ἰάματα. τὸ
μὲν γὰρ πλῆθος ἡ κένωσις ἰᾶται, τὴν δ' ἔνδειαν αἱ
τροφαί, καὶ ὅλως τὸ μὲν ὑπερβάλλον ἡ ἀφαίρεσις, τὸ
δ' ἐλλεῖπον ἡ πρόσθεσις. ὅθεν, οἶμαι, καὶ δι' ἑτέρου
λέλεκται γράμματος ὀρθῶς τῷ παλαιῷ· ἰατρικὴ γάρ
ἐστι πρόσθεσις καὶ ἀφαίρεσις· πρόσθεσις μὲν τῶν
ἐλλειπόντων, ἀφαίρεσις δὲ τῶν πλεοναζόντων. ἐλλείπει δὲ καὶ πλεονάζει τὰ μὲν κατὰ τὸ ποσὸν δηλονότι,
773K τὰ δὲ κατὰ τὸ ποιόν· κατὰ μὲν | τὸ ποσόν, ὅταν αἷμα
πλέον ᾖ τοῦ δέοντος ἢ πάλιν ἔλαττον γένηται· κατὰ δὲ
τὸ ποιόν, ὅταν ἤτοι θερμὸν ἢ ψυχρὸν ἢ παχύτερον ἢ
λεπτότερον. εἶτ' οὖν ἐναντίωσιν εἴτ' ἀντίθεσιν ὀνομάζειν ἐθέλοις, τοῦ ἐξ ἁπάντων τῶν παρὰ φύσιν
σκοποῦ τῆς ἰάσεως ἀεὶ μέμνησο καὶ πρόσεχε τὸν νοῦν
αὐτῷ, μάλιστα μέν, ὡς εἴρηται πρόσθεν, ἐν ταῖς διαθέσεσιν, ἤδη δὲ κἂν τοῖς συμπτώμασιν· ὧν οὔτε
πρώτως ἐστὶν ἴασις οὔτ' ἐναντίωσις κυρίως, ἀλλ' ἡ
μὲν ἴασις ἅμα ταῖς διαθέσεσι κατὰ συμβεβηκός,
ἐναντίον δὲ καταχρηστικῶς ὀνομαζόντων ἕνεκα σα-
φοῦς τε ἅμα καὶ συντόμου διδασκαλίας.

13. Ἀνέλθωμεν οὖν αὖθις ἐπὶ τοὺς θεραπευτικοὺς
σκοποὺς τοῦ προκειμένου γένους τῶν πυρετῶν, ἀνα-

makes a difference to the safety of people, and errors in it end in their destruction.[12]

Whether we use the names properly or improperly is neither here nor there to patients. You will understand more clearly if you don't say that opposites [are the cures] of their opposites, but that antitheticals (by which you please those men) are cures of each other. For evacuation cures abundance, nutriments cure deficiency and, in general, removal cures excess and addition cures deficiency. Whence, I think, it has been rightly said in another work by our ancient [teacher][13] that the art of medicine lies in adding to and taking away from—adding those things that are lacking and taking away those things that are in excess. Deficiency and excess quite clearly relate to quantity in some instances and to quality in others: to quantity whenever blood has become more than it ought to be or, on the contrary, less; and to quality whenever it has become hot or cold, thicker or thinner. Therefore, whether you wish to term this "opposition" or "contrariety," always bear in mind the indicator of the cure from all those things contrary to nature, and direct your attention to this particularly in the conditions and now even in the symptoms, as I said before. The cure of these is not primarily or properly opposition, but takes place contingently with the cure of the conditions when it is termed opposition catachrestically for the sake of clarity as well as brevity of teaching.

773K

13. Therefore, let me return once more to the therapeutic indicators of the class of the fevers before us, and let

12 See Plato, *Euthydemus*, 277e–78d.
13 See Hippocrates, *Breaths*, I, LCL, *Hippocrates*, vol. 2, p. 229.

μνήσωμέν τε τὰ γένη τῶν παρὰ φύσιν ἐν ἡμῖν τῶν τε
αἰτίων καὶ τῶν νοσημάτων καὶ τῶν συμπτωμάτων·
αἰτίων μὲν τοῦ πλήθους καὶ τοῦ πάχους καὶ τῆς
γλισχρότητος τῶν χυμῶν, αἰτίου δ' ἅμα καὶ συμπτώματος τῆς κατὰ τὴν διαπνοὴν ἐπισχέσεως, ὥσπερ
774K αὖ πάλιν αἰτίου τε ἅμα καὶ διαθέσεως | τῆς σήψεως,
νοσήματος δὲ τῆς ἐμφράξεως καὶ τοῦ πυρετοῦ, συμπτώματος δὲ τῆς ἐπισχέσεως τῆς διαπνοῆς. ἐνδείξεται
τοιγαροῦν, ὡς ἐλέχθη, τῶν μὲν προηγουμένων αἰτίων
ἕκαστον ἴδιόν τι· τὸ μὲν πλῆθος ἐν τῷ πρός τι κατὰ
τὴν τοῦ ποσοῦ προσηγορίαν⁴ τὴν κένωσιν, ἡ δὲ γλισχρότης καὶ τὸ πάχος κατὰ τὴν τοῦ ποιοῦ διάθεσιν
ὑπάρχοντα τὴν διὰ τῶν ἐναντίων ἴασιν.

ἅτε δὲ οὐκ ὄντων ἐν ταῖς πλείσταις ἀντιθέσεσι συν-
ηθῶν ὀνομάτων, ἀπατᾶσθαι συμβαίνει τοὺς ἀγυμνά-
στους τὴν περὶ τῶν πραγμάτων ἐπιστήμην, ὅταν ὀνο-
μάσαι μιᾷ προσηγορίᾳ μὴ δυνηθῶσι τὸ νοούμενον, ὡς
οὐδ' ὅλως ὄντος ἀφισταμένους· ὅπερ ἀμέλει καὶ κατὰ
τοὺς γλίσχρους καὶ παχεῖς χυμοὺς πεπόνθασιν. ὁ μὲν
γὰρ παχὺς χυμὸς ἀντικείμενον ἔχει τὸν λεπτόν, ὁ δὲ
γλίσχρος, ὡς μὲν ἐγὼ πρῶτον ὠνόμασα, τὸν ῥυπτι-
κόν, ὡς δ' ἄν τις ἴσως ἀπὸ στερεῶν σωμάτων ὁρμώμενος φαίη, τὸν κραῦρον. οὕτω γὰρ φαίνεται καὶ ὁ
Ἀριστοτέλης ἀεὶ ποιούμενος τὴν ἀντίθεσιν· ἀλλ' οὐκ
ἐπὶ τῶν χυμῶν, ὡς ἔφην, ἀλλ' ἐπ' αὐτῶν τῶν στερεῶν
775K σωμάτων. | εἰ δὲ νοήσαις τοῦ γλίσχρου χυμοῦ τὴν
φύσιν, ὡς παντὸς τοῦ ψαύσαντος ἀντέχεται δυσλύτως,
εἰκότως, οἶμαι, τὸν μὴ συμπλεκόμενον ἀπορρύπτοντά

170

me recall the classes of those things contrary to nature in us—causes, diseases and symptoms. Among the causes [of this kind of fever] are abundance, thickness and viscidity of the humors; what is both a cause and a symptom is the stoppage of transpiration just as, in turn, what is both a cause and a condition is putrefaction. The disease is the 774K blockage and the fever; the symptom is the stoppage of transpiration. So, as I said, each of the *proegoumenic* causes will indicate something specific: the abundance in that which pertains to something is designated as the evacuation of quantity, while the viscidity and thickness relate to the condition of quality, the cure of which is by opposites.

Inasmuch as customary terms do not exist in most oppositions, it happens that those unpracticed in the knowledge of the matters are deceived whenever they are unable to name the concept with a single term, because they dismiss it as not existing at all, which, of course, they also do in relation to those affected by viscid and thick humors. For the thick humor has what is thin as its opposite, while the viscid, as I first named it, has what is cleansing or—as someone might perhaps wish to say when impelled to do so by the solid bodies—the friable. Aristotle, too, always clearly makes the antithesis in this way, although not in the case of the humors, as I stated, but in the case of the solid bodies themselves. If, however, you understand the nature of the 775K viscid humor as adhering to everything it touches in a way that is difficult to loosen, you will say (reasonably I think) that what does not become adherent and also washes those

---

4 K; κατηγορίαν B

τε τοὺς συμπλεκομένους ἐναντίον εἶναι φήσεις αὐτῷ. κατὰ γοῦν τὰ προηγούμενα τῶν αἰτιῶν αἱ ἀντιθέσεις αὐτάρκως εἴρηνται· κατὰ δὲ τὴν ὀργανικὴν νόσον, ἥπερ ἐστὶν ἔμφραξις, τὸ ἐναντίον ταύτῃ ἡ ἔκφραξις ὑπάρχει· κατὰ δὲ τὸ σύμπτωμα τὸ τῆς ἐποχῆς τῶν διαπνεομένων ἡ ἔκκρισις αὐτῶν ἐστι τὸ ἐναντίον· κατὰ δὲ τὴν σηπεδόνα τῶν χυμῶν ἥ τε ἔκκρισις καὶ ἡ ἀνάψυξις καὶ ἡ ῥίπισις καὶ ἡ τῶν ἡμισαπῶν πέψις· αὐτῶν δὲ τῶν πυρετῶν ἡ ἀντίθεσις ψύξις ἐστίν.

ἐξ ὧν δ᾽ ἄν τις μεθόδῳ εὕροι τὴν ἐργαζομένην ἕκαστον τῶν εἰρημένων ὕλην ἤδη εἴρηται. μαχομένων οὖν αὐτῶν, ὥσπερ ἐλέγομεν, ἄλλοτ᾽ ἄλλη κρατεῖ τῶν ἐνδείξεων τε καὶ ὑλῶν. καὶ κατὰ τοῦτό τινες ἀγνοήσαντες ὡς ἀφ᾽ ἑκάστου τῶν ὄντων ἔνδειξίς ἐστιν ἀεὶ μία, καὶ ὡς εἰ μὲν εἴη τῶν παρὰ φύσιν τι τὸ ἐνδεικνύμενον, ἡ ἔνδειξις αὐτοῦ τὸ ἐναντίον ἐστίν, εἰ δὲ τῶν κατὰ φύσιν, οὐ τὸ ἐναντίον, ἀλλὰ τὸ ὅμοιον, οὐκ 776K ἠδυνήθησαν οὔτε θεραπευτικὴν | οὔτε ὑγιεινὴν συστήσασθαι μέθοδον. ὅπου γὰρ ἐν τοῖς πρώτοις ἐσφάλησαν σκοποῖς, πολὺ δήπου μᾶλλον ἔμελλον ἐν τοῖς μετ᾽ αὐτοὺς σφαλήσεσθαι. χρὴ γάρ, οἶμαι, τὰ θεμέλια τοῖς οἰκοδομήμασιν ἰσχυρὰ προκαταβεβλῆσθαι. καὶ τὴν τρόπιν τοῖς σκάφεσιν, εἰ μέλλει τι τῶν ἐπ᾽ αὐτοῖς οἰκοδομουμένων τε καὶ πηγνυμένων ἀσφαλὲς γενήσεσθαι· ὅπου δ᾽ ἂν ἐξ ἀρχῆς εὐθὺς ἡ πρώτη κρηπὶς σαθρὰ συμπαγῇ, τίς μηχανὴ τῶν ἐπ᾽ αὐτῇ τι γενήσεσθαι μὴ σαθρόν;

things that have become adherent is the opposite to it. At any rate, enough has been said about the oppositions pertaining to the *proegoumenic* causes. In relation to the organic disease, which is blockage, the opposite to this is removal of the blockage. In relation to the symptom, which is the retention of those things that undergo transpirations, the opposite is separation of these. In relation to the putrefaction of the humors, separation, cooling, ventilation and concoction of those things that are semiputrefied are the opposites. The opposite of the fevers themselves is cooling.

How someone might discover from these by method the effective material in respect of each of the things mentioned has already been stated. Therefore, when these are at odds, as I am wont to say, one or other of both the indications and the materials prevails at different times. In relation to this, some people, because they don't know that there is always one indication from each of those things existing, and they don't know that, if what has been demonstrated is among the things contrary to nature, the indication is the opposite of this, whereas if it is one of the things in accord with nature, it is not the opposite but the like, are rendered unable to establish either a therapeutic or 776K health-preserving method. For where they erred in the first indicators, by very much more, certainly, are they going to err in those that come after them. I believe it is necessary to have laid strong foundations for buildings beforehand, and the same with the keels of boats, if something that is erected on or attached to these is going to be safe. Where the primary foundation is made unsound right from the start, won't whatever structure that is created on it be unsound?

ἐπεὶ τοίνυν τῶν κατὰ τὴν προκειμένην στέγνωσιν
συνόχων, οὐ γὰρ ὀκνητέον αὖθις καὶ αὖθις ἀναλαμ-
βάνειν τὸν λόγον, ὑπὲρ τοῦ μαθεῖν τοὺς πολλοὺς τῶν
ἰατρῶν κἂν νῦν γοῦν τὴν ἀληθῆ μέθοδον, ἑπτὰ μὲν
ὑπέκειτο τὰ παρὰ φύσιν, ἀφ' ὧν αἱ ἐνδείξεις, πλῆθος
χυμῶν καὶ πάχος καὶ γλισχρότης ἔμφραξίς τε καὶ
τῶν ἀναπνεομένων ἐπίσχεσις καὶ σῆψις καὶ πυρετὸς
οὐκ ἔχων διάλειμμα· καὶ ταῦτ' ἐνεδείκνυτο πάντως τὸ
ἐναντίον. ἔξωθεν δ' αὐτῶν ἔνδειξις ἀπὸ τῶν κατὰ
777K φύσιν ἐκ τῆς κράσεως τοῦ τε νοσοῦντος | μορίου καὶ
τῶν ἄλλων ἁπάντων τῶν κυρίων ἐκ τοῦ ποσοῦ τῆς
οὐσίας αὐτῶν καὶ ἐκ τοῦ ποσοῦ τε καὶ ποιοῦ τῆς τοῦ
πνεύματος οὐσίας· αἱ δὲ ἀπὸ τούτων ἐνδείξεις οὐ τῶν
ἐναντίων ἦσαν, ἀλλὰ τῶν οἰκείων·⁵ εἰς δὲ τὴν διάγνωσιν αὐτῶν, ὡς ἐλέγομεν, ἥ θ' ἡλικία καὶ τὸ ἔθος
ἅπαντά τε τὰ προκατάρξαντα τῆς νόσου συντελεῖ·
προερχομένων δὲ κἀντεῦθεν ἐνδείξεων οὐκ ὀλίγων, εἰς
ἁπάσας χρὴ βλέπειν ἅμα τὸν ἰατρόν, ἵν' ἐξεύρῃ τίνι
μὲν πρῶτον καὶ μᾶλλον, τίνι δ' ἧττόν τε καὶ δεύτερον,
τίνι αὖθίς τε καὶ τρίτον καὶ οὕτως ἐφεξῆς ἄλλῳ μετ'
ἄλλο χρήσεται βοηθήματι.

14. Τῶν μὲν γὰρ δυνάμεων ἰσχυρῶν ὑπαρχουσῶν
τοῦ τὸν ἐπὶ σηπεδόνι πυρετὸν πυρέττοντος, ὡς ὑπόκει-
ται, φλεβοτομητέον αὐτίκα χωρὶς ἀπεψίας τῆς κατὰ
γαστέρα τοῦ νοσήματος ὑπαρξαμένου. τῆς δυνάμεως
δὲ ἀσθενεστέρας οὔσης, ἢ τῆς ἡλικίας κωλυούσης, οὐ

⁵ B, K; ὁμοίων conj. nos (cf. similicem KLat)

174

Therefore, since in the proposed stoppage of pores of the continuous fevers (I must not shrink from taking up the argument again and again so the majority of doctors may learn the true method even now), there are seven things postulated that are contrary to nature, from which the indications [are taken]—abundance, thickness and viscidity of humors, obstruction, suppression of transpiration, putrefaction and a fever which is without intervals. In every case these indicate their opposite. Apart from these, there is an indication from those things in accord with nature: from the *krasis* of the diseased part and all the other parts of importance, from the quantity of their substance, and from the quantity and quality of the substance of the *pneuma*. The indications from these are not of their opposites but of their similars. As I am wont to say, the age, custom, and all those things that stand in a *prokatarktic*[14] relation to the disease, jointly contribute to the diagnosis of these [fevers]. Since quite a number of indications also arise from that source, it is necessary for the doctor to look into all these at the same time, so that he may discover which one is primary and more [important], which one is less [important] and secondary, and which one is next and third. In this way, he will use one remedy after another in succession.

777K

14. When the capacities are strong in someone who is febrile due to putrefaction, as is postulated, you must carry out phlebotomy immediately the disease begins, except when there is failure of concoction in the stomach. However, when the capacity is weaker or when age prevents,

14 On this causal term, see note 9 above.

χρὴ τέμνειν φλέβα. λέλεκται δ᾽ ἔμπροσθεν ὡς οὐδὲ συνεισβάλλει τῷ τοιούτῳ πυρετῷ δύναμις ἄρρωστος· εὐεκτικῶν τε γάρ ἐστι σωμάτων καὶ ἡλικίας θερμῆς ὁ 778K σύνοχος πυρετὸς οἰκεῖος. ἀλλ᾽ | ἐὰν ἐν παιδίῳ γένηται μήπω τεσσαρεσκαιδέκατον ἔτος ἄγοντι, φλεβοτομεῖν οὐ προσήκει, διότι τοῖς τηλικούτοις γε θερμοῖς καὶ ὑγροῖς οὖσι καθ᾽ ἑκάστην ἡμέραν ἀπορρεῖ καὶ διαφορεῖται πάμπολυ τῆς τοῦ σώματος οὐσίας. ὥσθ᾽ ὅπερ ἂν ἐκ τῆς φλεβοτομίας ἐμηχανησάμεθα, τοῦτ᾽ αὐτόματον ἐκ τῆς τοῦ θεραπευομένου σώματος ὑπάρχει φύσεως.

εἰ δ᾽ ὑπὲρ τεσσαρεσκαιδέκατον ἔτος εἴη τὸ σῶμα, σκεπτέον αὐτοῦ τὴν σχέσιν ὁποία τίς ἐστιν, ἆρ᾽ ἰσχνὴ καὶ πυκνὴ καὶ σκληρὰ καὶ πολύαιμος ἢ τἀναντία· καὶ οὕτως ἐπὶ μὲν τῇ προτέρᾳ παραλήψῃ τὴν φλεβοτομίαν, ἐπὶ δὲ τῆς δευτέρας οὐκέτι. καὶ μὲν δὴ καὶ τὸ μέτρον σοι τῆς κενώσεως ἐκ τῶν αὐτῶν σκοπῶν λαμβανέσθω. καὶ γὰρ εἰ τριακονταετὴς μὲν ὁ φλεβοτομούμενος εἴη, πλαδαρός τε καὶ μαλακὸς καὶ πιμελώδης καὶ λευκὸς καὶ μικρὰς ἔχων τὰς φλέβας, ἤτοι γ᾽ οὐδ᾽ ὅλως φλεβοτομήσεις αὐτὸν ἢ ὀλίγον ἀφαιρήσεις· οὐδ᾽ ὅλως μὲν ἐν ὥρᾳ θερινῇ καὶ πνιγώδει χωρίῳ, θερμῆς καὶ ξηρᾶς οὔσης τῆς καταστάσεως, ὀλίγον δ᾽ ἐν ταῖς ἄλλαις ὥραις καὶ χώραις καὶ καταστάσεσιν.

εἰ μὲν γὰρ μήτ᾽ ἔμφραξις ὑπέκειτο μήτε σηπεδών, 779K ἀλλὰ μόνον ἦν | τὸ πλῆθος τῶν χυμῶν, εὐθὺς ἂν ἐπὶ τῇ φλεβοτομίᾳ κατὰ φύσιν ἔσχεν. ὥστε ἀφαιρεῖν ἦν προσῆκον ἐν ἁπάσῃ κράσει σώματος καὶ πάσῃ χώρᾳ

you must not open a vein. It has been said before that a weak capacity does not appear together with such a fever, since a continuous fever is characteristic of healthy bodies and an age that is hot. But if it occurs in a child who has not 778K yet reached the fourteenth year, it is not appropriate to carry out phlebotomy because, in ages that are hot and moist, a great part of the substance of the body flows away and is dispersed every day. As a result, what we might have achieved by phlebotomy occurs spontaneously by virtue of the nature of the body being treated.

If, however, the body is over fourteen years, you must consider what kind of state it is in—whether it is lean, dense, hard and blood-filled, or the opposites. Thus, if it is in the former state, you will undertake phlebotomy; if in the latter, you will not. You must also take the measure of the evacuation from the same indicators. So if the person being phlebotomized is thirty years of age, and is flabby, soft, fat and pale, and has small veins, you will either not phlebotomize him at all or you will take only a little [blood]. "Not at all" [applies] in summer and in a stifling place, and when climatic conditions are hot and dry; "a little" [applies] in the other seasons, places and climatic conditions.

For if there is no underlying blockage or putrefaction but only the abundance of humors, [the patient] would im- 779K mediately return to normal with the phlebotomy so that it is appropriate, in every *krasis* of the body, place, season

καὶ ὥρᾳ καὶ καταστάσει τοσοῦτον τῶν χυμῶν ὅσον
ὑπὲρ τὸ κατὰ φύσιν ηὔξητο. ἐπεὶ δ' οὔτε τὴν ἔμφραξιν
οὔτε τὴν σηπεδόνα δυνατόν ἐστι θεραπεῦσαι διὰ τῆς
φλεβοτομίας, ἑτέρων δεομένας βοηθημάτων, ὡς ἔμπροσθεν ἐδείκνυμεν, ἀποθέσθαι χρή τι τοῦ αἵματος
εἰς τὸν τῆς θεραπείας χρόνον, ὅπως μήποτ' ἀνάγκη
καταλάβῃ τις ἡμᾶς ἀκαίρως τρέφειν. ὅσον μὲν γὰρ
ἐπὶ τοιούτῳ πυρετῷ θρεπτέον οὐδ' ὅλως ἐστίν. ἐδείχθη
γάρ σοι καὶ πρόσθεν ὁ ἀπὸ τῆς δυνάμεως σκοπὸς
μόνος ἐνδεικνύμενος τὴν τροφήν· ὅσον δ' ἐπὶ τῷ κεκενῶσθαι τὸ σῶμα δεήσει τρέφειν. ὥστ' ἀναγκαῖον
ἔσται δυοῖν θάτερον, ἢ τρέφοντα τὸν πυρετὸν αὐξάνειν, ἢ μὴ τρέφοντα καταλύειν τὴν δύναμιν. ἄμεινον
οὖν, ὡς εἴρηται, καταλιπεῖν τι τοῦ αἵματος οἰκείαν
τροφὴν τοῖς τοῦ ζῴου μορίοις· αὐτὸν δ' ἀρκεσθῆναι
780K προσφοραῖς ὀλιγίσταις ῥοφημάτων καὶ ποτῶν, | ὧν
ὡς φαρμάκων μᾶλλον ἢ τροφῶν χρήζομεν. ἐδείχθη
γάρ σοι καὶ περὶ τοῦδε πρόσθεν, ὅθεν οὐ χρὴ μηκύνειν
ἔτι περὶ αὐτῶν.

15. Ἀλλὰ καὶ περὶ καταπλασμάτων τε καὶ περὶ τῆς
τῶν ὑποχονδρίων αἰονήσεως λεκτέον. ὁ μὲν γὰρ πολὺς
ὅμιλος τῶν ἰατρῶν ἕν τι καὶ τοῦτο τῶν ἐκ τοῦ νόμου
τῆς ἀμεθόδου θεραπείας ἔθετο. καὶ πάντας ὥσπερ
τρέφουσι μετὰ τὴν διάτριτον, οὕτω καὶ προκαταντλοῦ-
σιν ἐλαίῳ, κἄπειτα μίαν ὑπερβαλόντες αὖθις τρέφουσι προκαταπλάττοντες, εἰ δέ που γαστὴρ ἐπισχεθείη,
καὶ κλύζοντες· ὥστε τό γε κατὰ τούτους ὥρᾳ μιᾷ
δύνασθαι μαθεῖν τινα τέχνην διαιτητικήν. οὐ μὴν ὧδ'

and climatic condition, to take away as much of the humors as there is excess over what accords with nature. Since neither blockage nor putrefaction can be treated by phlebotomy, requiring other remedies as I previously demonstrated, the matter of the blood must be put aside to the time of treatment, so that at no point does necessity compel us to nourish [the patient] in an untimely manner. As far as it pertains to such a fever, you must not nourish at all. I also demonstrated to you earlier that the indicator from the capacity is the only one that indicates nourishment. However, to the extent to which the body has been evacuated, it will need to be nourished. And so, inevitably, there will be the two conflicting aspects: if you nourish, you increase the fever; if you don't nourish, you break down the capacity. It is better, then, as I said, to leave behind some of the blood as proper nourishment for the parts of the organism, and this will be adequately provided for by very small administrations of gruel and drink which we need as medications more than as foods. This I also showed you previously so it is not necessary to delay further on these matters.

780K

15. I must also speak about poultices and fomentation of the hypochondrium. For the great throng of doctors established this as one of the habitual practices of the amethodical treatment. And just as they nourish all [patients] after the three day period, so too do they give them a prior bathing with oil; and when they have gone beyond one day, they again nourish them after having applied poultices beforehand, while if the stomach is retaining to some degree, they also administer a clyster. And so, according to them at least, you are able to learn a kind of dietetic craft in one hour. The truth, however, is not like this.

ἔχει τἀληθές. ἀλλ᾽ ὅλης οὔσης μεγάλης τῆς τέχνης ἕν
τι τῶν μερῶν αὐτῆς οὐ τὸ φαυλότατόν ἐστιν ἐπιστήμη
διαίτης, ἥτις ἐκ τούτων μάλιστα συμπληροῦται τῶν
βοηθημάτων, τροφῶν καὶ πομάτων ἐπιβροχῶν τε καὶ
καταπλασμάτων καὶ κλυστήρων. ἡ γάρ τοι φλεβοτομία κατὰ τὴν ἀρχὴν καὶ αὐτὴ μόνη πρὸς τῶν αὐτῶν
ἰατρῶν παραληφθεῖσα τὴν ἐφεξῆς ἅπασαν ἴασιν τοῦ
781K νοσοῦντος ἐπιτρέπει τῇ διαίτῃ. | μιᾶς οὖν ὥρας, ὡς
ἔφην, ἐστὶν ἡ μάθησις τῆς διαίτης αὐτῶν. εἰ δ᾽ ὁ
μανθάνων εἴη συνετὸς οὐδὲ ταύτης ὅλης, ἀλλ᾽ ὀλιγο-
στοῦ μέρους αὐτῆς, ἐν ᾧ τούτων τῶν νῦν λεχθησομέ-
νων ἀκοῦσαι δυνατόν ἐστιν.

ἅπαντας τοὺς πυρέττοντας ἐν ἀρχῇ μὲν φλεβοτομήσεις, ἐὰν ἰσχυροὶ τὴν δύναμιν ὦσι· καταντλήσας δ᾽
ἐλαίῳ μετὰ τὴν διάτριτον, ἤτοι μελικράτῳ μετὰ χόνδρου θρέψεις ἢ ῥοφήματι. κἄπειθ᾽ ἑξῆς τρέφε παρὰ
μίαν, ἀρτομέλιτι προκαταπλάττων. εἰ δ᾽ ἐπισχεθείη
ποτὲ ἡ γαστήρ, ὑπάγειν αὐτὴν κλυστῆρι. τὴν μὲν οὖν
ἐκείνων διαιτητικὴν τέχνην ἤδη σύμπασαν ἀκήκοας
ἐν τοῖσδε τοῖς ῥήμασιν, οὐκ ἐν ἓξ μησίν, ἀλλ᾽ ἐν ἓξ
στίχοις. τὴν δ᾽ ὄντως διαιτητικὴν ἑξῆς ἄκουσον. θερμαίνειν ὑποχόνδρια καταπλάσμασιν ἢ αἰονήσεσιν οὐ
διὰ παντὸς ἀσφαλές, ἀλλ᾽ ἐπ᾽ ἐκείνων μόνων τῶν
ἀρρώστων, ἐφ᾽ ὧν οὐκέτ᾽ ἀλᾶται περιττὸν ἐν ὅλῳ τῷ
σώματι. τοῖς δ᾽ ἄλλοις ἅπασι κακὸν ἔσχατον. ἐάν τε
γὰρ ᾖ πλῆθος ὁποτερονοῦν, εἴτε τὸ πρὸς τὴν δύναμιν
εἴτε πρὸς τὴν εὐρυχωρίαν τῶν ἀγγείων, ἐάν τε περιτ-

Rather, the whole range of the craft is great; the knowledge of regimen is just one part of it, and by no means the most trivial. It comprises these remedies particularly: foods and drinks, bathings, poultices and clysters. Now when phlebotomy is employed at the beginning and alone, these same doctors entrust the whole remaining cure of the sick person to regimen. Learning about regimen for them takes one hour, as I said. If, however, the person learning is intelligent, it is not the whole hour but a very small part of it in which they can listen to those things I shall now speak about. 781K

If they are strong in terms of capacity you will phlebotomize all those who are febrile at the start. Having bathed [them] with oil, you will nourish them after the three-day period, either with melikraton along with gruel or with broth. And then, next, nourish them on alternate days, applying a poultice of bread and honey. If, however, the stomach is constrained at some time, purge it with a clyster. You have now heard the entire dietetic art of those people in these words—and not in six months but in six lines! Now listen to the real dietetics. To heat the hypochondrium with poultices or fomentations is not safe in every case, but only in those patients in whom superfluity has not yet spread to the entire body. In all others the bad effect is extreme. If the abundance is one of two things, either involving the capacity or involving the lumina of the

τώματα μοχθηρὰ κατά τι μόριον ἓν ἢ καὶ πλείω, ταῦθ᾽
782K ἕλκεται πάντα πρὸς | τὸ θερμαινόμενον.

ἐθεάσω γοῦν οὐκ ὀλιγάκις ἀφλέγμαντα τελέως
ὑποχόνδρια κατὰ τέσσαρας ἡμέρας ἀρχομένων αἰο-
νᾶν αὐτὰ τῶν ἀγελαίων ἰατρῶν φλεγμήναντα, αὐτὸ
τοῦτό σοι δείξαντος ἐμοῦ καὶ κελεύσαντος ἀκριβῶς
προσέχειν τὸν νοῦν τῇ μελλούσῃ γενήσεσθαι φλεγμονῇ. ἐφ᾽ ὧν καὶ μάλιστα ἔγνως πηλίκον ἐστὶ κακὸν
ἄλογος τριβή. διὰ παντὸς γὰρ αὐτοὶ γεννῶντες τὰς
φλεγμονὰς ἐξ ὧν πράττουσιν οὐ γινώσκουσιν, ἀλλ᾽
ἀεὶ τοῖς αὐτοῖς ἁμαρτήμασι περιπίπτουσιν, ὥσπερ
Ἱπποκράτης ἔλεγε, τοὺς ἔνθεν τε καὶ ἔνθεν ἐπιδοῦντας
τὰ μεθ᾽ ἕλκους κατάγματα. καὶ γὰρ κἀκεῖνοι φλεγμονὰς ἐργαζόμενοι κατὰ τὸ ἕλκος ὅμως οὐκ ἐπαύοντο,
νομίζοντες οὐ τὴν ἐπίδεσιν αἰτίαν ὑπάρχειν, ἀλλ᾽
ἄλλην τινὰ ἀποτυχίαν. πῶς δ᾽ ἂν καὶ μετέβησαν ἐφ᾽
ἕτερον ἰάσεως τρόπον οἱ τὴν ἄλογον πρεσβεύοντες
τριβήν, ἀφ᾽ ὧν αὐτοί τε διὰ παντὸς εἰθίσθησαν οὕτω
πράττειν, τούς τε διδασκάλους ἐθεάσαντο πρὸ αὐτῶν;
οὐδὲν γὰρ οὐδ᾽ ἐπιχειρεῖν ἀξιοῦσιν οἱ τοιοῦτοι δι᾽
ἐνδείξεως λαμβάνειν, ἀρκούμενοι μόνῃ τῇ πείρᾳ. ἀλλ᾽
783K Ἱπποκράτης ἀπὸ | τῆς τοῦ πράγματος ὁρμηθεὶς φύ-
σεως ἐξεῦρε τρόπον ἰάσεως ἐπιτήδειον. οὗ λοιπὸν εἰς
πεῖραν ἐλθόντος ἐγνώθη πόσον ὁ πρότερος ἐσφάλ-
λετο.

καὶ τοίνυν εἴ τις ἐθελήσειεν ἐπὶ τῶν τοιούτων διαθέσεων ἀποστὰς καταπλασμάτων τε καὶ τῆς δι᾽ ἐλαίου καταντλήσεως, ἐπὶ τὴν ἐκ τοῦ λόγου διδασκομένην

vessels, and if the abnormal superfluities involve one or more parts, all these superfluities are drawn toward the 782K heated part.

At any rate, you have frequently seen hypochondria that are completely free of inflammation become inflamed within four days when the common herd of doctors begin to apply fomentations to them, since I have shown you this very thing and directed you to focus your attention precisely on the possible future occurrence of inflammation. In these things, too, you know particularly well how great an evil irrational practice is. Time and again, when these very doctors create inflammation by those things they do in their ignorance, they always fall into the same errors, just as those men do who bandage compound fractures on each side, as Hippocrates said. For those men too, when they bring about inflammation in the wound, nevertheless do not stop since they think the bandaging is not the cause but some other failing. How could they change to another kind of cure, those men who cultivate an irrational practice by those things they are continually accustomed to do in this way and saw their teachers do before them? Such men do not think it worthwhile to attempt to take anything from an indication because they regard experience alone as sufficient. But Hippocrates, making a start from the nature of 783K the matter, discovered a suitable manner of cure. Thereafter, when he came to experience, he learned how much the previous way was in error.

Thus, if someone wished to steer clear of poultices and fomentation with oil in such conditions and to come to a

ἀφικέσθαι θεραπείαν, οὐ χαλεπῶς ἂν οὐδ' αὐτὸς τῇ
πείρᾳ κρίνειε πηλίκον ἁμαρτάνουσιν οἱ οὕτω θεραπεύοντες τά τε πληθωρικὰ καὶ τὰ μὴ καθαρὰ σώματα.
συμβαίνει δ' ἐν τοῖς τοιούτοις νοσήμασι τοὺς πλουσίους μᾶλλον τῶν πενήτων κακῶς θεραπεύεσθαι. ἀμφότερα γὰρ ἐπ' αὐτῶν ἁμαρτάνεται διὰ τὴν τρυφὴν οὐκ
ὀλιγάκις, ἥ τε τῆς φλεβοτομίας ἔνδεια καὶ ἡ περιτ-
τοτέρα δῆθεν ἐπιμέλεια τῶν ἰατρῶν, ὡς καθ' ἑκάστην
τι πράττειν ἡμέραν ἐπὶ τῷ τοῦ κάμνοντος σώματι. κατ'
ἀρχὰς μὲν οὖν ὑπὸ τρυφῆς οὐκ ἀνέχονται τῆς φλεβοτομίας οἱ πλείους αὐτῶν, καίτοι μᾶλλον τῶν πενήτων ἁλισκόμενοι ταῖς πληθωρικαῖς διαθέσεσιν, ὡς ἂν
καὶ μᾶλλον ἐμπιπλάμενοι καὶ βιοῦντες ἀργότερον. οἱ
θεραπεύοντες δ' αὐτοὺς ἰατροί, διότι παραλέλειπται
784K μέγιστον βοήθημα τῇ διὰ | τῶν ἄλλων χρήσει νομί-
ζουσιν ἀναπληρώσειν ὁπόσον ἐνδεῖ. καὶ μέντοι καὶ οἱ
κάμνοντες, οἰόμενοι τὴν μὲν ἐφ' ἡσυχίας δίαιταν ἀμέ-
λειαν εἶναι, τὸ δ' ὁτιοῦν πράττειν ἐπιμέλειαν, ἐπαναγ-
κάζουσιν αὐτοὺς ἑκάστης ἡμέρας προσφέρειν τι τοῖς
ὑποχονδρίοις· βουλομένοις δὲ δήπου καὶ τοῖς ἰατροῖς
ἔτι τοῦτο πρὸς τὸ δοκεῖν ἐνεργεῖν τι διαπαντός, ἐντεῦθεν γὰρ ἐλπίζουσι καὶ τὸν μισθὸν πλείω λήψεσθαι.

συμβαίνει τοιγαροῦν ἐν ἐκείνῳ τῷ χρόνῳ τὰς φλεγμονὰς ἄρχεσθαι τοῖς πλουσίοις, ἐν ᾧ πλησίον ἤδη
τῆς λύσεώς ἐστιν ἐπὶ τῶν πενήτων τὸ νόσημα. φλεβο-
τομηθέντες γὰρ ἐν ἀρχῇ κατὰ τοὺς προκειμένους ἐν
τῷ λόγῳ πυρετοὺς οὐ μόνον οἱ πένητες, ἀλλὰ καὶ οἱ
δοῦλοι τῶν πλουσίων, οἱ πλεῖστοι μὲν ἐν τῇ πέμπτῃ

treatment informed by reason, he would have no difficulty in judging by experience how greatly those men erred who treated the plethoric and nonpurified bodies in this way. However, what happens in such diseases is that the rich rather than the poor are treated badly because the former not infrequently go wrong in two ways due to their delicacy: by omitting phlebotomy and by what is really the over-attentiveness of the doctors insofar as they do something to the body of the patient every day. In the first place, then, the majority of these [rich people] do not tolerate phlebotomy because of their delicacy, and further, they are more susceptible to the plethoric conditions than poor people inasmuch as they stuff themselves with food and live a more idle life. Because they leave aside the most important remedy, those doctors who treat them think that by   784K the use of other remedies they will make good what is lacking. And, of course, those who are ill, because they think the regimen of quiet rest constitutes lack of care, whereas to do anything whatsoever constitutes care, force the doctors to apply something to the hypochondrium every day. And the doctors themselves, who no doubt wish to appear to be still doing something continually, hope they will receive a greater fee thereby.

So it happens that in the rich, inflammation begins at the time when the disease is already close to resolution in the poor. For when they are phlebotomized at the start in the fevers, which are the subject under discussion, not only the poor, but also the slaves of the rich in the majority of cases, come to a crisis on the fifth day, although some go

τῶν ἡμερῶν ἐκρίθησαν, ἔνιοι δὲ ἐς τὴν ἑβδόμην ἀφί-
κοντο, πορρωτέρω δ' οὐδείς. ἀλλ' οἵ γε πλούσιοι τὸ
μὲν πλῆθος καὶ τὰ περιττώματα πλείω τῶν πενήτων
καὶ δούλων ἔχοντες, παραλιπόντες δὲ τὸ τῆς φλεβοτομίας βοήθημα, κἄπειτα καθ' ἑκάστην ἡμέραν
αἰονώμενοί τε δι' ἐλαίου θερμοῦ καὶ καταπλαττόμενοι
τοῖς χαλαστικοῖς καταπλάσμασι φλεγμονήν τινα
συνισταμένην ἴσχουσιν, ἤτοι περὶ τὸ ἧπαρ ἢ τὴν
785K   κοιλίαν ἤ τι τῶν ἄλλων τῶν τῇδε | σπλάγχνων, ἔνιοι
μὲν ἐν τῇ τετάρτῃ τῶν ἡμερῶν, ἔνιοι δ' ἐν τῇ πέμπτῃ,
πάντες δ' οὖν ἐν τῇ ἕκτῃ. οἶσθα δὲ δήπου καὶ τὸν ἐν
αὐτῷ τῷ καταντλεῖσθαι δυσπνοήσαντα, τοῦ περιττοῦ
παντὸς ἐπὶ τὰς φρένας ἑλχθέντος ὑπὸ τῆς θερμασίας.
οὗτός τε οὖν ἀπέθανε καὶ ἄλλοι μυρίοι καθ' ἑκάστην
ἡμέραν ἀποθνήσκουσιν, ὥσπερ σικύας τοῦ καταπλάσματος ἕλκοντος ἐπὶ τὰ σπλάγχνα τὰ καθ' ὅλον τὸ
σῶμα περιττά. τινὲς δ' ἀμέλει καὶ αὐτῇ τῇ σικύᾳ κατ'
αὐτῶν ἐχρήσαντο, τοῦθ' ἓν μόνον ἐννοοῦντες κἀκεῖνοι,
τὸ φαίνεσθαί τι πράττοντες ἀεὶ περὶ τὸν κάμνοντα.

κάλλιστον μὲν οὖν, ὡς εἴρηται, φλέβα τέμνειν οὐ
μόνον ἐν τοῖς συνόχοις πυρετοῖς, ἀλλὰ καὶ τοῖς ἄλλοις ἅπασι τοῖς ἐπὶ σήψει χυμῶν, ὅταν γε ἤτοι τὰ τῆς
ἡλικίας ἢ τὰ τῆς δυνάμεως μὴ κωλύῃ. κουφισθεῖσα
γὰρ ἡ διοικοῦσα τὰ σώματα ἡμῶν φύσις, ἀποθεμένη
τε τὸ βαρῦνον αὐτὴν οἷόν πέρ τι φορτίον, ἐπικρατήσει
τοῦ λοιποῦ ῥᾳδίως. ὥστε καὶ πέψει τὸ πεφθῆναι δυνάμενον, ἐκκρινεῖ τε τὸ δυνάμενον ἐκκριθῆναι ἀναμνη-
σθεῖσα τῶν οἰκείων ἐνεργειῶν. ἡ γὰρ τῶν μέσων τοῦ

as far as the seventh day, but none longer. But the rich, who have an abundance and more superfluities than the poor and slaves, and who set aside the remedy of phlebotomy, then have both fomentations with warm oil every day and relaxing poultices applied, and so retain a coexisting inflammation involving either the liver or stomach, or one of the other internal organs, [come to a crisis] on the fourth day in some cases, on the fifth in some, and all by the sixth. And you know, I presume, that the patient has difficulty breathing at the very time of the fomentation, when all the superfluity is drawn to the diaphragm by the heating. Therefore, the person concerned dies, as do countless others every day when the poultice, just like a cupping glass, draws the superfluities in the whole body to the internal organs. Of course, there are also some who use the cupping glass itself on them; that is, those men who think about this one thing alone—to appear to be always doing something to the patient.

785K

It is best, then, as I said, to open a vein, not only in the continuous fevers but also in all the other fevers due to putrefaction of humors, at least whenever the factors of age and capacity do not prevent this. For Nature, which governs our bodies, is relieved when it has cast off a burden that is, as it were, weighing it down, and will prevail over what remains with ease. Consequently, it will concoct what can be concocted and will excrete what can be excreted, calling to mind the specific functions. Giving prior atten-

σώματος, ὡς οὗτοι καλοῦσι, πρόνοια μέγιστον μὲν
786K κακὸν ἐπὶ τῶν | μὴ φλεβοτομηθέντων ἐστίν, οὐ μέγι-
στον δὲ ἐπὶ τῶν φλεβοτομηθέντων. ἀλλ' ὅμως κἀπὶ
τούτων τι βλάπτειν πέφυκεν ἐν τοῖς συνόχοις πυρε-
τοῖς· διακαιόμενα γὰρ ὑπὸ τοῦ πλήθους τῆς θέρμης τὰ
σπλάγχνα προσέτι διακαίεται καὶ ἐξοπτᾶται.

μόνοι τοιγαροῦν ἐκεῖνοι δεήσονται προνοίας τοιαύτης, οἷς ἡ σῆψις ἐν ἑνὶ συνέστη μορίῳ, προφλε-
βοτομηθέντες δηλονότι καὶ οὗτοι. ποιησόμεθα δ' αὐτῶν τὴν πρόνοιαν οὐχ ὁμοίως τοῖς ἐκ τῆς τριόδου
τούτοις ἰατροῖς, εὐθέως ἐξ ἀρχῆς χαλῶντες, ἀλλὰ πᾶν
τοὐναντίον ἐπὶ τῶν πλείστων ἐργαζόμενοι. τὰ μὲν γὰρ
χαλαστικὰ τῶν βοηθημάτων ὥσπερ γε[6] διαφορεῖ τὸ
περιεχόμενον ἐν τοῖς μέλεσιν, οὕτως ἐπισπᾶται πλέον
ἀντ' αὐτοῦ, κατ' ἀρχὰς προσαγόμενα. φερομένων γὰρ
ἔτι τῶν ῥευμάτων ἐπὶ τὸ πεπονθός, ἀποκρούεσθαι
βέλτιόν ἐστι καὶ ἀναστέλλειν, οὐχ ἕλκειν ἐπ' αὐτά. τὰ
δ' ἀποκρουστικὰ καλούμενα, μετέχοντα δηλονότι τῆς
στυπτικῆς δυνάμεως, ἐπιτήδεια πρὸς τὰς ἀρχάς ἐστιν,
ἅμα μὲν ἐντιθέντα τόνον τοῖς πάσχουσι μορίοις, ὡς
μὴ ῥᾳδίως ὑποδέχοιντο τὰ ἐπιρρέοντα τῶν περιττω-
787K μάτων, ἅμα δὲ καὶ τῶν ἐν αὐτοῖς | ἤδη περιεχομένων
ἀντεκθλίβοντα τὸ λεπτότατον. εἰ δ' ἥ τε φορὰ παύ-
σαιτο διὰ τῶν τοιούτων βοηθημάτων, ἥ τ' ἐκ τῆς
στύψεως ἐγγενομένη πύκνωσις τοῖς πεπονθόσι κατ-
έχοι τὰ παχύτερα, καιρὸς ἤδη χαλᾶν ὑπὲρ τοῦ κενῶσαι τὰ περιεχόμενα. μάλιστα δ' ὅταν εἰς ἧπαρ ἢ
γαστέρα κατασκήπτῃ τὰ περιττὰ τοῖς στύφουσι χρή-

tion to the central parts of the body, as they call them, is very bad in the case of those who have not been phleboto- 786K mized, although it is not so bad in the case of those who have. Nevertheless, even in the latter, there is a natural tendency to harm in the continuous fevers; for the internal organs, having been overheated by the abundance of heat, are still further overheated and thoroughly baked.

For that very reason, only those in whom the putrefaction is established in one part will require such forethought, obviously after they have had prior phlebotomy. We shall provide forethought for them unlike that of these common doctors who cause relaxation right from the start, but in most cases bring about the complete opposite, because the relaxing remedies, just as they disperse what is contained in the parts, so they draw more compared to this when they are introduced at the beginning. When the fluxes are still being carried to the affected part, it is better to drive them away and repel them and not draw them to these parts. Those medications that are called repulsives, since they quite clearly partake of the astringent potency, are suitable at the beginning, simultaneously providing strength to the affected parts so they do not readily receive the fluxes of the superfluities, and squeezing out the thinnest of the superfluities already contained in them. If, 787K however, the movement ceases due to such remedies, or the condensation generated by the astringency retains those that are thicker in the affected [parts], it is already time to loosen those things that are contained, for the purpose of evacuation. Use the astringents especially whenever the superfluities rush down to the liver or stomach

---

6 K; γε *om.* B

σθαι· κύριά τε γὰρ ἱκανῶς τὰ μόρια καὶ πάντως
ἐργάζεσθαι τὸ σφέτερον ἔργον ἀναγκαῖα κἂν ταῖς
νόσοις. ἐστὶ δὲ οὐ σμικρὸν αὐτῶν τὸ ἔργον, οἷόν περ
ἑκάστου τῶν ἄλλων, ἃ μόνον ἐκεῖνο κατεργάζεσθαι
πέφυκεν ὑφ᾽ οὗ θρέψεται· ἀλλ᾽ οὕτως εἰς μέγα τῷ ζῴῳ
διαφέρειν, ὡς εἰ μήτε κατὰ γαστέρα πεφθείη καλῶς ἡ
τροφὴ μήθ᾽ αἱματωθείη κατὰ τὸ ἧπαρ, οἷόν περ ἐν
λιμῷ πάσχομεν ἀπορίᾳ βρωμάτων ἀσιτεῖν ἀναγκαζό-
μενοι, τοιοῦτόν τι καὶ νῦν συμβήσεται πᾶσι τοῖς τοῦ
ζῴου μορίοις· οὐ γὰρ ἐκ τῶν καταποθέντων, ἀλλ᾽ ἐκ
τῶν κατεργασθέντων ἐν τοῖς εἰρημένοις σπλάγχνοις ἡ
χορηγία τῆς τροφῆς ἐστιν ὅλῳ τῷ σώματι.

δὶα ταῦτα μὲν δὴ περιττότερον ἢ τὰ ἄλλα μόρια
γαστήρ τε καὶ ἧπαρ τῶν στυφόντων χρῄζουσιν. ἀμέ-
λει κἀπειδὰν καιρὸς | ᾖ διαφορεῖν τὰ στηριχθέντα, καὶ
τότε δεῖται στύφεσθαι τὰ μέτρια· τὸ γὰρ ἐξ ἐπιμέτρου
περὶ τὰ ἄλλα μόρια κατ᾽ ἀρχὰς αὐτοῖς προστιθέμενον
τῆς στύψεως ἀεὶ φυλάττεσθαι χρή, μενούσης γε τῆς
ἐνδειξαμένης αὐτὸ χρείας· ἡ χρεία δ᾽ ἐστίν, ὡς ὀλίγον
ἔμπροσθεν εἴπομεν, ἡ κατεργασία τῆς τροφῆς. ἐὰν
οὖν ἀτονήσαντα μὴ πέψῃ καλῶς, οὐ μόνον οὐδὲν
ὄφελος ἔσται τῶν καταποθέντων σιτίων, ἀλλ᾽ ἐπὶ
ταῖς διαφθοραῖς αὐτῶν πολλάκις ἐρεθισθείσης τῆς
γαστρός, συναπέρχεταί τι καὶ τῶν προϋπαρχόντων.
ὥστε διχόθεν ἤδη τὴν βλάβην γίνεσθαι τῷ παντὶ
σώματι, μήθ᾽ ὑπὸ τῶν σιτίων τραφέντι καὶ τῆς οἰκείας
παρασκευῆς τι προσαπολλύντι. καὶ χωρὶς δὲ τῶν
εἰρημένων αἱ πέψεις τῶν σιτίων, ἄν τε καθ᾽ ἧπαρ

788K

190

because these parts are sufficiently important to make it essential that they carry out their action at all times, even in diseases. Their action is not minor compared to each of the other parts whose only natural actions are to carry out that by which they will be nourished. But [the liver and stomach] do, in this way, make a great difference to the organism because, if concoction is not carried out properly in the stomach and the nutriment is not turned into blood in the liver, we shall suffer from a lack of food, since we are compelled to fast as in a famine, and such a thing will now occur in all the parts of the organism. The supply of nourishment to the whole body is not from those things swallowed down but from those things produced in the aforementioned viscera.

Now for this reason, the stomach and liver need the astringents to a greater extent than the other parts. Of course, whenever it is time to disperse those things that are firmly fixed, there is also, at that time, the need for moderate astringency. To begin with it is necessary to always keep the provision of astringency for these parts in excess of that for the other parts, their ongoing use being an indication of this. The use is, as I said a little earlier, the preparation of nutriment [by digestion]. If, because they are weak, they do not digest properly, not only will they derive no benefit from the foods swallowed, but often when the stomach is irritated by their corruption, some of the foods previously present are passed. Consequently, harm now occurs to the whole body in two ways: it is not nourished by the foods, and additionally it destroys some of what has been properly prepared. And apart from those things mentioned, the concoctions of the foods, should

788K

ἀτυχήσωσιν, ἄν τε κατὰ γαστέρα, τοὺς πυρετοὺς
αὐξάνουσι καὶ δριμυτέρους ἐργάζονται τῇ κακοχυμίᾳ.
διὰ τοῦτο τοίνυν πολλὴν χρὴ πεποιῆσθαι πρόνοιαν ἐν
ἅπασι πυρετοῖς πέψεως σιτίων. καὶ διὰ ταύτην τοῦ
τόνου τῶν πεπτικῶν ὀργάνων οὐ σμικρὰ φροντιστέον.
ἔνιοι δ᾽, ὥς φησιν Ἱπποκράτης, οὐ σμικρὰ κερδαίνου-
789K σιν ὅτι ἀγνοοῦσιν. οἴονται | γοῦν ἥπατος φλεγμονὴν
ἰᾶσθαι χαλαστικοῖς βοηθήμασιν, ἐξ ὧν ἐν τῇ διαγνώσει σφάλλονται, δόξαν ἑαυτοῖς ποριζόμενοι· φαίνεται
γὰρ ἐνίοτε τοῦ καθήκοντος εἰς ὑποχόνδρια μυὸς ἐκ
τῶν πλευρῶν τοῦ θώρακος ἡ περιγραφὴ παραπλησία
τῷ ἥπατι. θαυμαστὸν δ᾽ οὐδέν, οἶμαι, τοὺς μήτ᾽ ἄλλο
μηδὲν ἐν τῇ τέχνῃ καλῶς ἐκμαθόντας μήτ᾽ ἀσκηθέν-
τας ἐν διαγνώσεσι πεπονθότων μορίων τὰ τοιαῦτα
κερδαίνειν.

16. Ἐκείνους μὲν οὖν ἐατέον, ἡμεῖς δὲ κἂν ἐλαχί-
στην ποθ᾽ ὑπόνοιαν σχῶμεν ἥπατος ἢ γαστρὸς κακο-
πραγούντων, εὐθέως ἀψίνθιον ἐλαίῳ προσαφεψήσαν-
τες αἰονήσομεν τὰ μόρια. διττὴν δὲ ἅπαντος ἀψινθίου
ποιότητα καὶ δύναμιν ἔχοντος, ὡς κἂν τοῖς περὶ φαρμάκων εἴρηται, κατὰ μὲν τὸ Ποντικὸν ἡ στύφουσα
ποιότης ἐστὶν οὐκ ὀλίγη, τοῖς δ᾽ ἄλλοις ἅπασιν ἡ μὲν
πικρὰ ποιότης ὑπάρχει σφοδροτάτη· στύψεως δ᾽ ἤτοι
παντάπασιν ἀμυδρᾶς ἢ καὶ οὐδ᾽ ὅλως ἂν αἴσθοιο
γευόμενος αὐτῶν. ταῦτά τοι τὸ Ποντικὸν ἀψίνθιον

they fail in both the liver and the stomach, increase the fevers and make them more acute due to the *kakochymia*. Therefore, because of this, it is necessary to give much forethought to the concoction of foods in all the fevers. And through this forethought you must give no little attention to the strength of the digestive organs. However, as Hippocrates says, some [doctors] derive no small gain because they are ignorant. At any rate, they think to cure inflammation of the liver with relaxing remedies, establishing a reputation for themselves from instances where they fail in the diagnosis, for sometimes it seems that the contour of the muscle which passes down to the hypochondrium from the chest wall is like the liver. It is no wonder, I think, that those who learned nothing else properly in the craft, nor are practiced in the diagnoses of the affected parts, derive advantage from such things.

789K

16. Well then, we must let those men be. However, if I have even the slightest suspicion at any time that the liver or stomach is adversely affected, I will immediately foment the parts, after boiling down wormwood in oil. As I said in the writings on medications,[15] all wormwood has a twofold quality and capacity. In the Pontic (wormwood gentle) the astringent quality is not insignificant, whereas in all the others the bitter quality is very strong. However, the astringent quality is either quite indistinct or you do not perceive it at all when you taste them. As a result of these factors it is better to choose the Pontic wormwood

---

[15] On wormwood, see in particular *De simplicium medicamentorum temperamentis et facultatibus*, XI.798K ff. The different kinds of wormwood, including the Pontic, are considered in Dioscorides, III.26–28.

790K αἱρεῖσθαι βέλτιόν ἐστιν εἰς τὰς ἥπατος | καὶ γαστρὸς
φλεγμονάς· ἔστι δ' αὐτοῦ καὶ τὸ φύλλον καὶ τὸ ἄνθος
πολὺ σμικρότερον ἢ τῶν ἄλλων ἀψινθίων, καὶ ἡ ὀσμὴ
τούτῳ μὲν οὐχ ὅπως ἀηδής, ἀλλά τι καὶ τῶν ἀρωμάτων ἐμφαίνουσα· τοῖς δ' ἄλλοις δυσώδης ἅπασι·
φεύγειν μὲν οὖν ἐκεῖνα προσήκει, χρῆσθαι δ' ἀεὶ τῷ
Ποντικῷ. καὶ τῶν ἄλλων δέ τι φαρμάκων ἐμβαλὼν εἰς
τοὔλαιον, οἷς ἐπιμέμικται τῇ πικρᾷ ποιότητι δύναμις
στυπτική, κατὰ τὸν αὐτὸν τρόπον χρήσῃ. δέδεικται
γὰρ ἐν τοῖς περὶ φαρμάκων ἡ πικρὰ ποιότης ποδη-
γοῦσα τὴν στύφουσαν. διὸ καὶ κρεῖττόν ἐστι τὸ τοιοῦτον φάρμακον ἅπαν τοῦ στύφοντος μόνον. εἰ μὲν
οὖν ἰσχυροτέραν ἐθέλοις ἐργάσασθαι τὴν στύψιν,
ἔστω σοι καὶ τὸ ἔλαιον στῦφον, ὁποῖόν ἐστι τό τε
Ἱσπάνον ὀνομαζόμενον, ὅσα τ' ἄλλα σκευάζουσι μετὰ
θαλλῶν ἐλαίας· ἢ οἷόν πέρ ἐστι τὸ καλούμενον ὀμφάκινον.

εἰ δ' ἀσθενεστέρᾳ βούλοιο χρήσασθαι τῇ στύψει,
τῶν ἄλλων ἐλαίων τι παρασκεύαζε καὶ μάλιστα τῶν
λεπτομερῶν, οἷόν πέρ ἐστι τὸ Σαβῖνον. ἄριστον δὲ ἐν
791K οἷς στύψεως ἰσχυροτέρας | ἐστὶ χρεία, τὸ Ἰστρικὸν
ἔλαιον, ὡς ἂν ἑκατέρας ἔχον ἐν ἑαυτῷ τὰς ποιότητας,
στύφουσαν καὶ πικράν. ἀποροῦντι δ' ἀψινθίου μήλινον ἢ μαστίχινον ἢ σχίνινον ἀρκέσει τὴν πρώτην· εἰ
δὲ μικρὸς ὁ πυρετὸς εἴη, καὶ τὸ διὰ τῆς νάρδου μύρον.
ἔστω δὲ καὶ τοῦτ' ἄριστον· οὐ σμικρὰ γὰρ ἡ διαφορὰ
τοῦ τοιούτου πρὸς τὸ φαῦλον· ὁ γοῦν ἐν Νεαπόλει τῆς

for inflammation of the liver and stomach. Both its leaf   790K
and flower are very much smaller than those of the other
wormwoods and its smell is not as unpleasant but displays something of the aromatics, whereas all the others
are malodorous. It is therefore appropriate to avoid those
and always use the Pontic. And you will use it in the same
way after putting one of the other medications into the oil
in which the astringent potency has been mixed with the
bitter quality. For it has been shown in the works on medications that the bitter quality paves the way for the astringent.[16] Because of this, such a medication is altogether more powerful than the astringent [medication]
alone. Therefore, if you do wish to make the astringency
stronger, avail yourself of astringent oil, like the kind called
Hispanic, and other such things people prepare with the
young shoots of the olive—an example is the so-called
"omphacinum."[17]

If you wish to use a weaker astringency, prepare one
of the other oils, and particularly one that is fine-particled
like the Sabine. When there is need of stronger astrin-   791K
gency, the best of them is the Istrian oil as it has both qualities in itself—the astringent and the bitter. If wormwood
is not available, oil made from apple blossoms, mastich
or lentiscinum will suffice to begin with, as will the unguent made from spikenard if the fever is slight. And regard this as the best, for the difference in quality between
this and ordinary oil is significant. Anyway, that which they

16 On the astringent properties of wormwood, see *De simplicium medicamentorum temperamentis et facultatibus*, XI.844K.

17 See Dioscorides, V.6, although here the preparation is described as being from unripe grapes (see also I.75).

Ἰταλίας σκευάζουσιν ὄνομα μόνον ἐστὶ μύρου ναρδί
νου, παραβαλλόμενον τῷ κατὰ τὴν Ἀσίαν ἔμπροσθεν
μὲν ἐν Λαοδικείᾳ μόνῃ σκευαζόμενον, νυνὶ δὲ ἐν πολ
λαῖς ἤδη πόλεσιν. εἰ δ᾽ ἡ παρακμὴ τοῦ πυρετοῦ, καθ᾽
ἣν δηλονότι πράττεις τὰ τοιαῦτα, μὴ πάνυ τι πραεῖα
γίγνοιτο, φυλάττεσθαι μὲν τὴν νάρδον· ἄμεινον δὲ
χρῆσθαι τηνικαῦτα τῷ μηλίνῳ· μὴ παρόντος δὲ τού
του, τῷ σχινίνῳ· καὶ μετ᾽ αὐτὸ τῷ μαστιχίνῳ. πάντων
γὰρ τούτων τὸ νάρδινον μύρον μᾶλλον θερμαίνει· διὸ
καὶ μικτέον αὐτῷ τῶν ἄλλων ἀποροῦντι ῥοδίνου. ἡ δὲ
τοῦ μυρσίνου στύψις οὐκ ἐπιτήδειος· ἔστι γὰρ τοῦτο
παχυμερέστερον ἢ ὥστε διὰ βάθους ἱέναι. τὰ μὲν οὖν
792K τοιαῦτα παραδείγματος ἕνεκα | λέλεκται· πολὺ γὰρ
αὐτῶν ἐκλέξῃ πλῆθος ἐκ τῶν περὶ φαρμάκων ὑπομνη
μάτων.

ὡσαύτως δὲ καὶ τὸ κατάπλασμα σύνθετον ἐχέτω
τὴν δύναμιν ἔκ τε τῆς χαλαστικῆς λεπτομεροῦς καὶ
πικρᾶς καὶ στυφούσης. ἐπικρατείτω δ᾽ ἐν αὐταῖς, εἰ
μὲν ἀποκρούεσθαί τε καὶ τόνον ἐντιθέναι βουλοίμεθα,
τὸ στῦφον· εἰ δὲ τέμνειν καὶ διαρρύπτειν τὸ πικρόν. εἰ
δὲ διαφορεῖν, τὰ λεπτομερῆ τῶν χαλώντων· τὰ γὰρ
παχυμερῆ τοῖς ἐκπυήσουσιν ἐδείχθη χρηστά. διωρι
σμένων δ᾽ ἐν τοῖς περὶ φαρμάκων ὑπομνήμασι τῶν τε
μόνην ἐχόντων ἤτοι τὴν στύφουσαν ἢ τὴν πικρὰν
ποιότητα τῶν τε συναμφοτέρας, ἄριστον μὲν οὖν ἀεὶ
αἱρεῖσθαι τὰ συναμφοτέρας ἔχοντα· μὴ παρόντων δὲ

prepare in Naples in Italy is oil of spikenard in name only, if it is compared with that previously prepared only in Laodicea in Asia, but already nowadays in many cities. If the abatement of the fever, during which you are obviously doing such things, does not become completely mild, be on guard against the spikenard [oil]. Under these circumstances, it is better to use that from apple blossom, but if this is not available, use that from lentiscinum, and next after this use mastic; for the oil of spikenard heats more than all these. Accordingly, you must mix with it oil of roses should the others be lacking. The astringency of oil of myrrh is not suitable in that it is quite thick-particled so that it does not penetrate deeply. Such things have been stated by way of examples; you may choose the great majority of these from the treatises on medications.[18] 792K

Similarly, let the poultice have its potency compounded from a fine-particled relaxing agent that is both bitter and astringent. Let the astringency prevail in these qualities if your wish is to repel and build up strength. If, however, it is to cut and clean thoroughly, let the bitter prevail. If your wish is to disperse, let the fine-particled relaxing agents prevail, for the thick-particled ones were shown to be useful for suppurations. When a distinction was made in the treatises on medications between those having either the astringent or the bitter quality alone and those having both together, it was best always to choose those having both together. If these are not available, mix that

[18] This is taken to be a general reference to the three major treatises on simple and compound medications; see note 2 above.

τούτων, αὐτὸν μιγνύναι τὰ πικρὰ τοῖς στύφουσιν,
ὥσπερ ἐθεάσω ποθ᾽ ἡμᾶς κατάπλασμα συντιθέντας
ἥπατος φλεγμαίνοντος ἐκ τῶν παρόντων.

ἦν μὲν γὰρ ἰατρὸς ὁ κάμνων· ἐθεασάμεθα δ᾽ αὐτὸν
ἤδη λύχνων ἡμμένων, ὡς μηδὲν ἔτι δύνασθαι πρία-
σθαι παρὰ τῶν καπηλευόντων τὰ τοιαῦτα. παρακμὴν
οὖν εὑρόντες ἀξιόλογον, ἐσπεύσαμεν ὅτι τάχιστα χρή-
793K σασθαι τῷ καιρῷ | καὶ μάλισθ᾽ ὅτι τὴν ὕποπτον ὥραν
προσεδοκῶμεν ἀρχὴν οἴσειν ἑτέρου παροξυσμοῦ περὶ
τὰς τῶν ἀλεκτρυόνων ᾠδάς. ἀφεψήσαντες οὖν ἐλαίῳ
μὲν ἀψίνθιον, ἐν ὕδατι δὲ κυδώνιον μῆλον. ἐν ᾧ ταῦθ᾽
ἥψετο, μυροβαλάνου πίεσμα καὶ ἴρεως τὴν ῥίζαν
εὑρόντες ἔνδον καὶ κόψαντες καὶ διαττήσαντες, εἶτ᾽
ἐμβαλόντες λέβητι[7] τοῦ τ᾽ ἐλαίου καὶ τοῦ ὕδατος,
ὧνπερ ἐσκευάκειμεν, ἀφεψήσαντές τε μετρίως ἐπενε-
βάλλομεν ἅμα κηρῷ βραχεῖ τὴν μυροβάλανον καὶ
τὴν ἴριν. ἔψομεν δ᾽, ὡς οἶσθα, τὸ μῆλον οὐ δι᾽ ὕδατος
μόνον, ἀλλὰ καὶ δι᾽ οἴνου πολλάκις αὐστηροῦ· καὶ
μίγνυμεν ἐνίοτε καὶ αὐτῆς τι τοῦ μήλου τῆς σαρκὸς
ἅπασι τοῖς τοιούτοις καταπλάσμασιν. ὅταν μὲν γὰρ
τό τε ἐπιρρέον εἴη πλέον, ἀτονώτερόν τε τὸ μόριον,
αὐξάνομεν, ὡς οἶσθα, τὴν στύψιν· ὅταν δὲ τό θ᾽ ὅλον
σῶμα κενόν, οὐκ ἄρρωστόν τε τὸ μόριον, ἐπιρρέῃ τε
μηδὲν ἔτι, βραχύτατον ἔστω τὸ στῦφον, ἀξιολογώ-
τερον δὲ αὐτοῦ τὸ χαλαστικὸν καὶ πρὸς τούτοις τό τε
πικρὸν καὶ τὸ δριμὺ κατὰ τὴν τῶν ἐπεμβαλλομένων
ὕλην.

794K   εἰ δὲ καὶ μικρὸς ὁ πυρετὸς εἴη, καὶ | ἡ φλεγμονὴ μὴ
πάνυ θερμὴ καὶ ἀσιτίαν ὁ κάμνων ἐνεγκεῖν δυνάμενος,

which is bitter with those that are astringent, as you saw me do on one occasion when I made up a poultice for an inflammation of the liver from what was to hand.

The patient was a doctor. When I saw him the lamps were already lit, so it was no longer still possible to buy such things from traders. Therefore, since I found the abatement to be significant, I was eager to make use of the time as quickly as possible, particularly because the sus- 793K pected hour which I anticipated would bring the beginning of another paroxysm was around cockcrow. Therefore, I boiled up wormwood in oil and quince in water. While these were being boiled, I discovered inside the pressed juice of myrobalan and the root of iris, which I pounded and sifted. Then, when I put into a caldron some oil and water which I had prepared and boiled this up moderately, I put in, along with a little wax, the myrobalan and the iris. I will, as you know, boil the apple, not only with water but also often with astringent wine. And sometimes I also mix some of the actual flesh of the apple in all such poultices. For whenever the flow is greater and the part is weaker, I increase the astringency, as you know. However, when the whole body is empty and the part is not weak, and nothing is still flowing, let what is astringent be very slight but what is relaxing more substantial compared to this and, in addition to these, what is bitter and sharp according to the material of the ingredients.

If, however, the fever is slight, the inflammation not al- 794K together hot, and the patient able to tolerate fasting, you

---

7 K (cf. in lebetem KLat); μέλιτι B, *recte fort.* (*nisi* βέλιτι [*sic*])

αὐξανέσθω σοι τὸ διαφορητικὸν εἶδος τῆς θεραπείας,
ἐκλυομένου τοῦ στυπτικοῦ. οὐ γὰρ δὴ τὴν φλεγμονὴν
δυνατὸν ὠφελῆσαι τοῖς στύφουσιν, ἡνίκα μήτ᾽ ἐπιρρέῃ μηδέν, ἐστήρικταί τε δυσλύτως ἐν τῷ πάσχοντι
παχυμερὴς χυμός, ἀλλ᾽, ὡς εἴρηται, βραδύνειν αἱρούμεθα μᾶλλον ἐν τῇ θεραπείᾳ μιγνύντες τι τῶν στυφόντων ὑπὲρ τοῦ φυλάξαι τῶν μορίων τὸν τόνον. ὅθεν
οὔτ᾽ ἐπὶ κώλου πάσχοντος ἢ τῶν ἄλλων ἐντέρων τινὸς
οὔτε τῶν καθ᾽ ὑποχόνδρια μυῶν ἢ περιτοναίου μικτέον
τὰ στύφοντα· καθάπερ οὐδὲ κύστεως ἢ μήτρας, ὅταν
γ᾽, ὡς εἴρηται, μήτ᾽ ἐπιρρέῃ μηδὲν ἔτι μήτε πλῆθος ἢ
περιττώματα πολλὰ καθ᾽ ὅλον ὑπάρχῃ τὸ σῶμα.

νεφροὶ δὲ καὶ θώραξ ἐν μέσῳ τῶν εἰρημένων εἰσίν·
ὅσον γὰρ ἀπολείπονται γαστρὸς καὶ ἥπατος, τοσοῦτον τῶν ἄλλων πλεονεκτοῦνται· ἐπὶ μὲν γὰρ τοῦ ἥπατος ἢ τῆς γαστρὸς ἐκλυθῆναι τὸν τόνον ὀλεθριώτατον,
ἐπὶ δὲ τῶν ἄλλων ἀκινδυνότατον. ἐπὶ τούτων δὲ κιν-
δυνῶδες μέν, οὐ μὴν ἐκείνοις | γ᾽ ὁμοίως· πλὴν εἴ ποτε
πῦον ἐκκαθαίρειν δέοι περιεχόμενον ἐν θώρακι καὶ
πνεύμονι. περὶ δὲ τοῦ σπληνὸς οὐχ ἁπλῶς ἀποφήνα-
σθαι δυνατόν, ἀλλὰ μετὰ τοῦ διορίσασθαι τήν τε
φύσιν ὅλου τοῦ σώματος καὶ τὴν ἐν τῷ τῆς θεραπείας
χρόνῳ διάθεσιν. εἰ μὲν γὰρ ἤτοι φύσει τῶν ἀθροιζόν-
των εἴη περίττωμα μελαγχολικὸν ὁ κάμνων ἢ κατ᾽
ἐκεῖνον τὸν χρόνον ὁ τοιοῦτος ἐν αὐτῷ πλεονάζοι
χυμός, ἀναγκαῖόν ἐστιν ἐν τῇ θεραπείᾳ τοῦ σπληνὸς
ἐπιπλέκεσθαι τοῖς ἄλλοις τὰ στύφοντα, χάριν τοῦ
φυλάττεσθαι τοῦ σπλάγχνου τὸν τόνον. οὕτω γὰρ

must increase the discutient form of the treatment, leaving out the astringent. For certainly it is not possible to help the inflammation with astringents when there is nothing flowing and the thick-particled humor is held fast in the affected [part] in a way that is difficult to dislodge. Rather, I choose to proceed more slowly in the treatment, as I said, mixing one of the astringents with the aim of preserving the strength of the parts. For this reason, you must mix in the astringents if neither the colon, nor one of the other intestines, nor the muscles of the hypochondrium, nor the peritoneum is affected. The same applies if neither the bladder nor uterus is affected, at least whenever, as I said, there is nothing still flowing and there is neither an abundance nor many superfluities in the whole body.

The kidneys and chest are betwixt and between the structures mentioned in that, to the extent that they are less than the stomach and liver, they are greater than the other [structures]. In the case of the liver or stomach, it is absolutely fatal for the strength to be dissipated, whereas in the case of the other [structures], there is very little danger. It is dangerous in the latter but not like in the former, unless at some time you need to clear out pus contained in the chest and lungs. Regarding the spleen, the situation cannot be made clear in isolation, but only in conjunction with defining the nature of the whole body and the condition at the time of the treatment. If the patient is by nature one of those who gathers together melancholic superfluity, or such a humor is in abundance in him at that time, it is essential in the treatment of the spleen to mix astringents with the other [agents] for the sake of preserving the strength of the organ. In this way, it will draw the super-

795K

GALEN

ἕλξει τε πρὸς ἑαυτὸν τὰ περιττώματα, καὶ καθάρας τὸ
σῶμα, πάλιν αὐτὰ ἐκκρινεῖ διὰ τῆς γαστρός, ὡς ἐν
τοῖς φυσικοῖς περὶ τούτων ἐπιδέδεικται λόγοις. εἰ δὲ
μηδὲν οὐδ᾽ ὅλως ᾖ περίττωμα μελαγχολικὸν ᾖ οὐδ᾽
ὅλως στυπτέον ᾖ ὡς ἥκιστα.

προσέχειν δ᾽ ἀκριβῶς τὸν νοῦν οὐ κατὰ τὸν σπλῆνα μόνον, ἀλλὰ καὶ κατὰ σύμπαντα τἆλλα μόρια, καθ᾽
ἃ σήπεται χυμός, ὁποῖον μέν τοι τῶν συμπτωμάτων
ἐστὶ τὸ ἰσχυρότατον, ὁποῖον τὸ δεύτερον ᾖ τὸ τρίτον,
ἵν᾽ ἐξ αὐτῶν τὴν διάθεσιν ἀκριβῶς ἐξευρών, οἰκείαν
αὐτῇ καὶ τὴν | θεραπείαν ἁρμόσῃς. ἐνίοτε μὲν γὰρ ἡ
θέρμη κρατεῖ κατὰ τὸ πεπονθός, ἐνίοτε δ᾽ ὄγκος, ἤτοι
διὰ τὸ πλῆθος τοῦ χυμοῦ σκληρὸς ᾖ διὰ τὸ πάχος. εἰ
μὲν οὖν ἡ θέρμη κρατοίη, πειρᾶσθαι διαφορεῖν ἀτρέ-
μα, διὰ τῶν χλιαρὰν ἐχόντων θερμασίαν, οἷόν ἐστι τό
τε λινόσπερμον καὶ τὸ χαμαίμηλον· ἐπιμιγνύναι δ᾽
αὐτοῖς ἄλευρον ἐκ κριθῶν ᾖ κυάμων· οὐδέτερον γὰρ
αὐτῶν θερμαίνει. πολλάκις δ᾽, ὡς οἶσθα, τὸ καλούμενον ὑπὸ τῶν ἰατρῶν ἀρτόμελι παρασκευάζοντες εἰς
κατάπλασμα τοιαύτης φλεγμονῆς ὕδατος ἐπεμίξαμεν.
ἐπὶ δὲ σπληνὸς ἔστιν ὅτε καὶ ὄξους. εἰ δ᾽ ἡ θέρμη τοῦ
φλεγμαίνοντος μορίου μὴ πολὺ τοῦ κατὰ φύσιν ἐξ-
εστήκοι, μέγεθος δ᾽ ἀξιόλογον εἴη, διαφορεῖν θαρ-
ρούντως χωρὶς ὕδατος ἐπιμιγνύντα τοῖς εἰρημένοις
καὶ τῆλιν. εἰ δ᾽ οἷον σκιρρώδης τις φλεγμονὴ τύχοι
διὰ τὸ πάχος ᾖ τὴν γλισχρότητα τῶν ἐν αὐτῇ χυμῶν,
ὄξους τε ἅμα καὶ τι τῶν πικρῶν μικτέον φαρμάκων καὶ

796K

202

fluities to itself and, since it purifies the body, it excretes these again through the stomach, as has been shown in the physiological discussions about them.[19] If, however, there is absolutely no melancholic superfluity, you must either not use astringents at all or as little as possible.

Pay diligent attention, not only in relation to the spleen but also in relation to all the other parts in which a humor putrefies, to what kind the strongest of the symptoms is, and to what kind the second and third [symptoms] are, so that when you discover the condition accurately from them, you may also prepare the proper treatment for it. For sometimes heat prevails in the affected part, and 796K sometimes a swelling that is hard due to the abundance or thickness of the humor. If heat does prevail, attempt to disperse it gently with those things that are lukewarm in terms of heat, such as linseed and chamomile. Mix meal [made] from barley or beans with these as neither of them heat. As you know, I often mixed what doctors call artomeli[20] with water in preparing a poultice for such an inflammation. In the case of the spleen, sometimes vinegar is mixed too. If the heat of the inflamed part is not far removed from an accord with nature, but is nonetheless significant in amount, disperse it confidently without water, having mixed fenugreek with those things mentioned. If, for example, an inflammation should happen to be scirrhous due to the thickness or viscidity of the humors in it, you must mix in vinegar and, at the same time, one of the bitter medications, particularly in the case of the spleen.

[19] See Galen, *De naturalibus facultatibus*, Book 2, chapter 9 (II.134K).　　[20] This was a poultice or plaster made from bread and honey; see also 692K and Aëtius, 3.177.

GALEN

μάλιστ᾽ ἐπὶ σπληνός. τὸ γάρ τοι σπλάγχνον τοῦτο,
διὰ τὸ παχὺν ἐπισπᾶσθαι χυμὸν ἐξ ἥπατος, ὑπὸ τοῦ
τοιούτου καὶ νοσεῖ τὰ πολλά.

797K διὰ τοῦτ᾽ οὖν αὐτὸ τά τε δι᾽ | ὀξυμέλιτος καὶ
ἀψινθίου καὶ καππάρεως ἐπιτήδεια καταπλάσματα.
ταῦτ᾽ οὖν ἀγνοοῦντες οἱ πολλοὶ τῶν ἰατρῶν, ἅπαντας
αἰονῶσιν ἐλαίῳ καὶ καταπλάττουσι τοῖς χαλαστικοῖς
καταπλάσμασι, πρὶν κενῶσαι τὸ σῶμα, πάντων δ᾽, ὡς
εἴρηται, μάλιστα τοὺς πλουσίους, οἷς οὐ μόνον εἰς
τὰ τοιαῦτα ὑπηρετοῦσιν, ἀλλὰ καὶ λούεσθαι συγχω-
ροῦσιν. ἐγὼ δ᾽ οὔτε λούσαιμ᾽ ἂν οὐδένα τῶν ἐπὶ σηπεδόνι χυμῶν πυρεττόντων οὔτε χαλαστικοῖς χρήσομαι
καταπλάσμασι, πρὶν κενῶσαι τὰ περιττά. κενώσας δὲ
καὶ λούοιμ᾽ ἂν ἤδη θαρρῶν καὶ καταπλάττοιμι τὰ μὲν
ἄλλα μόρια τοῖς χαλαστικοῖς, ἧπαρ δὲ καὶ γαστέρα
μετὰ τοῦ τὰ μέτρια στύφειν.

17. Ἧττον δὲ ἐπὶ θώρακος τοῖς στύφουσι χρηστέον, ἀποκρούεται γὰρ εἰς πνεύμονα καὶ καρδίαν
ἐνίοτε τὰ τοιαῦτα τοὺς τὴν φλεγμονὴν ἐργαζομένους
χυμούς. ἀλλὰ διὰ τῶν ἐδεσμάτων αὐτοῖς φυλακτέον
τὸν τόνον· οὐ γὰρ ὥσπερ ἐπὶ γαστρὸς καὶ ἥπατος,
οὕτω καὶ ἐνθάδε λεπτότατα διαιτᾶν ἀναγκαῖον. ἐν
798K ἐκείνοις μὲν γὰρ ἡ τροφὴ πέττεται καὶ | κίνδυνός ἐστι
φλεγμαινόντων αὐτῶν μήτε πεφθῆναι καλῶς αὐξη-
θῆναί τε τὰς φλεγμονάς· εἰς θώρακα δὲ τοσοῦτον
ἀφικνεῖται τῶν πεφθέντων ὅσον ἱκανὸν αὐτῷ μόνῳ τῷ
τρεφομένῳ μορίῳ. πολὺ δὲ δὴ μᾶλλον ἐπὶ πνεύμονος
φλεγμαίνοντος ἀφεκτέον ἐστὶ τῶν στυφόντων, ὅπου

204

For indeed, due to its drawing the thick humor from the liver, this organ is very often diseased due to such a humor.

Because of this, the poultices made from oxymel, 797K wormwood and capparis are suitable.[21] The majority of doctors use fomentations with oil on everyone, as they are unaware of these things, and apply poultices with the relaxing agents before they purge the body, and, as I said, most of all in the rich, whom they not only minister to with such things but also agree to bathe. I, on the other hand, will not bathe anyone who is febrile due to putrefaction of humors, nor will I use relaxing poultices before purging the superfluities. After purging, I would now have the confidence to bathe and apply poultices with relaxing agents to the other parts, but to the liver and stomach, together with those agents that are moderately astringent.

17. However, in the case of the chest, you must use the astringents less, for sometimes such things drive the humors to the lungs and heart creating inflammation. But you must preserve their strength through foods; it is not the case, as it is with the stomach and liver, that a very meager diet is also essential here in this way. In those [organs] the nourishment is concocted, but there is a danger, when they 798K are inflamed, that it is not properly concocted and that inflammation is increased. As much of what is concocted comes to the chest as is sufficient for the nourishment of that part alone. In the case of a lung that is inflamed, you must avoid the astringents much more, whereas you must

[21] Oxymel was basically a mixture of vinegar and honey; for the precise composition, see Dioscorides, V.22. For wormwood, see note 15 above. For capparis, see Dioscorides, II.204.

γε καὶ τοῖς χαλαστικοῖς μικτέον, ἐπ' αὐτῶν τῶν δριμυ-
τέρων τι καὶ σαφῶς θερμαινόντων· ἕλκειν γὰρ ἔξω
μᾶλλον ἢ ἀποκρούεσθαι προσήκει. διὸ καὶ αἱ σικύαι
προκενωθέντων χρήσιμοι, πληθωρικῶν δὲ ὑπαρχόντων οὐ μᾶλλον ἐκ πνεύμονος εἰς θώρακα μεθιστᾶσί τι
τῶν περιττωμάτων ἢ ἐξ ὅλου τοῦ σώματος ἕλκουσιν
εἰς ἀμφότερα.

18. Τῷ δὲ αὐτῷ λόγῳ κἀπὶ τῶν κατὰ τὸν ἐγκέφαλον
καὶ τὰς μήνιγγας φλεγμονῶν οὐ χρησόμεθα σικύαις
ἐν ἀρχῇ τῶν παθῶν· ἀλλ' ὅταν μήτ' ἐπιρρέῃ μηδὲν ἔτι
καὶ προκενώσωμεν ὅλον τὸ σῶμα, κατὰ τὴν ἀρχὴν
δὲ κἀνθάδε τοῖς ἀποκρουστικοῖς ὀνομαζομένοις χρηστέον. ἐπεὶ δὲ τῶν ὀστῶν τοῦ κρανίου μέσων κειμένων
799K τὴν δύναμιν | αὐτῶν ἐξικέσθαι βουλοίμεθα πρὸς τὸ
βάθος, ἐπιμίξομεν τοῖς ἀποκρουστικοῖς ὀνομαζομένοις τῶν ποδηγεῖν τι δυναμένων, τουτέστι τῶν λεπτομερῶν κατὰ τὴν οὐσίαν. ὄξος μὲν οὖν οὐ μόνον ἐστὶ
λεπτομερές, ἀλλὰ καὶ αὐτῆς τῆς ἀποκρουστικῆς δυνάμεως οὐ μετρίως μετείληφεν. ὅθεν εἰκότως ἐν ἀρχῇ
τῶν παθῶν αὐτῷ χρῶνται, τῷ ῥοδίνῳ μιγνύντες. ἐπὶ
προήκοντι δὲ τῷ χρόνῳ καὶ σπονδυλίου καὶ ἑρπύλλου
μιγνύουσιν, ἤδη τι καὶ θερμαῖνον ἐχόντων, οὐ μόνον
λεπτομερές· ἐν ᾧ καιρῷ χρὴ μεταβαίνειν ἐπὶ τὰ δια-
πέττοντά τε καὶ διαφοροῦντα πλέον, ἢ κατὰ τὴν χρεί-
αν τῶν φλεγμαινόντων αἱρούμενον ἑκάτερον, ὡς ἂν
ἐκλυομένης αὐτῶν τῆς δυνάμεως ἐν τῷ μεταξὺ τετα-
γμένων ὀστῶν. οὕτως οὖν καὶ τῷ καστορείῳ χρώμεθα,
καίτοι γ' ἐπὶ τῶν ἄλλων οὐ χρώμενοι φλεγμονῶν, οὐδ'

also mix with the relaxing agents things that are sharper and clearly heating. It is more appropriate to draw away externally than it is to repel. And for this reason too, cupping glasses are useful after prior purging because, when people are plethoric, they do not transfer any of the superfluities from the lung to the chest more than they draw to both from the whole body.

18. For the same reason, in inflammations involving the brain and meninges, we will not use cupping glasses at the start of the affections. But whenever there is no longer anything flowing and we have previously purged the whole body, here too at the beginning, we must use the so-called repulsives. However, since we wish the potency of these to reach to the depths of those bones lying in the middle of the cranium, we shall mix with the so-called repulsives one of those things that is able to act as a guide—that is to say, one of those things that is fine-particled in terms of substance. Vinegar is not only fine-particled but also partakes of the repulsive capacity, and not just to a moderate degree. On this account people reasonably use this at the start of the affections after mixing it with rosaceum.[22] At the appropriate time, you also mix in spondylium and tufted thyme, since they already have something heating [in them] and are not just fine-particled. At this time, it is necessary to change to those things that are digestive and discutient to a greater degree, having chosen each in relation to the use of the inflamed [parts] so that their potency is released in the interval between the relevant bones. Therefore, we also use castor in this way, although in the case of other inflammations we do not, even if they are at

799K

---

[22] For the composition of this preparation, also called rhodinon, see Dioscorides, I.53.

ἢν ἐν ἐσχάτῳ τῆς παρακμῆς ὦσι· θερμότερον γάρ
ἐστιν ἢ ὡς ἐν ταῖς φλεγμοναῖς ἐπιτήδειον εἶναι. τοῖς
μέντοι κατὰ τὸν ἐγκέφαλον χωρίοις ἄριστον ἐν ταῖς
παρακμαῖς τῶν φλεγμονῶν, ὡς ἂν οὐκ εὐθέως αὐτοῖς
προσπῖπτον, ἀλλὰ διὰ μέσων τῶν ὀστῶν. ὑπάρχει δὲ
800K   τῷ τοιούτῳ φαρμάκῳ καὶ τὸ λεπτομερὲς | τῆς οὐσίας, ὃ
καὶ αὐτὸ τοῖς διὰ προβλημάτων στεγανῶν διαπέμψειν
μέλλουσι τὴν ἑαυτῶν δύναμιν ἐπιτήδειόν ἐστιν.

19. Οἶσθα δὲ δήπου καὶ τοὺς ἑκάστου τῶν μορίων
οἰκείους πόρους εἰς τὴν τῶν περιττωμάτων ἔκκρισιν
εὔρους ἡμᾶς παρασκευάζοντας· ἐντέρῳ μὲν καὶ γαστρὶ καὶ μεσαραίῳ καὶ τοῖς σιμοῖς τοῦ ἥπατος τὸν δι᾽
ἀπευθυσμένον· νεφροῖς δὲ καὶ κύστει καὶ τοῖς κυρτοῖς
τοῦ ἥπατος καὶ κοίλῃ φλεβὶ καὶ ἀρτηρίᾳ τῇ μεγάλῃ
καὶ πᾶσι τοῖς κατ᾽ ὀσφὺν τὸν τοῖς οὔροις ἀνακείμενον·
πνεύμονι δὲ καὶ θώρακι τὸν διὰ τραχείας ἀρτηρίας καὶ
φάρυγγος· ἐγκεφάλῳ δὲ καὶ μήνιγξι τὸν δι᾽ ὑπερῴας
καὶ ῥινός. ἡ δ᾽ εὔροια τοῖς μὲν δι᾽ ἕδρας ἐκκριθή-
σεσθαι μέλλουσιν ὑπό τε μελικράτου γενήσεται καὶ
τῶν ἐδεσμάτων ὅσα λαπάττει τὴν γαστέρα, καὶ φαρμάκων ὅσα μετρίως ἐρεθίζειν πέφυκε· φυλάττεσθαι
γὰρ χρὴ τὰ δριμύτερα, παροξύνοντα τὰς φλεγμονάς.
ἐπὶ δὲ γαστρὸς καὶ ἥπατος καὶ σπληνὸς φλεγμαινόντων, οὐδὲ τὸ μελίκρατον ἀγαθόν· ἐκχολοῦται γὰρ
801K   αὐτίκα καὶ τὰς φλεγμονὰς | αὐξάνει τῶν σπλάγχνων.
τοῖς δὲ διουρητικοῖς ὀνομαζομένοις φαρμάκοις ἐπ᾽
οὖρα προτρέψεις τὴν περιουσίαν· ὥσπερ γε καὶ τοῖς
βηχικοῖς τὰ κατὰ θώρακα καὶ πνεύμονα διὰ βηχῶν

the very end of the abatement, because they are hotter than is suitable in the inflammations. Nevertheless, in the places around the brain, it is best during the abatements of the inflammations because it would not fall upon them directly but through the intervening bones. On the other hand, in such a medication there is also the fine-particled substance, and this itself is suitable to those who intend to transmit their potency through obstructing coverings. 800K

19. You also know, of course, that we render the proper channels of each of the parts freely patent for the excretion of the superfluities. For the intestines, stomach, mesentery and concavities of the liver, this is through the rectum. For the kidneys, bladder, convexities of the liver, hollow vein, great artery and all those structures in the lower part of the loins, this is assigned to the urine. For the lungs and thorax, it is through the rough arteries (bronchial tree and trachea) and pharynx. For the brain and meninges, it is through the palate and nose. The free passage for those things that are going to be excreted via the rectum will come about due to melikraton, those foods that empty the stomach, and those medications that are moderately irritating in nature. You must, however, guard against those that are too sharp, since they provoke inflammation. In the case of the stomach, liver and spleen, when they become inflamed, melikraton is not good, because it immediately charges with bile and increases the inflammation of the viscera. You will impel the excess toward the 801K urine with the so-called diuretic medications, just as you will evacuate those things in the chest and lungs via coughing with the cough stimulants. In the writings on medica-

ἐκκενώσεις· ἔμαθες δ' ἐν τοῖς περὶ φαρμάκων τὰς ὕλας ἑκάστων οὐ μόνον τῶν εἰρημένων, ἀλλὰ καὶ ὅσα διὰ ῥινῶν ἐκκενοῖ τὰ κατ' ἐγκέφαλόν τε καὶ μήνιγγας περιττά.

20. Καὶ γὰρ καὶ τούτων ἁπάντων ἐν ἐκείνοις τοῖς ὑπομνήμασι τὴν ὕλην ἄφθονον ἔχεις. ὥστ' ἀρκεῖ πρός γε τὰ παρόντα περὶ τῆς τῶν φλεγμαινόντων μορίων διαφορᾶς ὅσα λέλεκταί μοι. μελλήσω γὰρ ἐπὶ πλέον ἐρεῖν ὑπὲρ αὐτῶν ἐν τῇ Μεθόδῳ τῆς θεραπείας ἁπάντων τῶν παρὰ φύσιν ὄγκων· οὐδὲ γὰρ οὐδὲ νῦν ὡς ἔργον τι καὶ σπούδασμα μετεχειρισάμην τόνδε τὸν λόγον, ἀλλ' ἐν παρέργῳ διῆλθον, ὡς ἐξ ἀκολουθίας τινὸς ἐμπεσὼν αὐτῷ. προὔκειτο γὰρ οὐχ ὡς ἄν τις ἄριστα θεραπεύσειεν ἓν ἕκαστον τῶν μορίων φλεγμαῖνον ὑποθέσθαί σοι κατὰ τὸν ἐνεστῶτα λόγον, ἀλλ' ἐνδείξασθαι τὴν βλάβην τῶν καταπλαττομένων ἐν πυρετοῖς ἄνευ τοῦ κενωθῆναι | τὸ σύμπαν σῶμα καὶ μάλισθ' ὅταν μὴ φλεγμαίνῃ τι τῶν ἔνδον· εἰ γὰρ εἴη τὸ περιέχον θερμὸν ὡς κεχύσθαι τὰς ὕλας ἱκανῶς, μείζων ἡ βλάβη τοῖς τοιούτοις ἕπεται καταπλάσμασιν ἤπερ τοῖς λουτροῖς. καίτοι δοκοῦσί γε οἱ καθ' ἑκάστην ἡμέραν αἰονῶντές τε καὶ καταπλάττοντες τὰ ὑποχόνδρια ἐκ τούτων μὲν οὐχ ὅπως βλάβην, ἀλλὰ καὶ μεγάλην ὠφέλειαν γίγνεσθαι τοῖς νοσοῦσιν, ἐκ λουτρῶν δ' οὐκ ὠφέλειαν γίγνεσθαι, ἀλλὰ καὶ μεγίστην βλάβην, ἀγνοοῦντες ὥσπερ τῶν ἄλλων ἁπάντων ὧν περὶ τὸν κάμνοντα πράττουσιν, οὕτω καὶ τῶν βαλανείων τὴν φύσιν.

802K

210

tions you learned not only the materials of each of the aforementioned medications, but also those that evacuate the superfluities in the brain and meninges through the nostrils.

20. And in fact, you have material in abundance on all these in those treatises.[23] As a result, what I have said about the variation among inflamed parts is enough, at least for our present purposes. For I do intend to say still more in the *Method of Medicine* about all the unnatural swellings themselves.[24] I am not taking this subject up now, as some task or pursuit, but to go through it as a secondary matter, having come upon it in the course [of the discussion]. For what I am setting out for you in the present work is not how someone might best treat each one of the parts when it is inflamed, but to show the harm caused by applying poultices in fevers without the whole body being evac- 802K uated, particularly when one of the internal parts is not inflamed. For if the ambient air is so hot that it has liquefied the materials excessively, a greater harm follows such poultices than it does baths. Furthermore, those who apply fomentations and poultices to the hypochondrium on a daily basis think that not only does no harm accrue to the patients from these in any way but actually very great benefit, whereas [they think] that not only does no benefit accrue from baths but actually very great harm, since they are ignorant about all the other treatments they perform for the patient, just as they also are about the nature of baths.

[23] Both here and in the previous sentence this is taken to be a general reference to Galen's three major pharmacological treatises; see note 2 above.   [24] Book 14 of the present work deals with this topic. There is also the specific work *De tumoribus praeter naturam*, VII.705–32K.

ἐγὼ δὲ περὶ μὲν ὅλης αὐτῶν τῆς δυνάμεως ἔμπροσθεν εἴρηκα τελεώτατα, νυνὶ δ' ἀρκεῖ μοι τό γε
τοσοῦτον εἰπεῖν, ὡς ἐπὶ τινῶν μὲν οὔ τι βλάψει τὸ
βαλανεῖον ὅλως, ἐπὶ τινῶν δὲ τοῦ καταπλάσματος
ἧττον. ὃ γὰρ ἐν τοῖς ὑποχονδρίοις ἐργάζεται τὰ χαλαστικὰ πρὸς αὐτῶν ὀνομαζόμενα βοηθήματα, τοῦτ'
ἐν ὅλῳ τῷ σώματι τὸ βαλανεῖον. ὥσθ' ὅταν ἐν τῇ
παρακμῇ παραλαμβάνηται, διαπνεομένου τοῦ σώματος ἤδη μετρίως, οὐ σμικρὸν ὄφελος ἐργάζεται, κενώ-
803K σαν ἅπαντ' ἐξ αὐτοῦ τὰ λιγνυώδη καὶ | καπνώδη
περιττώματα. καὶ εἴ γε μηδὲν εἴη τῶν σπλάγχνων
ἀσθενές, ὀνήσει τὸν κάμνοντα, χωρὶς τοῦ φρῖξαι τότε
γενόμενον.

τρεῖς γὰρ ἔχειν δεῖ τούτους σκοποὺς ἐπὶ τῶν ἐν
ἅπασι πυρετοῖς λουτρῶν· ἕνα μὲν εἰ χωρὶς τοῦ φρῖξαι
γένοιτο, δεύτερον δ' εἰ μηδὲν τῶν κυρίων μορίων
ἀσθενὲς ὑπάρχοι, καὶ τρίτον εἰ μὴ πλῆθος ὠμῶν εἴη
χυμῶν κατὰ τὰς πρώτας φλέβας. ἡ μὲν γὰρ φρίκη
λέλεκται πρόσθεν ὅπως οὐ μόνον αὐξάνειν τοὺς ὄντας
ἤδη πυρετούς, ἀλλὰ καὶ γεννᾶν ἐνίοτε πέφυκε τοὺς οὐκ
ὄντας· ἡ δ' ἀσθένεια τῶν μορίων χυθέντας τοὺς χυμοὺς ὑποδέχεται μᾶλλον ἢ πρὶν χυθῆναι· τὸ δὲ τῶν
ὠμῶν χυμῶν πλῆθος εἰς ὅλον ἀναδίδοται τὸ σῶμα.
μηδενὸς δὲ τούτων ἐμποδίζοντος ἐπὶ τῶν βαλανείων
δύο ἂν ταῦτα κερδαίνοιτο τῷ κάμνοντι, κενωθῆναί τι
τῆς περιουσίας τῶν χυμῶν καὶ διαπνεῦσαι τὸ πολὺ
τῆς τοῦ πυρετοῦ θερμότητος. εἰ δὲ πρὸς τῷ μηδὲν
ἀσθενὲς εἶναι τῶν κυρίων μορίων, ἔτι καὶ τῶν ἀκύρων

I have previously spoken in the greatest detail about the whole potency of baths, so now it is enough for me to say just this: in some cases bathing will do no harm at all while in other cases it will cause less harm than a poultice. What the relaxing agents, which they call remedies, bring about in the hypochondrium, this a bath does in the whole body. As a result, whenever it is undertaken in the abatement, if the body is already transpiring moderately, it brings about no little benefit, evacuating from the body all the sooty and smoky superfluities. And if none of the internal organs is weak, it will benefit the patient, provided he is without shivering fits at that time.

803K

You need to have these three indicators concerning baths in all fevers. One is if [the fever] occurs without shivering; the second is if none of the important parts is weak; and the third is if there is not an abundance of crude humors in the primary veins. For it has been said previously that shivering not only exacerbates fevers that already exist, but also is of a nature to generate sometimes those that do not [yet] exist. The weakened parts receive the humors in liquefied form more than before they were liquefied, while the abundance of crude humors is distributed to the whole body. If none of these [three indicators] is present as a contraindication to baths, they would offer two benefits to the patient: some of the surplus of the humors would be evacuated and there would be transpiration of the heat of the fever to a large extent. If, in addition to none of the important parts being weak, one of the unimportant parts is

ἀσθενές ὑπάρχει τι, καθάπερ ἐπὶ μὲν τῶν ποδαγρικῶν
οἱ πόδες, ἐπὶ δὲ τῶν ἀρθριτικῶν ἅπαντα τὰ τοῦ σώ-
ματος ἄρθρα, μεγίστη τις ἂν ἐκ τοῦ βαλανείου τῷ |
804K κάμνοντι πρὸς τὴν σωτηρίαν ἀπεργασθείη μοῖρα,
δεξαμένων τὰ περιττὰ τῶν ἀσθενῶν. εἴωθε μὲν γὰρ
ἐνίοτε καὶ χωρὶς τῶν βαλανείων ἐπιρρεῖν τοῖς ἀσθενέσι τὸ περιττόν· ἐπὶ βαλανείοις δὲ μᾶλλον, ὡς ἂν καὶ
τῶν ὑγρῶν χεομένων καὶ τῶν ὁδῶν αὐτοῖς παρασκευ-
αζομένων εἰς εὔροιαν· ἄμφω γὰρ ταῦτα θερμαινομέ-
νοις μετρίως ἀναγκαῖον ἐπακολουθεῖν. λέγω δὲ τήν τε
χύσιν τῶν ὑγρῶν καὶ τὴν εὐρυχωρίαν τῶν ὁδῶν.

ὅθεν ὅταν ἧπαρ, ἢ γαστήρ, ἢ πνεύμων, ἢ θώραξ, ἤ
τι τῶν οὕτω κυρίων ἀσθενὲς τύχῃ, μέγιστα βλάπτον-
ται λουσάμενοι πάντες οἷς ἂν ᾖ τι περιττὸν ἐν τῷ
σώματι· τοῖς δ' αὐτοῖς τούτοις καταπλασθεῖσιν ἡ
βλάβη γίγνεται διπλασία μετὰ τοῦ μηδὲν ὀνίνασθαι.
τὰ μὲν γὰρ βαλανεῖα καὶ κενοῖ τὸ σύμπαν σῶμα καὶ
τὸ καπνῶδες ἅμα τῷ λιγνυώδει διαφορεῖ, τὰ δ' εἰρημένα καταπλάσματα τὰ χαλαστικὰ μετὰ τῶν ὁμοίων
αἰονήσεων οὔτε τι τῶν ἀγαθῶν ἔχει καὶ πάνθ' ἕλκει τὰ
περιττὰ πρὸς τὸ τῶν μορίων ἀσθενέστερον, ὅ τί περ
ἂν ᾖ τοῦτο τῶν κατὰ τὰ μέσα τοῦ σώματος, εἴτ' οὖν
ἧπαρ, εἴτε γαστήρ, εἴτε φρένες, εἴτε μεσάραιον ἢ
νῆστις ἢ κῶλον ἢ νεφροί. |

805K εἰ δ' ἐπιμελέστεροι βουλόμενοι[8] φαίνεσθαι καὶ τὸν
θώρακα συνθερμαίνοιεν, ἤτοι γ' εἰς αὐτὸν ἐκεῖνον
ἕλξουσι τὸ περιττὸν ἢ εἰς τὴν καρδίαν ἢ εἰς τὸν
πνεύμονα. λέγω δὲ ἀσθενὲς μόριον ἤτοι τὸ φύσει

214

also still weak—for example, the feet in the case of gout or all the joints of the body in the case of the arthritis—a very significant component in terms of safety will be provided   804K
for the patient from the bath, since the weak parts receive the superfluities. It is also customary sometimes for the superfluity to flow to the weak parts without baths, although more flows with baths, as the fluids are dissolved and the paths are being prepared for them to flow easily. Both these things inevitably follow moderate heatings. I refer to the dispersion of the fluids and the patency of the pathways.

For this reason, whenever the liver, stomach, lungs or chest, or one of the other parts important in this way, happens to be weak, all those who bathe when there is some superfluity in their body harm themselves greatly; but for these same people, if poultices are applied, the injury becomes twofold along with their not bringing any benefit at all. The baths both empty the whole body and disperse the smoky along with the sooty [superfluity], while the aforementioned relaxing poultices together with similar fomentations do nothing good and draw all the superfluities toward the weaker of the parts if this should happen to be one of those in the middle of the body, whether it is the liver, stomach, diaphragm, mesentery, jejunum, colon or kidneys.

If, however, they wish to appear more careful and heat   805K
the chest thoroughly, they will in fact draw the superfluity either to that [structure] itself (i.e. the chest) or to the heart or lungs. I call a weak part either one that is very

---

8 B; βουλόμεθα K

215

δυσκρατότατον, ἄλλῳ γὰρ ἄλλο τοιοῦτον, ἢ τὸ κατά
τινα προηγησαμένην νόσον ἐν δυσκρασίᾳ γενόμενον,
ἢ κατὰ τὸν ἐνεστῶτα χρόνον εἰς αὐτὴν ἠγμένον. οὐ
γὰρ ἁπάντων εὐθέως τῶν φλεγμαινόντων μορίων ἢ
σηπομένους χυμοὺς περιεχόντων ἢ ἐμπεφραγμένων ἡ
δύναμις ἀρρωστεῖ· θᾶττον μέντοι κἀπὶ τούτων εἰς τὸ
πεπονθὸς φέρεται τὸ περιττόν, ὁπόσον ἂν ᾖ διὰ τὴν ἐκ
τοῦ παθήματος θέρμην· ἡ γὰρ οἷον ἑστία τοῦ πυρετοῦ
τὸ τὴν ἔμφραξιν ἢ τὴν σῆψιν ἢ τὴν φλεγμονὴν ἐκ-
δεξάμενόν ἐστι μόριον. ὅταν οὖν ἕλκοντι διὰ θέρμην
αὐτῷ προσέλθοι τις ἔξωθεν ἑτέρα πρόφασις, ἔκ τινος
τῶν δυνάμει θερμαινόντων ἀναγκαῖον αὐξάνεσθαι τὴν
διάθεσιν.

εἰ δ᾽, ὡς εἴρηται, προκεκενωμένος ὅλον τὸ σῶμα διὰ
τῶν χαλαστικῶν θεραπεύοιτο, χωρὶς τοῦ μέγιστον
εἶναι τὸν πυρετὸν ὀνίναιτ᾽ ἄν τι. τηνικαῦτα δ᾽, ὡς
ἐλέχθη, καὶ λούοιτ᾽ ἂν ἀβλαβῶς. χρὴ δὲ φυλάττεσθαι
806K τὴν | ψυχρὰν δεξαμενὴν τοὺς τοιούτους· εἰώθασι γὰρ
οἵ τ᾽ ὠμοὶ χυμοὶ καὶ οἱ σηπόμενοι δυσδιάπνευστοι
γίγνεσθαι χρονιζόντων ἐν αὐτῇ, βλάπτει δ᾽ αὐτούς,
κἂν ἰσχυρῶς εἴη τὸ ὕδωρ ψυχρόν, ἐπειδὴ διαπνεῖσθαι
τῶν οὕτω διακειμένων καὶ ὅλον μὲν τὸ σῶμα δεῖται,
μάλιστα δὲ τὸ πεπονθός. ὥστε ἤτοι περιχέειν αὐτοῖς,
ὡς εἰώθαμεν, ἐπιτήδειον ὕδωρ, ἢ εἰ μετρίως εἴη τὸ
κατὰ τὴν κολυμβήθραν ψυχρόν, ἐπιτρέψαντας ἐμ-
βῆναι διὰ ταχέων ἐξιέναι κελεύειν. ἐπὶ δὲ προήκοντι
τῷ νοσήματι, καθ᾽ ὃν ἤδη καιρὸν ἐκδεδαπάνηται μὲν

dyskratic in nature (for such parts are different in different ways), or one that has become *dyskratic* during some preceding disease, or one brought to a *dyskrasia* at the present time. For the capacity is not immediately weak in all parts that are inflamed which contain putrefying humors or are blocked. Nevertheless, the superfluity is carried more quickly to the affected part even in these cases, whatever the part might be, due to the heat from the affection; for the part which has received the blockage, putrefaction or inflammation is, as it were, the hearth of the fever. Therefore, whenever some other external cause comes near to the part as it is drawing up due to heat, the condition is inevitably increased by something that is heating by virtue of potency.

If, however, as was said, someone who has previously evacuated the whole body were to treat by means of relaxing agents, unless the fever is very great, he might help to some degree. Under these circumstances, as I said, he might also bathe [the patient] without harm. It is necessary to guard against the cold tank for such people because the crude and putrefying humors customarily become difficult to disperse when people spend a long time in this. And if the water is very cold, it harms them because, when people are in such a state, their whole body needs to transpire, but particularly the part that has been affected. Consequently, either pour suitable water over them, as I am accustomed to do, or if the water in the swimming pool is moderately cold and you do allow them to go in, direct them to get out quickly. In the progression of the disease, at the time when the superfluities have already been exhausted and

806K

τὰ περιττά, πέπτεται δὲ τὰ ὠμὰ καὶ τὰ σηπόμενα, καὶ καταπλάσμασι χαλῶσι καὶ βαλανείοις ἀκίνδυνον χρῆσθαι.

21. Καιρὸς δ᾽ ἐπιτηδειότατος ἐν τοῖς διαλείπουσι πυρετοῖς ὅταν, ὡς εἴρηται, διαπνεῖν ἄρξηται τὸ πυρετῶδες θερμόν. εὐθὺς γὰρ τῇ τροφῇ παρασκευάζει τὸ σῶμα, δυοῖν τούτοιν σκοπῶν καὶ τοῦ ταύτης καιροῦ δεομένου, τῆς τ᾽ ἀκμῆς τοῦ προγεγονότος παροξυσμοῦ καὶ τῆς ἀρχῆς τοῦ γενησομένου. πρὸς τούτους γὰρ ἀποβλέποντας χρὴ πειρᾶσθαι πορρωτάτω τρέ-

807K φειν ἑκατέρου. εἰ μὲν οὖν ἱκανὸν εἴη τὸ | ἐν μέσῳ τῆς τ᾽ ἀκμῆς τοῦ προτέρου παροξυσμοῦ καὶ τῆς ἀρχῆς τοῦ δευτέρου, ῥᾷστον ἐξευρεῖν τὸν καιρόν· εἰ δ᾽ ὀλίγον, ἀναγκαῖόν ἐστι δυοῖν θάτερον, ἢ θερμὸν ἱκανῶς ἔτι τὸν κάμνοντα τρέφειν ἢ μελλούσης ὅσον οὔπω τῆς εἰσβολῆς γενέσθαι τοῦ δευτέρου παροξυσμοῦ.

προγνωστικοῦ τοιγαροῦν ὁ τοιοῦτος πυρετὸς δεῖται ἰατροῦ, δυναμένου στοχάζεσθαι κατὰ τὰς πρώτας εὐθέως ἡμέρας εἰς πόσον ἐκταθήσεται χρόνον ὁ πυρετός· δῆλον γὰρ ὅτι σύνοχος ὑπάρχων ὀξεῖαν ἔχει τὴν λύσιν. εἰ μὲν οὖν μὴ πόρρω τῆς ἑβδόμης ἡμέρας μέλλοι προέρχεσθαι, παντάπασιν ἀσιτητέον ἐστὶ τῷ νοσοῦντι στοχασαμένων ἡμῶν εἰ ἡ δύναμις ἐξαρκέσει. εἰ δ᾽ ἤτοι μέχρι τῆς ἐννάτης ἢ ἑνδεκάτης ἢ τεσσαρεσκαιδεκάτης ἐκτείνεσθαι μέλλοι, ἢ καὶ περὶ μὲν τὴν ἑβδόμην λύεσθαι, τὴν δύναμιν δ᾽ ἀσθενεστέραν ἢ ὥστ᾽ ἐξαρκέσαι μέχρι τοσούτου χρόνου δόξειεν ἔχειν ὁ κάμνων, ἀναγκαῖον μὲν ἔσται τρέφειν, αἱρεῖσθαι δ᾽, ὡς ἐν ἀπόροις προσήκει, τὰ ἧττον βλαβερὰ

the crude and putrefying humors are concocted, use both relaxing poultices and baths without danger.

21. The most suitable time [for treatment] in the intermittent fevers is, as I said, whenever the feverish heat begins to disperse in vapor. This immediately prepares the body for nourishment, there being these two indicators of the required time for this: the peak of the previously occurring paroxysm and the beginning of the one that is about to occur. Focusing attention on these, you should attempt to nourish at the farthest point from each. So if the interval between the peak of the first paroxysm and the beginning of the next paroxysm is sufficient, it is very easy to identify the time. If, however, it is short, one of two things is necessary: either to nourish the patient when he is still excessively hot or when the onset of the second paroxysm is almost about to occur.

807K

Therefore, such a fever needs a doctor who is able to prognosticate—one who is able to calculate right from the first days how long the fever will last. For it is clear that when a fever is continuous it has an acute resolution. Therefore, if it is not going to advance beyond the seventh day, we must fast the sick person completely provided we have determined that the capacity will be sufficient. If, however, it is going to extend to the ninth, eleventh or fourteenth day, or if it is going to resolve around the seventh day, but the patient seems to have a weaker capacity than will be sufficient for such a long time, it will be essential to provide nourishment, but to choose, as is appropriate in difficult circumstances, those things that are less harmful, distinguishing and demarcating off those that cause less

διοριζόμενόν τε καὶ ἧττον ἀφορίζοντα τῶν μᾶλλον
βλαπτόντων αὐτὰ τῷ τε τόπῳ τῷ πεπονθότι σὺν τῇ
808K διαθέσει δηλονότι καὶ τῷ τοῦ | παροξυσμοῦ καὶ τῷ τῆς
ἀρχῆς καὶ τῷ τῆς ἀκμῆς ἤθει. φλεγμαίνοντος μὲν γὰρ
ἥπατος ἢ γαστρὸς ὀλεθριώτατον θρέψαι πρὸ τοῦ παρ-
οξυσμοῦ· χωρὶς δὲ φλεγμονῆς ἀρρωστούντων τὴν
δύναμιν ὠφελιμώτατον. εἰώθασι γὰρ τῶν οὕτως ἐχόντων ἐπὶ μὲν τῷ ἥπατι κακοπραγοῦντι διαχωρήσεις
τοῖς παροξυσμοῖς συνεισβάλλειν, ἐπὶ δὲ τῷ στόματι
τῆς γαστρὸς συγκοπήν.

τοῖς δ᾽ εἰρημένοις ὀργάνοις τοῦ ζῴου τὸ ἀνάλογον
ἐπὶ τῶν ἄλλων σκοπεῖν κατά γε τὴν θέσιν καὶ τὴν
ἀξίαν τοῦ μορίου· λεχθήσεται δὲ καὶ αὖθις ὑπὲρ αὐτῶν. ἦθος δ᾽ ἀκμῆς καὶ παροξυσμοῦ ἀρχῆς ἀξιῶ σε
σκοπεῖσθαι προσέχοντα τὸν νοῦν· ἐπὶ μὲν τῆς ἀκμῆς,
εἰ ξηρὰ καὶ αὐχμώδης ἄνευ μεγέθους τῆς θέρμης τῆς
πυρεκτικῆς ἢ διακαὴς χωρὶς αὐχμοῦ· τὴν μὲν γὰρ
προτέραν ὅτι τάχιστα χρὴ τέγγειν ὑγραινούσαις τρο-
φαῖς· ἐπὶ δὲ τῆς δευτέρας ἀναμένειν λωφῆσαι τὸ
πλεῖστον τῆς θερμότητος. οὕτω δὲ καὶ τῆς ἀρχῆς τοῦ
παροξυσμοῦ τὸ ἦθος ἐν τῷ καταψύχειν τὰ τοῦ σώματος ἀκρωτήρια καὶ πολλὴν εἴσω ποιεῖσθαι μετ-
άστασιν τοῦ αἵματος ἢ χωρὶς ἁπάσης γίγνεσθαι
θλίψεως ἐπισκέψῃ. τῆς μὲν γὰρ δευτέρας εἰρημένης |
809K τῆς ἀθλίπτου καταφρονεῖν ὡς ἐπιεικοῦς, ἐπὶ δὲ τῆς
προτέρας διορίσασθαι. χωρὶς μὲν γὰρ φλεγμονῆς
σπλάγχνου ἢ περιουσίας χυμῶν, ἐπικρατούσης τῆς

harm from those that cause more harm on the basis of the affected place, along with the condition, obviously, and the 808K character of the paroxysm, both its onset and its peak. When the liver or stomach is inflamed, it is absolutely fatal to nourish before the paroxysm. If, however, there is no inflammation, it is most beneficial with respect to the capacity of those who are weak [to nourish them]. Excretions in those who are so disposed due to the liver being adversely affected customarily occur coincidentally with the paroxysm, whereas those who are so disposed due to the opening of the stomach being adversely affected suffer syncope coincidentally with the paroxysms.

In the case of the other [organs], consider them analogously to the aforementioned organs of the person in relation to the position and importance of the part. There will also be further discussion about these. I think it worthwhile for you to consider and direct your attention to the character of the peak and onset of the paroxysm—in the case of the peak, whether it is dry and squalid without severe feverish heat, or burning hot without squalor. In the first instance, it is necessary to moisten as quickly as possible with foods that relax; in the second instance, it is necessary to wait for the greater part of the heat to abate. Also, in the same way, you will observe the character of the onset of the paroxysm—whether it cools the extremities of the body and produces a substantial transfer of the blood internally, or occurs without any affliction at all. Pay scant attention to the second, termed "nonafflicting," as being 809K mild. Make this distinction, however, in the case of the former. When movement to the depths prevails without inflammation of a viscus or a surplus of humors, you will do

εἰς τὸ βάθος κινήσεως, ἐν τοῖς παροξυσμοῖς οὐδὲν βλάψεις ὀλίγῳ θᾶττον τρέφων· εἰ δ' ἤτοι φλεγμονή τις ἢ πλῆθος εἴη, φυλακτέον τὴν πρὸ τοῦ παροξυσμοῦ τροφὴν ὡς βλαβερώτατον.[9]

[9] K; βλαβερωτάτον B, recte fort.

no harm if you nourish a little more quickly during the paroxysms whereas, if there is either some inflammation or abundance, you must guard against nourishment prior to the paroxysm as this is a very harmful thing [to do].

BIBΛION M

1. Οὔτε γένος οὔτε εἶδος οὔτε διαφορά τίς ἐστι πυρετῶν, ὡς ἔνιοι νομίζουσιν οἱ μετὰ συμπτωμάτων συνιστάμενοι, ἀλλ᾽ ὥσπερ ἄλλο τι ἄλλῳ νόσημά τε νοσήματι καὶ σύμπτωμα συμπτώματι καθ᾽ ἕνα χρόνον ἐπιπλέκεται περὶ ἓν σῶμα, κατὰ τὸν αὐτὸν τρόπον ἅμα πυρετῷ γίγνεταί τινα μείζω συμπτώματα, περιττοτέρας προνοίας δεόμενα. τῶν μὲν γὰρ μικροτέρων ἐξ ἀνάγκης ἑπομένων ταῖς γενέσεσι τῶν πυρετῶν οὐδὲ μέμνηνται τὴν ἀρχήν, ἀλλ᾽ ὡς οὐδ᾽ ὅλως ὄντα παρα-
λείπουσιν ἀνώνυμα· τὰ δ᾽ ἤτοι | κατὰ τὸ σπάνιον ἐπιγινόμενα τοῖς πυρέττουσιν ἢ τὸ σύνηθες ὑπερβάλλοντα μέγεθος ἐξαιρέτου παρὰ τἄλλα διδασκαλίας ἀξιοῦσιν.

εἰσὶ μὲν οὖν αὐτῶν αἱ πρῶται διαφοραὶ τρεῖς. ἤτοι γὰρ ἐν ταῖς τῶν ἐνεργειῶν βλάβαις εἰσὶν ἢ ἐν ταῖς τῶν ἐπεχομένων ἢ κενουμένων ἀμετρίαις, ἢ διαθέσεσι σωμάτων ὑπάρχουσι. οὐ γὰρ δὴ τό γε τέταρτον[1] γένος ἐν συμπτώμασι θετέον, εἰ καὶ ὅτι μάλιστα πάμπολλοι καὶ τοῦτο τοῖς συμπτώμασιν ἔμιξαν, οἷον ἐρυσιπέλατα καὶ ὀφθαλμίας καὶ ἕλκη καὶ παρωτίδας, ὅσα τ᾽ ἄλλα τοιαῦτα. τῶν δ᾽ εἰρημένων τριῶν αἱ μὲν τῶν

BOOK XII

1. There is not one class, kind, or differentia of fevers, as 810K
some [doctors] who conflate them with symptoms think.
Rather, just as one disease is combined with another disease, and one symptom with another symptom at one particular time in one body, in the same way, certain more se-
rious symptoms occur along with a fever, and require quite
considerable forethought. These doctors don't mention at
all the origin of the lesser symptoms that necessarily follow
the genesis of fevers; they leave them unnamed as if they
were altogether nonexistent. They do, however, think the 811K
symptoms that either seldom supervene, or that exceed
the customary magnitude in those with fever are worthy of
special teaching compared to the other symptoms.

There are three primary differentiae of these symptoms. They lie either in the injuries of functions, or in the
excesses of those things retained or evacuated, or they are
in the conditions of bodies. You must definitely not establish the fourth class in symptoms, even if it is the case that
very many [doctors] particularly have mixed this with the
symptoms. Examples are erysipelas, ophthalmia, wound/
ulcer, parotid gland swelling, and other such things. Of the
three differentiae mentioned, the injuries of functions, if

---

¹ B; οὐ γὰρ τὸ τέταρτον K

225

ἐνεργειῶν βλάβαι λυπηραὶ γενηθεῖσαι πρὸς ἑαυτὰς
ἐπιστρέφουσι τὸν ἰατρόν, ἀφίστασθαι τῶν διαθέσεων
ἀναγκάζουσαι, καθάπερ γε καὶ αἱ τῶν ἐκκρινομένων ἢ
ἐπεχομένων ἀμετρίαι. τὸ δ᾽ ἐν ταῖς διαθέσεσι γένος
τῶν συμπτωμάτων σημεῖον μέν ἐστιν ἑτέρας διαθέσεως, ἣν χρὴ θεραπεύειν, αὐτὸ δ᾽ οὐδεμίαν ἐξαίρετον
ἴασιν ἔχει, παυόμενον ἅμα τῇ ποιούσῃ διαθέσει.

καθόλου μὲν οὖν εἰπεῖν οὐδὲν τῶν συμπτωμάτων
ἢ σύμπτωμά ἐστιν οὔτ᾽ ἐνδείκνυται θεραπείαν οὔθ᾽
ὑπαλλάττει πρώτως. ἐδείχθη γὰρ ἐκ τῶν νοσημάτων |
812K ἡ ἔνδειξις τῆς θεραπείας γινομένη, καθάπερ ἐκ τῶν
αἰτίων ἡ προφυλακή. λέλεκται δὲ καὶ ὅπως ἡ προφυλακὴ τῷ τῶν ἰαμάτων μίγνυται γένει. κατὰ συμβεβηκὸς δέ ποτε καὶ τὰ συμπτώματα τὴν θεραπείαν
ἐξαλλάττει, λόγον αἰτίας ἔχοντα κατ᾽ ἐκεῖνον τὸν
χρόνον. ἐν δ᾽ αὐτῶν ἐστιν ἁπάντων κεφάλαιον, ὅταν γ᾽
ὡς αἰτία τὴν θεραπείαν ὑπαλλάττῃ· λέγω δὲ τὸ ἐν
τοῦτο τὴν βλάβην εἴτ᾽ οὖν τῆς δυνάμεως εἴτε τῆς
διαθέσεως· ὡς ὅταν γε βλάπτῃ μηδέν, οὐ διακόπτει
τὴν ἴασιν. βλάπτει δὲ τὴν μὲν δύναμιν ἀγρυπνία καὶ
ὀδύνη καὶ κένωσις ἄμετρος, τὴν διάθεσιν δ᾽ οὐκ ἀεὶ
μέν, ὡς τὰ πολλὰ δὲ καὶ μάλισθ᾽ ὅταν ἄμετρα γένηται. ἐπὶ τούτοις οὖν μόνοις ἀναγκαζόμεθα τὴν ἀγωγὴν
τῆς θεραπείας, ἣν ἐξ ἀρχῆς ἐνεστησάμεθα, μεταβα-
λόντες ἐν τῷ παρόντι πρῶτον ἐκκόψαι τὸ σύμπτωμα.

καταλυομένης οὖν τῆς δυνάμεως, οὐ μὴν ἤδη γε
ἰσχυρῶς καταλελυμένης, ἀλλ᾽ ἀντεχούσης ἔτι, πρὸς
τὸ σύμπτωμα μόνον ἀποβλέπομεν, ὅτι τάχιστα σπεύ-

they become distressing, divert the doctor toward themselves, compelling him to set aside the conditions, just as in fact the excesses of those things excreted or retained also do. However, the class of the symptoms in the conditions is a sign of another condition which it is necessary to treat, although it has no special cure, since it ceases along with the condition creating it.

To speak in general terms, there is no symptom *qua* symptom which either indicates treatment or changes it primarily. It was shown that the indication of the treatment   812K
arises from the diseases just as the prophylaxis arises from the causes. I have also stated how prophylaxis is connected with the class of cures. On some occasions—that is, when they have the ground of cause at that time—the symptoms do change the treatment contingently. There is one chief point of all these symptoms, at least whenever, by virtue of being causes, they change the treatment. I refer to the damage either of the capacity or the condition, since whenever the symptom damages nothing it does not impact on the cure. Insomnia, pain and excessive evacuation damage the capacity but not always the condition, as very many things do, and particularly whenever they are disproportionate. It is because of these alone that we are compelled, having changed the course of treatment which we instituted from the beginning, to eradicate the symptom first in the prevailing situation.

Therefore, if the capacity is being dissipated, although it has not yet been severely dissipated but is still holding out, we direct our attention to the symptom alone, hasten-

δοντες ἀναιρεῖν αὐτό· καταλελυμένης δ᾽ ἰσχυρῶς οὐ
μόνον τὸ σύμπτωμα παύειν, ἀλλὰ κἀκείνην ῥωννύειν
813K σπεύδομεν. εἰ δ᾽ ἑνὶ | βοηθήματι καὶ τὸ σύμπτωμα
παύεσθαι δύναιτο καὶ τὸ νόσημα θεραπεύεσθαι, τοῦτο
οὐκ ἂν εἴη πρὸς σύμπτωμα βλέπειν ἢ συμπτώματος
ἕνεκα βοήθημα παραλαμβάνειν, ἀλλ᾽ ἄντικρυς ἡ τοιαύτη πᾶσα τῶν βοηθημάτων ἰδέα νοσημάτων ἴασίς
ἐστιν. οὔτε γὰρ ὅταν ἄλγημα πλευρᾶς ἐν πλευρίτιδι
φλεβοτομήσαντες ἢ καθάραντες ἰασώμεθα, συμπτώ-
ματός ἐστιν ἡ τοιαύτη θεραπεία, οὔθ᾽ ὅταν ἀπόστημα
διατεινόμενον καὶ σφῦζον ὀδυνηρῶς διελόντες ἐκκρί-
νωμεν τὸ πῦον, ἀλλ᾽ ὅταν εἰ οὕτως ἔτυχεν ἐπὶ δυσεν-
τερικαῖς ἐκκρίσεσι δακνούσαις σφοδρῶς ἤτοι τράγου
χυλὸν ἐνίεμεν ἢ στέαρ αἴγειον ἢ ῥοδίνην κηρωτήν. ὑφ᾽
ὧν αὐτὴ² μὲν ἡ τῶν ἐντέρων ἕλκωσις οὐ θεραπεύεται·
καὶ μάλισθ᾽ ὅταν ἔχῃ τι σηπεδονῶδες· ἀναπαύεται δ᾽
ἡ δύναμις ἐν τῷ μεταξύ· καὶ τοῦτ᾽ ἐστὶ τὸ πρὸς τὸ
σύμπτωμα στῆναι, τοῦ νοσήματος ἀμελήσαντας παρ᾽
ἐκεῖνον τὸν καιρόν. ὡς ὅταν γε θαρρῶμεν τῷ τόνῳ τῆς
δυνάμεως, οὐδέποτε πρὸς τὸ σύμπτωμα ἐνιστάμεθα,
τοὐναντίον δ᾽ ἅπαν ἐνίοτε διὰ τῶν ὀδυνηρῶν βοηθημάτων τὰ νοσήματα θεραπεύομεν, ὥσπερ ἀμέλει
814K τὴν δυσεντερίαν αὐτὴν δηκτικωτάτοις | φαρμάκοις
τότε μάλιστα θεραπεύομεν, ὅταν ᾖ σφοδροτάτη.

πεντηκοντούτης γοῦν ἰατρός τις νοσῶν, ἑβδόμην
ἄγων ἐκείνην τὴν νύκτα, μὴ πάνυ τι τὴν δύναμιν
ἰσχυρός, ἀλγήματος αὐτῷ κεφαλῆς ἰσχυροῦ γενο-

---

² B; αὕτη K

228

ing to remove it as quickly as possible. However, when the capacity has been severely dissipated, not only do we hasten to put an end to the symptom but we also hasten to strengthen the capacity. But if it were possible to put an 813K end to the symptom and treat the disease with a single remedy, this would not be to focus on the symptom or to take up a remedy for the sake of the symptom. Rather, every kind of remedy of this sort is a cure of diseases outright. For whenever we cure pleuritic pain in pleurisy by phlebotomy or purging, this is not treatment of a symptom, nor is it when, having incised an abscess that is under tension and throbbing painfully, we evacuate the pus, but when, perchance, we introduce the juice of goatwort,[1] or goat fat, or a rosewater salve for the dysenteries with biting evacuations, [this is treatment of a symptom]. The actual ulceration of the intestines is not treated by these things, particularly when there is some putrefaction. However, the capacity is given respite in the meantime, and this is to take a stand against the symptom, having neglected the disease during that time. Even so, when I am confident about the strength of the capacity, I never take action against the symptom. Indeed, sometimes quite the opposite; I treat the disease with painful remedies just as I actually treat the dysentery itself with very mordant medica- 814K tions, especially during the time it is very severe.

Anyway, there was a fifty-year-old doctor whose illness had reached the seventh night, and who was not very strong in terms of capacity. When a severe pain arose in his

---

[1] This is taken to be a preparation from Tragian Allo (*Tragium calumnae*) which Dioscorides lists as useful for dysentery (IV.50).

μένου κατὰ τὸν δεξιὸν μάλιστα κρόταφον, οὐκ ἀνα-
μείνας ἀφικέσθαι τινὰ τῶν ἑταίρων ἑαυτὸν ἐφλεβο-
τόμησεν ἐν τῇ νυκτί. καὶ τὸ μὲν ἄλγημα διὰ ταχέων
ἐπαύσατο, μέχρι μέντοι χρόνου πολλοῦ κακόχρους
καὶ ἄρρωστος τὴν δύναμιν ἰσχνός τε καὶ ἄτροφος ἦν,
ὡς μόλις ἀναλαβεῖν τὴν ἐξ ἀρχῆς ἕξιν. ἔστι δὲ κἀνταῦθα πρόφασις διττὴ τῶν πρὸς ὀδύνην ἱσταμένων
ἰατρῶν, ἐνίοτε μὲν ὑπὲρ τοῦ χαρίσασθαι τῇ μαλακίᾳ
τῶν καμνόντων, ἐνίοτε δὲ κατὰ τὸν τῆς τέχνης λόγον.
εἰ μὲν γὰρ ἐνὸν ἀντισχόντα ταῖς ὀδύναις διὰ τῶν
ἰσχυρῶν βοηθημάτων ἡμέρᾳ μιᾷ θεραπευθῆναι, πρὸς
τὴν ὀδύνην ἵστατό τις τῆς ὅλης θεραπείας ἀμελῶν, οὐ
κατὰ τὸν λόγον τῆς τέχνης, ἀλλὰ τῷ κάμνοντι χαρι-
ζόμενος οὕτω πράττει. εἰ δὲ ὑπὸ τῆς ὀδύνης ἡ δύναμις
καταλύοιτο καὶ κίνδυνος ἐφεδρεύοι διὰ τοῦτο, πραΰνειν μὲν χρὴ τὸ ἄλγημα, ῥωννύναι δὲ τὴν δύναμιν,
815K ὅπως ἀντισχεῖν μὲν τῷ νοσήματι | δυνηθῇ, πρός τε
τὸν χρόνον ἐξαρκέσαι τῆς ἰάσεως.

οἶδα γὰρ ἐνίους τῶν γενναίων εἶναι προσποιουμέ-
νων ἰατρῶν τε καὶ καμνόντων ἀπολλυμένους δι᾽ αὐτὸ
τοῦτο τὸ καρτερῶς τε καὶ ἀνδρείως ὁμόσε χωρεῖν ἀεὶ
ταῖς ὀδύναις, οὐδὲν τῶν παρηγορικῶν ἑλομένους, ἀλλ᾽
ἐν τοῖς τραχέσι καὶ ὡς ἔλεγον αὐτοὶ τὴν διάθεσιν
ἀνασκευάζουσι διαγιγνομένους βοηθήμασιν, οὓς ἐν
χρόνῳ πλείονι θεραπεῦσαι βέλτιον ἦν ἢ τοὺς σπεύ-
δοντας ἀνδρείως ἀποθανεῖν. ἰώμενος δέ τις, ὡς αὐτὸς
ἐνόμιζε, γενναίως δυσεντερικοὺς ἰσχυροτάτῳ φαρμάκῳ πολλοὺς μὲν ἐθεράπευσεν ἐν ἡμέρᾳ μιᾷ, τινὰς δὲ

head, predominantly on the right side of the forehead, he did not wait for one of his associates to attend him but carried out a phlebotomy on himself during the night. This quickly put an end to the pain. However, he remained of bad complexion for a long time, was weak in terms of capacity, and was thin and wasted so that it was difficult to restore his original state. There is here also a twofold motivation in the position of doctors toward pain: sometimes it is about indulging the delicacy of patients and sometimes it relates to the theory of the craft. If it were possible for someone who has withstood the pains to be treated in a single day with strong remedies, and some [doctor] took a stand against the pain while neglecting the treatment as a whole, he would be acting like this, not in accord with the theory of the craft but because he wishes to please the patient. If, however, the capacity is dissipated by the pain and danger is imminent because of this, it is necessary to assuage the pain and strengthen the capacity so it can resist the disease and be strong enough for the time of the cure.   815K

For my part, I know of some who, considering themselves courageous, both doctors and patients, die for this very reason—that they choose to struggle continuously against the pains stoutly and manfully, taking none of the pain-relieving medications, but persevering with the remedies that are harsh and, as they themselves acknowledge, destructive of the condition. These are people whom it would be better to treat over a longer time rather than have them hasten to their death bravely. However, a certain person, when curing dysentery courageously, as he himself thought, with a very strong medication, treated many in

GALEN

ἀπέκτεινεν. ἦν δὲ ὁ τρόπος αὐτοῦ τῆς θεραπείας τοιόσδε. κρόμμυα τὰ καρτὰ³ καλούμενα μετὰ ἄρτου διδοὺς⁴ ἐσθίειν καὶ πίνειν ὀλίγον ἡμέρᾳ μιᾷ, κατὰ τὴν ὑστεραίαν ἔωθεν ἔκλυζεν ἅλμῃ δριμυτάτῃ, καὶ μετ' αὐτὴν ἐνίει φάρμακον ἰσχυρόν. ὅσοι μὲν οὖν ἀντισχεῖν ἠδυνήθησαν αὐτῷ τελέως ὑγιάσθησαν, ἔνιοι δὲ σπασθέντες ἢ μετὰ νοτίδος ὑπὸ τῆς ὀδύνης συγκοπέντες ἀπέθανον.

ὅρος οὖν ἐπὶ καμνόντων τῷ κατὰ τὸν λόγον τῆς τέχνης ἀγωνιζομένῳ γενναίως πρὸς τὸ νόσημα τὸ τῆς
816K ἰάσεως | ἀσφαλές· ὥσπερ γε καὶ τῷ πραΰνοντι τὰς ὀδύνας ἡ τῆς δυνάμεως φυλακή. τὸ⁵ δ' ἐπέκεινα τῶνδε σκαιοῦ μὲν ἀνδρὸς ἔργον ἐστίν, ἅμα τῷ νοσήματι καὶ τὴν ζωὴν ἀφελέσθαι τὸν ἄνθρωπον· κόλακος δὲ τὸ χαρίζεσθαι τῷ νοσοῦντι, σκοπὸν ὧν πράττει θέμενον ἡδονήν, οὐχ ὑγείαν. ἐμπίπτουσι δ' εἰς τὰς τοιαύτας ὑπερβολὰς ἐν πολλαῖς μὲν καὶ ἄλλαις ὕλαις βοηθημάτων οἱ ἰατροί, μάλιστα δ' ἐν τοῖς καλουμένοις ἀνωδύνοις φαρμάκοις, ὅσα δι' ὀποῦ μήκωνος, ἢ ὑοσκυάμου σπέρματος, ἢ μανδραγόρου ῥίζης, ἢ στύρακος, ἤ τινος τοιούτου συντιθέασιν. οἵ τε γὰρ χαριζόμενοι τοῖς νοσοῦσι πλεονάζουσιν ἐν τῇ χρήσει τῶν τοιούτων φαρμάκων, οἵ τ' ἀκαίρως καὶ ἀμέτρως γενναῖοι μηδ' ὅλως χρώμενοι διαφθείρουσιν ὀδύναις τοὺς κάμνοντας.

³ καρτὰ B; κάρτα K
⁴ B; ἐδίδουν K; ἐδίδου?(cf. dabat KLat)
⁵ Inter φυλακή (corr. ex ἀσφαλές) et τό add. B ὥσπερ γε καὶ

232

one day but killed some. The manner of his treatment was as follows: he was in the habit of giving so-called chopped onions to eat with bread and a small amount to drink on day one. On the following day, early in the morning, he purged [the patient] with very sharp brine and after that, inserted a strong medication *per rectum*. Those who could withstand this were restored to health completely. Some, however, when they suffered convulsions or fainted with sweating due to the pain, died.

Therefore, in those who are ill, the safety of the cure is the determining factor for someone struggling courageously against the disease in accord with the rationale of the craft, just as the preservation of the capacity is for the person soothing the pains. It is the action of an incompetent to exceed these limits and to take away the patient's life along with the disease while it is the action of a flatterer to gratify the sick person, establishing pleasure rather than health as the goal of what he does. Doctors fall into such excesses in many and various materials of remedies, but especially in the so-called anodyne medications which they compound from poppy juice, seed of henbane, root of mandrake, storax,[2] or some such thing. Those who indulge the sick go to excess in the use of such medications, and those inappropriately and immoderately courageous fellows who don't use them at all, destroy their patients through the pains.

816K

---

[2] Storax is the name given to the resin of any of the trees or shrubs of the genus *Styrax*. It had a range of medicinal uses in ancient times; see Dioscorides, I.79.

τῷ πραΰνοντι τὰς ὀδύνας ἡ τῆς, sed dittogr. statim recognovit scriba.

ὥσπερ οὖν ἐν ἁπάσαις ταῖς καθ᾽ ὅλον τὸν βίον
ἕξεσί τε καὶ πράξεσιν, οὕτω κἀνταῦθα τὸ μηδὲν ἄγαν
αἱρετέον, ὅρον ἔχοντα τὴν ὠφέλειαν τοῦ κάμνοντος. εἰ
μὲν γὰρ οἷόν τε εἴη τοῖς ἰωμένοις τὸ νόσημα βοηθήμασι χρώμενον ἀνύσαι τὸ δέον, ἀφίστασθαι χρὴ
817K τῶν | καρωτικῶν φαρμάκων, ἃ καλοῦσιν ἀνώδυνα. εἰ
δὲ ὑπό τε τῶν ἀγρυπνιῶν καὶ τοῦ καταλύεσθαι τὴν
δύναμιν εἰς κίνδυνον ὁ κάμνων ἥκει θανάτου, τότ᾽ ἂν
ἐν καιρῷ χρήσαιο καὶ τοῖς τοιούτοις φαρμάκοις, ἐπι-
στάμενος μὲν ὅτι βλάψεις τι τὴν ἕξιν τοῦ σώματος,
αἱρετώτερον δὲ εἶναι νομίζων τοῦ θανάτου τὴν βλάβην. εἰ μὲν γὰρ μὴ πάνυ τις αὕτη μεγάλη γένοιτο, κἂν
ἐκκόψαιμεν αὖθις αὐτὴν ἐπὶ μακροτέρας σχολῆς· εἰ δ᾽
οὕτως ἄμετρος ὡς μηδ᾽ ἐν χρόνῳ πολλῷ τελέως ἰαθῆναι, τοῦτο γοῦν αὐτὸ βέλτιον εἶναι νομίζομεν ἢ ἀπολέσθαι τὸν ἄνθρωπον· κατὰ τοῦτον τὸν λογισμὸν καὶ
ἡμεῖς αὐτοί, καίτοι μάλιστα πάντων ἐξιστάμενοι χρῆσθαι καρωτικοῖς φαρμάκοις, ἔσθ᾽ ὅτε καὶ κωλικοῖς
ἐδώκαμεν αὐτὰ καὶ ὀφθαλμοὺς ὀδυνωμένοις σφοδρότατα καὶ οὓς, ἕτερά τε μόρια κατὰ τὸν αὐτὸν τρόπον.
ἔσθ᾽ ὅτε καὶ διὰ κατάρρουν λεπτὸν ἀγρυπνοῦντός τε
καὶ βήττοντος ἰσχυρῶς τοῦ κάμνοντος ἐδώκαμεν ὀλίγον τι τοῦ τοιούτου φαρμάκου, ῥᾷστον εἶναι νομίζοντες ἐπανορθώσασθαι τὴν βλάβην τῷ χρόνῳ, εἴ τις
ἅπαξ χρήσαιτο καρωτικῷ φαρμάκῳ.

818K διαφέρει δὲ δήπου καὶ αὐτὰ τὰ φάρμακα | σφῶν
αὐτῶν. τὰ μὲν γὰρ πλεῖστον ἔχοντα τῶν καρούντων
εὐδοκιμεῖ μὲν ἐν τῷ παραυτίκα μᾶλλον, ἰσχυρὰν δὲ

Therefore, just like in all the states and actions in life as a whole, here too you must choose "nothing in excess,"[3] making the determining factor the benefit of the patient. If it is possible to achieve the desired result by using the remedies that are curative for the disease, you should keep away from the soporific medications which they call anodyne. If, however, due to wakefulness and destruction of the capacity, the patient faces the danger of death, you might, at that time, also use such medications appropriately, recognizing that you will harm the state of the body to some extent, but thinking it better to choose harm rather than death. If some harm which is not very great were to occur, we would also eradicate this in turn over a longer time. If it is so excessive as not to be completely cured even over a long time, I do think this is, at any rate, better than the man dying. By the same reasoning, although I abhor above all the use of soporific medications, there are times I have given them, both to those with colic and to those with very severe pains of the eyes and ears, and of other parts. There are also times when I have given a little of such a medication if the patient cannot sleep and has a violent cough due to a thin catarrh, thinking it easiest to correct the harm over time if someone were to use a soporific medication once only.

817K

Of course, the actual medications differ among themselves. Thus, those that are most soporific are more highly regarded in the short term but produce a coldness in the

818K

[3] The maxim attributed to Chilon; see Aristotle, *Rhetorica*, 1389b4.

καὶ δύσλυτον ἐναποτίθεται τὴν ψύξιν τῷ τοῦ κάμνοντος σώματι· τὰ δ᾽ ἔλαττον μὲν τούτων ἔχοντα, τῶν μιγνυμένων δὲ αὐτοῖς θερμαντικῶν πλεῖστον, ὅσον ἀπρακτότερα πρὸς τὸ παρὸν ὑπάρχει, τοσοῦτον ἀβλαβέστερα πρὸς τὸ μέλλον. ἅπαντα δ᾽ ἀμείνω τὰ τοιαῦτα μετ᾽ ἐνιαυτὸν τῆς συνθέσεως λαμβανόμενα, καθάπερ καὶ τὸ τοῦ Φίλωνος οὐδενὸς ἧττον ἔνδοξον ὀδύνας πραῦναι ναρκῶσαν τὴν αἴσθησιν. ἔστι δὲ καὶ ἄλλα πολλὰ φάρμακα διὰ τῶν σπερμάτων ὀνομαζόμενα καὶ τρίγωνα, μετριώτερα μὲν εἰς τὴν ἐν τῷ παραχρῆμα νάρκην, ἀκινδυνότερα δὲ εἰς τὸ μέλλον· ἅπαντα δ᾽ ἀμείνω καὶ ταῦτα μετ᾽ ἐνιαυτὸν τοῦ συντίθεσθαι λαμβανόμενα. τό γε μὴν τοῦ Φίλωνος εἰ καὶ δυοῖν ἢ καὶ τριῶν ἢ καὶ τεττάρων ἐτῶν εἴη, πολὺ δήπου βέλτιον· οὐκ ἄχρηστον δ᾽ οὐδ᾽ ἐν τοῖς ἐφεξῆς ἔτεσιν ἄχρι που ἐτῶν δέκα. τὸ γὰρ ἐπὶ πλέον κεχρονισμένον ἀβλαβέστερον μὲν εἰς τοσοῦτον εἰς ὅσον

819K χρονιώτερον · | ἤδη δ᾽ ἐξίτηλόν ἐστι τὴν δύναμιν, ὥστε ἀσθενῶς ἀνύειν οὗ χάριν ἐσκευάσθη. ὥσπερ οὖν χρὴ φυλάττεσθαι καθ᾽ ὅσον οἷόν τε τὴν τῶν τοιούτων φαρμάκων χρῆσιν, οὕτως ἀναγκασθέντας χρήσασθαι, μήτε νέα ἱκανῶς προσφέρειν ἰσχυρὰν ἔτι δύναμιν ἐχόντων τῶν ἐν αὐτοῖς ψυκτικῶν, μήθ᾽ οὕτως παλαιὰ ὡς ἁμαρτεῖν τοῦ σκοποῦ· ἀλλ᾽ ὡς εἴρηται, τοσοῦτον ἀναμένειν χρόνον ἐν ὅσῳ κολασθήσεται τὸ βίαιον τῶν ψυκτικῶν ὑπὸ τῆς τῶν θερμαινόντων δυνάμεως. ἡ μὲν δὴ τῶν τοιούτων φαρμάκων χρῆσις ὡς πρὸς σύμπτωμά ἐστιν ἱσταμένων καὶ μάλισθ᾽ ὅταν

patient's body that is severe and difficult to resolve. On the other hand, those that have less of these things but have the most calefacients mixed with them, are more ineffectual for the present purpose to the extent that they are less harmful for the future. All such medications are better when taken after they have been compounded for a year. The same applies to the medication of Philo which has no peer for soothing pains because it dulls sensation.[4] There are also many other medications named "made from seeds" (*diaspermaton*) and *trigona* which are more moderate in terms of narcosis in the immediate situation and less dangerous in terms of the future. All these too are better when taken after they have been compounded for a year. In fact, the medication of Philon would, I presume, also be much better if it were two, three, or four years old, and is not without effect in subsequent years, anywhere up to ten years. That which is even older is less harmful the older it is. It is, however, already declining in terms of potency so that it is [only] weakly effective for the purpose for which it was prepared. Therefore, just as it is necessary to guard against the use of such medications as far as possible, so too, if you are compelled to use them, do not apply them when they are new enough to still have a strong cooling in them, nor when they are so old as to fail to achieve their objective. But, as I said, await such a time that the strength of the cooling features will be mitigated by the potency of those features that are heating. Certainly, the use of such medications is characteristic of those who are taking a stand against symptoms, and particularly whenever the

819K

---

[4] Philo of Tarsus (AD 10–35). For a brief description of his medication and other references to this, see *EANS*, pp. 657–58.

ὑπὸ ψυχρᾶς αἰτίας ἡ ὀδύνη συνίστηται· βλάπτει γὰρ
ἰσχυρῶς τὴν διάθεσιν ὅλην ταῦτα καὶ δύσλυτον ἀπεργάζεται.

2. Φλεβοτομία δέ γε ἡ μὴ καταλύουσα τὴν δύναμιν
οὐ πρὸς τὸ σύμπτωμα βλεπόντων ἐστί, ἀλλὰ τὴν
διάθεσιν ὅλην ἐκκοπτόντων. οὕτω δὲ καὶ λουτρὸν καὶ
κάθαρσις καὶ πυρία καὶ οἴνου πόσις. ὧν ἁπάντων
Ἱπποκράτης ἐπὶ ὀφθαλμῶν ὀδυνωμένων ἐμνημόνευ-
σεν, ἐν Ἀφορισμοῖς λέγων ὧδε· Ὀδύνας ὀφθαλμῶν
820K ἀκρατοποσίη ἢ λουτρὸν ἢ | πυρίη ἢ φλεβοτομίη ἢ
φαρμακίη λύει. ἅπαντα γὰρ ταῦτα τὰς διαθέσεις ἰώ-
μενα συνεξιᾶται τὰς ὀδύνας αὐτῶν. οὕτω δὲ καὶ τροφή, καθάπερ καὶ ἐν τοῖς ἔμπροσθεν εἶπον, ξηρῷ[6] καὶ
αὐχμώδει σώματι[7] κάμνοντος στομάχου διδομένη τὴν
διάθεσιν ἰωμένη κωλύει συγκόπτεσθαι. οὐ μὴν ἥ γε
διὰ πλῆθος ψυχρῶν ἤτοι γ᾿ ἔκλυσις ἢ στομαχικὴ
συγκοπὴ τῆς τροφῆς ὡς ἰάματος δεῖται· καίτοι πολλάκις ἐν ταῖς τοιαύταις διαθέσεσιν οὐ τροφὴν μόνον,
ἀλλὰ καὶ οἴνου τι δοῦναι τοῖς κάμνουσιν ἀναγκαζόμεθα, τὸν ἐν τῷ παραχρῆμα κίνδυνον ἀποτρέποντες,
οὐ τὴν διάθεσιν ἰώμενοι. κένωσις γὰρ μᾶλλον ἢ πρόσθεσις ἡ τῶν τοιούτων ἐστὶν ἴασις· καὶ πλείστους ἰδεῖν
ἐστι τῶν οὕτως ἐχόντων ὁσημέραι διαφθειρομένους
ἀμαθίᾳ τῶν ἰατρῶν. ὑπὲρ ὧν μοι δοκεῖ καιρὸς εἶναι
τὸν λόγον ἐπιστήσαντα διελθεῖν.

3. Ἄρχονται τοίνυν πυρέττειν ἔνιοι, πάμπολυ πλῆ-

---

[6] K; εἴπομεν, ἐν ξηρῷ B   [7] K; νοσήματι B

pain arises due to a cold cause, because these pains harm the whole condition substantially and make it difficult to resolve.

2. Phlebotomy, since it does not dissipate the capacity, is not among those [treatments] directed against a symptom, but is one of those which eradicates the whole condition. The same applies to bathing, purging, the external application of heat and drinking wine. Hippocrates made mention of all these measures for pains in the eyes when, in the *Aphorisms*, he said this: "Drinking neat wine, bathing, the external application of heat, phlebotomy, or the use of drugs resolve pains of the eyes."[5] For all these things that cure the conditions jointly heal the pains that accompany them. In the same way too, as I said before, nourishment given to a dry and parched body of an affected cardiac orifice, having cured the condition, prevents the occurrence of syncope, for neither faintness nor gastric syncope arising from an abundance of cold humors requires nourishment as a cure. Nevertheless, in such conditions, we are often compelled to give patients not only nourishment but also some wine, not to cure the condition but to turn aside the immediate danger. Evacuation rather than administration of food is the cure of such conditions. It is possible to see very many of those patients, when they are in such a state, dying every day due to the ignorance of their doctors. It seems to me to be time now to go over these things, having initiated the discussion.

3. Thus some people begin to be febrile when they have

820K

---

[5] *Aphorisms* VII.46, *Hippocrates*, LCL, vol. 4, pp. 202–3. The text is somewhat different; see also Jones' note 10, p. 202.

θος ὠμῶν χυμῶν ἠθροικότες, ἅμα τῷ κεκακῶσθαι τὸ
στόμα τῆς γαστρός· ὃ δὴ καὶ στόμαχον ὀνομάζουσιν,
821K ἤτοι | γ᾽ ἐξ ἀπεψιῶν ἢ καὶ ἄλλως πως ἐμπεφύσηται
τούτοις τὸ ὑποχόνδριον, ἐν ὄγκῳ τε μείζονι τοῦ κατὰ
φύσιν ὅλον τὸ σῶμά ἐστι· καὶ ἡ χροιὰ τοῖς μὲν ἐπὶ τὸ
λευκότερόν τε καὶ ὑδαλεώτερον ἐκτέτραπται τοῦ κατὰ
φύσιν, ἔστι δὲ οἷς ἐπὶ τὸ μελάντερον ἢ πελιδνότερον,
οὓς ἔνιοι τῶν ἰατρῶν ὀνομάζουσι μολιβδοχρῶτας. οἱ
σφυγμοὶ δὲ ἁπάντων μικρότεροί τέ εἰσιν ἢ κατὰ
λόγον τῆς θέρμης, ἀμυδροὶ δὲ καὶ πάντως ἀνώμαλοι,
πολλάκις μὲν κατὰ τὴν συστηματικὴν ὀνομαζομένην
ἀνωμαλίαν, ἀεὶ δὲ τὴν κατὰ μίαν πληγήν· ἐπ᾽ οὐδενὸς
τῶν τοιούτων αἵματος ἀφαίρεσις ἄνευ μεγίστης εἴωθε
γίγνεσθαι βλάβης, καίτοι δέονταί γε κενώσεως· ἀλλ᾽
οὔτε φλεβοτομίαν οὔτε κάθαρσιν φέρουσιν, εἴ γε καὶ
χωρὶς τούτων ἐξαίφνης συγκόπτονται.

πῶς οὖν χρὴ τοὺς τοιούτους ἰᾶσθαι, δεομένους μὲν
κενώσεως, μὴ φέροντας δὲ τὰ κενωτικὰ βοηθήματα,
οὐδεμίαν ἄλλην εὗρον ἐπ᾽ αὐτῶν κένωσιν πλὴν τὴν
τῆς διατρίψεως. ἄρχεσθαι δ᾽ εὐθέως ἐν ἀρχῇ τῆς
νόσου προσήκει, τὸ μὲν πρῶτον ἀπὸ τῶν σκελῶν
ἄνωθεν κάτω διὰ σινδόνων μὴ πάνυ μαλακῶν, ἀλλά τι
822K μέτριον ἐχουσῶν τραχύ· χρὴ γὰρ | ἀμύττεσθαι πρὸς
αὐτῶν τὸ δέρμα, θερμαίνεσθαί τε ὁμοῦ αὐτὸ καὶ διαφορεῖσθαι δεομένου τοῦ κάμνοντος. ἐφεξῆς δὲ καὶ
ὅλας τὰς χεῖρας ἀπὸ τῶν ὤμων ἄνωθεν κάτω κατὰ τὸν
αὐτὸν τρόπον τρίβειν. ἐπειδὰν δὲ ἱκανῶς ἅπαντα τὰ
κῶλα φαίνηται θερμὰ καὶ φόβος ᾖ κοπώδη τινὰ

accumulated crude humors in great abundance at the same time as the the the opening of the stomach (which they also term *stomachos*) has been damaged, either from the humors being undigested or in some other way. In these [patients] the hypochondrium is inflated and the whole body is swollen to an abnormal degree. Also, in some the color is changed to become paler and more "dropsical" than normal, while in others, whom some doctors call "leaden colored," it is darker and more livid. In all cases, the pulse is smaller than would be expected on the basis of the heat, and is faint and altogether irregular, often in relation to what is termed a systematic irregularity, and is always of a single beat. In none of these patients is it customary for the withdrawal of blood to occur without the greatest harm, and yet they do in fact require purging. But they tolerate neither phlebotomy nor purging, even if they are liable to sudden fainting without these measures.

How, then, should we cure those patients who require evacuation but do not tolerate the evacuating remedies? I have found no other evacuation for them apart from that of rubbing. It is appropriate to start this immediately from the onset of the disease First [rub] the legs from above downward with linen cloths that are not very soft but have a moderate roughness, as it is necessary for the skin to be "scraped" by these, since it needs to be heated and, at the same time, made to perspire. Next, rub the arms in their entirety from the shoulders above downward in the same way. When all the limbs seem sufficiently hot and there is fear of some painful feeling arising in them because the

821K

822K

αἴσθησιν αὐτοῖς γενέσθαι, τεθλασμένης τῆς σαρκὸς ἐπὶ τῷ πλήθει τῆς τρίψεως ἐλαίῳ χαλαστικῷ χρηστέον, οἷόν ἐστι τὸ Σαβῖνον. ἀπέχεσθαι δὲ τῶν στυφόντων Ἰσπανοῦ καὶ Ἰστρικοῦ καὶ τοῦ μετὰ θαλλῶν ἢ ἐξ ὠμῆς ἔτι τῆς ἐλαίας ἐσκευασμένου. χειμῶνος δ' ὄντος οὐδὲν ἂν εἴη χεῖρον ἐπί τι τῶν διαφορητικῶν ἰέναι, οἷόν πέρ ἐστι τὸ Σικυώνιον ἔλαιον ἢ τὸ διὰ τοῦ χαμαιμήλου ἢ τὸ παρὰ τοῖς Αἰγυπτίοις εὐδοκιμοῦν.

εἰ δὲ μὴ παρείη ταῦτα, τῶν ἐλαίων τι τῶν χαλαστικῶν, οἷς οὐδεμία σύνεστί τις στύψις καὶ μάλιστα εἰ παλαιότερον εἴη λαβόντας ἐμβαλεῖν αὐτῷ σύμμετρον ἀνήθου. κἄπειτα, ὡς εἰώθαμεν, ἐπ' ἀγγείου διπλοῦ θερμήναντάς τε, ὥστε δέξασθαι μετρίως αὐτὸ τὴν ποιότητα τῆς βοτάνης, ἀλείφειν τε τούτῳ, τρίβοντας 823K καὶ νῦν | ἐπὶ πλεῖστον τὰ κῶλα· μετὰ δὲ ταῦτα τὸ μὲν ἔλαιον ἀπομάττειν, ἀσηρὸν γάρ, ἰέναι δὲ ἐπὶ τὴν ῥάχιν ὅλην καὶ ταύτην ἀνατρίβειν ὁμοίως, τὰ μὲν πρῶτα ξηρᾷ τρίψει, τῷ λίπει δ' ὕστερον. εἶτ' αὖθις ἐπὶ τὰ σκέλη μετιέναι, κἀκ τούτων αὖθις ἐπὶ τὰς χεῖρας, εἶτ' αὖθις ἐπὶ τὴν ῥάχιν, ὅλης τῆς ἡμέρας οὕτω πράττοντας ἐν οἴκῳ φῶς ἔχοντι καθαρὸν καὶ ἄνικμον· ἔστω δὲ δηλονότι καὶ τῇ θερμότητι σύμμετρος ὁ οἶκος.

ἐπιτηδειότατον δὲ τούτοις ἔστω τὸ μελίκρατον, ἐναφεψηθέντος ὑσσώπου. καὶ χρὴ μήτε σιτίον αὐτοῖς μήτε ῥόφημα μήθ' ὕδωρ διδόναι μήτ' ἐπιτρέπειν ὅλως πίνειν δαψιλῶς, ἀλλὰ τῷ μελικράτῳ μόνῳ χρῆσθαι κατὰ τὰς τρεῖς τὰς πρώτας ἡμέρας, ἐκ διαδοχῆς τρίβοντα καὶ μόνον ἐκεῖνον ἀνιέντα τὸν καιρόν, ἐν ᾧπερ

flesh has been bruised by the amount of rubbing, you must use a relaxing oil like the Sabine. Stay away from oils that are astringent like the Spanish and the Istrian,[6] and that which is prepared from young shoots, or from oil that is still raw. However, in winter, it would be no worse to go to one of the discutient oils such as the Sicyonian, or that made from chamomile, or that held in high repute among the Egyptians.

If these are not available, taking one of the relaxing oils in which there is no astringency, and particularly if you choose one that is older, throw into it a suitable amount of dill. Then, as I am accustomed to do, having heated it in a double vessel so that it receives the quality of the herb to a moderate degree, anoint with it, now also rubbing the limbs as much as possible. After this, wipe away the oil because it causes discomfort. Then proceed to the whole spine and rub this in the same way, first with a dry massage and subsequently with oil. Then, in turn, move to the legs, and from these back again to the arms, and then again to the spine, following this procedure for the whole day in a house that has pure light and is without moisture. And let the house also be quite clearly moderate in terms of heat.

Regard melikraton as the most useful [medication] for these [patients] after hyssop has been boiled down in it. You should give them neither food, gruel nor water, and you should not allow them to drink copiously in any way, but use melikraton alone during the first three days, rubbing [the parts mentioned] in turn, avoiding only that time

823K

[6] A region on the shore of the Adriatic Sea bordering on Illyria.

ἂν ὕπνος αὐτοὺς καταλάβῃ· συνεργεῖ δ᾽ εἰς τοῦτο
μάλιστα καὶ ἡ πλείων τρίψις· ὥστε καὶ κατὰ τοῦτ᾽ ἂν
εἴη βοήθημα χρηστὸν οὐ μόνον τὸ κενοῦν, ἀλλὰ καὶ
τὸν ὕπνον ἐμποιοῦν. οὐ μὴν οὐδ᾽ ὁ πλείων ὕπνος
ἀγαθὸς τοῖς τοιούτοις· οὔτε γὰρ ἐπιτρέπει διαφορῆσαι
τοὺς ὠμοὺς χυμοὺς καὶ βαρύνει τὰ σπλάγχνα· ἀλλ᾽
824K εἴπερ τις ἄλλη διάθεσις ἑκατέρων | δεῖται μετρίων,
ὕπνου τε καὶ ἐγρηγόρσεως, οὕτω καὶ ἥδε. πέπτει μὲν
γὰρ ὁ ὕπνος, διαφορεῖ δὲ ἡ ἐγρήγορσις· ἀμφοτέρων δ᾽
ἐστὶ χρεία τοῖς ἐκ πλήθους ὠμῶν χυμῶν νοσοῦσιν· ἐν
μέρει τοίνυν μετρίως γινόμενα τιμωρεῖν ἀλλήλοις δυ-
νήσονται χρηστῶς.

εἰ μὲν γὰρ ὁ σφυγμὸς αὐτοῖς μικρὸς ἱκανῶς εἴη καὶ
ἄρρωστος, ἢ καὶ πρὸς τούτοις ἀνώμαλος ἰσχυρῶς,
ἔσχατος ὁ κίνδυνος, καὶ χρὴ πράττειν τὰ λελεγμένα
μηδὲν ἄλλο περιεργαζόμενον. εἰ δὲ καὶ τόνου μετρίως
ἔχοι καὶ μεγέθους, ἄθλιπτός τε εἴη καὶ ὁμαλός, ἐπισκέπτου τὰ κατὰ γαστέρα καὶ κλύζε θαρρῶν, εἰ μὴ
καλῶς αὐτομάτως διεξέρχοιτο. συμβαίνει γὰρ ἐπὶ τῷ
μελικράτῳ κενοῦσθαι ταύτῃ χρηστῶς ἅπασαν τὴν
περιουσίαν τὴν ἐν ταῖς πρώταις τῶν φλεβῶν εἰωθυίαν
ἀθροίζεσθαι ταῖς καθ᾽ ἧπάρ τε καὶ μεσεντέριον. εἰ δὲ
πλείων τοῦ δέοντος ἡ ὁρμὴ τῶν περιττῶν ἐπὶ τὴν
γαστέρα γένοιτο, τὴν μὲν πρώτην ἐπὶ πλέον ἕψειν τὸ
μελίκρατον· ἧττον γὰρ ὑπάγον τὸ τοιοῦτον τρέφει
μᾶλλον· ὥσπερ γε τὸ ὠμὸν ἧττον μὲν τρέφει, μᾶλλον
δὲ ὑπάγει. μετὰ δὲ ταῦτα φερομένης ἐπὶ πλέον κάτω |

in which sleep descends upon them. More rubbing also helps, particularly with regard to this (i.e. sleeping) so that also, for this purpose, evacuation would not only be a useful remedy but also one that brings about sleep. Nevertheless, too much sleep is not good for such people as it does not allow the crude humors to be dispersed, and it weighs down the internal organs. But even if there is some other condition that requires moderate amounts of each, [that is] of sleep and wakefulness, this one also certainly does because sleep digests while wakefulness disperses, and there is a use for both to those who are ill from an abundance of crude humors. Accordingly, when sleep and wakefulness occur moderately in turn, they will be able to bring useful relief for each other.

824K

If the pulse in these patients is exceedingly small and weak, or if, in addition to these features, it is also strongly irregular, the danger is extreme and it is necessary to do those things that have been mentioned without wasting effort on anything else. But if the pulse is moderately strong and large, and is not oppressed, and is even, turn your attention to those things in the stomach and, if they don't pass through it spontaneously and well, wash the latter out with confidence. What happens after melikraton is that all the surplus which usually gathers in the first of the veins—that is, those of the liver and mesentery—is properly evacuated in this way. If, however, the impulse of the superfluities toward the stomach occurs more than is required, first boil the melikraton further, for the less it draws off the more it nourishes, just as what is undigested nourishes less but draws off even more. After this, if the surplus is car-

825K τῆς περιουσίας μηδ' οὕτως μὲν ἵστασθαι, διδόναι δ'
ἀντὶ τοῦ μελικράτου πτισάνης χυλόν.

εἰ δ' ἐπιμένει φερόμενα, τῷ ἐκ τοῦ χόνδρου ῥοφήματι τρέφειν, παρακολουθοῦντα δηλονότι τοῖς σφυ-
γμοῖς· ἔσθ' ὅτε γὰρ ἐξαίφνης μεταβάλλουσιν εἰς
ἀρρωστίαν ἢ ἀνωμαλίαν ἢ μικρότητα. καθ' ὃν καιρὸν
ἄρτον ἐξ οἴνου κεκραμένου διδόναι προσήκει μήτε
γαστρὸς δηλονότι μήθ' ἥπατος φλεγμαινόντων· ὡς εἴ
γε φλεγμαίνοιεν, ὠμῶν χυμῶν πεπληρωμένου τοῦ σώματος ἀνέλπιστος ὁ κάμνων ἐστί, τῶν σφυγμῶν οὕτω
τραπέντων. ἐπὶ μὲν δὴ τῶν τοιούτων ἀρρώστων προλέ-
γων τὸν θάνατον, ἀνέγκλητον φυλάξαις σεαυτόν, μηδενὶ βοηθήματι γενναίῳ χρώμενος. ἐφ' ὧν δ' ἐλπὶς
σωτηρίας, ἐπὶ πάντων δ' ἐστὶ τῶν χωρὶς φλεγμονῆς
οὕτω νοσούντων, ὅταν ἐξ ἀρχῆς παραλαμβάνωμεν
αὐτούς, αἰσχρότατον εἶναί μοι δοκεῖ συγκοπῆναι τὸν
κάμνοντα· καίτοι γ' ὁσημέραι γίγνεται τοῦτο, διὰ τὴν
ἀμαθίαν τῶν ἰατρῶν. ἀλλ' οὐ πρὸς τούτους χρὴ βλέπειν, οὐδὲ τούτοις παραβάλλειν ἑαυτόν, ἀλλ' ὅστις ἂν
ὄντως ᾖ τέχνης ἰατρικῆς ἐπιστήμων αἰσχρὸν εἶναι
826K νομίσει[8] | ἐξαίφνης ἄρρωστον συγκοπῆναι μήτε προ-
ειπόντος αὐτοῦ τὸ σύμπτωμα μήτε παρεσκευασμένου
πρὸς αὐτὸ μήτε κωλύσαντος. εἴρηται δέ μοι καὶ
πρόσθεν ὡς ἐν ταῖς τοιαύταις διαθέσεσιν αὐτοὶ κατα-
σκευάζουσι τὰς φλεγμονὰς οἱ διαπαντὸς τοῖς ὑποχονδρίοις λωβούμενοι διὰ καταπλασμάτων τε καὶ καταντλήσεων.

οὔκουν χρηστέον αὐτοῖς πρὶν ἢ τὸ πολὺ τοῦ πλή-

ried downward still more, don't stand in the way of this   825K
but give juice of ptisan instead of melikraton.

If, however, those things flowing continue, nourish with
porridge made from wheat, following the pulse closely of
course, because sometimes it suddenly changes to become
weak, irregular and small, at which time it is appropriate to
give bread soaked in wine provided, that is to say, neither
the stomach nor the liver is inflamed. If they are in fact
inflamed, when the body has been filled with crude humors, the patient is without hope, the pulse having been
altered as described. Certainly, in the case of such patients, if you predict death, you yourself will remain
blameless if you use no notable remedy. In those in whom
there is hope of safety, which is all those diseased in this
way without inflammation whenever we attend them from
the beginning, it seems to me to be most shameful for the
patient to suffer syncope. And yet this happens every day
due to the ignorance of doctors. But we should pay no attention to these doctors, nor should we compare ourselves
to them. Rather, anyone who is truly knowledgeable in the
medical art will think it disgraceful for the sick person to   826K
suddenly suffer syncope without he himself having pre-
dicted the symptom, or prepared [the patient] for it, or
prevented it. I said before that in such conditions, those
who continually irritate the hypochondrium with poultices
and fomentations themselves establish the inflammations.

You must not, therefore, use these [measures] before

---

8 K; νομίζειν B.

247

θους τὸ μὲν ἐκκενῶσαι, τὸ δὲ πέψαι· τηνικαῦτα δὲ ἤδη
χρηστέον αὐτοῖς οὐχ ἁπλῶς χαλῶντας, ἀλλ᾽ αἰονῶν-
τας δι᾽ ἀψινθίου μετρίως, ἢν ὁ πυρετὸς ὑπάρχῃ μὴ
πάνυ μέγας. εἰ δ᾽, ὡς εἴρηται, θεραπεύοντί σοι τὰς
τρεῖς τὰς πρώτας ἡμέρας μηδὲν ἀπαντήσειε χεῖρον,
ἐπὶ τῶν αὐτῶν ἄγειν ἄχρι τῆς ἑβδόμης, μελικράτῳ
μόνῳ χρωμένους ὕσσωπον ἔχοντι· μακροτάτην γὰρ οἱ
οὕτω διακείμενοι φέρουσιν ἀσιτίαν, ἐξ αὐτῶν ὧν ἔχουσιν ὠμῶν χυμῶν πεττομένων τρεφομένου τοῦ σώματος. εἰ δὲ μὴ μόνον ὠμοὶ τύχοιεν ὄντες, ἀλλὰ καὶ
διαφθορᾶς εἰς τοσοῦτον ἥκοντες ὡς μὴ δύνασθαι
πεφθῆναι τελέως, ἀδύνατον σωθῆναι τοὺς οὕτω νοσοῦντας.

ἀλλὰ τούς γε σωθῆναι δυναμένους, εἰ μηδὲν |
827K ἁμαρτάνοιτο, μέχρι τῆς ἑβδόμης ἐπὶ μελικράτῳ μόνῳ
διαιτᾶν· ἢ εἴ ποτ᾽ ἄρα δεήσειεν ἤτοι τῆς γαστρὸς
ἐκκρινούσης ἢ ἀποστραφέντων αὐτῶν τὸ μελίκρατον,
ἐπὶ τὸν τῆς πτισάνης ἰέναι χυλόν. εἰ δὲ καὶ τοῦτον
ἀποστρέφοιντο, χόνδρον ὁμοίως ἀρτύειν πτισάνῃ· καὶ
γὰρ οὐδὲ βλάπτει τοὺς οὕτω διακειμένους ὄξος· ἀλλ᾽
ἢν αἴσθοιό ποτε παχεῖς ἱκανῶς εἶναι τοὺς ὠμοὺς
χυμοὺς ὀξύμελι διδόναι διὰ παντός, ἀντὶ τοῦ μελικράτου. εἰ δ᾽ ἄχθοιντο τῷ συνεχεῖ τῆς δόσεως, ἐξαλ-
λάττειν μὲν ἐπὶ μελίκρατον καὶ πτισάνην· ἐπανέρ-
χεσθαι δ᾽ αὖθις ἐπ᾽ αὐτό· μάλιστα δὲ ἐπὶ τῶν μικρὸν
ἐχόντων καὶ ἀραιὸν καὶ βραδὺν τὸν σφυγμὸν ἡγεῖ-
σθαι ψυχροὺς εἶναι τοὺς χυμούς. εὐθὺς δὲ τούτοις
σύνεστι καὶ ἡ τοῦ παντὸς σώματος ἄχροια σαφῶς,

the greater part of the surplus has either been evacuated or digested. However, when these circumstances already obtain, you must use them, although if the fever is not very great, do not simply use those that are relaxing but those that are moderately moistening by means of absinthe. If, as I said, nothing worse should happen during the first three days of your treatment, continue with the same measures as far as the seventh day, using only melikraton which has hyssop in it, as patients in this state tolerate very long fasting because the body is nourished from the concoction of the crude humors which they themselves have. If, on the other hand, it should happen that the humors are not only crude but have also reached such a degree of corruption that they cannot be concocted completely, it is impossible for those who are diseased in this way to be saved.

But, with respect to those who can be saved if nothing goes wrong, feed them with melikraton alone up to the seventh day. Or if, at some time, the need arises, either if the stomach rejects [it] or if the patient is averse to melikraton, resort to the juice of ptisan. If they are also averse to this, prepare gruel in like manner to the ptisan, for truly vinegar does not harm those in such a state. But if you sense, at some time, that the crude humors are excessively thick, give oxymel throughout instead of melikraton. If, however, [the patient] is distressed by the continued administration [of the former], change to melikraton and ptisan and then return again to the oxymel, particularly in those who have a small, thin and slow pulse, considering the humors to be cold. In these [patients], there is also present, immediately and obviously, a decoloration of the

827K

οἷα περ πρόσθεν εἴρηται. δόξει δέ τισιν ἴσως ἀδύνατον εἶναι τὸ λεγόμενον οὐδενὶ τῶν πυρεσσόντων οὔτε βραδυτέρων οὔτε ἀραιοτέρων γινομένων τῶν σφυγμῶν.

ἀλλὰ κατὰ διττόν γε τρόπον ἀληθὲς ὑπάρχει τὸ εἰρημένον· ἕνα μὲν ἐπειδὴ σαφῶς ἐν τοῖς παροξυσμοῖς

828K καὶ βραδύτεροι καὶ ἀραιότεροι | γίνονται τοῦ κατὰ φύσιν· ἕτερον δέ, διότι καὶ κατὰ τοὺς ἄλλους καιροὺς τοῦ παροξυσμοῦ πάντων τῶν ὁπωσοῦν πυρεττόντων ἀραιότερόν τε καὶ βραδύτερον αἱ τούτων ἀρτηρίαι σφύζουσιν. εἰ μὲν γὰρ δὴ θέρος εἴη καὶ ψυχροπότης ὁ νοσῶν, ψυχρὸν τούτῳ διδόναι τὸ ὀξύμελι, χειμῶνος δ᾽ ὄντος ἅπασι θερμόν, ὡσαύτως ψυχροπόταις τε καὶ θερμοπόταις· ὥσπερ εἰ καὶ ἄμετρον καῦμα θέρους ὥρᾳ ψυχρὸν καὶ τοῖς θερμοπόταις, εἰ μηδὲν εὐπαθὲς ἔχοιεν σπλάγχνον. ἐναντιώτατα δὲ τούτοις ἐστὶ βαλανεῖα καὶ ὁ περιέχων ἀήρ, εἰ θερμὸς ἱκανῶς ἢ ψυχρός. ὅθεν οὔτε λούειν προσήκει καὶ κατακλίνειν ἐν οἴκῳ ὥρᾳ μὲν θέρους ψυχεινῷ, χειμῶνος δ᾽ ἀλεεινῷ. βλάπτονται γὰρ ἱκανῶς ὑφ᾽ ἑκατέρας τῆς ἀμετρίας ἐν μὲν τοῖς βαλανείοις, ὡς κἂν τῷ θέρει καὶ πάσῃ τῇ κατὰ τὸ θερμὸν αὐξήσει χεομένων τῶν ὠμῶν χυμῶν καὶ πάντῃ τοῦ σώματος ῥεόντων, ὡς κίνδυνον εἶναι καὶ πρὸς τὸν πνεύμονα καὶ τὴν καρδίαν ἀφικέσθαι καί ποτε καὶ εἰς τὸν ἐγκέφαλον ἀναδραμεῖν, ἄμεινον ὑπάρχον μακρῷ

829K περί τε τὸ ἧπαρ αὐτοῖς ἔτι | μένειν καὶ τὰς πρώτας φλέβας.

ἐν δὲ τοῖς ἀμέτρως ψυχροῖς οἴκοις καὶ χειμῶνι καὶ

whole body, such as was spoken of before. Perhaps what is being said will seem to some to be impossible, for in none of those who are febrile does the pulse become slower or thinner.

But what was said is, in fact, true in two situations. One is in the paroxysms when the pulse clearly does become 828K slower and thinner than normal. The other is because, at the other times besides the paroxysm, the arteries of all those who are febrile in any way whatsoever, pulsate more thinly and slowly. So then, if it is summer and the sick person is accustomed to cold drinks, give him the oxymel cold whereas, if it is winter, make the drink hot for everyone, both those accustomed to cold drinks and those accustomed to hot drinks. Similarly, if there is also excessive heat in summer, give it cold, even to those who are accustomed to hot drinks, if they have no internal organ that is easily affected. Most inimical to these measures are baths and the ambient air, if it is excessively hot or cold. On which account, it is not appropriate for patients to bathe or to recline in a cool house in summer, or in a warm one in winter, for they are harmed unduly by each extreme in the baths since, in the summer, and in every increase in heat, the crude humors liquefy and flow to every part of the body, so there is a danger of them reaching even the lungs and heart and sometimes also of passing up to the brain. It is better by far for them (i.e. the crude humors) to remain 829K near the liver and the primary veins.

In houses that are excessively cold, in winter, and in

GALEN

ὅλως ψυχρῷ τῷ περιέχοντι δύσπεπτοί τε μένουσιν οἱ
ὠμοὶ χυμοὶ καὶ τὰς κατὰ τὸ ἧπαρ ἐμφράξεις, εἰ μὲν
εἶεν μικραί,[9] παραύξουσιν· εἰ δ᾽ οὐκ εἶεν, γεννῶσι διὰ
πάχος ἰσχόμενοι καὶ σφηνούμενοι. ταῦτα μὲν οὖν
οὕτω πρακτέον, ὅταν ἀπὸ πρώτης ἡμέρας αὐτὸς ἄρχῃ
τῆς θεραπείας αὐτῶν. εἰ δὲ συγκοπτομένων ἤδη κλη-
θείης, ἀφλεγμάντων ὄντων ὧν εἶπον χωρίων, διδόναι
μὲν αὐτοῖς ἄρτου μὴ πολύ, δι᾽ οἴνου δὲ τῶν ἀνάδοσιν
ἐργαζομένων ταχεῖαν· εὐθέως δ᾽ ἐπὶ τὴν τρίψιν ἰέναι
καὶ χρῆσθαι κατὰ τὸν εἰρημένον ὀλίγον ἔμπροσθεν
τρόπον. εἰ μὲν οὖν εἴη θέρος καὶ τὸ χωρίον φύσει
θερμὸν καὶ πνιγῶδες ἢ ἡ κατάστασις ἱκανῶς θερμή,
μεθ᾽ ὕδατος ψυχροῦ δοτέον τὸν οἶνον· εἰ δὲ μηδὲν τῶν
τοιούτων εἴη, μετὰ θερμοῦ. τὸ μέντοι δεύτερον καὶ
τρίτον ἐκ παντὸς τρόπου πειρᾶσθαι θερμὸν διδόναι
ποτόν. εἰ γὰρ τὴν ὅλην θεραπείαν, ἧς ἕνεκα καὶ τὰς
τρίψεις παραλαμβάνομεν, ἄμεινον τὸ θερμὸν ταῖς
πέψεσι τῶν ὠμῶν χυμῶν συνεργοῦν. |

830K   4. Ἐφ᾽ ὧν δὲ διὰ χολὴν ξανθὴν ἀδικήσασαν τὸ
στόμα τῆς γαστρὸς ἡ συγκοπὴ γένοιτο, ψυχρὸν τούτοις χρὴ προσφέρειν τὸ ποτόν. οἶνον μέντοι τῇ φύσει
θερμὸν εἰς ἀνάδοσιν ὁρμῶντα τοῖς συγκοπτομένοις
ἅπασι δοτέον· ἀναδοθῆναι γὰρ βουλόμεθα τὴν λη-
φθεῖσαν τροφήν, οὐκ ἐν τῇ γαστρὶ μεῖναι. δῆλον δὲ
ὅτι τοὺς οἴνους κιρροὺς μὲν τῇ χροιᾷ, λεπτοὺς δὲ τῇ
συστάσει, παλαιοὺς δὲ τὴν ἡλικίαν ἐκλεκτέον, εὐθὺς
δ᾽ ἂν εἶεν εὐώδεις οἱ τοιοίδε. καὶ μέντοι καὶ αὐτῷ σοι

252

general, in cold surroundings, the crude humors remain difficult to concoct, and they increase blockages in the liver if the veins are small. If, however, they are not, the crude humors create obstructions when they adhere and plug up due to their thickness. You must, therefore, do these things in this way whenever you yourself begin their treatment from the first day. If, however, you are called in when they are already suffering syncope, if the places which I mentioned are without inflammation, give the patient bread (not much), with one of the wines that make distribution swift, and proceed immediately to the rubbing, using this in the manner spoken of a little earlier. If it is summer and the place is naturally hot and stifling, or the climatic conditions are excessively hot, you must give the wine with cold water. If, however, none of these things obtain, [give it] with hot water. Nevertheless, attempt to give the second and third drink hot as far as possible because, in respect to the whole treatment, for the sake of which we also undertook the rubbings, heat is better to help the concoction of the crude humors.

4. In the case of those in whom syncope occurs due to yellow bile having harmed the cardiac orifice of the stomach, it is necessary to administer a cold drink. Of course, we must give wine that is hot in nature to all those affected with syncope, as it stimulates distribution, because we want the nourishment that has been taken to be distributed and not to remain in the stomach. It is clear that we must select wines that are tawny yellow in color, thin in consistency, and old in age—such wines would be naturally fragrant. Furthermore, it is also possible for you your-

830K

---

9 B (cf. leves KLat); μακραί K, Linacre, p. 611.

πάρεστι καταπιόντι βραχὺ κραθέντων αὐτῶν αἰσθάνεσθαι θερμασίας εἰς ὅλον ἰούσης τὸ σῶμα. μὴ μέντοι πικροί γε ἔστωσαν ὑπὸ παλαιότητος οἱ οἶνοι καὶ μάλιστα ἐφ᾽ ὧν ὁ τοῦ κάμνοντος στόμαχος ὑπὸ τῆς ξανθῆς ἀδικεῖται χολῆς. οὔτε γὰρ ἔτι τὸ τῶν οἴνων εὔχυμον καὶ τρόφιμον ἔχουσιν οἱ τοιοῦτοι καὶ ἀηδεῖς ὄντες αὐτῷ τούτῳ τὸν στόμαχον ἀνιῶσιν.

ἄριστοι τοίνυν εἰσὶν ὅσοι τῶν οἴνων αὐστηροὶ τὴν φύσιν ὄντες αἰσθητὴν μὲν οὐκέτι τὴν στύψιν ἔχουσι διὰ τὴν παλαιότητα, σαφῆ δ᾽ ἱκανῶς τὴν θερμασίαν. ἅπαντα γὰρ ὧν δεόμεθα ποιήσουσιν, ἐφ᾽ ὧν ὁ στό-
831K μαχός ἐστι | πικρόχολος, ἡδέως τε λαμβανόμενοι καὶ πέψει καὶ ἀναδόσει συνεργοῦντες καὶ τὴν κακίαν τοῦ χυμοῦ πραΰνοντες καὶ θερμὴν τὴν ἕξιν ἐργαζόμενοι καὶ ῥωννύντες τὸν στόμαχον. ἄριστοι δ᾽ εἰσὶ τῶν τοιούτων οἴνων Σουρρεντῖνος, Σιγνῖνος, Σαβῖνος, Τιβουρτῖνος, Μάρσος, Ἰταλιῶται πάντες οὗτοι, στύφοντες μέν, ἀλλ᾽ οὐχ ὁμοίως ἅπαντες. ὁ μὲν γὰρ Σουρρεντῖνος μετρίως τε στύφει καὶ θερμότερος τῶν ἄλλων καὶ ἡδίων ἐστί. ἐφεξῆς δ᾽ ὁ Τιβουρτῖνος ὁ εὐγενής· ἔστι γάρ τις καὶ ἄλλος Τιβουρτῖνος ἄτονος ὁμοίως Σαβίνῳ, βραχείας μετέχων στύψεως. ὁ δὲ Σιγνῖνος αὐστηρότερος· καὶ τούτου πολὺ μᾶλλον ὁ Μάρσος. πάντες οὗτοι χρηστοὶ πικροχόλῳ στομάχῳ. καὶ διὰ τοῦτ᾽ αὐτοῖς χρηστέον, ὅταν γ᾽ ἤτοι ὑπὸ καυσωδῶν πυρετῶν ἀδικηθεὶς ἐπιφέρῃ συγκοπήν, ἢ καὶ ἄλλως πολλῆς εἰς αὐτὸν ἤτοι ῥυείσης ἢ ἀναποθείσης χολῆς.

self, if you drink a little of these [wines] when they are mixed, to sense the warmth as it spreads to the whole body. Do not, however, let the wines be bitter due to great age, and particularly in those instances where the patient's cardiac orifice is adversely affected by yellow bile; such wines are no longer wholesome and nutritious, and being unpleasant, harm the cardiac orifice of the stomach for this very reason.

Therefore, the best wines are those that are naturally sharp and no longer have any perceptible astringency due to their age, but have a heat that is very clear, for they will do everything we require for those in whom the cardiac orifice of the stomach is picrocholic, since they are pleasant to take, facilitate digestion and distribution, mitigate the badness of the humor, bring about a state that is hot, and strengthen the cardiac orifice of the stomach. The best of such wines are the Sorrentine, Signine, Sabine, Tiburtine, and Marsyan; all these are Italian and astringent, but not all in like manner.[7] The Sorrentine is moderate in terms of astringency and more warming and more pleasant than the others. Next in order is the Tiburtine of high quality; for there is another Tiburtine which is weak like the Sabine and partakes of little astringency. The Signine is more harsh and the Marsyan much more than this. All these are useful for the cardiac orifice of the stomach that is picrocholic and, because of this, we must use them whenever someone is either adversely affected by a burning fever which brings syncope, or otherwise when there is much bile either flowing or attracted to the cardiac orifice of the

831K

---

[7] See Galen's *De sanitate tuenda*, Book 5, chapter 5, for a more detailed consideration of the various wines.

οἷς μέντοι δι᾽ ὠμοὺς χυμοὺς ὁ κίνδυνος τῆς συγκοπῆς, ὁ Φαλερῖνος ἀμείνων τῶν εἰρημένων εἰς τοσοῦτον εἰς ὅσον εὐχυμότερός τέ ἐστι καὶ θερμότερος· ἀναδοθή-
832K σεται γὰρ αὐτῶν θᾶττον καὶ τῇ πέψει | τῶν ὠμῶν χυμῶν συνεργήσει.

τῶν δ᾽ εἰρημένων ἕκαστος ὁ μὲν μᾶλλον, ὁ δ᾽ ἧττον, ἅπαντες δ᾽ οὖν Φαλερῖνοι τονοῦσι μᾶλλον τὸν στόμα-χον. ἐπεὶ δὲ καὶ αὐτῶν τῶν Φαλερίνων ὁ μέν τις ἱκανός ἐστι γλυκύς, ὃν ὀνομάζουσι Φαυστῖνον, ὁ δ᾽ ὡς μὲν πρὸς ἐκεῖνον αὐστηρός, ὡς δὲ πρὸς τὸν Σιγνῖνόν τε καὶ Μάρσον γλυκύς, ἐκλεκτέον εἰς τὰ τοιαῦτα τὸν δεύτερον. εἰ δ᾽ ὁ κάμνων εἴθιστο χρῆσθαι περὶ τὸν τῆς ὑγείας χρόνον οἴνοις γλυκέσιν, οὐδὲν κωλύει τούτῳ δοῦναι τὸν γλυκύτερον Φαλερῖνον. ἐπὶ μέντοι τῆς Ἀσίας καὶ τῆς Ἑλλάδος, ὅσα τε τούτοις ἔθνη γειτνιᾷ, μὴ παρόντων τῶν εἰρημένων οἴνων, ἔσθ᾽ ὅτε κάλλι-στος πρὸς ἀνάδοσιν τῶν ἐν ἐκείνοις τοῖς χωρίοις εὐπορουμένων οἴνων ὁ Ἀρούίσιός τε καὶ ὁ Λέσβιος· ὁ μὲν οὖν Λέσβιος εὔδηλός ἐστι κἀκ τῆς προσηγορίας ἐν Λέσβῳ γινόμενος, ὁ δ᾽ Ἀρούίσιος ἐν χωρίοις τισὶ τῆς Χίου γεωργεῖται. τριῶν δὲ οὐσῶν ἐν Λέσβῳ πό-λεων, ἧττον μὲν εὐώδης καὶ γλυκὺς ἐν Μιτυλήνῃ γεννᾶται, μᾶλλον δ᾽ εὐώδης καὶ γλυκὺς ἐν Ἐρεσσῷ, κἄπειτα ἐν Μηθύμνῃ. λαμβάνειν δ᾽ ἀπαραχύτους,
833K οὕτω δ᾽ ὀνομάζουσιν οἷς οὐ μέμικται | θάλασσα, μεγί-στην βλάβην ἡγουμένους ἐφ᾽ ὧν μέμικται γενήσε-σθαι. οὐ μὴν οὐδὲ εἰώθασι τοῖς εὐγενέσιν οἴνοις, ὑπὲρ ὧν ὁ λόγος ἐστί, μιγνύναι τῆς θαλάσσης ἐν Λέσβῳ,

stomach. However, in those in whom there is danger of syncope due to crude humors, the Falernian is better than those mentioned to the extent that it is more wholesome and hotter, and so will be distributed quicker than the others and will help with the concoction of the crude humors.   832K

However, each of the wines spoken of does strengthen the cardiac orifice of the stomach to a greater or lesser degree, although more so all the Falernian wines. And since, of the Falernian wines themselves, one is excessively sweet (the one they call Faustian), while another is harsh if you compare it with that, although it is sweet in comparison to the Signine and the Marsyan, you must choose the second for such matters. But if the patient had been accustomed to using sweet wines during a time of health, there is nothing to prevent you giving him the sweeter Falernian. Nevertheless, in the case of Asia and Greece, and those nations adjoining them, if the wines spoken of are not available, sometimes the best of the wines in abundance in these regions in respect to distribution are the Ariusian and Lesbian. Obviously from its name the Lesbian is produced in Lesbos while the Ariusian is cultivated in certain regions of Chios. Since there are three cities on Lesbos, [you should know that] the less fragrant and sweet [wine] is produced in Mitylene, the more fragrant and sweet [wine] in Eressos, and that in between in Methymna. Find those that are unmixed, for this is how those who think that the   833K
greatest harm will be generated from wines that have been mixed describe the wines with which seawater has not been mixed. In Lesbos they are not accustomed to mix seawater with the high-quality wines, about which the discus-

257

καθάπερ οὐδ' ἐν Χίῳ τῷ Ἀρουισίῳ. γεννᾶται δὲ καὶ κατὰ τὴν Ἑλλησποντιακὴν Μυσίαν οἶνος ὅμοιος τῷ Σουρρεντίνῳ, Μύσιος ὀνομαζόμενος κατ' ἐξοχήν.

οὗτοι πάντες οἱ οἶνοι κιρροὶ καὶ θερμοὶ καὶ εὐώδεις εἰσὶ καὶ τῇ συστάσει μέσοι τῶν θ' ὑδατωδῶν καὶ τῶν παχέων. ὑδατώδεις μὲν οὖν εἰσι τὴν σύστασιν ὅ τε Ἀδριανὸς[10] καὶ Σαβῖνος καὶ Ἀλβανὸς καὶ Γαυριανὸς καὶ Θοῦσκος, ὅ τε Νεαπολίτης ὁ Ἀμιναῖος, ἐν τοῖς περὶ Νεάπολιν χωρίοις γενόμενος· ὥστε καὶ ὀνομάζουσιν οὕτως αὐτόν. ἐπὶ δὲ τῆς Ἀσίας ὅ τε Τιβηκῖνος καὶ ὁ Τιτακαζηνὸς καὶ τρίτος μετ' αὐτοὺς ὁ Ἀρσυϊνός. ἔμπαλιν δὲ παχεῖς οἶνοι, καθάπερ τὸ σίραιον, ὃ παρ' ἡμῖν ὀνομάζουσιν ἕψημα, ὁ Σκυβελλίτης καὶ ὁ Θηραῖος καὶ Ἀλβάτης· ἐπὶ δὲ τῆς Ἀσίας Αἰγεάτης τε καὶ Περπερῖνος.[11] ἐφ' ὧν οὖν διὰ πλῆθος ὠμῶν χυμῶν ἡ συγκοπὴ γίγνεται, τοὺς μὲν παχεῖς φευκτέον ὡς βλά-

834K πτοντας, τοὺς δὲ ὑδατώδεις ὡς | ἀπράκτους· αἱρετέον δὲ τοὺς μέσους αὐτῶν κιρρούς, ὡς εἴρηται, καὶ θερμοὺς ὄντας. οὐδὲ γὰρ ἂν εὕροις οὐδένα τῶν λευκῶν οἴνων θερμόν, ὅπου γε καὶ οἱ αὐστηροὶ καὶ μετρίως λευκοὶ παλαιούμενοι κιρρότεροί πως γίνονται. εἰ δ' ἄλλως ἐθέλεις ὀνομάζειν τὸ κιρρὸν χρῶμα, δύναιο ἂν λέγειν πυρρὸν ἢ ὠχρόν. ὅσοι δ' ἐν αὐτοῖς εἰσι θερμότατοι, ξανθοὶ πάντες εἰσίν· ὁποῖος καὶ ὁ Καίκουβος ἐπὶ τῆς Ἰταλίας. ὃς οὐχ ἕν τι γένος ἐστὶν οἴνου τοιούτου ἐξ ἀρχῆς, ὡς ἔνιοι νομίζουσιν, ἀλλ' ὑπὸ παλαιότητος εἰς τοῦθ' ἥκων, ὡς πυρρὰν ἔχειν χρόαν, ὅθεν περ καὶ τοὔνομα αὐτῷ· ἥκουσι δ' οὐ πάντες εἰς

sion is, just as they are not in Chios with the Ariusian. Also, in Hellespontian Mysia, there is a wine produced like the Sorrentine that is called Mysian by virtue of its excellence.

All these wines are tawny yellow, warm and fragrant, and midway in consistency between those that are watery and those that are thick. The wines that are watery in consistency are the Adrian, the Sabine, the Alban, the Gaurian, the Tuscan, and the Neapolitan Aminean produced in those regions around Naples, so that they also name it accordingly. In Asia, there are the Tibecine, the Titacazene and third, after them, the Arsynian. Conversely, thick wines, like new wine boiled down—which our folk call *hepsema*—are the Scybelline, the Theraean and the Albate. In the province of Asia there are the Aegeate and the Perperine. In those [patients] in whom syncope occurs due to an abundance of crude humors, you must avoid the thick wines as harmful and the watery wines as ineffectual. You must choose those wines that lie 834K between these [extremes], which are, as I said, [the wines that are] tawny yellow and hot. You would find none of the white wines hot since both the harsh and the moderately white as they age somehow become more tawny yellow. If you wish to name the tawny yellow color in another way, you could speak of yellowish red or yellow. Among these, those that are hottest are all yellow. In Italy the Caecubine is a wine of this kind. This is not one sort of such wine from the beginning, as some think, but has come to this by being old, so as to have a yellowish red color, from which perhaps the name for it also comes. However, not all these wines

---

10 K; Ἀρσήνιος B.
11 K; Περπερηνός B

τοῦτο φθάνοντες παχύνεσθαι, πρὶν ἐν τῷ διαυγεῖ τὸ
ξανθὸν λαβεῖν χρῶμα. τῶν δ' ἄλλων οἴνων τῶν αὐστηρῶν ὅσοι μετρίως λευκοὶ καὶ παχεῖς, οὐδεὶς ἐπιτήδειος εἰς ἀνάδοσιν· εἰ μέντοι παλαιωθεῖεν ἱκανῶς,
χρήσαιτ' ἄν τις αὐτοῖς ἑτέρων μὴ παρόντων· εὐστό-
μαχοι δ' εἰσὶν οἱ τοιοῦτοι πάντες παλαιούμενοι.

ὅπως δὲ καὶ τούτων ἔχοις τι παράδειγμα, τοῦ τ'
ἀπὸ τῆς Νικομηδίας ἀναμιμνήσκω σε πᾶσιν ἀνθρώ-
835K πoις γνωρίμου | καὶ τοῦ Σικελοῦ τοῦ Ἀμιναίου τοῦ ἐν
τοῖς μεγάλοις κεραμείοις· ὁ γὰρ ἐν τοῖς μικροῖς λα-
γυννίοις ἐναντιώτατος τῷδε, κακοστόμαχος ἅμα καὶ
κεφαλαλγὴς ὑπάρχων. πειρᾶσθαι μὲν οὖν ἀεὶ τὸν
ἄριστον αἱρεῖσθαι· μὴ παρόντος δὲ τὸν ὁμοιότατον
ἐκείνῳ κατὰ τὰ λελεγμένα γνωρίσματα. κάλλιστοι
γοῦν[12] εἰς ἀνάδοσιν ὅ τ' Ἀριούσιός[13] ἐστι καὶ ὁ
Λέσβιος· ἀλλ' εἰ μὴ παρεῖεν, ἐν Ἀσίᾳ δὲ εἴημεν, ἤτοι
τὸν Ἀφροδισιαῖον ληπτέον ἢ τὸν αὐστηρὸν Τμωλίτην
ἢ τινα τῶν ὁμοίων αὐτοῖς, κιρρὸν τῇ χρόᾳ καὶ θερμαίνοντα σαφῶς, εἰ ποθείη παραχρῆμα. ἐπισκέπτεσθαι
δὲ χρὴ πρὸ πάντων ὁποία τις ἡ τοῦ κάμνοντός ἐστι
φύσις. εἰ μὲν γὰρ ἀσθενὴς τὴν κεφαλὴν εἴη καὶ
ῥᾳδίως ὑπὸ τῶν θερμῶν οἴνων πληρούμενος, ἐπὶ τοὺς
ἧττον θερμοὺς ἔρχεσθαι μᾶλλον χρή· εἰ δ' ἰσχυρὸς
τὴν κεφαλὴν ᾖ καὶ χαίρων οἴνοις τοῖς τοιούτοις ἢ
εἰθισμένος ἀλύπως φέρειν, χρηστέον αὐτοῖς θαρρούν-
τως. κεφαλαλγεῖς δέ εἰσιν ὅσον ἐφ' ἑαυτοῖς οἶνοι
πάντες οἱ εὐώδεις καὶ κιρροί· διὰ τοῦτο καὶ ὁ Φαλερῖνος καὶ ὁ Τμωλίτης καὶ ὁ Λέσβιος ἀσθενεῖ κεφαλῇ

come to this, since they are thickened first before they take a yellow color in translucency. Of the other wines that are harsh, none of those that are moderately white and thick are useful to distribution. Nevertheless, if they are aged sufficiently, you might use them when others are not available; all such wines are good for the stomach as they age.

So that you may have an example of these, I remind you of the wine from Nicomedia, which is known to all men, and the Sicilian Aminaean in the large earthenware jars; 835K for what is kept in small flasks is most contrary to this since it is bad for the stomach and at the same time provokes headache. Attempt always to choose the best. If this is not available, choose the one most similar to it according to the indicators mentioned. At any rate, the best with regard to distribution are the Ariusian and the Lesbian. But if they are not available, and we are in Asia, we must choose either the Aphrodisian, or the harsh Tmolitan, or one of those like them, tawny yellow in color and clearly heating, if we need it immediately. It is necessary, above all, to consider what kind of nature the patient has. For if he is weak with respect to the head and easily filled by the hot wines, it is necessary instead to go to those wines that are less hot. If, however, he is strong in respect to the head and enjoys such wines, or is accustomed to tolerate them without distress, you must use them with confidence. All wines that are fragrant and tawny yellow are to some degree headache-producing in themselves. Because of this, the Falernian, the Tmolitan and the Lesbian are harmful to a

---

12 K; δ' οὖν B.

13 B; Ἀρουίσιος K (cf. Aruisium KLat).

836K βλαβεροί. ὁποία δὲ ἐν τοῖς Φαλερίνοις οἴνοις | εἴρηται διαφορά, τοιαύτη κἂν τοῖς Τμωλίταις· ὁ μὲν γὰρ ἕτερος ἱκανῶς γλυκύς, ὁ δὲ ἕτερος αὐστηρός, ὡς ἐκείνῳ περιβάλλειν· θερμοὶ δ' ἱκανῶς ἀμφότεροι καὶ διὰ τοῦτο κεφαλαλγεῖς.

οἱ δὲ αὐστηροὶ τήν τε γαστέρα ῥωννύουσι καὶ ἄλυποι τῇ κεφαλῇ πάντων οἴνων μάλιστ' εἰσίν· εἰς ἀνάδοσιν δὲ οὐχ ὁρμῶσιν· ὥστε σοι φευκτέον αὐτοὺς ἐπὶ τῶν συγκοπτομένων οὐδὲν ἧττον ὕδατος. ἐπιτήδειος δὲ εἰς ἀνάδοσιν καὶ ὁ Ἱπποδαμάντειος παλαιωθεὶς ἐκ τῆς ἰδέας ὢν τοῦ τε Φαλερίνου γλυκέος καὶ τοῦ Τμωλίτου· πρόσεστι δὲ αὐτῷ στύψις σαφής, ἥπερ ἐκείνων οὐδετέρῳ, καὶ διὰ τοῦτό ἐστιν εὐστομαχώτερος. ἀποροῦντι δέ σοι τῶν οἰκείων οἴνων ἐν ἑκατέρᾳ τῇ συγκοπῇ, λέγω δὲ τῇ τε διὰ πλῆθος ὠμῶν καὶ τῇ διὰ χολήν, ἀναγκαζομένῳ τε πάντως οἴνῳ χρήσασθαι, φευκτέοι μέν, ὡς εἴρηται, πάντες οἱ αὐστηροὶ καὶ νέοι· φευκτέοι δὲ καὶ οἱ παχεῖς· αἱρετέοι δὲ οἱ ὑδατώδεις καὶ μᾶλλον εἰ παλαιοὶ τύχοιεν ὄντες· οὐ θερμαίνουσι μὲν γὰρ ἱκανῶς οἱ τοιοῦτοι, ῥᾳδίως δὲ ἀναδίδονται.

ὥστε τοῦ τάχους τῆς ἀναδόσεως κοινοῦ τοῖς | 837K τοιούτοις οἴνοις πρὸς τοὺς κιρροὺς ὄντος, ἡ διαφορὰ γένηται ἂν[14] ἐν τοῖσδε· πέψει μὲν τῇ κατὰ τὴν γαστέρα καὶ φλέβας οἱ κιρροὶ συναίρονται μᾶλλον, ὅτι καὶ

[14] K; γένοιτο ἂν B, recte fort.

weak head. The kind of difference that was spoken of in the Falernian wines is the same with the Tmolitan; for the one is excessively sweet while the other is harsh when compared to that. Both, however, are excessively hot and, for this reason, headache-producing.[8]

836K

The harsh wines strengthen the stomach and are the least headache-producing of all the wines. They do not, however, stimulate distribution, so that you must avoid them in those who are syncopal no less than you must avoid water. The Hippodamantine wine that has been aged is also suitable for distribution, and is of the same kind as the sweet Falernian and the Tmolitan wines. In addition to this, it has a clear astringency which neither of the others have and, because of this, it is better for the stomach. If you are in doubt about the proper wines in either [kind of] syncope—I refer to that due to the abundance of the crude humors and that due to bile—and if you are absolutely compelled to use wine, you must avoid, as I said, all the harsh and young wines. You must also avoid those that are thick. The ones to choose are the watery wines, and more so if they happen to be aged, for such wines do not heat excessively and are easily distributed.

Consequently, since the swiftness of distribution is common to such wines compared with the tawny yellow, the difference would arise in the following: the tawny yellow wines are more conducive to digestion in the stomach and the veins because they are also more heating, and are

837K

[8] Apart from the chapter in *De sanitate tuenda* referred to in the previous note, information on the various wines can be found in modern works such as P. E. McGovern (2003), *Ancient Wines*, and C. Sellman (1957), *Wines in the Ancient World*.

μᾶλλον θερμαίνουσιν, ἐπικεραστικοὶ δ' εἰσὶ καὶ θρε-
πτικοί· καὶ διὰ τοῦτο καὶ ταῖς εὐχυμίαις συντελοῦσιν.
ὧν οὐδὲν ὑπάρχει τοῖς ὑδατώδεσιν, ἥκιστα γὰρ ἐκ τῆς
τούτων οὐσίας ὁμοιοῦταί τι τῷ αἵματι. πληττόντων δὲ
τῶν κιρρῶν τὴν κεφαλήν, ἀλυπότατοι πεφύκασιν οἱ
ὑδατώδεις· οὐρητικοὶ δέ εἰσιν ἁπάντων οἴνων μάλιστα. δεύτεροι δ' ἐπ' αὐτοῖς ὅσοι λεπτότατοι τῶν κιρρῶν, οὓς μάλιστ' ἄν τις εἰς συγκοπὰς αἱροῖτο. βρα-
δυπορώτεροι δὲ τούτων οἱ κιρροὶ καὶ παχεῖς· ἀλλ'
ὅμως τῶν αὐστηρῶν ἁπάντων πορμώτεροι. τρέφουσι
δὲ μᾶλλον τῶν λεπτῶν, κακοχυμίαν τε τάχιστα πάντων τῶν οἴνων ἐπανορθοῦνται, χρηστὸν αἷμα γεννῶντες. ἐπ' ἀρχὴν οὖν αὖθις ἐπαναγάγωμεν τὸν λόγον.

5. Ὅτι μὲν οὖν ἡ συγκοπὴ κατάπτωσίς ἐστιν ὀξεῖα
δυνάμεως εἴρηται τοῖς πρὸ ἡμῶν. ἐπεὶ δὲ ἡ οὐσία τῶν
διοικουσῶν ἡμᾶς δυνάμεων ἔν τε τῷ πνεύματι καὶ τῇ
τῶν στερεῶν σωμάτων ἐστὶ κράσει, ταῦθ' ἡμῖν φυλακτέον ἐστὶ παρόντα καὶ ἀνασωστέον διαφθειρόμενα.
πῶς μὲν οὖν χρὴ φυλάττειν αὐτὰ κατὰ τὸν τῆς ὑγείας
χρόνον, ἐν τοῖς Ὑγιεινοῖς δεδήλωται· πῶς δὲ ἐν ταῖς
νόσοις, εἴρηται μὲν ἤδη καὶ διὰ τῶν ἔμπροσθεν ὑπομνημάτων, εἰρήσεται δὲ καὶ νῦν. οὐ μὴν ἤδη γέ πω τὸ
σύμπαν εἰς πρέπουσαν ἥκει σύνοψιν, οὐδὲ τοιαύτην
ἔχει μέθοδον οἵαν τἆλλα τὰ πρότερα. ὅπερ οὖν ἐνδεῖ,
προσθεῖναι καιρός. τὴν τοῦ πνεύματος οὐσίαν ἅμα
τοῖς στερεοῖς σώμασι φυλακτέον ἡμῖν ἐστιν ἐν ταῖς

tempering of the humors and nourishing. And because of this, they also contribute to [a state of] *euchymia*. None of these characteristics exists in the watery wines, for least of all is something like blood made from the substance of these. However, of the wines that are overpowering to the head, the watery wines are by their nature the least pain-producing, whereas they are the most diuretic of all the wines. Second to these are the thinnest of the tawny yellow wines which one might particularly choose for syncope. Of these, the tawny yellow are slower to pass and thick. Nevertheless, they are more able to find a passage than all the harsh [wines]. They nourish more than those that are thin, and of all the wines they correct *kakochymia* most rapidly because they generate useful blood. Let us, then, bring the discussion back to the beginning.

5. That syncope is an acute collapse of capacity has been stated by my predecessors. Since, however, the substance of the the capacities controlling us lies in the   838K
pneuma and in the *krasis* of the solid bodies, what we must do is preserve these when they are present and restore them when they are weakened. How we must preserve them in a time of health has been shown in my work *On the Preservation of Health*.[9] How we must preserve them in the diseases has already been stated in my earlier works and will also be stated now. Up to this point, however, the whole overview has not yet arrived at an appropriate general statement, nor does it have the kind of method that pertains to the other things considered earlier. Therefore, it is time to add what is lacking. What we must do is preserve the substance of the *pneuma* along with the solid

[9] This is taken to be a general reference to *De sanitate tuenda*, VI.1–452K (English translation by R. M. Green, 1951).

νόσοις, ὥστε καὶ τῇ ποιότητι καὶ τῇ ποσότητι καθ᾽
ὅσον ἐνδέχεται κατὰ φύσιν ἔχειν αὐτά. εἰ μὲν οὖν
ἐνεχώρει τοῦ τε μὴ κενοῦσθαί τι τῆς οὐσίας αὐτῶν καὶ
τοῦ μηδ᾽ ὅλως ἀλλοιοῦσθαι προνοήσασθαι, τοῦτο ἂν
ἦν ἄριστον.

ἐπεὶ δ᾽ ἐν τοῖς Ὑγιεινοῖς ὑπομνήμασιν ἀδύνατον
ἐδείχθη τὸ τοιοῦτον ὑπάρξαι ποτὲ τῷ γεννητῷ σώματι,
πειρᾶσθαι χρὴ τὸ μὲν ἐκρέον τῆς οὐσίας ἐπανορθοῦ-
σθαι προσθέσει, τὸ δὲ ἀλλοιούμενον εἰς εὐκρασίαν |
839K ἐπανάγειν δι᾽ ἐναντίας ἀλλοιώσεως. εἰ μὲν οὖν ἥ τε
κένωσις ἥ τε ἀλλοίωσις γίγνοιτο κατὰ βραχύ, καὶ ἡ
ἐπανόρθωσις ἀμφοῖν ἔσται κατὰ βραχύ· καὶ τοῦτ᾽
ἐστὶν ἔργον, ὡς ἐδείχθη, τῆς ὑγιεινῆς τέχνης. εἰ δὲ καὶ
ἀθρόως καὶ κατὰ μεγάλα μὴ μόνον αἱ κενώσεις, ἀλλὰ
καὶ αἱ προσθέσεις ποτὲ γένοιντο, νόσος ἂν οὕτω γε
συσταίη, θεραπευτικῆς δὲ ἂν εἰς τὴν ἴασιν αὐτῶν
δεοίμεθα μεθόδου. ὥσπερ οὖν τῆς ὑγιεινῆς ἦν ἔργον
τὸ κατὰ μικρὰ τὴν εἰς τὸ παρὰ φύσιν ἐκτροπὴν ἐπανορθοῦσθαι, οὕτω τῆς θεραπευτικῆς ἐστι τὸ κατὰ
μεγάλα. τοῦ μὲν δὴ ψυχικοῦ πνεύματος ἐναργῶς ἐδείξαμεν οἷον πηγήν τινα οὖσαν τὸν ἐγκέφαλον, ἀρδο-
μένου καὶ τρεφομένου διά τε τῆς εἰσπνοῆς καὶ τῆς ἐκ
τοῦ δικτυοειδοῦς πλέγματος χορηγίας. τοῦ δὲ ζωτικοῦ
πνεύματος οὐχ ὁμοίως μὲν ἐναργῶς ἡ ἀπόδειξις ἦν, οὐ
μὴν ἀπίθανόν γε κατά τε τὴν καρδίαν αὐτὸ καὶ τὰς
ἀρτηρίας δοκεῖν περιέχεσθαι, τρεφόμενον καὶ τοῦτο
μάλιστα μὲν ἐκ τῆς ἀναπνοῆς, ἤδη δὲ καὶ τοῦ αἵματος.
840K εἰ δέ ἐστί τι καὶ φυσικὸν πνεῦμα, περιέχοιτ᾽ | ἂν καὶ

bodies in diseases so that, in terms of both quality and quantity, they are in accord with nature as far as possible. If it is possible to make provision so that nothing of the substance of these is evacuated or changed in any way at all, this would be best.

Since, however, in *On the Preservation of Health* it was shown to be impossible that such a thing should ever exist in the mortal body, what we must attempt is to correct the outflow of the substance by an addition and restore what is being changed to *eukrasia* through an opposite change. If 839K both the evacuation and the change occur gradually, then the correction of both will be gradual, and this is the work of the craft of health, as was shown. If, however, not only the evacuations but also the additions occur all at once and in large amounts, this would constitute a disease and we would require a therapeutic method for the cure of these. Just as the correction of a small deviation to a contrariety to nature is the task of the craft of health, so too the correction of a large deviation is the task of the craft of therapeutics. Now I showed clearly that the brain is a fount, as it were, of the psychic *pneuma* which is refreshed and nourished by the inspiration of air and from what the netlike plexus arrangement (*rete mirabile*) provides. My exposition of the vital *pneuma* was not, however, similarly clear. It is certainly not implausible that it seems to be contained in the heart and arteries, this too being nourished particularly from the respiration, but now from the blood as well. If there is a physical *pneuma*, it too would be contained 840K

τοῦτο κατά τε τὸ ἧπαρ καὶ τὰς φλέβας. εἴρηται δὲ ἐπὶ πλεῖστον ὑπὲρ οὐσίας δυνάμεων ἐν τοῖς Περὶ τῶν Ἱπποκράτους καὶ Πλάτωνος δογμάτων.

ἡ δὲ τῶν στερεῶν σωμάτων οὐσία δεῖται μὲν δήπου ποσή τις ὑπάρχειν· καὶ διὰ τοῦτο αἱ τροφαὶ δια-σώζουσι τὸ θνητὸν γένος. οὐ μὴν ἧττον εὐκρασίας χρῄζει τῶν συνθετικῶν αὐτῶν στοιχείων· καὶ λέλεκται πολλάκις ἤδη δι' ὧν χρὴ φυλάττειν αὐτὴν εὔκρατον. ἀλλ' ἥ γε κατάπτωσις τῆς δυνάμεως, ὑπὲρ ἧς νῦν πρόκειται λέγειν, ἐπὶ μὲν τῇ στερεῶν οὐσίᾳ διαφορου-μένῃ κατά γε τὰ χρονιώτατα νοσήματα γίγνεται πολ-λάκις, ἀτροφίᾳ λεπτυνθέντος τοῦ ζῴου καὶ τῶν ὀξέων ἐν τοῖς συντηκτικοῖς πυρετοῖς. ἐπὶ δὲ τῇ κράσει μετα-βαλλούσῃ θερμανθέντων αὐτῶν ἀμέτρως ἢ ψυχθέν-των ἢ ὑγρανθέντων ἢ ξηρανθέντων ἢ κατὰ συζυγίαν τι τούτων παθόντων. ἡ δὲ τοῦ πνεύματος ἀλλοίωσις διά τε μοχθηροὺς γίγνεται χυμοὺς καὶ τὴν τοῦ περι-έχοντος ἀέρος κακίαν ἄλλοτε ἐξ ἄλλης αἰτίας εἰς τοῦτ' ἀχθέντος, ἔτι τε τὰς δηλητηρίους δυνάμεις ἢ τοὺς τῶν ἰοβόλων ζῴων ἰούς.

841K ἀλλ' | ἐξηρήσθω γε ταῦτα τῆς ἐνεστώσης ὑποθέ-σεως. ἡ δὲ τῆς οὐσίας τοῦ πνεύματος φθορὰ γενήσε-σθαί τε κινδυνεύουσα καὶ ἡ ἤδη γινομένη διά τε πάθος ψυχικὸν καὶ ὀδύνην ἰσχυρὰν καὶ κινήσεις πλεί-ους· ἐξ οὗ γένους ἐστὶ καὶ ἀγρυπνία καὶ προσέτι δι' ὑπερβάλλουσαν αὐτοῦ λεπτότητα καὶ τῶν περιεχόν-των αὐτὸ σωμάτων ἀραίωσιν καὶ δὴ καὶ πρὸ τούτων

in the liver and the veins. There was a very full discussion about the substance of capacities in my work *On the Opinions of Hippocrates and Plato*.[10]

The substance of the solid bodies needs, of course, to be of a certain quantity, and because of this, the nutriments preserve the class of living things. No less does it need the *eukrasia* of the actual constituent elements that compose it, and it has been stated often already by what means we must preserve these in a *eukratic* state. But the collapse of the capacity, about which I now propose to speak, frequently occurs after the destruction of substance of solid [bodies] in the course of the most chronic diseases, when the organism is thinned by atrophy, and in the colliquative (*syntectic*) fevers that are acute. It arises after a change in *krasis* when [the solid bodies] have been disproportionately heated, cooled, made moist or dry, or are affected by some conjunction of these. The change of the *pneuma* occurs due to humors in a bad state and the bad quality of the ambient air when it has been brought to this on some occasion for one reason or another, and further, due to the noxious potencies or poisons of venomous animals.

But let these things be put aside from the present purpose. The destruction of the substance of the *pneuma* requires a more precise restoration, both when it is at risk of occurring and when it has already occurred due to a psychical affection, severe pain and much movement (insomnia is also from this class) and besides this, due to excessive thinness of it and rarefaction of the bodies containing it,

841K

---

[10] See *De placitis Hippocratis et Platonis libri ix*, V.181–805K—in particular, Book 6 (P. H. De Lacy, 1984, vol. 2, pp. 360–427).

GALEN

ἁπάντων δι' ἐπίσχεσιν τῆς ἀναπνοῆς καὶ τροφῆς ἀπορίαν ἐπανορθώσεως ἀκριβεστέρας δεῖται.

τὰ μὲν οὖν πάθη τὰ ψυχικὰ φόβοι τέ εἰσιν ἐξαιφνίδιοι καὶ σφοδροί, οὓς ἐκπλήξεις καλοῦσιν· αἵ τ' ἐναντίαι τοῖς φόβοις ἡδοναὶ μέγισται, καλοῦσι δὲ καὶ ταύτας περιχαρείας. ἴσμεν γὰρ ἐξ ἀμφοῖν ἀποθανόντας τινάς. ἤδη δὲ καὶ οἱ ἄλλοι φόβοι πάντες οἱ μεγάλοι σὺν ταῖς μεγίσταις ἡδοναῖς εἰ καὶ μὴ ἀπέκτειναν, ἀλλ' ἔκλυτον ἐργάζονται καὶ ἄτονον τὸ πνεῦμα. καὶ λῦπαι δὲ καὶ ἀγωνίαι καὶ θυμοὶ καὶ φροντίδες, ἐν οἵῳ τρόπῳ καὶ αἱ πλείους ἀγρυπνίαι, βλάπτουσι καταλύουσαι τὴν δύναμιν. ἐν μὲν δὴ τούτοις ἅπασιν αὐτὴ καθ' ἑαυτὴν ἡ ψυχὴ κινεῖται, κατὰ δὲ τὰς πρακτικὰς ἐνεργείας τὸ σῶμα κινεῖ· καταλύει δὲ αὐτῆς τὸν |
842K τόνον ἑκάτερα τὰ γένη τῶν οἰκείων κινήσεων εἰς ἀμετρίαν ἐκταθέντα. περὶ δὲ πόνων τί δεῖ λέγειν; ὅπως ἀνεῖλον μέν τινας αὐξηθέντες ἐπὶ τὸ σφοδρότερον, ἔβλαψαν δὲ πάντας, εἰ καὶ μὴ τύχοιεν ἀποκτείναντες. ἐν τοῖς τοιούτοις ἅπασιν ἐναργῶς φαίνονται λειποδρανοῦντές τε καὶ καταλυόμενοι τὴν δύναμιν οἱ κάμνοντες· ἔνιοι δ', ὡς εἴρηται, καὶ ἀποθνῄσκοντες, ἔσθ' ὅτε μὲν ἀπολλυμένου τοῦ πνεύματος, ἔσθ' ὅτε δὲ ἀλλοιουμένου, ποτὲ δὲ ἐκ μέρους διαφορουμένου, πολλάκις δ' ἀμφότερα ταῦτα πάσχοντος. διῄρηται γὰρ ἐπὶ πλέον ὑπὲρ αὐτῶν ἐν ἄλλοις τέ τισι κἂν τοῖς Περὶ τῶν Ἱπποκράτους καὶ Πλάτωνος δογμάτων.

270

and certainly also, before all these things, due to stoppage of the breathing and lack of nourishment.

The affections that are psychic—sudden and strong fears, which people call terrors, and the extreme pleasures, the opposite to the fears, which they call excessive joys—are both things we know some people have died from. And all the other great fears as well as the extreme pleasures, even if they have not caused death, at least create *pneuma* that is dissipated and weakened. Grief, anguish, outbursts of strong anger and anxiety, and in like manner frequent episodes of insomnia also cause harm by dissipating the capacity. Now in all these the *psyche* moves itself by itself and, in terms of its practical functions, moves the body. However, each class of its specific movements, if extended to an excessive degree, dissipates its own strength. What needs to be said about sufferings? How they have destroyed some people when increased to a more violent level, and have harmed everyone if they don't happen to have killed them. In all such instances, the patients quite obviously fail in strength and suffer dissipation of the capacity. As I said, some people obviously also die, sometimes when the *pneuma* is being destroyed and sometimes when it is being changed and in part destroyed, often suffering both these things. A more detailed analysis has been made of these matters in certain other [writings] and in that *On the Opinions of Hippocrates and Plato*.[11]

842K

11 The other writings are presumably those on the soul—*De propriorum animi cuiuslibet affectuum dignotione et curatione*, V.1–57K, and *De animi cuiuslibet peccatorum dignotione et curatione*, V.58–103K. The theories of Hippocrates and Plato are given detailed consideration in the work referred to in note 10 above.

εἰς δέ γε τὰ παρόντα τὸν μὲν τρόπον ᾧ βλάπτειν ἕκαστον πέφυκεν οὐκ ἀναγκαῖον ζητεῖν· τὸ δὲ ὅτι βλάπτει, λαβόντα παρὰ τῆς ἐναργείας χρῆσθαι συμφερόντως εἰς φυλακὴν δυνάμεως ἐν νόσοις. οὕτω δ' εἰ καὶ λεπτυνθείη ποτὲ περαιτέρω τοῦ δέοντος ἡ τῶν διοικούντων ἡμᾶς πνευμάτων οὐσία, γίγνοιτ' ἂν ἀσθενὴς κατ' ἀμφότερα, καὶ ὡς ἠλλοιωμένη τὴν κρᾶσιν καὶ ὡς εὐσκέδαστος ἢ εὐδιαφόρητος ἢ ὅπως ἄν τις ὀνο-

843K μάζειν ἐθέλῃ γεγενημένη. | κατὰ δὲ τὸν αὐτὸν τρόπον εἰ καὶ τὸ σῶμα σύμπαν εἰς ἄμετρον ἐκτραπείη μανότητα, διαφοροῖτ' ἂν ἡ τῶν πνευμάτων οὐσία ῥᾳδίως, αὐτή τε λεπτομερὴς ὑπάρχουσα καὶ τῶν σωμάτων μὴ στεγόντων αὐτήν. οὔκουν οὐδ' ἀραιοῦν ἀμέτρως χρὴ τοῦ νοσοῦντος τὸ σῶμα, συνέχειν ἐν ἑαυτῷ τὸ πνεῦμα προαιρούμενον, οὔτε λεπτύνειν ἰσχυρῶς τὰ κατ' αὐτὸ διὰ τῶν ἐσθιομένων τε καὶ πινομένων. ὅτι δὲ καὶ αὐτῶν τῶν τροφῶν ἅμα τοῖς ποτοῖς οὐ σμικρόν ἐστι φροντιστέον εἰς ῥώμην δυνάμεως, ὅπως τῇ τε ποσότητι σύμμετροι καὶ ταῖς ποιότησιν ἠκριβωμένοι εἶεν, πρόδηλον εἶναι νομίζω. πρόδηλον δὲ οὐδὲν ἧττον, ὡς καὶ τῆς εἰσπνεομένης οὐσίας φροντιστέον, ὅπως εὐκρατοτάτη τ' εἴη καὶ παντὸς τοῦ μιαίνοντος αὐτὴν καθαρωτάτη, μήτ' ἐκ μετάλλων ἢ καμίνων ἢ βαράθρων ἐπιμιξίαν τινὰ λαμβάνουσα, μήτε ἐκ σηπεδόνος ὀσπρίων ἢ λαχάνων ἢ ζῴων ἢ ὁτουδήποτε, μήτ' ἀτμοὺς ἐκ λιμνῶν ἢ ἑλῶν ἢ ποταμῶν δεχομένη. ταῦτ' οὖν ἅπαντα προνοεῖσθαι χρή, διασῴζειν βουλόμενον τὴν δύναμιν ἀβλαβῆ·

For present purposes, it is not necessary to seek the manner by which each thing naturally does harm, but it is necessary to accept the fact that it does do harm and use [this knowledge] expeditiously for the preservation of capacity in diseases. And if, in this way, the substance of the *pneumas* governing us were to be at some time thinned further than it should be, it would become weak in both respects, having been changed in terms of its *krasis* and having become easy to disperse or easy to evaporate, or whatever someone might wish to call it. In the same way 843K too, if the whole body were to change toward an disproportionate looseness of texture, the substance of the *pneuma* would be easily dispersed, since it is fine-particled and the containing bodies do not retain it. We must not, therefore, thin the body of the sick person disproportionately if we are proposing to preserve the *pneuma* in it, nor must we thin excessively those things in it through what is eaten and drunk. I think it is clear too, that we must take heed of the nutriments themselves along with the drinks to no small extent in regard to the strength of the capacity in order that they may be moderate in quantity and precise in quality. It is no less clear that we must also take heed of the substance of [the air] being inspired in order that it may be most *eukratic* and extremely pure without anything defiling it, not taking on any admixture from mines, kilns or pits, nor receiving [anything] from the putrefaction of pulses or herbs, or animals, or anything whatsoever, or of vapor from stagnant pools, marshes or rivers. You must, then, give forethought to all these things, if you wish to preserve the capacity unharmed.

844K ἔτι τε πρὸς | τούτοις αὐτῶν τῶν στερεῶν σωμάτων,
ἐν οἷς δὴ καὶ μάλιστα ἔοικεν ἡ οὐσία τῶν δυνάμεων
ὑπάρχειν, οὐ σμικρῶς προνοητέον, ὅπως ὑγιεινὰ ταῖς
κράσεσιν εἴη· τοῦτο μὲν οὖν ἁπάντων μορίων κοινὸν
ἀγαθόν. ἐξαίρετον δὲ εἰς ῥώμην δυνάμεως καὶ προφυ-
λακὴν τοῦ μή ποτ᾽ ἐξαιφνίδιον ἐπιπεσεῖν παροξυσμὸν
συγκοπτικὸν ἡ φυλακὴ τῆς εὐκρασίας ἐστί, πρῶτον
μὲν τῶν τριῶν ἀρχῶν, ἔπειτα δὲ καὶ τῶν ἄλλων μορίων
ὅσα τὰς ἀρχὰς εἰς συμπάθειαν ἐπισπᾶται ῥᾳδίως,
οἷόν πέρ ἐστι καὶ τὸ τῆς γαστρὸς στόμα τῷ περιττῷ
τῆς αἰσθήσεως, ἄλλα τε πολλὰ συμπτώματα καὶ συγκοπὰς ἐπιφέρον. ἡ μὲν οὖν προειρημένη διάθεσις τῶν
ὠμῶν χυμῶν, εἰ μὲν καὶ τῷ πλήθει βαρύνοιεν οὗτοι
τὴν δύναμιν, ἐμφράττοιέν τε διὰ πλῆθος καὶ πάχος
τοῦ ζῴου τοὺς πόρους, κατὰ πολλὰς προφάσεις ὀλέ-
θριός τέ ἐστι καὶ συγκοπτική, καὶ τῷ μὴ τρέφεσθαι τὸ
ζῷον καὶ τῷ καταπνίγεσθαι καὶ ἀλλοιοῦσθαι καὶ δια-
φθείρεσθαι τῆς κράσεως τὴν συμμετρίαν. οἱ μὲν γὰρ
ὠμοὶ τρέφειν οὐ δύνανται πρὶν πεφθῆναι, οἱ δὲ πολλοὶ |
845K βαρύνουσιν· εἰ δ᾽ ἐμφράττουσι τὰς διαπνοάς, σβεννύ-
ουσι τὸ θερμόν· εἰ δὲ μήτ᾽ ἐμφράττοιεν μήτε βαρύνοιεν, οὐ συγκοπὰς οὗτοί γε φέρουσιν, ἀλλὰ τὰς
καλουμένας λειποψυχίας· ἐκλύονται γάρ, εἰ μὴ τρέ-
φοιντο συνεχέστερον, οἱ τοιοῦτοι.

διδόναι δὲ οὐ χρὴ πλῆθος εἰς ἅπαξ αὐτοῖς, οὐ μὴν

---

12 Taken to be the three components of the *psyche* according
to Plato; see *Timaeus*, 89e.

And in addition to these things, we must give consider-   844K
able forethought to the solid bodies themselves in which,
above all, it seems that the substance of the capacities lies,
because health lies in the mixtures. This is a common good
for all parts. Of particular importance for strength of capacity and prophylaxis against a sudden paroxysmal syncopal attack at any time is the preservation of *eukrasia*, first
of the three "principles"[12] and then also of those other
parts [of the body] that easily draw the "principles" on to a
sympathetic affection.[13] For example, the mouth of the
stomach, by virtue of its excess of sensitivity, brings many
other symptoms as well as syncope. Thus, the aforementioned condition of the crude humors, if they were also to
weigh down the capacity by their abundance and block up
the pores of the organism by their amount and thickness,
is deadly for many reasons and predisposes to syncope,
both by the fact that the organism is not nourished and by
stifling, changing and corrupting the balance of the *krasis*.
For the crude [humors] are not able to nourish before they
have been concocted and, if there are many [such hu-
mors], they weigh down [the organism]. But if they ob-   845K
struct the transpirations, they quench the heat. If they neither obstruct nor weigh down, they do not in fact bring on
syncope but the so-called "swooning."[14] Such patients are
weakened if they are not nourished quite frequently.

We should not, however, give them a large amount all at

[13] On sympathetic affection see *De morborum causis*,
VII.127–8K, VII.136–7K and I. Johnston (2006), pp. 226, 231.

[14] It is not clear what this distinction is in modern terms. KLat
has *syncopen* and *deliquia*; Peter English has "swooning" and
"raving."

οὐδὲ παχείας ἢ ψυχρὰς τὴν δύναμιν τροφάς, ἀλλ' ὡς
εἴρηται πρόσθεν, ἐκ τοῦ γένους τῶν λεπτυνουσῶν τε
καὶ θερμαινουσῶν. οὕτω δὲ αὐτοῖς καὶ τὰ φάρμακα
δοτέον, ὅσα γε λαβεῖν δύναται πυρέττων ἄνθρωπος,
φάρμακα λεπτύνοντα καὶ θερμαίνοντα. τούτοις ὁ ὑδατώδης οἶνος εὐθέως ἐξ ἀρχῆς ἐπιτήδειος, εἰ μὴ σφοδρῶς πυρέττοιεν, ὅπερ οὐ πάνυ τι συμβαίνει κατὰ τὴν
εἰρημένην διάθεσιν· ἔτι δὲ μᾶλλον εἰ πρεσβῦται τύχοιεν ὄντες, οἶνον αὐτοῖς δοτέον ἐφ' ἑκάστῃ τροφῇ καὶ
μᾶλλον εἰ τὰ διαλείμματα εἴη μέτρια. παροξύνονται
δὲ ἐπὶ τοῖς τοιούτοις χυμοῖς καθ' ἑκάστην ἡμέραν καὶ
μᾶλλον εἰς ἑσπέραν τε καὶ νύκτωρ, οὐχ ἕωθεν δὲ οὐδ'
ἄχρι μεσημβρίας. περὶ μὲν δὴ τούτων καὶ ταῦθ' ἱκανά
ἐστι. |

846K   6. Μεταβῶμεν δὲ ἐπὶ τοὺς ἐναντίαν μὲν ἔχοντας
αὐτοῖς τὴν διάθεσιν, ὁμοίως δὲ τούτοις συγκοπτο-
μένους, ἂν μὴ προσχῇ τις ἀκριβῶς αὐτοῖς. εἰσὶ δ'
οὗτοι λεπτοὺς ἱκανῶς ἔχοντες τοὺς χυμούς, ὡς διαφορεῖσθαι τάχιστα. καὶ δὴ καὶ θεραπευτέον αὐτοὺς
ἐναντίως τοῖς προειρημένοις. ἐκείνους μὲν γὰρ ἐκκενοῦμεν κατὰ βραχὺ συνεχῶς διὰ τὸ μὴ φέρειν ἀθρόαν
τὴν κένωσιν· τοὺς δ' οὕτως ἔχοντας θρέψομεν κατὰ
βραχὺ συνεχῶς. εἰ δ' ἐξ ἀρχῆς παραλάβωμεν ἔτι τῆς
δυνάμεως ἐρρωμένης, οὐδὲ κατὰ βραχύ· δύνανται γὰρ
οἱ τοιοῦτοι καὶ πλείονος εἰς ἅπαξ δοθείσης περιγίνε-
σθαι τροφῆς. ὥσπερ δὲ τοὺς ἐκ πλήθους ὠμῶν χυμῶν
κινδυνεύοντας συγκοπῆναι παντάπασιν ἐλέγομεν ὀλε-
θρίως ἔχειν, ὅταν τὸ ἧπαρ αὐτῶν ἢ ἡ κοιλία φλεγμαί-

once, nor should we give them those nutriments that are thick and cold in terms of potency but, as I said before, nutriments from the class of those that are thinning and heating. In the same way too, we must give them the medications that a person who is febrile is able to take; that is, medications that are thinning and heating. Among these, watery wine is useful right from the start, if they are not severely febrile, which is not very likely to happen in the condition mentioned. Still more, if they happen to be old, we must give them wine on the occasion of each nourishment, and more so if the intervals are moderate. There are paroxysms daily due to such humors, and especially toward evening and during the night, but not early in the morning, nor up until midday. This, then, is enough about these things.

6. Let us pass on to those [patients] who, although they have a condition opposite to the ones [mentioned above], nevertheless suffer syncope in like manner to these if someone does not attend to them assiduously. These are people who have the humors excessively thinned so as to be very quickly dissipated. In particular, we must treat them oppositely to those previously spoken of. Those we evacuate gradually and continuously due to their being unable to tolerate a concentrated evacuation, whereas these, when so affected, we will nourish gradually and continuously. If, however, we were to undertake their care from the beginning, when the capacity was still strong, it would not be gradually because such people are able to survive, even when too much food has been given all at once. Just as I said that those who are in danger of syncope from an abundance of crude humors are undoubtedly affected fatally whenever their liver or stomach is inflamed, in the

νη, κατὰ τὸν αὐτὸν τρόπον καὶ τοὺς νῦν ὑποκειμένους
ἐν τῷ λόγῳ νομιστέον ἀνιάτως ἔχειν, ἢν φλεγμήνῃ τι
τούτων σπλάγχνον, καμνούσης ἤδη τῆς δυνάμεως·
οὔτε γὰρ οἷόν τε τρέφειν αὐτοὺς συνεχῶς, ἀλλ' ἐν ταῖς
παρακμαῖς μόνον, οὔτε μὴ τρέφοντας οἷόν τε δια-
σῴζειν αὐτούς ἐστιν.

847K   ἐπὶ μὲν τῶν τοιούτων διαθέσεων | ἡ νεκρώδης ἐν τῷ
προσώπῳ κατάστασις ἐν τάχει γίνεται· ῥὶς ὀξεῖα,
ὀφθαλμοὶ κοῖλοι καὶ τἆλλα ὅσα τούτοις ἐφεξῆς λέλεκται. ἐάν τε γὰρ ἐκκρίνωσι πλέον, ἐάν τε ἀγρυπνή-
σωσιν ἢ λυπηθῶσιν, ἢ μὴ θᾶττον τραφῶσιν, αὐτίκα
τοιοῦτον ἴσχουσι τὸ πρόσωπον. εἰ μὲν οὖν σύν τινι
τούτων ἤ τισιν ὀφθεῖεν οἱ τοιοῦτοι, μικρότερον εἶναι
νόμιζε τὸ κακὸν ἢ εἰ χωρὶς τούτων ἐλεπτύνθησαν· εἰ δ'
ἄνευ τούτων, ὀλέθριον· ἔτι δὲ μᾶλλον, εἰ κατ' ἀρχὰς
τοῦ νοσήματος οὕτως ἔχοιεν· οὐδὲ γὰρ ἐξαρκέσουσιν
εἰς τὴν πέψιν αὐτοῦ. ταῦτ' ἄρα καὶ ὁ Ἱπποκράτης ἔφη·
Καὶ ἢν μέν τι τουτέων ὁμολογῇ, ἧττον νομίζειν δεινὸν
εἶναι. κατ' ἀρχὴν μὲν γὰρ τοῦ νοσήματος, ὡς προ-
εῖπον, εἰ καὶ μετά τινος τῶν προειρημένων οὕτως
ἔχοιεν, οὐκ ἀγαθόν· ἧττον μὲν τοῦτο δεινὸν ἢ εἰ καὶ
μηδὲ μετὰ τούτων τινός· ἀλλὰ τῷ διαφορεῖσθαι
ῥᾳδίως χαλεπόν. οὐ μόνον δὲ ἀπολεπτυνθέντος οὕτω
τοῦ τε πνεύματος αὐτοῖς καὶ τοῦ αἵματος ἡ διαφόρη-
σις εἴωθε γίνεσθαι πολλάκις, ἀλλὰ καὶ τῆς ἀποκρι-
τικῆς δυνάμεως τῆς καθ' ὅλον τὸν ὄγκον ἀμετρότερον |
848K   κινηθείσης. ὅπερ εἴωθε μάλιστα γίνεσθαι διὰ τὴν
ἀμετρίαν τῆς καθεκτικῆς.

278

same way too, one must deem those now under consideration in the discussion to be affected incurably, if one of their internal organs is inflamed when the capacity is already suffering. It is not possible to nourish them continuously but only in the abatements, and yet is it not possible to save them without nourishing them.

In such conditions the cadaveric appearance in the face   847K
occurs very quickly. The nose is sharp, the eyes sunken and there are, successively, those other [features] that have been described in them. If they expel more, are sleepless or distressed, and are not quickly nourished, they immediately have such a facies.[15] If such people are seen with one or several of these features, consider the harm to be less than if they are emaciated apart from them. If, however, they are without these features, [the disease] is fatal. Still more is this so if they are like this at the beginning of the disease, for they will not be strong enough for the concoction of it. This is why Hippocrates said: "If any one of these is consistent consider it to be less dire."[16] At the start of the disease, as I said before, if the patient has one of the previously mentioned features, the situation is not good. It is less dire than if it is not with one of these, but the difficulty lies in [the *pneuma*] being easily dispersed. However, not only when the *pneuma* and the blood have been attenuated in this way in these patients is it customary for dissipation to occur often, but also when the separative capacity is activated more immoderately in the whole mass of the body. This is something which customarily occurs through   848K
an imbalance of the retentive [capacity] in particular.

[15] The Hippocratic *facies*; see *Prognostic*, II.
[16] Ibid.

τὸ μὲν δὴ τρέφειν ἀναγκαῖον, ὅ τι περ ἂν ᾖ τούτων·
ἀναγκαῖον δὲ καὶ τὸ πυκνοῦν τὴν ἐπιφάνειαν, ὥσπερ
τῶν προτέρων ἀραιοῦν· ἐκ διαμέτρου γὰρ ἀλλήλαις αἱ
διαθέσεις ὑπάρχουσαι καὶ τῆς τῶν βοηθημάτων ἰδέας
ἐναντιωτάτης δεήσονται. ψυχρὸν οὖν ἐπὶ τούτων τὸν
ἀέρα καὶ στύφοντα ποιητέον· εἴρηται δὲ ἔμπροσθεν
ὅπως χρὴ ταῦτα πράττειν. ἀλειπτέον δὲ αὐτοὺς ὁμοίως
τοῖς στύφουσιν ἀλείμμασι. καὶ τροφὰς δοτέον οὐ
πάνυ τι διαρρεούσας ἑτοίμως, οὔτε μελίκρατον οὔτε
χυλὸν πτισάνης, ἀλλ᾽ ἄρτους καὶ τὰ διὰ τοῦ χόνδρου
ῥοφήματα καὶ ὀπώρας αὐστηρὰς καὶ δυσφθάρτους,
αὐτάς τε καθ᾽ ἑαυτὰς καὶ μετ᾽ ἄρτου καὶ χόνδρου δι᾽
ὕδατος ἑψημένου. δοτέον δὲ τούτοις ἔστιν ὅτε καὶ ᾠὰ
καὶ μᾶλλον τὰς λεκίθους αὐτῶν, δύσπεπτον γὰρ τὸ
λευκόν. ἀλεκτρυόνων τε τῶν γάλακτι τεθραμμένων
τοὺς ὄρχεις· δοτέον δὲ καὶ τοὺς ἐγκεφάλους τοὺς
ὑείους, ἤτοι γ᾽ ὀπτηθέντας ἀκριβῶς· ἔναιμοι γὰρ δο-
θέντες ἐσχάτως βλάπτουσιν· ἢ διὰ τοῦ λευκοῦ ζωμοῦ
καλῶς ἑψημένους, ἐμβεβλημένου πράσου καὶ ἀνήθου. |

849K καὶ ὅλως ἅπαντα πρακτέον ὑπὲρ τοῦ παχυτέραν μὲν
ἐργάσασθαι τὴν ὕλην τῶν χυμῶν, πυκνῶσαι δὲ τὸ
δέρμα καὶ διακωλῦσαι τὰς διαπνοάς. ἀναγκαῖος δὲ
τούτοις ὁ ὑδατώδης οἶνός ἐστιν, εὐθὺς ἐπὶ τοῖς σιτίοις
ἐξ ἀρχῆς λαμβανόμενος. εἰ δ᾽ ὡς ἐλπίζομεν ἅπαντα
γίγνοιτο, καὶ κρεώδους τροφῆς ἐπιδώσομεν αὐτοῖς τι
καὶ μάλιστα μετὰ τὴν τετάρτην ἡμέραν, ἤδη τῶν τοὺς
πυρετοὺς γεννώντων χυμῶν εἰς πέψιν ἡκόντων.

ἐν τοιαύτῃ διαθέσει γενομένου νεανίσκου κατὰ τὸν

So then it is necessary to nourish whichever of these there should happen to be. It is necessary also to thicken the outer surface [of the body], just as it is necessary to rarefy those of the former sort. Because the conditions are diametrically opposite to each other, they will also need the form of the remedies to be entirely opposite. You must, therefore, make the air cold and astringent in these cases. I said before how you should do these things. You must smear them similarly with the astringent unguents. And you must give nutriments that are not very readily diffluent; [that is], neither melikraton nor juice of ptisan, but bread and porridge made from gruel and astringent fruits which are not easily corrupted, [giving] these either by themselves or with bread and gruel, when it has been boiled with water. You must also give them eggs sometimes, and particularly their yolks (for the white is difficult to digest), and the testicles of cocks that have been nourished with milk. You must also give sows' brains, either roasted thoroughly (if blood-filled ones are given they are extremely harmful) or well boiled by means of white broth and with leek and dill added. And, in general, you must 849K do everything to make the material of the humors thicker, to thicken the skin and to prevent transpirations. Watery wine is necessary for these [patients], taken from the beginning immediately after food. If all these things happen as we hope, we will give them some meat and food too, particularly after the fourth day when the humors which create the fevers have already come to concoction.

I made mention, in the tenth book, of a young man who

δέκατον ἐμνήσθην λόγον, ὃν ἐτρέφομεν, ὡς εἶπον, ἑκάστης ἡμέρας καὶ μάλιστά γε πρὸ τοῦ παροξυσμοῦ. καὶ ὡς ἐξεπίτηδες ἅπαξ μου μὴ διδόντος αὐτῷ τροφὴν ἡ εἰσβολὴ τοῦ παροξυσμοῦ μετὰ συγκοπῆς ἐγένετο. κατὰ τὰς τοιαύτας οὖν θεραπείας ἁπάσας ἤδη μὲν συγκοπτομένου τοῦ κάμνοντος ὁ πρὸς τὸ σύμπτωμα κρατεῖ σκοπός· οὐδέπω δὲ εἰς τοῦθ' ἥκοντος ἀμφοτέρων στοχαζόμεθα, τοῦ τε μὴ γενέσθαι τὸ σύμπτωμα καὶ τοῦ κατὰ βραχὺ λύειν τὸ νόσημα. θᾶττον δ' ἂν ἐλύσαμεν αὐτό, μὴ δεδιότες τὸ σύμπτωμα. τὸν μὲν γὰρ ὑπὸ τοῦ πλήθους τῶν ὠμῶν χυμῶν βαρυνόμενον, 850K ἅπαξ ἂν οἶμαι | κενώσαντες ἀπηλλάξαμεν τῆς διαθέσεως, εἴπερ οἷός τε ἦν ἐνεγκεῖν ἀθρόαν κένωσιν καὶ μηδεὶς φόβος ἐφέδρευε συγκοπῆς· τὸν δ' εὐδιαφόρητον οὐκ ἂν πρὸ τοῦ παροξυσμοῦ τρέφειν ἀναγκαζόμενοι μακρότερον εἰργαζόμεθα τὸ νόσημα. περὶ μὲν δὴ τούτων ἀρκείτω τὰ λελεγμένα.

7. Τῶν δ' ἄλλων προφάσεων ἐφ' αἷς συγκόπτονται μνημονεύσομεν ἐφεξῆς. εἰσὶ δὲ τέσσαρες, ἄλγημά τε σφοδρὸν καὶ ἀγρυπνία καὶ κένωσις ἄμετρος γαστρός· ἐπὶ δὲ τῶν παραπαιόντων καὶ ἡ κίνησις ἔστιν ὅτε. καὶ εἰ βούλει καὶ πέμπτην προστίθει τὴν δυσκρασίαν τῶν ἀρχῶν, ὑπὲρ ἧς ἐπὶ τελευτῇ τοῦ λόγου διαλεξόμεθα, καὶ μάλισθ' ὅτι διττῶς γίγνεται, ἤτοι κατὰ ψιλὰς τὰς ποιότητας ἀλλοιουμένων τῶν μορίων ἢ κακοχυμίας τινὸς ἐν αὐτοῖς παραλαμβανομένης. ἑξῆς οὖν ὑπὲρ ἁπάντων ῥητέον τῶν διῃρημένων, ἀρξαμένους ἀπὸ τῶν ἀλγημάτων. εἰ μὲν οὖν δυναίμεθα τὴν αἰτίαν

was in such a condition, and whom I nourished each day, as I said, and particularly before the paroxyxm. And when, on one occasion, I deliberately did not give him nourishment, an attack of the paroxysm occurred with syncope. In relation to all such treatments, when the patient is already suffering syncope, the aim is directed to the symptom. But if he has not yet come to this, we aim at both—that the symptom should not occur and that the disease should gradually resolve. And we would resolve this quicker if we did not fear the symptom because, in the case of someone weighed down by the abundance of crude humors, if we had carried out evacuation, I think we would have removed the condi- 850K tion once and for all, should such a person be able to bear a sudden evacuation and no fear of syncope was lying in wait. If we are compelled to nourish someone who concocts easily before the paroxysm, we would not make the disease more protracted. Let what has been said be enough on these matters.

7. I shall make mention, in order, of the other causes due to which people suffer syncope. There are four: severe pain, insomnia, excessive evacuation of the stomach and sometimes, in addition, movement in those who are delirious. And if you wish, also add as a fifth, *dyskrasia* of the "principles," which I shall go over at the end of the discussion, and particularly because it occurs in a twofold manner: either when the parts are changed in terms of their qualities being simple or when some *kakochymia* is taken up in them. Therefore, I must speak next about all these features that have been distinguished, starting with the pains. If we could eradicate the actual cause due to which

αὐτὴν ἐκκόψαι δι᾽ ἣν ἀλγοῦσιν, οὕτως ἂν οὐ πρὸς
σύμπτωμα εἴημεν ἐνεργοῦντες, ἀλλ᾽ ἰώμενοι τὸν κά-
μνοντα· κωλυόμενοι δὲ διά τινα πρόφασιν ἐπὶ τὴν
851K διάθεσιν ἰέναι πραΰνομεν τὸ τῶν | ἀλγημάτων σφο-
δρόν. ἐπεὶ δ᾽ οὔτε ἀναιρεῖν οἷόν τέ ἐστι τὰς τῶν
ἀλγημάτων αἰτίας οὔτ᾽ ἀμβλύνειν τὰς σφοδρότητας
ἄνευ τοῦ γνῶναι τουτὶ τὸ ἄλγημα τὸ περὶ τὸν κάμνοντα συνεστηκὸς ὑπὸ τίνος αἰτίας γίγνεται, προεσκέ-
φθαι χρὴ δηλονότι τὰς αἰτίας καὶ τὰ σημεῖα δι᾽ ὧν
ἄν τις αὐτὰς γνωρίσειεν· ὡς τὰ πολλὰ γὰρ ἄδηλοι
ταῖς αἰσθήσεσιν ὑπάρχουσαι, γεγυμνασμένου δέον-
ται περὶ τὴν τοιαύτην θεωρίαν ἀνδρός.

εἴρηται μὲν οὖν ἡμῖν οὐκ ὀλίγα δι᾽ ἑτέρων πραγμα-
τειῶν ὑπὲρ ἀλγημάτων γενέσεως, ἁπάντων δὲ αὐτῶν
τοῦ καιροῦ νῦν ἥκοντος, ἀναγκαῖον εἶναι νομίζω διελθεῖν ἐπὶ κεφαλαίων ὅλον τὸν λόγον. ἐπεὶ τοίνυν ἐν τῷ
τοῦ ζῴου σώματι κατὰ πάσας τὰς αἱρέσεις οὐ πάντα
ἐστὶν αἰσθητικὰ τὰ μόρια, δῆλον ὡς οὐκ ὀδυνήσεται
τὰ τελέως ἀναίσθητα. μόνοις οὖν τοῖς αἰσθητοῖς τῆς
ὀδύνης ὑπαρχούσης, ἐφεξῆς σκεπτέον· ἆρά γε διὰ
παντὸς ὀδυνηθήσεται τὰ τοιαῦτα τῶν σωμάτων ἢ καθ᾽
ἕνα τινὰ χρόνον ἐγγινομένης αὐτοῖς διαθέσεως ὀδυνηρᾶς; ἀλλ᾽ εἴπερ αὐτὸ τὸ φαινόμενον ἐναργῶς ἡμᾶς
διδάσκει πολὺν εἶναι χρόνον ἐν ᾧ μηδ᾽ ὅλως ὀδυνώμε-
852K θα, | δῆλον ὡς κατά τινα χρόνον ἐγγινομένης ἑκάστῳ
τῶν ὀδυνωμένων ὀδυνηρᾶς διαθέσεως ἀναγκαῖον ἀλ-
γεῖν ἐστιν. ἐὰν οὖν εὕρωμεν ἥτις ποτ᾽ ἐστὶν ἡ τοῖς
αἰσθητοῖς σώμασιν ἐγγινομένη διάθεσις, καθ᾽ ἣν ἀνι-

patients suffer pain, we would be acting in this way, not against the symptom but curing the patient. However, when we are prevented for some reason from proceeding against the condition, we mitigate the severity of the pains. 851K But since it is not possible to remove the causes of pains or dull their severity without knowing what has caused this particular pain that has become established in the patient, it is clearly necessary to consider beforehand the causes, and the signs by means of which someone might recognize these [causes]. As they are obscure to the senses in the majority of cases, they require a man practiced in such speculation.

I have said quite a lot in other treatises about the genesis of pains and, since the right time has now come for all these [pains to be considered], I think it is necessary to go over the discussion as a whole under the chief points. Accordingly, since it is accepted by all the sects that not all the parts in the body of the organism have sensation, it is clear that those entirely without sensation will not feel pain. Therefore, as pain exists only in those parts with sensation, what I must consider next is: will such parts of the body feel pain continuously, or only at one particular time when a painful condition arises in them? But if what is actually apparent clearly teaches us that there is a long time in which we do not feel pain at all, it is obvious that at the par- 852K ticular time when a painful condition is engendered in each of the parts that feel pain, there is inevitably pain. If we should discover what, at some time, the condition is which is engendered in the perceiving bodies, by virtue of

ᾶται τὸ ζῷον, ἑαλωκὸς ἂν εἴη ἡμῖν τὸ ζητούμενον.
ἐνταῦθ᾽ οὖν ἀπὸ τῶν κατὰ φύσιν ἐννοιῶν, αἳ δὴ καὶ
κοιναὶ πᾶσιν ἡμῖν εἰσιν, ἰέναι χρὴ μεθόδῳ πρὸς τὸ
σκέμμα. διδάσκει δὲ ἡ κατὰ φύσιν ἔννοια τὸ πάθος
χρῆναι τοῦ σώματος, ᾧ μέλλει πονήσειν, ἤτοι συνεχείας λύσιν ἢ ἀλλοίωσιν εἶναί τινα.

διὸ καὶ καλῶς ἐλέγετο τοῖς ἐξ ἀτόμων ἢ ἀναισθήτων ἢ ἀνάρμων στοιχείων συντιθεῖσι τὸ σῶμα παντάπασιν ἄπορον εἶναι διάθεσιν εὑρεῖν, ὑφ᾽ ἧς ὀδύνη
γίγνεται. φαίνεται γὰρ οὐδὲν τῶν ἁπλῶς ψαυόντων
ὀδυνηρόν· ἁπλῶς δ᾽ ἀλλήλων ψαύει τὰ τοιαῦτα σώματα. κἂν εἰ μὴ ψαύοι δὲ μόνον, ἀλλὰ καὶ θραύοι κατὰ
τὴν πρόπτωσιν ἄλληλα, πλέον οὐδὲν εἰς ἀλγήματος
γένεσιν, ἀναισθήτων ὑποκειμένων τῶν θραυομένων· εἰ
μή τι καὶ τοὺς λίθους ὀδυνᾶσθαι φήσομεν διαιρου-
853K μένους. ἀλλὰ περὶ | μὲν τούτων ἐπὶ πλέον ἐν ἄλλοις τέ
τισι κἂν τῷ πέμπτῳ Περὶ τῶν Ἀσκληπιάδου δογμάτων ἐσκεψάμεθα. νυνὶ δὲ ἀρχὴν ὁμολογουμένην λαμ-
βάνοντες, ὡς ἀναγκαῖον ἤτοι διαίρεσιν εἶναι τὴν
διάθεσιν τῆς ὀδύνης ἢ ἀλλοίωσιν, ἀναμνησθῶμεν
αὖθις ὡς οὐδ᾽ ἡ ἀλλοίωσις ἡ κατὰ βραχὺ γινομένη
δύναιτ᾽ ἄν ποτε ὀδυνῆσαι τὸ ζῷον, ἀλλ᾽ ἀναγκαῖον
ἀθρόαν τε ἅμα καὶ βιαίαν γίνεσθαι τὴν μεταβολήν, εἰ
μέλλει τις ὀδύνη γενήσεσθαι. φαίνεται γὰρ οὐ μόνον
εἰς τὸ παρὰ φύσιν ἀθρόως ἀγόμενα τὰ σώματα λυπη-
ρὰν ἴσχειν τὴν μεταβολήν, ἀλλὰ καὶ τὴν εἰς τὸ κατὰ

---

17 See Galen's *De elementis secundum Hippocratem*, I.30.

which the organism is distressed, we would have found what it is we were seeking. Here then, from the notions of those things in accord with nature, which are also common to all of us, we must proceed by method toward the matter in question. However, the notion of accord with nature teaches us that the affection of the body due to which there is going to be pain must be either a dissolution of continuity or some change.

For this reason, it was said correctly by those who would put together the body from atoms, or impassible and indivisible elements, that it is quite impossible to find a condition from which pain arises.[17] For it is obvious that nothing that simply touches is painful, yet such bodies simply touch each other. Even if they don't only touch but also break each other in the collision, there is nothing more for the creation of pain if the things broken exist without sensation, unless we shall say that stones too feel pain when they are divided. But I gave consideration to these matters   853K
at greater length in certain other [works] and in the fifth book of *On the Opinions of Asclepiades*.[18] For the present, if we take it as an agreed principle that it is necessary for the condition producing pain to be either a division or a change, let us remind ourselves again that it is not the change which occurs gradually that is able at any time to cause pain to the organism; the change must be, at one and the same time, sudden and violent, if some pain is going to be produced. For it seems that not only do bodies, when suddenly brought to [a state that is] contrary to nature, have a change that is painful, but so too is the actual return

18 On the lost work *On the Opinions of Asclepiades*, see Galen's *De libris propriis*, XIX.38K.

GALEN

φύσιν ἐπάνοδον αὐτήν, εἰ μὴ κατὰ βραχὺ λαμβάνοι,
κἂν τούτῳ πονοῦντα. τοὺς γοῦν ὁδοιπορήσαντας ἐν
σφοδρῷ κρύει θερμαινομένους ἀθρόως ἴσμεν ἀλγοῦν-
τας οὕτως ἰσχυρῶς τὰ περὶ τὰς ῥίζας τῶν ὀνύχων, ὡς
μὴ δύνασθαι φέρειν. εἰ μὲν οὖν καὶ αὐτὸ τὸ σφοδρῶς
θερμαίνεσθαι ἢ ψύχεσθαι διαίρεσίν τινα τῆς συνεχείας ἐργάζεται, πλείονος ἴσως δεῖται λόγου, καὶ δέ-
854K δεικται καὶ τοῦτο ἐν τοῖς Περὶ τῆς | τῶν ἁπλῶν φαρ-
μάκων δυνάμεως ὑπομνήμασιν.

ἀλλ᾽ εἴς γε τὰ παρόντα τί ποιοῦν τὸ θερμαῖνον
ἰσχυρῶς ἢ ψῦχον σφοδρῶς ὀδύνην ἐργάζεται, ζητεῖν
οὐκ ἀναγκαῖον· εἴτε γὰρ ὡς τέμνον τὸ συνεχὲς τῆς
οὐσίας τῶν αἰσθητῶν, εἴθ᾽ ὡς ἀλλοιοῦν αὐτὴν ἀθρόως,
οὐδὲν διαφέρει πρὸς τὴν εὕρεσιν τῶν ἰασομένων τὰς
ὀδύνας, ἐάν γε μόνον εἰδῶμεν ὡς τὸ θερμαῖνον ἢ
ψῦχον βιαίως ὀδυνηρὸν γίνεται τοῖς αἰσθανομένοις
σώμασιν. εἴπερ γοῦν ἄρχεσθαι μὲν ἀπὸ τῶν ἐναργῶν
δεῖ, φαίνεται δὲ ἐναργῶς ταῦτα μόνα τῶν πλησιαζόντων ἡμῖν ὀδυνηρά, τά τ᾽ ἰσχυρῶς θερμαίνοντά τε καὶ
ψύχοντα καὶ τὰ διαιροῦντα τὸ συνεχὲς τῆς οὐσίας,
ἅπερ ἐν τοῖς θλῶσί τε καὶ τέμνουσι καὶ τείνουσι
περιέχεται, ζητήσωμεν εὑρεῖν τὴν ἴασιν αὐτῶν ὁδῷ
τινι προϊόντες ἀπὸ τῆς τῶν ζητουμένων φύσεως, ὅπερ
ὀνομάζεται μέθοδος. ἄνωθεν οὖν αὖθις ἀρξάμενοι λέγωμεν. ἐπειδὴ ζητοῦμεν ὅπως ἐν νόσοις ὀδυνώμεθα
μηδενὸς ἡμῖν ἔξωθεν ὀδυνηροῦ πλησιάζοντος, ἀναγ-

to an accord with nature [painful], if they do not receive this gradually, and that they are distressed also in this. At any rate, we know that when those who have walked in severe cold are heated suddenly, they feel pain around the roots of the nails so strongly that they cannot bear it. Therefore, whether the act of being strongly heated or strongly cooled itself also creates some division of continuity perhaps requires a longer discussion. This has also been shown in the treatise *On the Mixtures and Potencies of* 854K *Simple Medications*.[19]

But for present purposes at least it is not necessary to inquire into what it is that creates the severe pain of strongly heating or cooling. Whether the continuity of the substance of those [parts] capable of sensation is divided, or this substance is suddenly changed, makes no difference to the discovery of those things that will cure the pains, as long as we know that what heats or cools violently creates pain in bodies capable of sensation. At any rate, since we ought to begin from those things that are obvious, and since it seems clear that of those things that approach us, the only ones that are painful are those that heat or cool strongly and those that divide the continuity of the substance, which are encompassed in the bruisings, cuttings and stretchings, let us seek to discover their cure, proceeding by some path from the nature of what is sought, which is called method. Let me say again what I said above, when I began. Since what we are inquiring into is how we suffer pain in diseases when there is nothing pain-producing near us externally, it is essential to seek out what the cause 855K

[19] *De simplicium medicamentorum temperamentis et facultatibus*, XI.379–892K and XII.1–377K.

855K καῖον ἐν τῷ σώματι ζητῆσαι τί τῶν τοιούτων | αἴτιον, ὅ θερμαῖνον σφοδρῶς ἢ ψῦχον ἢ διαιροῦν τὴν οὐσίαν ἑνός γέ τινος αἰσθητοῦ σώματος ἀνιαρὸν ἡμῖν γίνεται.

καὶ πρῶτόν γε θεασώμεθα τίνος εἶναι χρὴ φύσεως ὅ διαιρήσει τὴν συνέχειαν. ἀναγκαῖον οὖν ἤτοι ῥῆξιν ἢ θλάσιν ἢ διάβρωσιν εἶναι τὴν διάθεσιν ἐν τῷ διαιρεῖσθαι τὸ συνεχές. ἀλλ' ἡ μὲν ῥῆξις ἐκ τάσεώς τινος, ἡ δὲ θλάσις ἐκ βάρους, ἡ δὲ διάβρωσις ἐκ δακνώδους γίνεται ποιότητος. ὅταν οὖν ἔξωθεν μὲν μηδὲν ἦ τὸ τεῖνον ἢ βαρῦνον ἢ δάκνον, ἐξ ἑαυτοῦ δέ τι πάσχῃ τούτων τὸ σῶμα, παντί που δῆλον ὡς ἡ μὲν τάσις ὑπό τινος ἔνδον οὐσίας πολλῆς διατεινούσης τὸ περιέχον αὐτὴν σῶμα γίνοιτ' ἂν ἢ ἀμέτρου ξηρότητος, ἡ δὲ θλάσις ἔξωθεν ἐπιπεπτωκυίας οὐσίας σκληρᾶς ἢ μεγάλης ἢ βαρείας ἢ συναμφοτέρων, ἡ δὲ δῆξις ὑπὸ χυμοῦ δακνώδη φύσιν ἔχοντος· ἐπισκεπτέον οὖν ἡμῖν ἐν ταῖς σφοδραῖς ὀδύναις ἤτοι χυμὸν πολὺν ἢ πνεῦμα διέξοδον οὐκ ἔχον, ἢ βαρὺν ὄγκον, ἢ δακνῶδες

856K ὑγρὸν ἢ διάθεσιν ξηράν· ἔτι δὲ πρὸς τούτοις, ὡς | ἐν ἀρχῇ διῄρηται, τὸ θερμαῖνον ἰσχυρῶς ἢ ψῦχον. ὧν πάλιν αὐτῶν τὰς αἰτίας ἢ ἐν τῷ πνεύματι θετέον ἢ ἐν τοῖς ὑγροῖς ἢ ἐν τοῖς στερεοῖς· ὅ τι γὰρ ἂν αὐτῶν ἰσχυρῶς ψυχθὲν ἢ ὑποθερμανθὲν ψαύῃ τῶν πλησιαζόντων, ὀδυνηρὸν αὐτοῖς γίγνεται.

πῶς μὲν οὖν χρὴ διαγινώσκειν ἕκαστον τῶν εἰρημένων καὶ ὡς οὐχ ἁπάντων ἔχομεν ἐναργῆ σημεῖα καὶ ὡς εἴπερ τι ἄλλο καὶ ἡ διάγνωσις τῶν ἀδήλων πρὸς

of such things is in the body—whether strong heating or cooling, or division of the substance of one of the bodies capable of sensation has occurred to distress us.

First, let us look into what the nature of that which will divide continuity must be. It is, of necessity, a rupture, contusion or erosion—that is, a condition in which continuity is divided. But rupture arises from some tension, contusion from a weighing down and erosion from a biting quality. Therefore, when there is nothing external stretching, weighing down or biting and yet, of itself, the body suffers one of these, it is surely clear to everyone that tension arises when the body is stretched significantly by some large substance which it contains within, or by excessive dryness; contusion arises when a hard, substantial, or heavy substance, or all of these together, has impinged on it from without; gnawing arises due to a humor which has a biting nature. What we must consider, then, in severe pains is either abundant humor, or *pneuma* not having a path out, or a heavy mass, or a biting fluid, or a dry condition. Over and above these are severe heating or cooling, as I distinguished at the start. We must, in turn, establish the 856K causes of these same things, either in the *pneuma*, or in the fluids, or in the solid [parts] because, if one of those that has been strongly cooled or heated touches the parts adjacent, it generates pain in them.

So it has been demonstrated elsewhere how we should diagnose each of the things mentioned, and how we do not possess clear signs of them all, and how—if indeed there is anything else—even our diagnosis of causes and condi-

αἴσθησιν αἰτιῶν τε καὶ διαθέσεων ἐκ πολλῆς μὲν
γυμνασίας, ἐπιμελοῦς δ' ἀπάντων περισκέψεως ἡμῖν
περιγίνεται, δι' ἑτέρων δεδήλωται. νυνὶ δὲ οὐ τοῦτο
πρόκειται σκοπεῖν, ἀλλ' αὐτὴν μόνην διέρχεσθαι τὴν
θεραπευτικὴν μέθοδον. ἐξ ἧς ἐπιστημονικῶς δια-
γνωσθείσης ἡλίκη βοήθεια καὶ ὠφέλεια γίνεται τοῖς
ἰατροῖς οὐ μόνον εἰς βοηθημάτων τε καὶ ἰαμάτων
εὐπορίαν, ἀλλ' ἔστιν ὅτε καὶ εἰς τὴν τῆς διαθέσεως
γνῶσιν, ἐναργῶς ὑμῖν ἔδειξα πολλάκις ἐπ' αὐτῶν τῶν
καμνόντων, ὧν ἤδη μνημονεύσω δυοῖν ἢ τριῶν οὐ πρὸ
πολλοῦ θεραπευθέντων.

857K   ὁ μέν γε τεσσαρακοντούτης ἦν, ὡς οἶσθα, | κωλι-
κὸς εἶναι νομιζόμενος, οὐ μόνον οὐδὲν ὀνινάμενος ὑπὸ
καταντλήσεων καὶ πυρίας καὶ καταπλασμάτων καὶ
κλυσμάτων, οἷς συνήθως εἰώθασιν ἐπὶ τῶν τοιούτων
χρῆσθαι διαθέσεων, ἀλλὰ καὶ παροξυνόμενος ὑπὸ τῶν
πλείστων. ἐπὶ γοῦν ἐλαίῳ πηγανίνῳ διὰ τῆς ἕδρας
ἐνεθέντι χείρων ἐγένετο καὶ αὖθις ἐπὶ καστορίῳ· καὶ
μέντοι καὶ μέλι ποτὲ προσενεγκάμενος ἐφθὸν ἔχον
πέπερι ἐσχάτως ὠδυνήθη· καὶ τὸν χυλὸν δὲ τῆς ἐφθῆς
τήλεως ἅμα μέλιτι λαβὼν ἱκανῶς παρωξύνθη. στοχα-
σάμενος οὖν ἐγὼ χυμοὺς δακνώδεις ἐν αὐτοῖς τοῖς
χιτῶσι τῶν ἐντέρων ἀναπεπόσθαι, συνδιαφθείροντας
ἑαυτοῖς τά τε κάτωθεν ἐνιέμενα καὶ τὰ διὰ τοῦ στόματος λαμβανόμενα, δύσφθαρτον αὐτῷ τροφὴν δούς.
εἶτ' ἰδὼν ὀδυνώμενον ἔγνων χρῆναι τὴν κακοχυμίαν
ἐκκαθαίρειν. ὄντος δ' ἀρίστου πρὸς τὰς τοιαύτας
κακοχυμίας φαρμάκου τοῦ διὰ τῆς ἀλόης, ὃ καλοῦσιν

tions not apparent to the senses is the result of much practice and our careful consideration of everything. I do not, however, propose to consider this now, but to go over the therapeutic method itself alone. I have clearly shown you on many occasions in actual patients how great a help and benefit arises for doctors from this when we diagnose scientifically, not only for the easy means of providing remedies and cures, but sometimes also for the recognition of the condition. I shall now call to mind two or three of those [patients] I treated not long ago.

One was a man forty years of age, as you know. Although he was thought to be suffering from colic, he not only derived no benefit from fomentations, applications of heat, poultices and clysters, which [doctors] are habitually accustomed to use in such conditions, but he also suffered exacerbations with most of them. At any rate, after oil of rue was inserted *per rectum*, he became worse, and worse again after castor. Furthermore, when he employed honey which had been boiled with pepper, he was afflicted with extreme pain; and when he took the juice of boiled fenugreek along with honey, he suffered severe irritation. Therefore, I guessed that there had been biting fluids absorbed in the actual walls of the intestines, and these were corrupting along with themselves those medications that were being inserted *per rectum* and also those things being taken *per os*, and I gave him nourishment that was not easily corrupted. Then, when I saw him in pain, I knew there was a need to purge the *kakochymia*. Although the best medication for such *kakochymias* is that made of bitter aloes, which people are now accustomed to call "higry-

857K

ἤδη συνήθως πικράν, ἀθρόως μὲν οὐκ ἐτόλμησα καθαίρειν αὐτὸν τὸν ἄνθρωπον, ὑπό τε τῆς ὀδύνης καὶ τῆς ἐνδείας καθῃρημένον ἤδη που δυοῖν μηνῶν. ἐκ δια-

858K στημάτων δέ τινων σύμμετρον | τοῦτ' ἐργαζόμενος ἡμέρας ὡς οἶσθά που πεντεκαίδεκα τελέως ἰασάμην αὐτὸν οὐδὲν οὐκέτι αὐτῷ προσαγαγὼν ἄλλο βοήθημα. οὗτος μὲν οὖν ἐν ἐκείνῳ τῷ χρόνῳ πρῶτον οὕτως ἠνωχλεῖτο, μηδέπω πρότερον ἀλγήσας ἔντερα.

νεανίσκος δέ τις ἐγγὺς ἐκείνῳ τὴν ἡλικίαν, οὐκ ὀλιγάκις ἔμπροσθεν ἠνωχλημένος ὑπὸ κωλικῶν ἀλγημάτων, ἐκαθάρθη λαβὼν σκαμμωνίας ὀπόν. ἀξιολόγου δὲ τῆς καθάρσεως γενομένης, ἐν μὲν τῇ πρώτῃ τῶν ἡμερῶν λουσάμενος εἰς ἑσπέραν καὶ λαβὼν πτισάνης χυλὸν πρῶτον, εἶτ' ἐπιφαγὼν ἰχθύας ἤμεσε τὰ ληφθέντα διὰ νυκτός. ἐν δὲ τῇ δευτέρᾳ λουσάμενος ἔφαγε· πρῶτον μὲν θριδακίνης, εἶτα κρεῶν ὀρνιθείων ἑψομένων ἐν λευκῷ ζωμῷ· κἄπειτα προσηνέγκατο χόνδρον, ἐξ ὕδατος ἐπιβαλὼν οἶνον αὐστηρόν. μετρίως δὲ διάγειν δόξας ἐπ' αὐτοῖς, διῃτήθη μὲν καὶ τῇ τρίτῃ τῶν ἡμερῶν παραπλησίως, ἐξέδωκε δὲ ἡ γαστὴρ αὐτοῦ μετὰ τοῦ δηχθῆναι πλείω τῆς τῶν ἐδεσμάτων ἀναλογίας. εἶτα κατὰ τὴν τετάρτην ἡμέραν ἔτι καὶ μᾶλλον ὠδυνήθη τὰ κατὰ τὴν γαστέρα· καὶ δόξας ἐν τῷ βαλανείῳ που λεληθότως ἐψῦχθαι, πηγάνινον

859K ἔλαιον ἐνθείς, | ὠδυνήθη τε σφοδρότερον καὶ μετὰ ταῦτα ἐξέκρινεν ὑγρὸν διαχώρημα πάμπολυ, σαφῶς ἐνδεικνύμενον ὡς κἀκ τοῦ σώματος ὅλου τι φέρεται

pigry" (*hiera picra*),[20] I did not dare purge this particular man suddenly because he had already been affected by pain and lack of food for, I suppose, two months. When I did purge him moderately, and at certain intervals for a   858K fortnight or so, I completely cured him, as you know, without applying any other remedy to him. This was the first time he was being troubled in such a way, never having experienced pain in the intestines previously.

[Another] young man, close to the previous one in age, who had been troubled with colicky pains on a number of previous occasions, was purged by taking the juice of scammony. When a significant purging had occurred, he bathed toward evening on the first day and initially took the juice of ptisan. Then, after eating dried fish, he vomited during the night those things he had taken. On the second day, having bathed, he ate: first wild lettuce and then the flesh of birds boiled in white broth. And then he took gruel made from water, having added bitter wine. When he seemed to manage moderately well as a result of these things, he was also nourished similarly on the third day. However, his stomach gave out a greater than expected proportion of the foods in conjunction with being "bitten." Then, on the fourth day, he suffered even more pain in the stomach, and when he thought he had been cooled in the bath somehow without realizing it, having inserted oil of rue *per rectum*, he suffered stronger pains,   859K and after these, he expelled watery excrement in a great amount, which clearly revealed that something was being borne from the whole body toward the lumen of the stom-

[20] A purgative medication made from aloes and canella bark.

πρὸς τὰ κατὰ τὴν γαστέρα χωρία. κἄπειθ' ἑξῆς ἐκ
περιόδων ὁμοίως ἠνωχλεῖτο.

πυθόμενοι[15] δ' ἡμεῖς αὐτῷ τὰ συμπτώματα, τὴν
βλάβην ἐκ τῆς σκαμμωνίας ἔγνωμεν ἐν ἐκείνοις μάλιστα γεγονέναι τοῖς ἐντέροις, ἃ καὶ πρότερον ἦν ἀσθενῆ, ὥσθ' οἷον ῥευματικήν τινα συνίστασθαι διάθεσιν.
οἶσθ' οὖν ὅπως ἰασάμην καὶ τοῦτον ἀπορρῖψαι μὲν
κελεύσας τὰ κωλικὰ βοηθήματα, τραφῆναι δὲ χόνδρῳ
θερμῷ, δι' ὕδατος ἡψημένῳ, κόκκους ῥοιᾶς ἐμβαλὼν
αὐτῷ. κοιμηθεὶς δ' ὅλῃ τῇ νυκτὶ χωρὶς ὀδύνης ἐθαύ-
μασε δήπου τὸ παράδοξον τῆς βοηθείας, ἠρώτα δ' ὅ τι
χρὴ πιεῖν. ἐδώκαμεν οὖν αὐτῷ πιεῖν ὑδαρῆ ῥοῦ χυλόν,
ὅπως εἴτε τις ἐπιπολῆς εἴη γεγενημένη περὶ τὸ ἔντερον ἑλκώδης διάθεσις ἀποστύψειεν, εἴτε καὶ ῥέῃ τι
κατὰ τὸ σύνηθες ἐκ τῶν ἄνω χωρίων εἰς τὸ πεπονθὸς
ἀναστείλειεν· ἀρίστῳ τε αὐτῷ χρήσασθαι προσετάξα-
μεν οὕτως· εἶτ' εἰς ἑσπέραν δειπνεῖν οἴνων τινὰ τῶν
860K αὐστηρῶν μὲν φύσει, παλαιῶν δέ, διαβραχέντος | ἄρ-
του καθαροῦ. συνεχωρήσαμεν δὲ καὶ τῶν αὐστηρῶν
ὀπωρῶν λαβεῖν ἧς ἂν αὐτὸς ἐθέλῃ μήλων ἢ ἀπίων ἢ
ῥοιᾶς. ὁμοίως δὲ καὶ τῇ τρίτῃ τῶν ἡμερῶν διαιτηθεὶς
ἐν τῇ τετάρτῃ πιὼν τῆς θηριακῆς, ὑγιὴς τελέως ἐγένετο, καὶ τοῦ λοιποῦ τοῖς συνήθεσι χρώμενος οὐδὲν
ἐβλάπτετο.

παραπλησίως δὲ ἕτερόν τινα διακείμενον ὥρᾳ θέρους, ἐπειδὴ διψώδης ἱκανῶς ὑπῆρχε πρὸς τοῖς ἄλλοις
τοῖς εἰρημένοις καὶ ψυχρῷ ποτῷ συνεχώρησα χρῆσθαι. μὴ τοίνυν ζήτει σημεῖα τοιαῦτα καθ' ἑκάστην

ach. Thereafter, he was periodically troubled in the same
way.

When I learned of his symptoms, I recognized that
injury from the scammony had occurred, particularly in
those intestines that were also previously weak, so that a
certain flux-inducing condition, as it were, was developing. You know how I cured this man also, having directed
him to cast aside the remedies for colic and to be nourished with warm gruel which had been boiled in water into
which he had put pomegranate seeds. When he had slept
the whole night without pain he was amazed, presumably
at the unexpected [efficacy] of the remedy, and asked what
he ought to drink. I gave him the watery juice of sumac to
drink in order either to provide astringency, if some superficial ulcerated condition had occurred involving the intestines, or to check anything that was flowing in the customary way from the places above to the affected place. I
directed him, in this way, to use it for lunch and then, toward evening, to have for his evening meal one of the
wines that was harsh in nature, but old, after soaking pure   860K
bread [in it]. I also allowed him to take whichever one of
the bitter fruits he wished—apples, pears or pomegran-
ates. After also observing this regimen in the same way on
the third day, when he had drunk theriac on the fourth day,
he became completely healthy, and thereafter, when he
used the customary things, he came to no harm.

In the case of another person in a similar state in summer, since he was excessively thirsty in addition to the
other things mentioned, I allowed him to use cold water.
Thus, do not seek in each condition the kind of signs that

---

15 B (cf. audientes KLat); πειθόμενοι K

διάθεσιν, οἷα πλευρίτιδός ἐστιν ἢ δυσεντερίας. ἐπιστημονικὴ μὲν γὰρ ἡ τῶν τοιούτων νοσημάτων διάγνωσις, ὡρισμένοις σημείοις γνωριζομένη, στοχαστικὴ δὲ ἡ τῶν ἀρτίως εἰρημένων καὶ μόνοις τοῖς ἀκριβῶς ἐπισταμένοις ἑκάστου νοσήματος τὴν οἰκείαν θεραπείαν εὑρίσκεσθαι δυναμένη. ὥστ' ἐπεὶ περὶ ταύτης πρόκειται νῦν λέγειν, αὐτὴ καθ' ἑαυτὴν περαινέσθω, μὴ προσαπτομένων ἡμῶν τῆς διαγνώσεως τῶν διαθέσεων· καὶ μάλισθ' ὅταν ἐκ στοχαστικῶν ἀρχομένη σημείων συνάπτηται τῇ τῶν ὠφελούντων ἢ βλαπτόντων διαγνώσει. τὴν γὰρ ἐν τούτοις μέθοδόν τε καὶ
861K γυμνασίαν ἐν ταῖς | τῶν νοσημάτων διαγνώσεσιν ἰδίᾳ ποιούμεθα καθ' ἑαυτήν, ἵνα τεχνωθείς τις εὑρίσκῃ καθ' ἕκαστον ἄρρωστον, ὥσπερ καὶ ἡμεῖς τά τε πεπονθότα μόρια καὶ τὰς διαθέσεις αὐτῶν.

8. Ἐπανέλθωμεν οὖν αὖθις ἐπὶ τὰς τὴν ὀδύνην ἐργαζομένας διαθέσεις. εἰ μὲν γὰρ αἵματος πλῆθος εἴη τὸ διατεῖνον, ὥσπερ ἀμέλει καὶ ἐν τοῖς φλεγμαίνουσι γίγνεται, φλεβοτομητέον αὐτίκα τῆς δυνάμεως ἰσχυρᾶς οὔσης· εἰ δ' ἤτοι φοβοῖτο τὴν φλεβοτομίαν ὁ ἄνθρωπος ἢ ἀρρωστότερος εἴη τὴν δύναμιν, ἀντισπαστέον τε καὶ παροχευτέον· εἴρηται δὲ ἔμπροσθεν ὅπως χρὴ ταῦτα ποιεῖν. εἰ δὲ καὶ τούτων γενομένων ἔτι μένοι τὸ ἄλγημα, δῆλον μὲν δήπου κατὰ τὸ πεπονθὸς μόριον ἐσφηνῶσθαι τὸ λυποῦν. εὔδηλος δὲ ἡ θεραπεία τοῖς διαφορητικοῖς γινομένη φαρμάκοις. ὡσαύτως δὲ καὶ τὰς διὰ φυσῶδες πνεῦμα γινομένας ὀδύνας ἰασόμεθα, προσβοηθοῦντες ἐπ' αὐτῶν μᾶλλον τοῖς λεπτύ-

occur, for example, in pleurisy or dysentery. The diagnosis of such diseases is scientific, recognized through determinate signs, but the diagnosis of those things just mentioned is conjectural, and can only be discovered by those with a precise knowledge of the specific treatment of each disease. Consequently, since I propose to speak about this matter now, let it be brought to conclusion of itself, lest we become fixed upon the diagnosis of the conditions, and particularly whenever, beginning from the conjectural signs, it links with the diagnosis of those things that help or harm. We create the method by these things and the practice in the diagnoses of the diseases specific to itself, so that   861K
someone who is practiced may discover, in each person who is ill, the affected parts and their conditions, just as we do.

8. Therefore, let us return once again to the conditions that bring about pain. If what distends is an abundance of blood, such as occurs, of course, in the inflammations, we must carry out phlebotomy immediately, if the capacity is strong. If, however, the person either fears phlebotomy or his capacity is rather weak, we must draw off by another outlet and divert [the blood]. I have said before how we ought to do these things. If they have also been done and the pain still remains, it is, I presume, clear that what is causing pain is plugged up in the affected part. Obviously the treatment is by the discutient medications. In like manner too, we shall cure the pains arising due to vaporous

νουσι ἐδέσμασί τε καὶ πόμασιν ἐνέμασί τε καὶ καταπλάσμασιν αἰονήσεσί τε καὶ πυριάμασιν. εἰ δ' ὄγκος
βαρύνων ἢ θλῶν ὀδύνην ἐργάζοιτο, τὸν ὄγκον ἰατέον.
862K   εἰ δὲ δακνῶδες ὑγρόν, | ἐναντιώτατα τούτοις ἐστὶ τὰ
λεπτύνοντα καὶ θερμαίνοντα. χρῄζει γὰρ ἡ τοιαύτη
διάθεσις, εἰ μὲν ἰαθήσεσθαι μέλλει, τῆς κενώσεως τῶν
λυπούντων· εἰ δ' εἴη τοῦτο ποιεῖν ἀδύνατον, τῆς ἐπι-
κράσεως· εἰ δὲ καὶ τοῦτο ἀδύνατον, διὰ τῆς τῶν ναρκωτικῶν φαρμάκων προσφορᾶς· ἥτις ἐστὶ μὲν κοινή,
βλάπτει δὲ ἥκιστα τῶν ἄλλων τὰς τοιαύτας διαθέσεις.
λεπτὰ γὰρ ὑπάρχει ταῖς συστάσεσι καὶ θερμὰ ταῖς
δυνάμεσι τὰ πλεῖστα τῶν τοιούτων ὑγρῶν· ὅσα δὲ δι'
ὀπίου καὶ ὑοσκυάμου καὶ τῶν οὕτως ψυχόντων σκευ-
άζεται φάρμακα, ψύχει τε ἅμα καὶ ξηραίνει πάντως·
κατὰ τοῦτο γοῦν οὐ μόνον ὡς αἰσθήσεως ναρκωτικὰ
χρήσιμα καθέστηκεν, ἀλλὰ καὶ ὡς συνιστάντα καὶ
παχύνοντα τὴν τῶν ὑγρῶν λεπτότητα καὶ προσέτι καὶ
τὴν θερμότητα σφοδρὰν ὑπάρχουσαν ἐμψύχοντα.

παχέων δὲ ὑγρῶν ἢ γλίσχρων ἐπικρατούντων ἐναν-
τιώτατα ὑπάρχει τὰ ναρκωτικὰ φάρμακα καὶ πάνυ
χρὴ φυλάττεσθαι τὴν χρῆσιν αὐτῶν ἐπὶ ταῖς τοιαύταις διαθέσεσιν. οὐ μὴν οὐδ' ὀδύνη τις ἕπεται σφο-
863K   δρὰ τοῖς τοιούτοις χυμοῖς μόνοις οὖσιν. εἰ δέ | ποτε
συμπλακείη κατά τι συμβεβηκός, ἀτμῶδες πνεῦμα
τῶν περιεχόντων ἑαυτὸ σωμάτων ἀδυνατοῦν διεξιέναι,
σφοδρότατα πάντως οἱ οὕτως ἔχοντες ὀδυνῶνται. ἕπεται δ' αὐτοῖς τοῦτο κατὰ διττὴν αἰτίαν, ἤτοι γ' ἔμφραξιν ἢ θερμασίαν· ἥ τε γὰρ ἔμφραξις ἴσχει τὸ πνεῦμα

300

pneuma, helping in this case particularly with thin foods
and drinks, enemas, poultices and moist vapor baths. If,
however, a swelling brings about pain by weighing down or
bruising, we must cure the swelling. On the other hand, if
it is a biting fluid, those agents that are thinning and heat-   862K
ing are most inimical to these. If the pain is going to be
cured, such a condition requires the evacuation of those
things that are pain-producing; but if it is impossible to do
this, what is required is a "tempering" [of humors]. If this
too is impossible, it requires the administration of [one of
the] narcotic medications—one that is common but in-
jures such conditions less than the others. The majority of
such humors are thin in consistency and hot in capacity,
whereas those medications that are prepared from poppy
juice and henbane, and the things cooling in this way, cool
and dry completely at the same time. At all events, in relation to this, they are useful, not only as establishing a dull-
ing of sensation, but also as compounding and thickening
the thinness of the humors as well as cooling the strong
heat that exists.

On the other hand, when thick and viscous humors prevail, narcotic medications are most inimical, and we must
particularly guard against their use in such conditions. Severe pain does not follow such humors when they exist
alone. If, however, at some time, contingently, vaporous   863K
pneuma which is unable to pass through the bodies containing it is intermingled, those so affected assuredly suffer very severe pain. This follows in them for two reasons:
either blockage or heating. The blockage holds back the

καὶ κωλύει διεξιέναι· θερμαινόμενά τε τὰ παχέα καὶ
γλίσχρα φυσῶδες πνεῦμα γεννᾶν πέφυκε. τοιαῦται
μάλιστα γίνονται διαθέσεις τοῖς ἐμπεπλησμένοις ἐδεσμάτων, ψυχρῶν μὲν ταῖς κράσεσι, παχέων δὲ καὶ
γλίσχρων ταῖς συστάσεσιν, ὅταν ἐν αὐτοῖς περι-
έχηται τοῖς χιτῶσι τῶν ἐντέρων τὸ φυσῶδες πνεῦμα.

δυοῖν γὰρ ὄντοιν αὐτῶν, ἐπειδὰν ἐν τῇ μεταξὺ
χώρᾳ τοιοῦτός τις ἀθροισθῇ χυμός, εἰς φυσῶδες μετα-
βάλλει πνεῦμα. παχὺ δὲ δή που τοῦτ' ἔστι καὶ ψυχρὸν
καὶ βραδύπορον. ὅταν οὖν ἴσχηταί τε καὶ διατείνῃ
τοὺς χιτῶνας, ὅ τε χυμὸς ἐξ οὗ τὴν γένεσιν ἔχει ψύχει
σφοδρῶς τὰ ψαύοντα μόρια τῶν ἐντέρων, κατὰ δύο
προφάσεις οἱ οὕτω κάμνοντες ὀδυνῶνται. τὸ μὲν οὖν
παραυτίκα τελέως ἀνώδυνοι γίνονται, πιόντες τι τῶν
ναρκωτικῶν φαρμάκων, αὐξάνεται δὲ ἡ διάθεσις αὐ-
τοῖς παντοίως· | οἵ τε γὰρ χιτῶνες τῶν ἐντέρων πυκνό-
τεροι καὶ δυσδιαπνευστότεροι γίγνονται τῇ ψυχρότητι
τῶν φαρμάκων, ὅ τε χυμὸς παχύτερος καὶ δυσκινητό-
τερος. ὥστε καὶ χρόνου πλέονος ἀναγκαῖον ἔσται
δεηθῆναι τὸν κάμνοντα καὶ τοῦ θεραπεύοντος εἰς
ἄκρον γεγυμνασμένου κατὰ τὴν τέχνην. εἰ γὰρ ὡς
ἔτυχεν αὖθις ἐγχειρήσειε τῇ θεραπείᾳ, κίνδυνος ἐκ
δευτέρου πνευματωθέντας τοὺς χυμοὺς τὴν αὐτὴν ὀδύνην ἐργάσεσθαι· εἶτ' ἐκ δευτέρου δηλονότι δοθῆναί τι
τῶν ναρκωτικῶν φαρμάκων δεῖ, τῆς αὐτῆς ἀνάγκης
καταλαβούσης· ὡσαύτως δὲ κἀκ τρίτου καὶ τετάρτου
καὶ πολλάκις ἐφεξῆς, ἄχρις ἂν ὁ ἄνθρωπος ἀνίατος

864K

pneuma and prevents it passing through and those things that are heating are wont to create a vaporous *pneuma* which is thick and viscous in nature. Such conditions occur particularly in those who are filled full of foods that are cold in terms of *krasis* and thick and viscous in terms of consistency, whenever the vaporous *pneuma* is contained in the actual walls of the intestines.

There are two of these walls, and when some such a humor is collected in the space between them, it changes to vaporous *pneuma*. It may be that this is thick, cold and slow to pass. Therefore, whenever it is retained and distends the walls, and the humor from which it has its genesis strongly cools the parts of the intestines that it touches, those who suffer in this way experience pain for two reasons. Thus, when they drink one of the narcotic medications, they immediately become completely pain-free, but their condition is increased in every way because the walls of the intestines become thicker and are less able to disperse [the liquid] due to the coldness of the medications, and the humor becomes thicker and more difficult to move. Consequently the patient will necessarily require a longer time and someone to treat him who is practiced in the craft to the highest degree. For if, as may happen, he should attempt the treatment again, there will be a danger of the humors, having been turned to a vapor a second time, bringing about the same pain. Then, on this second occasion, quite clearly one of the narcotic medications needs to be given, since the same necessity compels this. The same also applies on a third, or fourth, or frequent occasions in succession, until the man becomes incurable

864K

γενόμενος εἰς καχεξίαν τε τοῦ παντὸς σώματος ἀφίκηται καὶ πολλῷ χρόνῳ κακοπαθήσας ἀποθάνῃ.

πῶς οὖν χρὴ τὰς διὰ τοὺς ψυχροὺς χυμοὺς ἐν τῷ μεταξὺ τῶν ἐντέρων ἠθροισμένας ὀδύνας ἰᾶσθαι; οὐ θερμαίνοντες σφοδρῶς αἰονήσεσιν ἢ καταπλάσμασιν· ὑπὸ γὰρ τῶν θερμαινόντων, εἰ μὴ καὶ διαφορηθεῖεν ἱκανῶς, χέονται καὶ πνευματοῦνται πάντες οἱ γλίσχροι καὶ παχεῖς καὶ ψυχροὶ χυμοί. προσήκει τοίνυν

865K αὐτοὺς τέμνειν τε | ἅμα καὶ πέττειν, ὅπερ ἐκ τῶν λεπτυνόντων γίνεται φαρμάκων· καὶ μάλισθ᾽ ὅταν μὴ θερμαίνωσι σφοδρῶς. ὅτι δὲ καὶ τούτων αὐτῶν ὅσα ταῖς δυνάμεσίν ἐστιν ἀφυσότερα καὶ ξηραντικώτερα, ταῦθ᾽ αἱρεῖσθαι χρὴ μᾶλλον εὔδηλον παντί. καὶ πολλῶν ἰδιωτῶν ἐστιν ἀκοῦσαι πολλάκις ἧττον ἀλγεῖν φασκόντων τὸ κῶλον, ὅταν μήτε καταπλάττωνταί τινι μήτε καταντλῶνται δι᾽ ἐλαίου μήτε κλύζωνται. καὶ ὅσοι γενναῖοί εἰσι καὶ ἰσχυροί, μετρίως διαιτηθέντες ἐν χρόνῳ πλέονι συντομώτερον καὶ ἀκινδυνότερον ἐκθεραπεύονται τῶν βιαιοτέρως ὡς εἴρηται θερμαινόντων. οὐδὲν γὰρ οὕτω χρὴ δεδιέναι κατὰ τοὺς τοιούτους χυμοὺς ὡς θερμότητα χέουσαν μὲν καὶ πνευματοῦσαν αὐτούς, διαφορεῖν δ᾽ ἀδυνατοῦσαν. ἐθεασάμην γοῦν τινα τῶν κατ᾽ ἀγρὸν ἐργατῶν, ὃς ἐπειδὴ ᾔσθετο κωλικῆς ὀδύνης, ἐζώννυντο μὲν αὐτίκα, πρότερον οὐ ζωννύμενος, ἤσθιε δὲ μετ᾽ ἄρτου σκόροδα, μηδεμίαν ἐκλείπων συνήθη πρᾶξιν, ἔπινε δὲ δι᾽ ὅλης μὲν ἡμέρας οὐδέν, εἰς ἑσπέραν δὲ ἀκρατέστερον· εἶτα κοιμώμενος δι᾽ ὅλης νυκτὸς ἔωθεν ἀνίστατο παντάπα-

and reaches a cachectic disposition of the whole body and, after suffering over a long period, dies.

How, then, must we cure the pains due to cold humors that have collected together in the middle of the intestines? It is not by heating vigorously with fomentations and poultices, because due to those things that are heating, unless they are also dispersed sufficiently, all the viscid, thick and cold humors flow and turn into vapor. Therefore, it is appropriate to cut and concoct them at the same time—something which occurs through the thinning medications, particularly when they do not heat strongly. It is clear to everyone that, of these very things, you must choose especially those that are more flatus-dispelling and more drying in their capacities. And you often hear many ordinary people saying there is less pain in the colon whenever they are neither treated with poultices, nor fomentations with oil, nor clysters. Those who are excellent [in health] and strong, if they have followed a moderate regimen over a longer time, are cured more quickly and less dangerously than those who are heated vigorously, as I said. For in relation to such humors, we should fear nothing so much as heat which causes them to flow and be converted to wind, but is not able to disperse them. At any rate, I saw a certain man, a farmworker who, when he felt a colicky pain, immediately girded himself, not having done so before, and ate garlic with bread, leaving out no customary activity; and he drank nothing through the whole day but toward evening he drank more unrestrainedly. Then, having slept through the whole night, he rose

865K

866K σιν ἀνώδυνος. ἔστι γὰρ ἀμέλει καὶ | τὰ σκόροδα τῶν
ἀφύσων καὶ ἀδίψων βρωμάτων.

ἔνιοι δὲ ἀπείρως ἔχοντες αὐτῶν ὑπολαμβάνουσι
κρομμύων εἶναι διψωδέστερα τοῦ παντὸς ἁμαρτάνον-
τες· οὐ μόνον γὰρ οὐκ ἔστι διψωδέστερα κρομμύων,
ἀλλ᾽ οὐδὲ διψώδη τὴν ἀρχήν, ἀφυσότατά τε πάντων
βρωμάτων. ὥστ᾽ ἔγωγε τῶν ἀγροίκων θηριακὴν ὀνο-
μάζω τὸ βρῶμα. καὶ εἴ τις ἢ Θρᾷκας ἢ Κελτοὺς ἢ
ὅλως τοὺς ψυχρὰν γοῦν οἰκοῦντας εἴρξειεν ἐσθίειν
σκορόδων, οὐ σμικρὰ βλάψει τοὺς ἀνθρώπους. ὅσοι
μὲν οὖν ἄνευ πυρετῶν ἀλγοῦσι σφοδρῶς ἔντερα, διὰ
τὴν εἰρημένην αἰτίαν καὶ σκορόδων ἐγχωρεῖ τούτους
ἐσθίειν καὶ τῆς διὰ τῶν ἐχιδνῶν πίνειν, ὠφελούσης
ἄκρως τὰς τοιαύτας διαθέσεις. ἐν πυρετοῖς δὲ εἴ ποτε
γένοιτο, πειρατέον μὲν πρῶτον εἰ προσίενται τὴν διὰ
τῶν κέγχρων πυρίαν ξηράν· εἰ δὲ ἐπὶ τῆσδε μὴ παύ-
σαιντο, τῶν ἀφύσων τι σπερμάτων ἑψήσαντας ἐν
ἐλαίῳ λεπτομερεῖ, κἄπειτα δι᾽ ὀθόνην διηθήσαντας ὡς
καθαρὸν γενέσθαι τοὔλαιον, εἶτ᾽ ἐν αὐτῷ τήξαντας
στέαρ χηνὸς ἐνιέναι. μὴ παρόντος δὲ τοῦ χηνείου
867K στέατος ὀρνιθείῳ | χρῆσθαι· πρὸ παντὸς δὲ ἔστω τὸ
καλούμενον ἄναλον· ἔστω δὲ μὴ πάνυ παλαιόν· ἐγχωρεῖ δὲ καὶ προσφάτῳ χρῆσθαι καὶ μάλιστα τῷ
χηνείῳ. εἰ δὲ μηδ᾽ ἐπὶ τούτῳ καθίστανται, δεύτερον
αὖθις ἐνιέναι ταὐτό, βραχὺ προσεπεμβάλλοντας
καστορίου καὶ ὀπίου, μέγεθος δὲ ἑκατέρου μὴ μεῖζον
κυάμου· τὸ δὲ τοῦ ἐλαίου πλῆθος ἔστω κοτύλη. ἀνα-
λαμβάνεται δὲ ἐκ τούτου τοῦ φαρμάκου καὶ κροκιδίου

early in the morning completely pain-free. Garlic is, of 866K
course, also among the foods that dispel flatus and do not
cause thirst.

There are some who, being without experience, assume
garlic causes more thirst than onions, but they are altogether mistaken. Not only is garlic not more thirst-producing than onions, but it is not thirst-producing fundamen-
tally and is the most flatus-dispelling of all foods, so that I
myself call the food a "theriac for rustics." And if someone
were to prevent Thracians or Celts, or those who live in
cold regions generally, from eating garlic, it would harm
those people to no small extent. Therefore, it is permissible for those who, without fevers, feel severe pain in the intestines due to the cause spoken of, to eat garlic and drink
what is made from theriac, since it greatly benefits such
conditions. But if such a pain should occur at some time in
a fever, you must first try a dry fomentation made of millet
if the pains allow it. If, however, the pains do not settle with
this, you must try one of the flatus-dispelling seeds that has
been boiled in thin oil and then strained through fine
linen, so that the oil becomes pure, and having dissolved
goose fat in this, insert it *per rectum*. If goose fat is not
available use that of a cock; but above all let it be what is 867K
called unsalted and not very old. It is also permissible to
use what is recent and especially that of the goose. If the
pains do not settle after this, insert the same thing again for
a second time, putting in a little castor and the juice of
poppy, the amount of each not more than a bean. Make the
amount of oil a kotyle. Take up some of this medication on

τι· χρὴ δὲ ἐξάψαντας αὐτὴν ἰσχυροῦ νήματος ἐπὶ
πλεῖστον εἴσω κατὰ τὸ ἀπευθυσμένον ἔντερον ἐντιθέναι, πρὸς τῷ καὶ ῥαδίως ὅτε βούλει κομίζεσθαι
καὶ τὴν ὠφέλειαν ἐναργῆ παρέχεσθαι.

κατὰ δὲ τὸν αὐτὸν τρόπον καὶ τὰ τῶν ὀφθαλμῶν
καὶ τὰ τῶν ὤτων ἀλγήματα διὰ τοιοῦτον χυμὸν ἢ
πνεῦμα φυσῶδες ἐν πυρετοῖς γινόμενα πραΰνειν
προσήκει, τῇ τε διὰ τῶν κέγχρων πυρίᾳ χρώμενον,
ἐπειδὴ καὶ κουφότατόν ἐστιν ἁπάντων τοῦτο καὶ
ξηραντικώτατον, ἀλύπους τε τὰς ἀπορροίας ἔχει καὶ
ἀδήκτους, ἐγχέοντά τε τοῖς ὠσὶ τὸ διὰ καστορίου καὶ
ὀπίου φάρμακον, ᾧ συνήθως χρώμεθα πρὸς τὰ τοιαῦτα. τὸ δὲ ὑγρὸν ᾧ ἀναδεύσεται ταῦτα, τὸ παρ᾽ ἡμῖν
868K καλούμενον | ἕψημά ἐστιν· οἱ πλεῖστοι δὲ τῶν ἰατρῶν
ὀνομάζουσιν αὐτὸ σίραιον. εἰ δὲ καὶ καταπλάττειν
ποτὲ δεήσειε, κωδίας ἐν ὕδατι καθέψοντα,[16] δι᾽ ἀλεύ-
ρου τήλεως ἢ κριθῶν ἢ λινοσπέρμου ἐμβαλλομένου
τῷ ὕδατι, τὸ κατάπλασμα συνθετέον. εἰδέναι δὲ χρὴ
τὸ διὰ καστορίου φάρμακον οὐ μόνον ὤτων ἄλγημα
πραΰνειν, ἀλλὰ καὶ ὀφθαλμῶν καὶ ὀδόντων ἐνστα-
ζόμενον τοῖς ὠσίν. ἴσασι δὲ δήπου πάντες ἤδη καὶ τὰ
δι᾽ ὀπίου κολλύρια σφοδροτάτας ὀδύνας ὀφθαλμῶν
πραΰνοντα. χρηστέον δ᾽ αὐτοῖς, ὡς εἴρηται, μεγίστης
ἀνάγκης καταλαβούσης, εἰδότα μὲν ὅτι βλαβήσεταί
τι τὰ μόρια καὶ ἀσθενέστερα πρὸς τὸ λοιπὸν τοῦ βίου
γενήσεται ψυχθέντα, τὸ δὲ σωθῆναι τὸν ἄνθρωπον ἐν
τῷ παραχρῆμα τῆς εἰς ὕστερον ἀκολουθούσης βλά-

a wool suppository. It is necessary to insert this into the rectum as far as possible, having attached a strong thread to it so as to recover it easily when you wish to, and to provide obvious benefit.

In the same way too, it is appropriate to soothe the pains of the eyes and ears that occur due to such a humor or vaporous *pneuma* in fevers, using the fomentation made from millet, since this is both the lightest of all things and the most drying and has harmless and nonbiting emanations, and to pour into the ears the medication made from castor and poppy juice which we customarily use in such cases. The fluid with which these things are mixed into a paste I call "hepsema" although the majority of doctors call it *siraeon*.[21] If at some time there is also need to apply a poultice, you must compound this, boiling down poppy heads in water using meal of fenugreek, barley or linseed put in the water. You should know that the medication made from castor not only soothes pains of the ears, but also of the eyes and teeth when instilled into the ears. Everyone already knows, of course, that the collyriums made from the juice of poppy soothe very severe pains of the eyes. You must use them, as I said, when constrained by the greatest necessity, knowing that the parts will be harmed in some way and, having been cooled, will become weaker for the rest of the person's life, although he will be saved in the short term, preferring this despite the later,

868K

---

[21] Here the terms are taken to refer to what is boiled down and not necessarily wine; see Galen's *De compositione medicamentorum secundum locos*, XIII.8K.

---

16 B; καθέψοντας K

βης προαιρούμενον. ἐπεί τοί γ᾽ ὅτι πολλοὶ τῶν χρησαμένων αὐτοῖς ἐγγὺς ἧκον, οἱ μὲν τοῦ μηδ᾽ ὅλως ὁρᾶν,
οἱ δὲ τοῦ κωφωθῆναι, γινώσκεται πᾶσιν. ὅθεν ἡμεῖς εἴ
ποτε ἀναγκασθείημεν αὐτοῖς χρήσασθαι, μετὰ ταῦτα
κατὰ τὴν τῆς ὑγείας καιρὸν ἐκθερμαίνομεν τὰ μόρια,
τοῖς μὲν ὠσὶν ἐγχέοντες τὸ διὰ μόνου τοῦ καστορίου,
869K   τοῖς δ᾽ ὀφθαλμοῖς | τὰ θερμαίνοντα κολλύρια μόνα·
μάλιστα δὲ ἐπαινοῦμεν εἰς ταῦτα τὸ διὰ κινναμώμου.

περὶ μὲν οὖν ὀδύνης τῆς ἐπὶ παχέσιν ἢ γλίσχροις
ἢ ψυχροῖς χυμοῖς ἢ φυσώδει πνεύματι καὶ ταῦθ᾽
ἱκανά. παραπλησία γὰρ ἁπάντων αὐτῶν ἐστὶν ἡ θεραπεία διὰ τῶν εἰρημένων ὑλῶν περαινομένη, πλὴν ἤ γε
διὰ πνεῦμα φυσῶδες ἐξαίρετον ἴαμα κέκτηται τὴν
σικύαν πολλάκις προστιθεμένην ἅμα δαψιλεῖ φλογί.
καὶ δόξει σοι τὸ βοήθημα τοῦτο μαγείᾳ τινὶ παραπλήσιον ἐργάζεσθαι κατὰ τὰς τοιαύτας διαθέσεις, εἴτε
κατά τι τῶν ἐντέρων, εἴτε κατ᾽ ἄλλο τι τοῦ σώματος
γένοιτο μόριον. αὐτίκα γὰρ ἀνώδυνοί τε ἅμα καὶ εἰς
τέλος ὑγιεῖς οἱ διὰ πνεῦμα φυσῶδες ὀδυνώμενοι
γίγνονται, προσβληθείσης σικύας. εἰ δὲ μὴ μόνον εἴη
πνεῦμα φυσῶδες, ἀλλὰ καὶ χυμὸς ἐξ οὗ τοῦτο γεννᾶται, παραχρῆμα μὲν ἀνώδυνοι γίνονται, πάλιν δ᾽
αὐτοῖς ἤτοι διὰ τῆς ἐπιούσης νυκτὸς ἢ κατὰ τὴν
ὑστεραίαν ἡμέραν, ἢ καὶ διὰ τρίτης, ὅμοιαι συμβαί-
νουσιν ὀδύναι, καὶ μάλισθ᾽ ὅταν ἐξαμαρτάνωσί τι
870K   περὶ τὴν δίαιταν ἢ φιλοτιμότερον | ἐκθερμαίνωσι τὰ
μόρια. σοὶ δὲ καὶ τοῦτο μὲν αὐτὸ μέγιστον ἔστω
γνώρισμα τῆς διαθέσεως. ἐπέσθω δὲ καὶ ἡ θεραπεία

310

consequent damage. Mark you, everyone knows that many people, when they use these things, in some cases come close to losing their sight altogether, and in other cases to being made deaf. On which account, if we should ever be compelled to use them, after their use, in a time of health, we heat the parts, pouring into the ears [oil] made from castor alone, and into the eyes the heating collyriums 869K alone. I particularly commend those made from cinnamon for these purposes.

This is enough on the subject of pain due to thick, viscous or cold humors, or to vaporous *pneuma*. The treatment of all these things is similar because it is accomplished by the aforementioned materials. A notable exception is the cure of pain due to vaporous *pneuma* which is achieved with a cupping glass, often when applied along with abundant heat. And it will seem to you that this remedy acts like magic in such conditions, whether they occur in some part of the intestines or in some other part of the body. Those who are suffering pains due to vaporous *pneuma* become pain-free immediately and, at the same time, completely healthy when the cupping glass has been applied. If, however, it is not only vaporous *pneuma* but also a humor from which this pain is generated, they immediately become pain-free, but similar pains recur in them, either during the following night, or the next day, or the third [day], and particularly whenever they err in some respect in their regimen or heat the parts too liberally. Let 870K this also be an important sign of the condition for you. And

προσήκουσα μὴ θερμαίνοντι μὲν ἐπιφανῶς τὸ μόριον,
ἀγωγῇ δὲ ἐπιμελείας χρωμένῳ λεπτυνούσῃ. εἰ δὲ καὶ
κατὰ γαστέρα τῶν τοιούτων τι συμβαίνει, θαυμαστῶς
ὑπὸ κλυσμάτων ὀνίνανται δριμέων. χρὴ δὲ πρῶτον
μὲν αὐτῶν ἰᾶσθαι τὸν παροξυσμὸν τῆς ὀδύνης, προσ-
βάλλοντα σικύαν ὑπὲρ τοῦ διαπνεῦσαι τὸ φυσῶδες
πνεῦμα· μετὰ ταῦτα δὲ ἐκκενοῦν τὸν χυμόν, ἐνιέντα
τῶν τοιούτων τι φαρμάκων. ἐγὼ δὲ εἴωθα χρῆσθαι τῶν
λεπτομερῶν ἐλαίων τινί, πήγανον ἐναφεψῶν. ἔνιοι δὲ
ὑπὸ τὴν τοιαύτην θεραπείαν ἀχθέντων, ὅταν παρα-
δέξωνται τοὔλαιον, ὀδυνῶνται σφοδρότατα· κἄπειτ᾿
ὀλίγον ὕστερον ἐκκρίνουσιν ὑαλώδη χυμόν· ἐφ᾿ ᾧ
παραχρῆμα τήν τ᾿ ὀδύνην ἅμα καὶ τὴν διάθεσιν ἐκ-
θεραπεύονται· κενωθέντος γὰρ τοῦ τὸ φυσῶδες πνεῦμα
γεννῶντος αἰτίου πάντα παύεται.

τοιούτοις χυμοῖς ἐναντιωτάτην μὲν ἔχουσι φύσιν οἱ
λεπτοὶ καὶ δριμεῖς, ὁμοίαν δὲ τὴν ὀδύνην· ἐνίοτε δὲ καὶ
871K σπασμοὺς συντόνους ἐπιφέρουσιν ἐν τῷ | στόματι τῆς
γαστρὸς ἀθροισθέντες· ὥσπερ καὶ πρώην τῷ μικρὸν
ὕστερον ἐμέσαντι τὸν ἰώδη χυμόν. ἄμεινον δὲ οὐκ
ἰώδη λέγειν αὐτόν, ἀλλ᾿ ἀκριβέστατον ἰόν· ἦν γὰρ δὴ
τοιοῦτος οἷος ὁ κάλλιστος ἰός. ἀλλὰ τούτῳ γε τῷ
νεανίσκῳ μετὰ τοῦ σπᾶσθαι καὶ συγκόπτεσθαι καί
τινες ἐγίνοντο νοτίδες ψυχραὶ καὶ ὁ σφυγμὸς ἐσχάτως
μικρὸς ἦν. ἐξ ὧνπερ καὶ τεκμηράμενος ἐν τῷ στόματι
τῆς γαστρός, ὃ συνήθως ὀνομάζομεν στόμαχον, εἶναί
τινα δακνώδη χυμόν, ἔδωκα πιεῖν αὐτῷ ὕδατος χλια-
ροῦ· μεθ᾿ ὃ παραχρῆμα τοιοῦτον ἤμεσεν οἷόν περ εἰ

let the appropriate treatment follow, not significantly heating the part but using the treatment that is thinning. If one of these pains also occurs in the stomach, it is benefited marvelously by bitter clysters. But first it is necessary to cure the paroxysm of their pain, applying the cupping glass for dissipating the vaporous *pneuma* and, after this, evacuating the humor by inserting one of these medications *per rectum*. It is my custom to use one of the thin oils, boiling it down with rue. Some who are treated in this way, when they have received the oil, suffer very strong pains and then, a little later, evacuate a glassy humor. After this, they are immediately and effectively cured in terms of the pain and the condition at the same time because, when the cause generating the vaporous *pneuma* has been evacuated, everything settles down.

The humors that are thin and bitter have a nature that is absolutely opposite to such humors, but the pain is similar. Sometimes, they also bring on strong spasms when they are gathered in the opening of the stomach, just as [happened] to the person who, a little while later, vomited rust-colored humor. It is better not to speak of this as rust-colored but more precisely as verdigris, for it was like the most beautiful rust. But in this particular young man certain moist coolings occurred along with spasms and syncope, and the pulse was extremely small. Having conjectured from these things that there was also evidence of a certain biting humor in the opening of the stomach, which we customarily call the cardiac orifice, I gave him lukewarm water to drink which he immediately vomited— the kind of effect for which, if perchance you wished to

καὶ σὺ βουληθείης ἐργάσασθαι, μίξας ὕδατι τὸν εὐαν-
θέστατον ἴον.

ὅταν μὲν οὖν ἐν τῇ γαστρὶ συνίσταται τοιοῦτος
χυμός, ἐμέτοις ἐκκαθαίρειν αὐτόν· ὅταν δὲ ἐν τοῖς
ἐντέροις, ἐνιέναι διὰ τῆς ἕδρας ἐπιτήδειόν τι τῶν
τοιούτους χυμοὺς κατακλύζειν δυναμένων. εἶναι δὲ
χρὴ τοῦτο ῥυπτικὸν μὲν πάντως. ἀλλ' ἐπειδὴ τὰ πλεῖστα τῶν τοιούτων δάκνει, κάλλιστα ἂν εἴη τῶν ἀδήκτων τι ῥυπτικὸν ἐκλέγεσθαι· τοιοῦτον δέ ἐστιν ἐν τοῖς
μάλιστα πτισάνης χυλός. ἐδέσματα δὲ αὐτοῖς εὔχυμά
872K   τε καὶ δύσφθαρτα δοτέον, ὧν | εἴρηται καὶ παραδεί-
γματα κατὰ τὸν ἔμπροσθεν λόγον.

ἐπεὶ δὲ καὶ τῶν διὰ ξηρότητα σφοδρὰν τεινομένων
τε ἅμα καὶ ὀδυνωμένων ἐμνημόνευσα, προσθεῖναί τι
καὶ περὶ τούτων ἄμεινον. εἰδέναι γὰρ χρὴ τὴν τοιαύτην διάθεσιν ἐνδεικνυμένην μὲν εἰ μέλλει θεραπεύειν
τις αὐτήν, ὑγρότητα, χαλεπὴν δ' οὖσαν ἢ καὶ παντάπασιν ἀδύνατον ἐκθεραπεύεσθαι, λόγῳ πυρετοῦ γενομένην. ἕπεται δὲ μάλιστα ταῖς ὀλεθρίαις φρενίτισι καὶ
σωθέντα τινὰ τῶν οὕτω σπασθέντων οὔτ' αὐτὸς εἶδον
οὔτ' ἄλλου λέγοντος ἤκουσα. τὰ πολλὰ γὰρ οἱ σπασμοὶ γίγνονται διά τε πλήρωσιν τῶν νευρωδῶν μορίων, ᾧ λόγῳ καὶ τοῖς φλεγμαίνουσιν ἰσχυρῶς ἕπονται·
καὶ προσέτι καὶ διὰ δακνώδη χυμὸν λεπτόν, ἀναβι-
βρώσκοντα τὰ νευρώδη μόρια· καὶ ψύξιν ἰσχυράν,
ὅμοιόν τι πήξει δρῶσαν. οὗτοι μὲν οἱ εἰρημένοι τρεῖς
σπασμοὶ θεραπεύονται πολλάκις· ἀνίατος δὲ ὁ διὰ
ξηρότητα τῶν νευρωδῶν μορίων γιγνόμενος.

314

bring it about, you would mix the brightest verdigris with water.

Whenever such a humor exists in the stomach, purge it by vomiting. On the other hand, whenever it exists in the intestines, it is useful to insert *per rectum* one of those things capable of washing away such humors. It is necessary for this to be thoroughly cleansing. But since the majority of such things are biting, it would be best for one of those that is nonbiting to be chosen as a cleansing agent. Of such a kind, particularly, is juice of ptisan. And to these [patients] you must give foods productive of good humors and not easily spoiled, examples of which have also been mentioned in the previous discussion. 872K

Since I also made mention of those who lie stretched out and in pain due to severe dryness, it is better that I add something about them too. It is necessary to know, if one intends to treat such a condition, that it indicates [the need for] moisture, although it is difficult or altogether impossible to treat successfully if it has occurred by reason of a fever. It particularly follows the deadly phrenitides, and I myself have not seen anyone who has been saved after having suffered convulsions in this way, nor have I heard another say [he has]. In the majority of cases the convulsions occur due to the repletion of the nervous parts, for which reason they also follow in those with severe inflammation. And in addition, they may also occur due to a thin biting humor which erodes the nervous parts, and due to a strong cooling which acts in a similar way to freezing. The three [kinds of] convulsion mentioned are often treatable, whereas that occurring due to dryness of the nervous parts is incurable.

315

ἐδείχθη γάρ μοι κἂν τῷ Περὶ μαρασμοῦ λόγῳ
παντάπασιν ἀθεράπευτος ἡ τῶν στερεῶν σωμάτων
ξηρότης. ὥστ᾽ οὐδὲν ἔτι χρὴ περί γε τῶν τοιούτων
873K λέγειν συπτωμάτων· οὐ μὴν | οὐδὲ περὶ τῶν διὰ
κένωσιν ἀμέτρων[17] ἤτοι διὰ γαστρὸς ἢ δι᾽ ἐμέτου[18] ἢ
δι᾽ αἱμορραγίας· εἴρηται γὰρ ὑπὲρ αὐτῶν ἤδη μετρίως
ἐν τοῖς ἔμπροσθεν αὖθίς τε διελθεῖν ἀναγκαῖον ἔσται
καὶ μάλισθ᾽ ὅταν ὁ λόγος μοι γίγνηται περὶ τῶν παρὰ
φύσιν ὄγκων. καταπαύσω τοιγαροῦν ἤδη τὸν ἐνεστῶτα λόγον, ἐπειδὴ περὶ τῶν ἀναγκαιοτάτων συμπτωμάτων καὶ μάλιστα τῶν συνεζευγμένων αὐταῖς ταῖς
διαθέσεσι τῶν πυρετῶν αὐτάρκως διῆλθον.

[17] K; ἄμετρον B
[18] K; ἐμέτων B, recte fort.

I demonstrated in the work *On Marasmus*[22] that dryness of the solid parts is altogether untreatable. As a result, there is nothing I still need to say, at least about such symptoms; and certainly not about immoderate ones due to 873K evacuation, whether through the stomach, or through vomiting, or through hemorrhage. For I already spoke, within limits, about these matters in what has gone before, and it will be necessary to go over them again, and especially when my discussion turns to swellings contrary to nature.[23] So then, I shall put an end to the present discussion now, since I have gone over the most necessary symptoms sufficiently, and particularly those connected with the actual conditions of the fevers.

[22] *De marcore*, VII.666–704K; English translation by Th. C. Theocharides (1971).

[23] Reference to Books 13 and 14 of the present work. See also Galen's *De tumoribus praeter naturam*, VII.705–32K.

ΒΙΒΛΙΟΝ Ν

1. Δύο μὲν ἤδη γένη νοσημάτων ὅπως ἄν τις ἰῷτο μεθόδῳ δεδήλωται· τὸ μὲν ἕτερον, ἡ δυσκρασία, παλαιὰν ἔχουσα προσηγορίαν, τὸ δ᾽ ἕτερον ὑφ᾽ ἡμῶν ὠνομασμένον, ἡ τῆς συνεχείας λύσις. ὑπὲρ ἧς πρώτης γράψαντες ἐν τῷ τρίτῳ καὶ τετάρτῳ καὶ πέμπτῳ καὶ ἕκτῳ τῶνδε τῶν ὑπομνημάτων, ἐφεξῆς αὐτῇ τὰ κατὰ δυσκρασίαν γιγνόμενα μέχρι τοῦ δωδεκάτου διήλθομεν. ἐν δὲ τῷδε τῷ τρισκαιδεκάτῳ τῆς ὅλης πραγματείας ὄντι περὶ τῶν παρὰ φύσιν ὄγκων ἀρξόμεθα λέγειν, ἐν οἷς δηλονότι κατὰ μέγεθος ἐξίσταται τὰ |
μέλη[1] τοῦ κατὰ φύσιν. ὑγείαν δὲ καλεῖν ἢ κατὰ φύσιν οὐ διοίσει πρός γε τὰ παρόντα. πολλῶν δὲ κατ᾽ εἶδος ὄντων ἐν αὐτοῖς παθῶν, περὶ πρώτης ἐροῦμεν τῆς φλεγμονῆς.

ἄμεινον γὰρ ἀπὸ ταύτης ἄρξασθαι διά τε τὸ συνεχέστατα γίγνεσθαι καὶ πυρετούς τε καὶ ἄλλα συμπτώματα ἐργάζεσθαι σφαλερώτατα. λεγόντων δὲ πολλάκις τῶν παλαιῶν φλεγμονὴν τὴν φλόγωσιν, ἰστέον νῦν ἡμᾶς οὐ περὶ ταύτης διέρχεσθαι τῆς φλεγμονῆς, ἀλλ᾽ ἥτις ἅμα τῇ φλογώσει καὶ τάσιν ἔχει περὶ τὸ

---

[1] K, B; μέρη conj. Boulogne (cf. partes KLat)

BOOK XIII

1. It has now been shown how someone might cure two classes of diseases by method. One, *dyskrasia*, has an ancient name; the other, dissolution of continuity, I myself named. Having first written about the latter in the third, fourth, fifth and sixth books of these treatises, after that I went through the diseases occurring as a result of *dykrasia*, as far as the twelfth book. In this, which is the thirteenth [book] of the whole treatise, I shall begin to discuss swellings contrary to nature[1] in which the parts manifestly differ in magnitude from what accords with nature. Whether we call the latter "health" or "accord with nature" will make no difference, at least for our present purposes. Since, however, there are many kinds of affection among these [swellings], I shall speak first about inflammation.

It is better to begin with this because it occurs most frequently and gives rise to fevers and other very dangerous symptoms. Although the ancients often used *phlegmonē* for *phlogōsis*, we must now realize that we are not discussing this particular kind of *phlegmonē* but that which, together with *phlogōsis*, also holds the part in a state of ten-

---

[1] Throughout this book the translation "swelling" is used for *onkos*, although "tumor" in the general sense of "any swelling or tumefaction" (S) could also be used.

μόριον, ἡμῖν θ' ἁπτομένοις φαινομένην αὐτῷ τε τῷ
κάμνοντι διὰ τῆς ἰδίως ὀνομαζομένης συναισθήσεως.
οὐδὲν δ' ἧττον τῆς τάσεως ἀντίτυπόν ἐστι τὸ φλεγμαῖνον μόριον ἐν ὄγκῳ τε μείζονι τοῦ κατὰ φύσιν·
ὀδύνη δ' αὐτῷ σύνεστιν ἤτοι γ' ἐλάττων ἢ μείζων·
ἐνίοτε δὲ καὶ μετὰ σφυγμοῦ συναισθήσεως, ὅταν ἐπιπλέον αὐξηθῇ τὸ νόσημα, καὶ μάλισθ' ἡνίκα ἐκ-
πυΐσκεται. οὕτω δὲ καὶ τὸ καλούμενον ἔρευθος ἤτοι γ'
ἧττον ἢ μᾶλλον. ἀεὶ δὲ πάντως ἐστὶν ἐν τοῖς φλεγμαί-
876K νουσι μορίοις· ὥστε κἂν ἐν τῷ τοῦ ποδὸς | ἴχνει, κἂν
κατὰ τὸ τῆς χειρὸς ἔνδον γένηται μεγάλη φλεγμονή,
καὶ ταῦτα φαίνεσθαί πως ἑαυτῶν ἐνίοτε ἐρυθρότερα.

2. Δέδεικται γάρ τοι πᾶσα φλεγμονὴ δι' ἐπιρροὴν
αἵματος γιγνομένη τισὶ μὲν εὐθέως θερμοῦ πλέον ἢ
κατὰ φύσιν ἦν θερμόν, ἅπασι δ' οὖν ἐν τῷ φλεγμαίνοντι μορίῳ θερμοτέρου γιγνομένου. καὶ τοῦτο κοινὸν
ἁπάσαις ταῖς αἱρέσεσίν ἐστιν, εἴτε σφήνωσιν μόνην
αἰτιῶνται κατὰ τὰ πέρατα τῶν ἀγγείων, εἴτε παρέμ-
πτωσιν τοῦ αἵματος ἐν μόναις ταῖς ἀρτηρίαις, εἴτ'
ἔμφραξίν τινα, εἴτε ἔνστασιν ἐν λόγῳ θεωρητοῖς
ἀραιώμασιν. ὥστε καὶ ὁ τῆς ἰάσεως σκοπὸς ἁπάσαις
κοινὸς ἡ κένωσις τοῦ πλεονάζοντος αἵματος ἐν τῷ
φλεγμαίνοντι μορίῳ. γιγνομένης δ' ἔτι τῆς φλεγμονῆς
διττὸς ὁ σκοπὸς ὥσπερ καὶ τῶν ἄλλων ἁπάντων ἐδεί-

---

2 LSJ offers several meanings for the term "synesthesia,"
which has come to have a specific technical sense in modern usage. The sense here is an accompanying sensation of the disease
which the patient himself has, as found in Aretaeus, *On Causes*

sion which is apparent not only to us when we palpate it but also to the patient himself through what is termed, in a particular sense, "accompanying sensation."[2] In respect of the inflamed part, firmness of the tension is no less than in a swelling that is greater than normal. Pain is present in it, either more or less, sometimes with an accompanying throbbing sensation when the disease is increased still more, and especially when there is suppuration. Similarly too, the so-called redness is either less or more. Under any circumstances, redness is always present in inflamed parts so that, even if severe inflammation occurs in the sole of the foot or the palm of the hand, these parts also appear redder than normal in some way.

876K

2. So then, it has been shown that every inflammation arises through an influx of blood that in some cases is immediately hotter than accords with nature, but in all cases becomes hotter in the inflamed part itself. This [view] is common to all the sects, whether they blame obstruction at the ends of the vessels alone, or *paremptosis* of blood in the arteries alone, or some blockage, or impaction in the "theoretical" pores.[3] As a result, the aim of treatment, which is actually common to all [sects], is the evacuation of the excess blood in the inflamed part. However, when the inflammation is still in the process of occuring, there is a

and Signs of Acute Diseases, 2.9, and *On Causes and Signs of Chronic Diseases*, 2.2.

[3] *Paremptosis* is a concept particularly associated with Erasistratus and indicates a passage (spillover) of blood from veins to arteries; see Galen, *De plenitudine*, VII.542K. "Theoretical" pores (literally, pores visible to reason) are considered in the Introduction, section 2, on Methodic theory.

321

χθη νοσημάτων, ὅσα τὴν γένεσιν ἐνεστῶσαν ἔτι καὶ
μήπω συμπεπληρωμένην ἔχοι. τὸ μὲν γὰρ γεγονὸς
αὐτῶν ἤδη τῷ θεραπευτικῷ μέρει τῆς ἰατρικῆς ὑπο-
πέπτωκε, τὸ δ' ἔτι γιγνόμενον τῷ προφυλακτικῷ. καὶ
877K διὰ | τοῦτο ἔφαμεν οὐχ ἁπλῆν, ἀλλὰ σύνθετον εἶναι
τὴν ὅλην ἐπιμέλειαν τῶν ἔτι γιγνομένων παθῶν ἐκ
προφυλακτικῆς τε καὶ θεραπευτικῆς.

ὥσπερ γε καὶ εἰ μηδ' ὅλως ἄρχοιτο φλεγμαίνειν
μηδέπω, φαίνοιτο δὲ τὸ τῶν γεννῆσαι δυναμένων
αἰτίων εἶδος ἤδη κατὰ τὸ σῶμα, σκοπὸς κἀπὶ τούτων
ἁπάντων ἡ προφυλακὴ μόνη. μηδέπω δ' αἰτίας μηδὲ
μιᾶς ὑποτρεφομένης ἐν τῷ σώματι, τὸ καλούμενον
ὑγιεινὸν μέρος τῆς τέχνης προνοεῖται καὶ οὕτως ἐχόντων. ὅσα τοίνυν αἴτια τὴν φλεγμονὴν ὁρᾶται γεννῶν-
τα, ταῦτα ὅταν μὲν ἤδη πως ᾖ κατὰ τὸ σῶμα, μικρὰ δ'
ἔτι καὶ ἀρχόμενα, κωλύειν αὐτὰ δεῖ μείζω γενέσθαι,
καὶ τοῦτ' ἐστὶν ἡ προφυλακὴ τῆς φλεγμονῆς. εἰ δὲ
τηλικοῦτον ἔχει τὸ μέγεθος ὡς ἤδη ποιεῖν φλεγμονήν,
ἐκκόπτειν μὲν χρὴ ταῦτα, τὸ δὲ ἤδη γεγονὸς αὐτῆς
ἰᾶσθαι.

3. Γένεσις μὲν οὖν κοινὴ πάσαις ταῖς φλεγμοναῖς
ἐξ αἵματος ἐπιρροῆς ἐστι πλείονος ἢ ὅσου δεῖται τὸ
μέρος, ὡς ἔν τε τῷ Περὶ τῶν παρὰ φύσιν ὄγκων
878K ἐδείχθη | καὶ τῷ Τῆς ἀνωμάλου δυσκρασίας. ἐπιρρεῖ δὲ
πλέον, ἐνίοτε μὲν ἑτέρου τινὸς ἢ ἑτέρων τινῶν μορίων
εἰς αὐτὸ πεμπόντων, ὑποδεχομένου δὲ τοῦ φλεγμαίνειν
ἀρχομένου, ποτὲ δὲ ἕλκοντος ἐφ' ἑαυτὸ τοῦ πάσχον-

---

4 *De tumoribus praeter naturam*, VII.705–32K (see particu-

twofold objective, as was also shown for all other diseases that are still in the evolving phase and are not yet fully established. For what has already occurred of these diseases falls under the therapeutic part of medicine, whereas what is still coming into existence falls under the prophylactic part. And because of this, I said that the complete care of affections that are still in evolution is not simple but compound, involving both prophylactic and therapeutic elements. 877K

In fact, in similar fashion, if inflammation is not yet beginning at all, but the kinds of causes that can generate [inflammation] are already apparent in the body, the aim, in all such instances, is prophylactic alone. And even if no single cause has yet arisen in the body, the part of our craft that is referred to as "pertaining to health" provides also for people in this condition. Therefore, regarding those causes that are seen to generate inflammation, whenever these are already in the body in some way, but are still small and incipient, we must prevent them becoming greater, and this is the prophylaxis of inflammation. If, on the other hand, these causes have such magnitude that they already bring about inflammation, we ought to eradicate them and cure what has already occurred of the inflammation itself.

3. Therefore, a common genesis for all inflammations is from a flow of blood greater in amount than the part requires, as I demonstrated in the work *On Abnormal Swellings* and in *On Anomalous Dyskrasias*.[4] Sometimes, a greater amount [of blood] flows when one or several different parts send it and the parts beginning to become inflamed receive it, and sometimes when the affected part 878K

larly p. 723), and *De inaequali intemperie*, VII.733–52K (see particularly p. 738).

τος. τὰ μὲν οὖν πέμποντα ποτὲ μέν, ὡς τῷ πλήθει περιττὸν ἢ ἀνιαρὸν τῇ ποιότητι, διωθεῖται τὸν χυμόν, ἐνίοτε δὲ καὶ δι᾽ ἄμφω· τὰ δὲ ἕλκοντα διὰ θερμότητα νοσώδη. κατὰ δὲ τὰς ὀδύνας ἄρχεται μὲν ἐκ τοῦ τὴν ὀδύνην ἔχοντος ἡ αἰτία, τὰ δ᾽ ὑπερκείμενα τὸ σύμπαν ἐργάζεται τῆς φλεγμονῆς. τὸ μὲν οὖν ἐπὶ τὸ θερμαινόμενον ἤτοι γ᾽ ἕλκεσθαι τοὺς πλησιάζοντας χυμούς, ὡς ἡμεῖς φαμεν, ἢ ὡς Ἀσκληπιάδης ἐνόμιζε, ῥεῖν, ἐναργῶς φαίνεται καὶ φυλαττέσθω τῷ λόγῳ κἀνταῦθα τὸ ἀληθές ἐξ αὐτοῦ τοῦ βλέπεσθαι.

τά γε μὴν ὀδυνώμενα φαίνεται μὲν καὶ ταῦτα φλεγμαίνοντα διὰ τὴν ὀδύνην, ἡ δ᾽ αἰτία τισὶ μὲν οὐδ᾽ ὅλως εἴρηται, τισὶ δ᾽ οὐδαμῶς πιθανή. καθ᾽ ἡμᾶς δ᾽ ἐστὶ τοιάδε. δέδεικται δὲ κατὰ τὴν πραγματείαν ἣν Περὶ τῶν φυσικῶν δυνάμεων ἐποιησάμεθα μία καὶ ἥδε
879K τῆς φύσεως δύναμις, ἣν ἀποκριτικὴν | ὀνομάζομεν. ἐνεργεῖ δ᾽ αὕτη κατ᾽ ἐκείνους τοὺς καιροὺς ἐν οἷς ἂν αἴσθηται λυπούντός τινος. ἐν δέ τι τῶν λυπούντων αὐτήν ἐστι καὶ τὸ τὴν ὀδύνην ἐργαζόμενον αἴτιον, ὅ τί ποτ᾽ ἂν ᾖ. τοῦτ᾽ οὖν ἀποτρίψαι σπεύδουσα φλεγμονὴν ἔστιν ὅτε κατὰ τὸ μέρος ἐργάζεται. ὅταν γὰρ ταῖς πρώταις ἑαυτῆς κινήσεσι μηδὲν ἀνύσῃ, τηνικαῦτ᾽ ἤδη σφοδρότερον ἐπιχειροῦσα τὸ λυποῦν ἀποτρίψασθαι συνεκθλίβει τι πρὸς τὸ μέρος ἐκ τῶν ὑπερκειμένων αἷμα καὶ πνεῦμα. κἀντεῦθεν ἐπὶ ταῖς ὀδύναις εἰς ὄγκον αἴρηται[2] τὸ μέρος ἀνάλογον τῷ πρὸς αὐτὸ ῥυέντι χυμῷ.

[2] B (cf. attollatur KLat); εἴρηται K

draws the blood to itself. The parts that send [the blood] expel the humor, sometimes because there is excess in quantity, sometimes because it is distressing in quality, and sometimes for both reasons. Those parts which attract do so because of a morbid heat. As for pain, the cause arises from what has the pain whereas the overlying parts bring about all the inflammation. Therefore, it is clearly apparent that there is either a drawing of the adjacent humors to the part that is heated, as I say, or a flow, as Asclepiades thought; and here too the truth in the theory must be maintained by what is actually observed.

At any rate, in parts that sense pain, those that are inflamed are apparent because of the pain, although the cause is not stated at all in some, and not at all reliably in others. My view is as follows: it was shown in the treatise I wrote *On the Natural Capacities* that there is this one capacity of Nature which I called separative.[5] This functions on those occasions when there is a perception of something distressing. One of the things that is distressing is the cause which brings about the pain itself, whatever that might be. Therefore, when the separative capacity hastens to get rid of this, it sometimes brings about an inflammation in the part. Whenever nothing is achieved in the first actions of this [capacity], since it now, under these circumstances, attempts to get rid of what is distressing more vigorously, it squeezes out some blood and *pneuma* from what is overlying the part. And here, due to the pains, the part is raised up into a swelling in proportion to the humor flowing to it.

879K

[5] See *De naturalibus facultatibus*, II.1–214K, in particular II.149K ff.

4. Καὶ μέντοι καὶ πάντων τῶν παρὰ φύσιν ὄγκων ἡ ποικιλία τῆς διαφορᾶς ἕπεται τῇ τῶν ἐπιρρεόντων φύσει. πνευματωδέστεροι μὲν γάρ, ὅταν ἡ πνευματώδης οὐσία πλείων ἀφίκηται, γίνονται· φλεγμονωδέστεροι δέ, ὅταν ἡ τοῦ αἵματος· ἐρυσιπελατώδεις δὲ ὅταν ὁ τῆς ξανθῆς χολῆς χυμός· οἰδηματώδεις δέ, ὅταν ὁ τοῦ φλέγματος, ὥσπερ γε καὶ σκιρρώδεις, ὅταν ἤτοι παχὺς ἢ καὶ γλίσχρος ἱκανῶς ὁ κατασκήψας εἰς τὸ μόριον ᾖ χυμός. ὁ μὲν οὖν παχὺς ἤδη πώς ἐστι μελαγχολικὸς καὶ ἤτοι γε ἧττον ἢ μᾶλλον. ὁ δὲ 880K γλίσχρος ἔκ | τε γλίσχρων ἐδεσμάτων γίγνεται, καί ποτε καὶ αὐτῶν τῶν νευρωδῶν μορίων περίττωμα πολὺ γεννησάντων. ἀλλὰ περὶ μὲν τῶν ἄλλων ὄγκων ἐφεξῆς εἰρήσεται.

περὶ δὲ τῆς φλεγμονῆς τὰ κοινὰ πάντων λαβόντες εἰς τὸν λόγον οὕτως αὐτοῖς προσθῶμεν, ὅσα μόνης αὐτῆς ἐστιν ἴδια. ὅταν οὖν ἄρχηταί τι φλεγμαίνειν μόριον, ἐπισκεπτέον εἴτε διὰ θερμασίαν τινὰ παρὰ φύσιν ἐν αὐτῷ γενομένην εἴτε δι᾽ ὀδύνην ἤτοι γ᾽ οἰκείαν ἤ τινα τῶν πλησιαζόντων εἰς τοῦθ᾽ ἧκεν· ἵνα σοι παύοντι τὴν αἰτίαν ἡ φλεγμονὴ μηκέτ᾽ αὐξάνηται. μετὰ δὲ τήνδε τὴν ἐπίσκεψιν ἐφεξῆς θέασαι μή τι τῶν πλησιαζόντων μορίων ἐπιπέμπει πλέον αἷμα τῷ φλεγμαίνοντι· καὶ μετὰ τοῦτο μὴ καὶ σύμπαν τὸ σῶμα πληθωρικῶς διάκειται.

5. Μεμνῆσθαι δ᾽ οἶμαί σε καὶ τούτου τοῦ δεδειγμέ-

---

[6] The problems associated with the term *neuron* and its cog-

4. But, in truth, the diversity in the differentiae of all the swellings contrary to nature follows from the nature of the inflow. For the more airlike swellings arise whenever an airlike substance arrives in greater amount; more inflammatory swellings arise whenever the inflow is of blood; erysipelatoid swellings arise whenever the humor is yellow bile; edematous swellings arise whenever the inflow is of phlegm, just as scirrhous swellings arise whenever a thick or excessively viscid humor rushes down to the part. The thick humor, then, is already in some way "melancholic," and is in fact either less or more. However, the viscid humor arises from viscid foods, and sometimes also from those "nervelike" parts[6] when they generate a large amount of superfluity. But I shall say more about the other swellings in due course.

880K

On the matter of inflammation, having dealt with the generalities, let me advance the argument in the same way to those things that are specific to inflammation alone. Thus, whenever some part begins to become inflamed, what must be considered is whether the inflammation is due to some abnormal heat arising in it, or due to pain, either of the part itself or coming to it from one of the adjacent parts so that, if you put a stop to the cause, the inflammation does not increase any further. The next consideration after this is to see whether one of the adjacent parts is sending excess blood to the inflamed part and, after this, whether or not the whole body is in a plethoric state.

5. I think I should remind you also of what has been

nates have been mentioned earlier; see vol. I, 160K, note 3.

νου πολλάκις, ὡς ἐκ τῶν ἰσχυροτέρων μορίων ὠθού-
μενα τὰ περιττὰ κατὰ πλῆθος ἢ ποιότητα τοῖς ἀσθενε-
στέροις ἐγκατασκήπτει· καὶ διὰ τοῦτό γε καὶ οἱ ἀδένες
ἑτοίμως δέχονται τὸ ῥεῦμα καὶ μάλισθ᾽ ὅσοι μανώτε-
881K ροι φύσει. | σφοδρότερος μὲν γὰρ ὁ τῶν ἀρτηριῶν καὶ
φλεβῶν καὶ νεύρων καὶ μυῶν ἐστι τόνος· ἀσθενέστε-
ρος δὲ καὶ ἴσως οὐδ᾽ ὅλως ὁ τῶν ἀδενωδῶν σωμάτων.
οὕτως οὖν καὶ δι᾽ ἕλκος ἐν δακτύλῳ γενόμενον ἤτοι
ποδὸς ἢ χειρὸς οἱ κατὰ τὸν βουβῶνα καὶ τὴν μασχά-
λην ἀδένες ἐξαίρονταί τε καὶ φλεγμαίνουσι, τοῦ
καταρρέοντος ἐπ᾽ ἄκρον τὸ κῶλον αἵματος ἀπολαβόν-
τες πρῶτοι. καὶ κατὰ τράχηλον δὲ καὶ παρ᾽ ὦτα
πολλάκις ἐξήρθησαν ἀδένες, ἑλκῶν γενομένων ἤτοι
κατὰ τὴν κεφαλὴν ἢ τὸν τράχηλον ἤ τι τῶν πλησίων
μορίων· ὀνομάζουσι δὲ τοὺς οὕτως ἐξαρθέντας ἀδένας
βουβῶνας. εἰ δὲ σκιρρωδεστέρα ποτ᾽ αὐτῶν ἡ φλεγμονὴ γένοιτο, δυσίατός τέ ἐστι καὶ καλεῖται χοιράς.
ἥτις μὲν οὖν ἐστιν ἡ τῶν χοιράδων ἴασις ἰδία, κατὰ
τὸν ἑξῆς λόγον εἰρήσεται.

νυνὶ δὲ περὶ τῶν φλεγμονῶν ἐπειδὴ περὶ τούτων
πρόκειται διελθεῖν, ἀναλαβόντες περὶ τούτων αὖθις
λέγωμεν· ὡς τὸ κωλύειν αὐτὰς ἀρχομένας ἐκκοπτόντων τὴν γεννῶσαν αἰτίαν γίγνεται. καὶ πρῶτόν γε περὶ
τῶν ἐφ᾽ ἕλκεσι φλεγμονῶν εἴπωμεν· ἐπειδὰν γὰρ ἐγγὺς ἀρτηρίας μεγάλης ἢ φλεβὸς ἕλκος γένηται, τάχι-
882K στα | μὲν οἱ βουβῶνες ἀνίστανται. φαίνεται δ᾽ ἐνίοτε
καὶ ἡ φλὲψ αὐτὴ καθ᾽ ὅλον τὸ κῶλον ἐρυθρά τε καὶ
θερμὴ καὶ τεταμένη, καὶ εἰ θίγῃς αὐτῆς ὀδυνωμένη.

demonstrated often—that the superfluities thrust out from the stronger parts on account of amount or quality pass down to the weaker [parts] and, because of this, the glands also readily receive the flux, especially those that are by nature looser in texture. The strength of the arteries, veins, nerves and muscles is quite considerable whereas that of the glandular bodies is weaker, and perhaps not there at all. So it is that, due to a wound [or ulcer] occurring in a digit of either a foot or hand, the glands in the groin and armpit respectively are swollen and inflamed because they are first to receive the blood flowing to the farthest point of the limb. And the glands in the neck or behind the ears are often swollen when wounds [or ulcers] occur in the head or neck, or one of the adjacent parts. They call glands swollen in this way "buboes." If, however, the inflammation arising in them becomes harder at some point, it is difficult to cure and is called "scrofulous."[7] What the specific treatment of the scrofulous swellings is, I shall speak about in the discussion that follows.

881K

Now, about inflammations, since they are what I propose to go over, having brought them to the fore, let me say again about these that preventing them arising lies in eradicating the generating cause. Let me speak first about inflammation due to wounds [or ulcers] because, whenever they occur near a major artery or vein, glandular swellings very quickly build up. Sometimes it seems that the vein itself is also red, hot and tense through the whole limb, and painful if you touch it. If the whole body is either plethoric

882K

[7] Both these terms remain in use, albeit uncommonly. *Bubo* retains its ancient meaning; *scrofula* has become restricted to tuberculous glandular swellings, particularly in the neck, which were also called "King's-evil" (see, e.g., Peter English).

πληθωρικοῦ μὲν οὖν ὄντος ἢ κακοχύμου τοῦ παντὸς
σώματος ἡ θεραπεία δύσκολος γίνεται, ὑγιεινοῦ δ'
ἀκριβῶς ῥᾳδία. θερμαίνειν τε γὰρ καὶ ὑγραίνειν χρὴ
μετρίως ὅλον τὸ κῶλον, ὅπως ἀνώδυνον γένοιτο,
γινώσκεις δὲ δήπου τὴν τῶν τοιούτων ὕλην, αὐτῷ μὲν
οὖν τῷ ἕλκει τῆς τετραφαρμάκου δυνάμεως ἐπιτιθε-
μένης ἐν μοτῷ· λύεται δὲ ῥοδίνῳ μὲν μάλιστα, μὴ
παρόντος δὲ αὐτοῦ, τῶν χαλαστικῶν ἐλαίῳ τινί· τῷ δὲ
ὅλῳ κώλῳ περιελιττομένου πιλήματος ἐλαίῳ θερμῷ
βεβρεγμένου.

καὶ μέντοι καὶ αὐτῷ τῷ ἕλκει τὸ φάρμακον ἐπιτιθέναι χρὴ θερμόν, ἔξωθέν τε καταπλάττειν αὐτῷ θερμῷ καταπλάσματι τὸ μὲν ἄλευρον ἤτοι κρίθινον ἢ
πύρινον ἢ μικτὸν ἐξ ἀμφοῖν ἔχοντι, τὸ δὲ ὑγρὸν ὕδωρ
μετ' ἐλαίου βραχέος. οὕτω δὲ καὶ αὐτῷ τῷ ἀδένι τῷ
φλεγμαίνειν ἠργμένῳ παρηγορικῶς χρὴ προσφέρεσθαι τήν γε πρώτην ἡμέραν ἐξ ἐλαίου θερμοῦ διάβρο-
χον ἔριον ἐπιτιθέντας, οὐχ ὥς τινες εὐθέως μεθ' ἁλῶν· |
883K   ὕστερον γὰρ ἐκείνοις χρησόμεθα, τοῦ τε καθ' ὅλον τὸ
κῶλον ὄγκου παρηγορηθέντος, ἀνωδύνου τε τοῦ ἕλκους γενομένου. καὶ μέντοι γε καὶ τὸ Μακεδονικὸν
καλούμενον φάρμακον, ὡσαύτως τῇ τετραφαρμάκῳ
δυνάμει κατὰ τῶν ἑλκῶν ἐπιφέρειν προσήκει· καὶ γὰρ
καὶ παραπλήσιά πως ἀλλήλοις ἐστί, μόνῳ τῷ λιβα-
νωτῷ πλεονεκτοῦντος τοῦ Μακεδονικοῦ.

πληθωρικοῦ δ' ὄντος ἢ κακοχύμου τοῦ σώματος ἡ
διὰ τῶν οὕτω θερμαινόντων ἀγωγὴ ῥευματίζει τὸ κῶ-

or *kakochymous*, treatment becomes difficult, whereas if it is healthy, treatment is easy, it being necessary to heat and moisten the whole limb to a moderate degree, so that it becomes pain-free—I presume you know the material for such things, which is to place the "tetrapharmaceutical potency"[8] in lint on the wound itself. It is best if this is dissolved in oil of roses or, if this is not available, in some oil of a relaxing kind, bandaging the whole limb with compressed wool soaked in warm oil.

However, it is also necessary to apply the warm medication to the wound or ulcer itself, and to cover it externally with a warm poultice made of meal, either barley or wheat, or a mixture of both, the moisture of the water having with it a little oil. Similarly, it is also necessary, at least on the first day, to gently apply wool moistened with warm oil to the actual gland that has begun to become inflamed and not, as some do, immediately with [added] salt. Later, we shall use salt when the swelling involving the whole limb has settled and the wound or ulcer has become pain-free. It is, of course, also appropriate to apply the so-called "Macedonian" medication[9] to wounds or ulcers in the same way as the "tetrapharmaceutical potency" because, somehow, there are similarities between them, [the only difference being that] the Macedonian [medication] has a greater amount of frankincense.

883K

If, however, the body is plethoric or *kakochymous*, treatment with things that heat in this way creates flow in

---

[8] On the composition of this (from wax, tallow, pitch, and resin), see Galen, *De elementis secundum Hippocratem*, I.452K, and *De simplicium medicamentorum temperamentis et facultatibus*, XII.328K.

[9] There is no other reference to this in the Kühn index.

λον. οὐ μὴν οὐδ' ἄλλῃ τινὶ χρῆσθαι δυνατόν. ἀναγκαζόμεθα τοιγαροῦν ἐνίοτε κενοῦν αἵματος ἤτοι φλέβα
τέμνοντες ἢ ἀποσχάζοντες τὰ μὴ πεπονθότα κῶλα.
χειρὸς μὲν γὰρ κακῶς ἐχούσης τὰ σκέλη, τοῦ δ'
ἑτέρου τῶν σκελῶν πεπονθότος τὸ λοιπόν. ταύτας γὰρ
τὰς κενώσεις ἐνδείκνυται τὸ πλῆθος, ὥσπερ γε καὶ ἡ
κακοχυμία τὴν τοῦ πλεονάζοντος χυμοῦ κάθαρσιν. ὡς
τὰ πολλὰ μὲν οὖν ἐπὶ τοῖς εἰρημένοις βοηθήμασι
παύεται τῶν ἀδένων ἡ φλεγμονή. πολλάκις δ' ἤτοι τοῦ
θεραπεύοντος βραδύνοντος περὶ τὴν τοῦ παντὸς σώματος κένωσιν ἢ αὐτοῦ τοῦ κάμνοντος ὑπὸ μαλακίας
884K αὐτὴν οὐ προσιεμένου, μείζων ἡ φλεγμονὴ | γίνεται
τῶν ἀδένων, ὡς εἰς ἐκπύησιν ἔρχεσθαι.

καὶ μέντοι καὶ τὰ καλούμενα φύματα κατ' αὐτοὺς
τοὺς ἀδένας συμβαίνει, διὰ ῥεῦμα κατασκῆψαν ἄνευ
τῆς ἕλκους προφάσεως. ὅταν οὖν ποτε διατείνωνται
σφοδρῶς ἀδένες ἢ ἁπλῶς ὁτιοῦν μόριον ἄλλο φλεγμαῖνον, ἀναγκαζόμεθα προκενώσαντες τὸ ὅλον ἀποσχάζειν αὐτό. κενοῦμεν δὲ τὸ ὅλον, ὡς κἂν τῷ Περὶ
πλήθους ἐδείξαμεν, οὐ μόνον ἐν πληθωρικῇ διαθέσει
γιγνόμενον, ἀλλὰ καὶ διὰ μέγεθος τοῦ πάθους, ἐν
συμμετρίᾳ χυμῶν καθεστηκότος τοῦ παντὸς σώματος.
ἡ γὰρ ὀδύνη καὶ ἡ θερμασία τοῦ φλεγμαίνοντος
μέλους αἴτιαι ῥεύματος γίνονται, κἂν ἀπέριττον ᾖ τὸ
σύμπαν σῶμα. χρὴ τοίνυν ἐνδεέστερον αὐτὸ ποιεῖν
τηνικαῦτα κενοῦντα κένωσιν, ἥτις ἂν ἁρμόττειν φαίνηται μάλιστα τῇ θ' ἡλικίᾳ καὶ τῇ φύσει τοῦ κάμνον-

the limb. And yet it is not possible to use any other treatment. Therefore, we are sometimes forced to drain blood, either by opening a vein or by scarifying the nonaffected limbs. When the hand is badly affected, scarify the legs; when one of the legs is involved, scarify the other leg. Abundance indicates these evacuations, just as *kakochymia* indicates purging of the excessive humor. For the most part, then, after the aforementioned remedies, the inflammation of the glands ceases. Often, when either the person carrying out the treatment is tardy in purging the whole body, or the patient himself does not respond to the purging because of weakness, the inflammation of the glands becomes more marked such that it goes on to suppuration.

884K

And the so-called "phymata"[10] also occur in relation to the glands themselves due to a flux falling on them without a causative wound or ulcer. Whenever glands are markedly distended on any occasion, or simply when any other part whatsoever is inflamed, we are forced to scarify the whole [body] after prior purging. We evacuate the whole [body], as I showed in the work *On Plethora*,[11] not only when it is in a plethoric condition, but also because of the magnitude of the affection when the whole body is in a state of humoral balance. The pain and the heat of the inflamed part are the causes of flux, even if the whole body is free of superfluities. Therefore, it is necessary for the one who carries out the purging to make the body deficient in these circumstances to the extent that seems to accord best with the age and nature of the patient, attention also being

[10] It is not possible to identify this with a specific lesion; see Celsus, V.28 (9) for a description.

[11] *De plenitudine*, VII.513–83K.

τος, ἐπισκοποῦντα καὶ τὴν ὥραν καὶ τὴν χώραν καὶ τὰ
ἔθη τοῦ νοσοῦντος· ὑπὲρ ὧν ἤδη πολλάκις ἐν πολλοῖς
εἴπομεν, ὥστε κἂν μὴ προσκέηταί ποτε τῷ λόγῳ,
συνυπακούειν αὐτὰ χρή. ὅταν τὸ οἷον ζέον τῆς φλεγμονῆς παύσηται, τῶν παρηγορικῶν ἀποχωροῦντα
885K καταπλασμάτων ἐπὶ | τὰ διαφορητικὰ χρὴ μετα-
βαίνειν κατὰ βραχύ, πρῶτον μὲν τοῖς παρηγορικοῖς
μιγνύντα μέλιτος ὀλίγον, εἶτ᾽ ἀφαιροῦντα μὲν ὅλον τὸ
πύρινον ἄλευρον, ἀρκούμενον δὲ τῷ κριθίνῳ μετὰ τοῦ
καὶ τὸ μέλι προσαύξειν, εἶθ᾽ ἑξῆς ἐπί τι τῶν διαφορούντων ἰέναι φαρμάκων ὅσα ταῖς συστάσεσιν ἤτοι
γ᾽ ὑγρὰ τοῖς ἐμμότοις ὁμοίως ἐστὶν ἢ κηρωτοειδῆ.

ἀφίστασθαι δὲ τῶν σκληρῶν, οἷα πολλὰ τῶν ἐμ-
πλαστῶν ἐστι· συντείνει τε γὰρ τὰ λείψανα τῶν φλεγμονῶν, αὖθίς τε φλεγμαίνειν ἀναγκάζει τὰ πεπονθότα μόρια. κἂν εἰ πῦον δέ τι κατὰ τὸ διαπυῆσαν
ἀξιόλογον εἴη περιεχόμενον, οὐ χρὴ τέμνειν αὐτίκα,
καθάπερ ἔνιοι πράττουσιν, ἀλλὰ διαφορεῖν ἐπιχειρεῖν
τοῖς τοῦτο δρᾶν πεφύκοσι φαρμάκοις, ὧν ἡ χρῆσις
ἐστοχάσθω τῆς διαθέσεως· ὅταν μὲν γὰρ ἔτι φλεγμο-
νῶδές τι κατὰ τὸ μόριον ᾖ, τὰ δριμέα τῶν φαρμάκων
ἐρεθίζει μᾶλλον ἢ διαφορεῖ. ὅταν δὲ φαίνηταί σοι τὸ
τῆς φλεγμονῆς λείψανον σκιρρῶδες γινόμενον, θαρ-
ρεῖν ἤδη τοῖς ἰσχυροῖς φαρμάκοις, ἐπιβλέποντα δὶς
τῆς ἡμέρας ὁποῖόν τι δρᾷ κατὰ μὲν τὴν ἕω τὸ πρότερον, εἰς τὴν ἑσπέραν δὲ τὸ δεύτερον. εἰ δὲ καὶ
886K βαλανείῳ χρῷτο, καὶ κατὰ | τὸν ἐκείνου καιρόν.

given to the time of year, the region, and the habits of the person diseased. I have already spoken frequently about these things in many [places], so that even if they don't, at some particular time, come into the discussion, we ought to take them as understood. Whenever the fiery heat, as it were, of the inflammation has subsided and the pain-relieving cataplasms have been withdrawn, it is necessary to 885K make the transition gradually to those things that disperse, first mixing a little honey with the paregorics. Next, avoiding wheaten flour altogether, we should satisfy ourselves with barley flour, enhancing this with honey. Then, next, we should move on to one of the discutient medications such as those that are moist or waxy in composition like salves.

But we should keep away from hard things like many of the plasters because they draw together what remains of the inflammation and force the affected parts to become inflamed once again. Even if a significant amount of pus is contained in the suppurating part, it is not necessary to make an immediate incision, as some do; attempt, rather, to disperse it with strong medications, letting the use of these be estimated from the condition, for whenever there is still some inflammation in the part, the sharp medications irritate more than they disperse. On the other hand, whenever it seems to you that what remains of the inflammation is scirrhous, have the confidence now to use strong medications, examining twice a day what it (i.e. the inflammation) is doing, first in the morning and second in the evening. And if the patient also uses bathing, do so at the 886K time of that as well.

ὅταν οὖν ἴδῃς ποτὲ διὰ τὴν τοῦ φαρμάκου δριμύτητα τὸ πεπονθὸς μέρος ἠρεθισμένον ὡς ὀγκωδέστερον ἢ ἐρυθρότερον ἢ ὀδυνωδέστερον γεγονέναι, παρηγόρει μεταξὺ τῇ διὰ τῶν σπόγγων πυρίᾳ. καὶ αὕτη δέ σοι ποτὲ μὲν ἐξ ὕδατος ἔστω ποτίμου, ποτὲ δ' ἁλῶν ἔχοντός τι κατὰ τὰς σκιρρωδεστέρας δηλονότι φλεγμονάς. εἰ δὲ καὶ νικηθείη ποτὲ τὰ φάρμακα πρὸς τοῦ πλήθους τοῦ πύου καὶ φαίνοιτο μὴ δυνάμενα διαφορῆσαι πᾶν αὐτό, τέμνειν χρὴ τὸ οὕτως ἀφιστάμενον, ἔνθα μάλιστά ἐστιν ὑψηλότατον ἑαυτοῦ· καὶ γὰρ λεπτότατον εὑρήσεις ἐνταῦθα τὸ δέρμα. μέμνησο δὲ καὶ θατέρου σκοποῦ τοῦ τῆς ὑπορρύσεως ἐν τῇ τομῇ· καὶ πρὸς ἀμφότερα ἀποβλέπων οὕτως σχάζε τὸ διαπυῆσαν· ἐπιτίθει τε φάρμακον ἐφεξῆς τῶν ξηραινόντων ἀδήκτως. εἰ δὲ καὶ σεσηπέναι φαίνοιτό τινα τοῦ διαπυήσαντος, ἐκκόπτειν ἀναγκαῖον αὐτά. τινὲς δ' ἐπὶ τῶν κατὰ μασχάλην καὶ βουβῶνα διαπυΐσκόντων ἀεὶ κελεύουσι μυρσινοειδῶς ἐκτέμνειν τοῦ δέρματος, ἐπειδὴ φύσει χαλαρὸν ἐν αὐτοῖς ἐστι καὶ διὰ τοῦτο δεχόμενον ἑτοίμως πᾶν τὸ παραγινόμενον ἐπ' αὐτό· καὶ |

887K φλεγμαίνουσι ῥᾳδίως ἐπὶ σμικραῖς προφάσεσι. καὶ μεγίστας γ' ἔνιοι τὰς περιτομὰς εἰώθασι ποιεῖσθαι, δι' ἃς αἴσχιστόν τε τὸ μέρος εἰς οὐλὴν ἀχθὲν γίνεται καὶ προσέτι καὶ ἀσθενέστερον ἐμποδίζει τε πολλάκις εἰς τὰς κινήσεις. ταῦτ' οὖν ἡμεῖς φυλαττόμενοι τὰ μὲν πλεῖστα μόνῃ τῇ τομῇ μετὰ φαρμάκων ξηραινόντων ἱκανῶς ἰασάμεθα τὰς τοιαύτας διαθέσεις· εἰ δέ ποτε καὶ περιτέμνειν ἐδέησε, διὰ τὸ πλῆθος οὐ τοῦ πύου

Therefore, whenever you see the affected part irritated at any time by the sharpness of the medication such that it becomes more swollen, reddened or painful, you will alleviate it in the meantime with a fomentation applied by sponges. Sometimes you should make this from fresh water and sometimes from [water] containing some salts when the inflammation is obviously harder. If on occasion the medications are overwhelmed by the large amount of pus and seem unable to disperse this completely, it is necessary to incise what is prominent in this way, for here, particularly, is the highest part of it (i.e. the inflammatory swelling), so you will find the skin thinnest at that point. However, in the incision, you must also keep in mind the other objective which is that of drainage. Directing your attention to both aspects, incise the suppuration in this way. Next, apply one of the medications that is drying without being stinging. But if some part of what is suppurating seems to have putrified, it is necessary to cut this out. With suppuration in the axilla or groin, some direct that the skin incision should always be shaped like a myrtle leaf since the skin is, by nature, loose in these places and, because of this, readily receives anything that comes to it; [these sites] 887K are easily inflamed due to minor causes. Some customarily make very large encircling incisions, as a result of which the part becomes ugly when a scar is formed, and in addition, is also weaker and often acts as an impediment to movement. Therefore, I guard against these things for the most part, using a single incision and treating such conditions vigorously with drying medications. If, on the other hand, it is sometimes necessary to make an encircling incision, not only because of the amount of pus but also be-

μόνον, ἀλλὰ καὶ τῶν ἐφθαρμένων σωμάτων, ἤρκεσεν
ἡμῖν οὐ πάνυ μεγάλη μυρσινοειδὴς περιαίρεσις. ἐχούσης δὲ τῆς τοιαύτης τὸ μῆκος μεῖζον τοῦ πλάτους,
ἐγκάρσιον ἔστω τὸ μῆκος ἐπὶ τοῦ βουβῶνος, οὐ κατ'
εὐθὺ τοῦ κώλου. καὶ γὰρ κατὰ φύσιν οὕτως ἐπιπτύσ-
σεται τὸ δέρμα ἑαυτῷ, καμπτόντων τὸ κῶλον.

ἐπὶ δὲ τῇ περιαιρέσει πληροῦν χρὴ τὸ πεπονθὸς
μόριον τῇ καλουμένῃ μάννῃ· ἔστι δὲ ὑπόσεισμα λιβανωτοῦ τὸ φάρμακον τοῦτο, στύψεώς τε μετέχον ὀλίγης
καὶ κατὰ τοῦτο καὶ αὐτοῦ τοῦ λιβανωτοῦ πρὸς ἔνια
βέλτιον. ἐκεῖνος γὰρ ἐκ τῆς πυητικῆς δυνάμεώς ἐστι
μόνης, ὡς ἂν μὴ μετέχων στύψεως, καὶ μᾶλλον ὁ
888K λιπαρώτερος | ἐν αὐτῷ καὶ τῇ χροιᾷ λευκότερος, ὥσπερ
γε ὁ τοῦδε ξανθότερος ξηραντικώτερός ἐστι. τῇ δὲ
μάννῃ καὶ φλοιοῦ τι λιβανωτοῦ μέμικται σμικρόν, ἀφ'
οὗ τὸ στῦφον ἔχει. τοῦτο δ' αὐτὸ τὸ φάρμακον ὁ
φλοιὸς τοῦ λιβανωτοῦ καὶ στύφει καὶ ξηραίνει γενναίως. διὸ καὶ πρὸς τὰς μετριωτέρας αἱμορραγίας
αὐτῷ χρώμεθα μόνῳ, καθάπερ γε καὶ πρὸς τὰς σφο-
δροτέρας καυθέντι μόνῳ καὶ τῷ τε διηθημένῳ δηλονότι καὶ χνοώδει γεγονότι. καὶ μὲν δὴ καὶ παρηγορῆσαι χρὴ πρότερον τὸ τμηθέν, ὡς εἴρηται, μέρος εἰς
ὅσον ἂν φαίνηται δεόμενον, ἐπιβροχῆς μὲν πρῶτον,
εἶτα καταπλάσματος, εἶτα τῶν ὑγραινόντων τινὸς
φαρμάκων ἢ μὴ ξηραινόντων, ἔξωθεν ἐπιτιθεμένων
δηλονότι τούτων. κατ' αὐτοῦ γὰρ τοῦ ἡλκωμένου τήν
τε μάννην, ὡς εἴρηται, καὶ τῶν ἐμμότων φαρμάκων τὰ
διαπυΐσκοντα μὲν πρῶτον, εἶτ' ἀνακαθαίροντα θετέον

cause of the amount of the corrupted bodies, I find an encircling incision in the shape of a not particularly large myrtle leaf is enough. In such incisions, the length is greater than the width; in the groin make the long incision transverse and not vertical along the line of the limb, for when we bend the limb, the skin produces natural folds in this way.

After the encircling cut, it is necessary to fill the affected part with so-called "manna,"[12] a medication which is the dust of frankincense and partakes of a little astringency and, by virtue of this, is in some instances better than frankincense itself. For the latter is based on the pus-forming capacity alone inasmuch as it does not possess astringency and, in particular, and the more oiliness there is, the whiter it is in color, just as the yellower form of this is more drying. When a little of the bark of frankincense is mixed with manna, it acquires astringency from this. The bark of frankincense, as a medication in its own right, is markedly astringent and drying, and because of this we use it on its own in moderate hemorrhages, just as, for more severe hemorrhages, we use it on its own after burning it, but filtered obviously and made into fine powder. Nevertheless, it is first necessary to soothe the part that is cut, as was said, to the extent that there seems to be a need for fomentation first, then a poultice, then one of the moistening or nondrying medications, applying these externally, obviously. On the wounded part itself you must place manna, as I said, and then, of the salvelike medications, first those that bring about suppuration, and then those that cleanse

888K

---

12 On manna, see Dioscorides, I.83, and Galen, *De compositione medicamentorum secundum locos*, XII.722 and 845K.

ἐστίν· ἐφ᾽ οἷς εἰ μὲν εἴη κοιλότης ἔτι, τὰ σαρκοῦντα προσφέρειν, εἰ δ᾽ οὐκ εἴη, τὰ συνουλωτικά τε καὶ ἐπουλωτικὰ καλούμενα, καθάπερ καὶ τὸ διὰ τῆς καδμίας.

889K ἐπεὶ δὲ καὶ | κατὰ τὴν τούτων χρῆσιν οὐ σμικρόν τι παρορᾶται τοῖς πλείστοις τῶν ἰατρῶν, ἄμεινον ἂν εἴη καὶ περὶ τούτων δηλῶσαι. τηνικαῦτα γὰρ εἰς οὐλὴν ἄγειν ἄρχονται τὰ ἕλκη, τοῖς ἐπουλωτικοῖς χρώμενοι φαρμάκοις, ὁπόταν ἀκριβῶς ἀναπληρωθῇ καὶ μηδὲν ἔτ᾽ ἔχῃ κοῖλον, εἶτ᾽ αὐτοῖς συμβαίνει τὰς οὐλὰς ἐργάζεσθαι τοῦ πέριξ δέρματος ὑψηλοτέρας. ὅπως οὖν ἐκείνῳ γίγνοιντο ἴσαι, τοῖς τοιούτοις φαρμάκοις χρῆσθαι προσήκει, πρὶν ἀκριβῶς ὁμαλὲς ἀποδειχθῆναι τὸ ἕλκος, ἐπὶ μὲν τὰ χείλη διὰ μήλης πυρῆνος ἐπιτιθέντας τῶν ξηρῶν τι φαρμάκων, ὧν ἐν τῇ τῶν ἑλκῶν ἐμνημονεύσαμεν θεραπείᾳ· τῷ δ᾽ ἄλλῳ μοτῷ σκέποντας, κεχρισμένῳ τῶν ἐπουλωτικῶν τινι φαρμάκων, ὑγρῶν τῇ συστάσει. προκοπτούσης δέ σοι τῆς θεραπείας καὶ τοῦτο ἀφαιρήσεις ὕστερον, μόνῳ τῷ ξηρῷ φαρμάκῳ χρώμενος ἐφ᾽ ὅλου τοῦ ἕλκους, ἐπικυλιομένου τοῦ τῆς σπαθομήλης πυρῆνος. ἔξωθεν δ᾽ ἀρκεῖ μοτὸς ἤτοι ξηρὸς ἢ ἐξ οἴνου. καὶ μᾶλλον ὁ ἐκ τῶν μαλακῶν ἐλλυχνίων, οἷά πέρ ἐστι τὰ Ταρσικά· καὶ γὰρ καὶ αὐτὰ ἔχει τι καθαιρετικὸν τῶν ὑπερσαρ-

890K κούντων | ἑλκῶν. ταῦτα μὲν οὖν εἴρηται τῇ κοινωνίᾳ τῶν πραγμάτων ἀκολουθήσαντός μου.

completely. If, after these, there is still a cavity, apply enfleshing medications and, if there is not, apply the medications that are called cicatrizing, such as that made from
calamine.

However, since there is much that is overlooked by   889K
most doctors in the use of these things, it would be better
to provide clarification about them. For in those circumstances they begin by bringing the wound to a scar, using
the cicatrizing medications whenever [the wound] has
been completely filled in and there is no longer any cavity
left, and then they find that these medications make the
scars more raised than the surrounding skin. Therefore, so
that they become level with the skin, it is appropriate to
use such medications before the wound is made entirely
level, placing one of the drying medications, which I mentioned in the treatment of wounds, on the margins [of the
wound] with the round head of a probe, covering it with
another lint pledget soaked with one of the cicatrizing
medications that is fluid in consistency. When your treatment has made progress, you will subsequently remove
this, making use of the drying medication alone over the
whole wound, rolling [the medication] over it with the
round head of the flat probe. Externally, a lint pledget, either dry or moistened with wine, is sufficient, and particularly one of the soft dressings like those from Tarsus,[13]
since these have a reducing effect on exuberant wounds
[ulcers]. I have said these things, inferring from what is   890K
common to the matters.

[13] A form of surgical dressing described by Soranus, 2.11, and
Aëtius, 15.1, and also mentioned again by Galen in the next book
(984K).

341

6. Πάλιν δ᾽ ἐπὶ τὸν περὶ τῆς φλεγμονῆς λόγον
ἀφικόμενοι λέγωμεν ὡς κοινὸς μὲν ἁπασῶν σκοπὸς ἡ
κένωσις, ὅσαι δ᾽ ἔτι γίγνονται, πρότερον τοῦ κενοῦν
ἐστι κωλῦσαι τὸ αἷμα ῥεῖν ἐπὶ τὸ φλεγμαῖνον. ἔσται
δὲ τοῦτο καλῶς, εἰ τοῦ ῥεύματος αἰτίαν εὕροιμεν. ἔστι
δὲ καὶ αὕτη διττή· ποτὲ μὲν ἐξ αὐτοῦ τοῦ φλεγμαίνοντος ὁρμωμένη μορίου, ποτὲ δὲ ἐξ ἄλλου τινὸς ἢ
ἄλλων. ἐξ αὐτοῦ μέν, ὅταν ἤτοι θερμότερον ἢ ὀδυνώμενον γένηται, καθότι καὶ πρόσθεν εἴρηται· οὐκ ἐξ
αὐτοῦ δέ, ὅταν ἤτοι γ᾽ ἐξ ἑτέρου τινὸς ἢ ἑτέρων αὐτῷ
πέμπηται τὸ περιττόν, ἢ καὶ τῆς καθ᾽ ὅλον τὸ σῶμα
διαθέσεως. θερμότερον μὲν οὖν γίνεται διὰ κίνησιν
ἀμετροτέραν ἤ τινα θάλψιν ἐξ ἡλίου καὶ πυρός, ἢ διὰ
δριμὺ φάρμακον. ὀδυνᾶται δὲ διά τε δυσκρασίαν καὶ
τραῦμα καὶ θλάσμα καὶ στρέμμα καὶ τάσιν, ἔτι τ᾽
ἔμφραξίν τινα καὶ πνεῦμα φυσῶδες. ἡ δυσκρασία δὲ
891K ποτὲ μὲν ἔξωθεν αὐτῷ γίγνεται, ποτὲ δὲ ἐκ τῶν | κατὰ
τὸ σῶμα χυμῶν. ἔξωθεν μὲν ἐπί τινι τῶν ἰοβόλων
ὀνομαζομένων[3] ζῴων ἢ φαρμάκῳ θερμαίνοντι σφοδρῶς ἢ ψύχοντι, κἀκ τοῦ περιέχοντος ἐνίοτε, διὰ δὲ τὸ
σῶμα τοῦ κάμνοντος αὐτὸ μοχθηροὺς ἀθροῖσαν χυμοὺς ἀνομοίους ταῖς δυνάμεσι.

ταῦτ᾽ οὖν ἅπαντα διασκεψάμενος, ὅσαι μὲν ἔτι
γίγνονται φλεγμοναί, τὰς αἰτίας αὐτῶν ἔκκοπτε πρότερον, ὅσαι δ᾽ ἤδη γεγόνασιν, αὐτὰς μόνας θεράπευε.
πῶς οὖν χρή σε τοῦ παντὸς σώματος ἐπιμελεῖσθαι
μοχθηρῶς διακειμένου, λέλεκται μὲν οὖν οὐκ ὀλίγα

6. Returning once again to the discussion on inflammation, let me state that the common objective for all cases is evacuation. However, for those in whom the inflammation is still in evolution, the aim is to prevent blood flowing to the inflamed part prior to evacuation. If we discover the cause of the flow, this will be [done] properly. This, however, is twofold as well. Sometimes it arises from the inflamed part itself, and sometimes it arises from one or several other parts. It is from the inflamed part itself whenever this becomes either hotter or painful, as I said before. It is not from the inflamed part itself whenever the superfluity is sent to it from one or several different parts, or it arises from the condition of the whole body. Thus, it becomes hotter due to immoderate movement or to some kind of heat from the sun or a fire, or due to a bitter medication. It becomes painful due to a *dyskrasia*, wound, contusion, sprain or tension, and also due to some obstruction or flatulent *pneuma*. The *dyskrasia* sometimes arises from something external, and sometimes from the humors in 891K the body. It is external when caused by one of the so-called venomous animals or by a strongly heating or cooling medication, or sometimes by the surroundings; it is due to the actual body of the sick person when it collects bad humors which are dissimilar in their capacities.

Having considered all these things, the first priority in those inflammations that are still in evolution is to eradicate their causes, while in those that are already established it is to treat the inflammations alone. How you must take care of the whole body when it is in a pathological state is something I have spoken about at length, both

3 K; ὀνομαζομένων om. B, recte fort.

καὶ διὰ τῶν ἔμπροσθεν, εἴρηται δὲ κἂν τῷ Περὶ
πλήθους γράμματι· καὶ νῦν δ' εἰρήσεται τὰ κεφάλαια
τῶν λόγων. ὅταν μὲν γὰρ ὁμοτίμως ἀλλήλοις αὐξη-
θῶσιν οἱ χυμοί, πλῆθος τοῦτο καὶ πληθώραν ὀνομάζουσιν. ὅταν δ' ἤδη ξανθῆς χολῆς ἢ μελαίνης ἢ
φλέγματος ἢ τῶν ὀρρωδῶν ὑγρῶν μεστὸν γένηται τὸ
σῶμα, κακοχυμίαν, οὐ πληθώραν καλοῦσι τὴν τοιαύτην διάθεσιν. ἡ μὲν οὖν πληθώρα διά τε τῆς τοῦ
αἵματος ἀφαιρέσεως θεραπεύεται καὶ διὰ λουτρῶν
πλεόνων καὶ γυμνασίων καὶ τρίψεων, ἔτι δὲ καὶ[4] φαρ-
892K μάκων διαφορούντων, | καὶ πρὸς τούτοις ἅπασιν ἀσι-
τίαις, ὑπὲρ ὧν ἐν τοῖς ὑγιεινοῖς ὑπομνήμασιν εἴρηται
τελέως. ἡ κακοχυμία δὲ διὰ τῆς οἰκείας ἑκάστου τῶν
πλεοναζόντων χυμῶν καθάρσεως. εἴρηται δὲ καὶ περὶ
ταύτης ἐν τῷ προφυλακτικῷ μέρει τῆς Ὑγιεινῆς
πραγματείας. ἐκεῖθεν οὖν αὐτὰ μεταφέρειν ἐνταῦθα
σκοπούμενον ὅτῳ βέλτιον ἐξ αὐτῶν[5] χρῆσθαι.[6]

πυρέττοντος γὰρ ἤδη τοῦ κάμνοντος οὔτε γυμνασίοις ἔτι δυνατὸν ἐκκενῶσαι τὸ πλῆθος οὔτε θερμαίνουσι χρίσμασιν οὔτε τρίψει πολλῇ, καθάπερ οὔτε
τοῖς λουτροῖς· ἀφαιρέσει δ' αἵματος ἅμα ταῖς ἀσιτίαις
ἢ καθάρσει τινί. μηδέπω δὲ πυρέττοντος ἅπασι τοῖς
εἰρημένοις χρῆσθαι, τὸ βέλτιον εἰς τὰ παρόντα προαιρούμενον. εὔδηλον γὰρ δήπου, κἂν ἐγὼ μὴ λέγω, περὶ
μὲν τῆς ἐν σκέλει φλεγμονῆς ὡς οὐ προσήκει διὰ
περιπάτων ἢ δρόμων γυμνάζειν· ἀλλ' οὐδ' ἑστάναι
τούτῳ κάλλιον· ἄμεινον δὲ καθήμενον ἐπὶ πολὺ τρίψα-

throughout what has gone before and also in the work *On Plethora*.[14] Now I shall speak of the chief points of the discussions. Whenever the humors are increased to an equal degree to each other, [doctors] call this "abundance" or "plethora." On the other hand, whenever the body is already full of yellow or black bile, or phlegm, or the serous humors, they call such a condition *kakochymia* and not plethora. Plethora is treated by the letting of blood, and by numerous baths, exercises and rubbings, as well as by discutient medications, and in addition by all fastings, which I covered comprehensively in the treatises on health. *Kakochymia*, however, is treated by the specific evacuation appropriate for each of the humors in excess. There was also discussion about this in the section on prophylaxis in the work *On the Preservation of Health*. I ought then to transfer those things from that place to consider here which of them is better to use.

892K

If the patient is already febrile, it is no longer possible to purge the abundance with exercises, heating ointments, much rubbing, or baths, or, indeed, with bloodletting along with fasting and some purging. However, if the patient is not yet febrile, it is possible to use all the measures spoken of, choosing beforehand what is better for the existing circumstances. For it is, I presume, clear, even if I do not say so, that with inflammation in the leg it is not appropriate to exercise by walking or running. Instead, not standing is better for this. What is best is much massage

14 *De plenitudine*, VII.513–83K.

---

4 B; καὶ *om*. K   5 B; ἑαυτῶν K

6 χρή *fort. omissum est ante* χρῆσθαι *per haplogr.*

σθαι, κἄπειτα διὰ τῆς τῶν χειρῶν κινήσεως γυμνάσασθαι. τῷ δ᾽ ἐν τοῖς ἄνω μέρεσιν ἔχοντι τὸ φλεγμαίνειν ἀρχόμενον ἡ διὰ περιπάτων ἢ δρόμων κίνησις ὠφέ

893K λιμος. | οὕτω δὲ καὶ ἡ τρῖψις τούτοις μὲν ἡ τῶν σκελῶν μᾶλλον, οἷς δ᾽ ἐν σκέλεσι τὸ φλεγμαῖνον ἡ τῶν ἄνω· τὸ γὰρ τῆς ἀντισπάσεως παράγγελμα κοινὸν ἐπὶ τοῖς τοιούτοις ἅπασιν· οὔκουν οὐδ᾽ ὅταν ἤτοι κατὰ τὴν ἕδραν ἤ τι τῶν πλησίων μορίων ἀρχὴ φλεγμονῆς γίγνηται, γαστέρα λαπάξεις, ὥσπερ οὐδ᾽ εἰ κατὰ κύστιν ἢ αἰδοῖον ἢ νεφροὺς οὐρητικὰ φάρμακα καταπίνειν κελεύσεις· οὐδ᾽ εἰ γυναικὶ κατὰ μήτραν ἢ αἰδοῖον ἔμμηνα κινήσεις, ἀλλ᾽ ἐπὶ τὰ πορρώτατα τὴν ἀντίσπασιν ἀεὶ ποιήσεις, προσέχων δηλονότι καὶ τῷ τῆς φλεγμονῆς μεγέθει καὶ τῇ τοῦ παντὸς σώματος διαθέσει.

παμπόλλου μὲν γὰρ ὄντος τοῦ πλήθους οὔτε γυμνασίοις οὔτε λουτροῖς ἀκίνδυνον χρῆσθαι, βραχέος δ᾽ ὑπάρχοντος ἐγχωρεῖ καὶ διὰ τούτων κενοῦν. ἀλλ᾽ ὅπερ ἔφην, ἔν τε τῷ προφυλακτικῷ μέρει τῆς Ὑγιεινῆς πραγματείας γέγραπται ταῦτα κἂν τῷ Περὶ πλήθους γράμματι, κἂν τοῖς περὶ φλεβοτομίας, ἔτι κἂν τῷ Περὶ τῆς τῶν καθαιρόντων δυνάμεως. ὅσον δ᾽ ἀναμνῆσαι μόνον αὐτῶν, αὐτάρκως εἴρηται καὶ νῦν.

894K ἐπὶ | τοὺς ἰδίους οὖν μόνης τῆς φλεγμονῆς ἀφικώμεθα λόγους. ὧν εἰσιν εἰκότως πρῶτοι γιγνομένης αὐτῆς

---

15 The four works referred to are the *De sanitate tuenda*, VI.1–

while sitting and then to exercise through movement of the arms. For someone who has inflammation beginning in the upper parts, movement through walking or running is beneficial. Similarly, rubbing of the legs is better for them than rubbing of the upper parts whereas, for those with inflammation in the legs, rubbing of the upper parts is better. The precept of "revulsion" is common to all such cases. You will not, therefore, empty the stomach whenever the start of the inflammation occurs in the anus or one of the adjacent parts, just as you will not order the drinking of diuretic medications if it involves the bladder, genitals, or kidneys. Nor, in women, will you provoke menstruation if there is involvement of the uterus or genitals. Rather, you will always create a "revulsion" toward the most distant parts, paying attention, obviously, to both the magnitude of the inflammation and the condition of the whole body.

893K

If the abundance is very great, it is not possible to use either exercise or baths without risk whereas, if it is slight, it is possible to effect evacuation through these measures. But, as I said, these things have been covered in the section on prophylaxis in the work *On the Preservation of Health* and even in the work *On Plethora*, and in the writings on venesection, and further, in the work *On the Potencies of Purgative Medications*.[15] I shall call to mind only as much of these matters as is sufficient for my present subject. Let me proceed, then, to the arguments specific to inflammation alone. Appropriately, the first of these is when the

894K

452K (particularly Book 6), *De plenitudine*, VII.513–83K, the three works on venesection (XI.147–86K, XI.187–249K, and XI.250–316K, all translated into English by P. Brain, 1984), and *De purgantium medicamentorum facultate*, XI.323–42K.

διὰ τὴν ἐν αὐτῷ τῷ φλεγμαίνοντι μορίῳ διάθεσιν·
εὔδηλον γὰρ ὡς ἐκείνην μέν σοι πρότερον θεραπευτέον, ἐφεξῆς δὲ αὐτὸ τὸ γεγενημένον ἤδη τῆς φλεγμονῆς. ἐνίοτε δὲ διὰ τῶν αὐτῶν ἀμφοτέρων καθισταμέ-
νων· οἷον ὅταν ἐπὶ φυσώδει πνεύματι καὶ πυκνώσει
τοῦ μορίου γένηταί τις ὀδύνη. τηνικαῦτα γὰρ ἡ τῶν
θερμαινόντων μετρίως, ἃ δὴ καὶ χαλαστικὰ προσαγο-
ρεύομεν, ἁρμόττει χρῆσις, ἅμα μὲν ἀραιοῦσα τὰ με-
μυκότα τοῦ σώματος, ἅμα δὲ λεπτύνουσα τὸ φυσῶδες
πνεῦμα καὶ διαφοροῦσα τὸ γεγενημένον ἤδη τῆς φλεγμονῆς.

οὕτω δὲ κἂν τοῦ ψυχροῦ κρατοῦντος ἡ δυσκρασία
γίγνηται· θερμαίνων γὰρ καὶ τότε τὴν δυσκρασίαν
ἅμα καὶ τὴν φλεγμονὴν ἐκθεραπεύσεις, ὥσπερ γε καὶ
εἰ διὰ θερμασίαν πλείονα τοῖς ψύχουσιν ἰάμασιν
ἄμφω καταστήσει· ἡ μὲν γὰρ δυσκρασία τῶν ἐναντίων ἀεὶ δεῖται. κενοῦται δὲ τὸ πεπληρωμένον οὐ
μόνον τοῖς διαφορητικοῖς φαρμάκοις, ἀλλὰ καὶ τοῖς
895K στύφουσι | καὶ τοῖς ψύχουσι. καὶ μᾶλλόν γ' ἐπὶ τῶν
ἀρχομένων φλεγμονῶν τοῖς ψύχουσι καὶ στύφουσι
χρηστέον ἤπερ τοῖς διαφοροῦσιν· ἔτι δὲ μᾶλλον ὅταν
μὴ παχὺ τὸ ἐπιρρέον ᾖ. σφοδρᾶς δὲ τῆς τῷ φλεγμαίνοντι μορίῳ σφηνώσεως γεγονυίας οὐκ ἔθ' οἷόν τε
τοῖς ἀποκρουομένοις χρῆσθαι, ἀλλ' ἐπὶ τὸ διαφορεῖν
ἰέναι καιρός. ὅταν δὲ ἐπὶ θηρίῳ νύξαντί πως ἢ
δάκνοντι τὴν ὀδύνην γίνεσθαι συμβαίνῃ, διττὸς τῆς
ἀνωδυνίας ὁ σκοπός, ἢ κενῶσαι τὸν ἰὸν ἢ ἀλλοιῶσαι
τὸ τὴν ὀδύνην ἐργαζόμενον. κενοῦται μὲν οὖν διὰ τῶν

inflammation occurs due to the condition in the inflamed part itself, for clearly it is that condition which you must treat first, and next, what has already occurred of the inflammation. Sometimes both are removed by the same [remedies]—for example, whenever pain occurs due to flatulent *pneuma* and thickening of the part. Under these circumstances, the use of those agents that are moderately heating is fitting; i.e. those we also call relaxing, and which simultaneously rarefy what has been made dense in the body, thin the flatulent *pneuma*, and disperse what has already occurred of the inflammation.

And even the *dyskrasia* when cold prevails is like this, for by heating you will completely cure both the *dyskrasia* and the inflammation at the same time, just as you will also settle both with cooling remedies, if [the *dyskrasia*] is due to excess heat, because the *dyskrasia* always requires opposites. However, what has been filled is evacuated not only by discutient medications but also by those that are astringent and cooling. And, in fact, particularly in the case of incipient inflammations, you must use those medications that are cooling and astringent rather than those that are dispersing. Still more is this so whenever what flows in is not thick. If a severe impaction (of blood) has occurred in the inflamed part, it is no longer possible to use the "repulsives"; it is time to go to the discutients. Whenever it should happen that the pain arises due to a stinging or biting animal, there is a twofold objective of pain relief—either to evacuate the poison or to change what brought about the pain. Evacuation is through strongly drawing medications while change is through opposites, either in terms of qualities or in the whole substance. It was shown

895K

349

σφοδρῶς ἑλκόντων φαρμάκων, ἀλλοιοῦται δὲ διὰ τῶν
ἐναντίων, ἤτοι κατὰ τὰς ποιότητας ἢ καθ' ὅλην τὴν
οὐσίαν. ἐδείχθη γὰρ ἐν τοῖς περὶ φαρμάκων ἔνια μὲν
ὅλαις ταῖς οὐσίαις ἀλλήλοις ἐναντία, τινὰ δὲ ταῖς
ποιότησι μόναις. ἐδείχθη δὲ καὶ ὡς ἐπὶ μὲν τῶν ἐναντίων κατὰ ποιότητα μέθοδός τίς ἐστιν, ἐπὶ δὲ τῶν κατὰ
τὴν οὐσίαν οὐκ ἔστιν, ἀλλ' ἐκ πείρας εὕρηται πάντα·

896K καὶ σὺ τοίνυν ὅσα μὲν ἐκ Μεθόδου θεραπευτικῆς
ἐντεῦθεν μάνθανε, τὰ δ' ἐκ μόνης τῆς | πείρας ἐγνω-
σμένα κατὰ τὰς περὶ τῶν φαρμάκων πραγματείας
ἔχεις ἠθροισμένα, μίαν μὲν τὴν περὶ τῆς δυνάμεως
αὐτῶν, ἑτέραν δὲ τὴν περὶ τῆς συνθέσεως, καὶ τρίτην
τὴν περὶ τῶν εὐπορίστων ὀνομαζομένων, ἐν αἷς ἐπι-
δέδεικταί μοι τίνα μὲν ἐκ μόνης τῆς πείρας εὕρηται
φάρμακα, τίνα δὲ ἐκ μόνου τοῦ λόγου, τίνα δ' ἐξ
ἀμφοτέρων. ἡ τοίνυν μέθοδος, ὑπὲρ ἧς ἐν τῇδε τῇ
πραγματείᾳ πρόκειται λέγειν, ἐπὶ ταῖς ὀδύναις ἁπάσαις, ὅσαι διὰ θηρίων ἢ φαρμάκων γίνονται, διττὸν
ἔχει τὸν σκοπόν, κένωσίν τε καὶ ἀλλοίωσιν τοῦ τὴν
ὀδύνην ἐργαζομένου. κενοῖ μὲν οὖν τὰ θερμαίνοντα
πάντα καὶ τὰ χωρὶς τοῦ θερμαίνειν ἕλκοντα σφοδρῶς,
ὥσπερ αἵ τε σικύαι καί τινα τῶν κοίλων κεράτων, οἷς
ὡς σικύαις ἔνιοι χρῶνται. τινὲς δὲ καὶ δι' αὐτοῦ τοῦ
στόματος ἕλκουσι τὸν ἰόν, αὐτοὶ προσπίπτοντες τῷ
πεπονθότι μορίῳ καὶ περιλαμβάνοντες αὐτὸ τοῖς χεί-
λεσιν. ἔχεταί γε μὴν καὶ τοῦ προειρημένου σκοποῦ τὸ
καυτήριον, ὅσα τε φάρμακα παραπλησίως τοῖς καυ-

in the writings on medications[16] that some things are opposites to others in terms of the whole substance whereas others are opposites in terms of qualities alone. However, it was also shown what the method is in the case of opposites in terms of qualities, but not what it is in the case of those pertaining to substance. Instead, in all instances it is discovered by experience.

Accordingly, also among those things you must learn here from the *Method of Medicine* are those known from experience alone, collected together for you in the writings on medications; the one on their potency, the second on their composition, and the third on what is called their ease of procurement. In these works, I have been shown what medications are discovered by experience alone, what by reason alone, and what by both. Therefore, the method, which is what I propose to speak about in this particular work, in all pains which occur due to animals or medications has a twofold objective—evacuation and change of what has brought about the pain. All the medications that are heating, evacuate, as do those things that are strongly drawing apart from heating, such as cupping glasses and certain hollow horns which some use as cupping glasses. There are also some who draw the poison out with their own mouth, applying it to the affected part and surrounding the part with their lips. In fact, the cautery also achieves the aforementioned objective as do those

896K

16 These are the three works frequently referred to: *De simplicium medicamentorum temperamentis et facultatibus*, XI.379–892K and XII.1–377K, *De compositione medicamentorum secundum locos*, XII.378–1007K and XIII.1–361K, and *De compositione medicamentorum per genera*, XIII.362–1058K.

τηρίοις ἐσχάραν ἐργάζεται. ταῦτα μὲν οὖν ἐκκενοῖ
897K πάντα τὴν | οὐσίαν ὅλην τοῦ λυποῦντος. ἕτερον δὲ
γένος ἐστὶ βοηθημάτων ἀλλοιούντων τὴν ποιότητα
διὰ τῶν ἐναντίων, εἰ μὲν θερμασίας ὁ κάμνων αἰσθά-
νοιτο σφοδρᾶς ἤτοι κατ' αὐτὸ τὸ δεδηγμένον ἢ καθ'
ὅλον τὸ σῶμα, τὰ ψύχοντα φάρμακα προσφερόντων
ἡμῶν· εἰ δὲ ψύξεως, τὰ θερμαίνοντα. μεμάθηκας δὲ ἐν
ταῖς περὶ τῶν φαρμάκων πραγματείαις ἑκάτερα. τοιαῦται μὲν οὖν αἱ κοιναὶ πάσης φλεγμονῆς ἰάσεις·
ὑπαλλάττονται δὲ κατὰ τὰ πεπονθότα μόρια. δέδεικται
γὰρ ἤδη κἀν τῇ τῶν ἑλκῶν θεραπείᾳ τοῦτο· καὶ πολλῶν οὐ δεῖ λόγων τῷ μεμνημένῳ τῶν ἐν ἐκείνοις
εἰρημένων, ἀλλ' ἀρκέσει διὰ βραχέων ἐπελθεῖν αὐτά.

7. Μία μὲν οὖν ἔνδειξις ἐκ τῶν ὁμοιομερῶν καλουμένων γίνεται μερῶν τοῦ σώματος· ἑτέρα δὲ ἐκ τῶν
ὀργανικῶν. ἡ μὲν οὖν ἐκ τῶν ὁμοιομερῶν τὸ ποσὸν τοῦ
θερμαίνειν, ἢ ψύχειν, ἢ ξηραίνειν, ἢ ὑγραίνειν διο-
ρίζει, ἡ δ' ἐκ τῶν ὀργανικῶν τὸν τόπον δι' οὗ χρὴ
κενῶσαι καὶ προσέτι τὸν τρόπον τῆς κενώσεως· ἔτι τε
πρὸς τούτοις τὸ μᾶλλόν τε καὶ ἧττον ἐν τῇ τῶν
898K ὁμοειδῶν φαρμάκων χρήσει. | περὶ μὲν οὖν τῆς ἀπὸ
τῶν ὁμοιομερῶν ἐνδείξεως λέλεκται πρόσθεν ἐν τῇ τῶν
ἑλκῶν ἰάσει, περὶ δὲ τῆς τῶν ὀργανικῶν ἐν τῷδε
λεχθήσεται.

8. Τῆς γάρ τοι φλεγμονῆς κατὰ διττὸν τρόπον
ἐξεστώσης τοῦ κατὰ φύσιν, ὅτι τε πεπλήρωται τὸ
μόριον αἵματος πολλοῦ καὶ ὅτι θερμότερόν ἐστιν, ὁ

medications that create an eschar like cauteries. Thus, all
these things evacuate the whole substance of what is dis-   897K
tressing. There is also another class of remedies which
change the quality through the opposites—for example,
when we apply cooling medications, if the patient senses
strong heat either in the part that has been bitten or in the
whole body, or conversely, when we apply heating medications, if the patient senses cooling. You have learned each
[of these] in the works on medications. Such things are,
then, the common cures of every inflammation; the differences lie in the affected parts. This has already been shown
in the treatment of wounds.[17] There is no need for prolonged discussion to recall what was said about those—a
brief run-through will suffice.

7. Thus, one indication arises from what are called the
homoiomerous parts of the body and another from the organic parts. The indication from the *homoiomerous* parts
determines the quantity of heating, cooling, drying or
moistening, while the indication from the organic parts determines the place through which it is necessary to evacuate as well as the manner of the evacuation and, in addition
to these factors, it determines the issue of more or less in
the use of similar kinds of medications. I have spoken   898K
about the indication from the *homoiomeres* previously in
relation to the cure of wounds and ulcers; I shall speak
about the indication in relation to the organic parts here.

8. Now since inflammation departs from an accord with
nature in a twofold manner—because the part has been
filled with a lot of blood and because it is hotter—the indi-

17 This is taken as a general reference to Books 3–6 of the present work.

τῆς κενώσεως σκοπὸς ἐπικρατεῖ μᾶλλον τοῦ τῆς ἐμψύξεως, οὐχ ὡς ἐν τοῖς ἐρυσιπέλασιν· ἐπ' ἐκείνων γὰρ
ὁ τῆς ἐμψύξεως ἐπείγει πρὸ τοῦ τῆς κενώσεως. καίτοι
τό γε κεφάλαιον τῆς θεραπείας ἀμφοτέρων τῶν παθῶν
κοινόν ἐστιν ἡ κένωσις τοῦ λυποῦντος χυμοῦ. καὶ διὰ
τοῦτο μετὰ τὴν ἔμψυξιν τῶν ἐρυσιπελάτων ἐπὶ τὰ
διαφορητικὰ παραγινόμεθα φάρμακα. ἐπὶ τοίνυν τῆς
φλεγμονῆς ζεούσης εἰς τοσοῦτον ψυκτέον ἐστίν, εἰς
ὅσον ἐκκόψαι τε καὶ κωλῦσαι τὴν αὔξησιν αὐτῆς
συμφέρει. καὶ γὰρ ὀδυνώσης τῆς πλέονος θερμασίας
καί τι καὶ πρὸς τὸ πεπονθὸς ἑλκούσης, ἐξ ἀμφοῖν
αὐξάνεσθαι συμβαίνει τὴν φλεγμονήν. ὅσον μὲν οὖν
ὡς θερμῷ νοσήματι τῇ φλεγμονῇ τῆς ψύξεως ἁρμότ-
899K τει, τοῦτο κωλυτικόν ἐστι τῆς αὐξήσεως. | ὡσαύτως δὲ
καὶ καθ' ὅσον ἀναστέλλει τὸ ἐπιρρέον. ὅσον δὲ τοῦ
περιεχομένου κατὰ τὸ φλεγμαῖνον ἀποκρουστικὸν εἰς
τὰ πλησιάζοντα μόρια, θεραπευτικὸν τοῦτ' ἔστι τῆς
οὔσης ἤδη φλεγμονῆς. ὡσαύτως δὲ καὶ τὰ θερμαίνοντα μετρίως ἐνίοτε κατ' ἀμφοτέρους τοὺς τρόπους
ὀνίνησι, ὅταν τὴν ὀδύνην παύῃ καὶ διαφορῇ τὸ περιεχόμενον ἐν τοῖς πεπονθόσι μορίοις· ἐν μὲν τῷ παύειν
τὴν ὀδύνην κωλύει τὴν αὔξησιν, ἐν δὲ τῷ διαφορεῖν
ἰᾶται τὸ γεγονὸς ἤδη τῆς φλεγμονῆς.

9. Ἐπεὶ τοίνυν τὸ κῦρος ἅπαν ἐστὶ τῆς τῶν
φλεγμαινόντων θεραπείας ἐν τῷ κενῶσαι τὸ περιττὸν
αἷμα τοῦ φλεγμαίνοντος μορίου, κένωσις δὲ ἐπινοεῖται διττὴ τῶν οὕτως ἐχόντων, ἢ μεθισταμένου πρὸς
ἕτερα χωρία τοῦ περιεχομένου κατὰ τὸ φλεγμαῖνον

cator of evacuation takes precedence over that of cooling, unlike in erysipelas, for in that [disease] the objective of cooling is more pressing than that of evacuation. And yet, the chief point of treatment, common to both affections, is the evacuation of the distressing humor. Because of this, after the cooling of the erysipelas, we proceed to the discutient medications. Therefore, in the case of a seething inflammation, it must be cooled to a degree that is appropriate to eradicate it altogether or to prevent it from increasing. And because too much heat excites pain and also draws something to the affected part, what happens is that the inflammation is increased by both these factors. Therefore, the amount of cooling that is fitting for inflammation as for a hot disease is the amount that prevents the increase, and the same also applies to the amount that keeps   899K back the flux. As much of what is contained in relation to what is inflamed that can be repelled to the adjacent parts is therapeutic of the inflammation that already exists. Similarly, things that are moderately heating are sometimes beneficial in both ways when they stop the pain and disperse what is contained in the affected parts; on the one hand, by stopping the pain, they prevent the increase and, on the other, by dispersing, they cure what already exists of the inflammation.

9. Therefore, since the whole principle of the treatment of inflammations lies in evacuating the excess blood of the inflamed part, this evacuation is thought of in a twofold way: the transfer of blood contained in what is

αἵματος, ἢ ἔξωθεν τοῦ σώματος ἐκκρινομένου, βέλτιόν ἐστιν ἀμφοτέροις χρῆσθαι προσέχοντα τὸν νοῦν,
μὴ κατὰ συμβεβηκὸς ἔπηταί τις βλάβη. διττῆς τοίνυν ἑκατέρας τῶν εἰρημένων κενώσεως οὔσης, εἰς
τέσσαρας τὰς πάσας ἡ τομὴ γίνεται τῶν κενωτικῶν
900K ἁπάντων | βοηθημάτων, τῆς μὲν εἰς τὰ ἄλλα μόρια
μεταρρύσεως τοῦ αἵματος ἡ μὲν ἑτέρα διωθουμένων
αὐτὸ τῶν φλεγμαινόντων μορίων, ἡ δ᾽ ἑτέρα τῶν
ἀπαθῶν ἑλκόντων γίνεται· τῆς δὲ ἔξω τοῦ σώματος
κενώσεως ἡ μία μὲν αἰσθηταῖς ἐκροαῖς, ἡ δὲ ἑτέρα
λόγῳ θεωρηταῖς ἐπιτελεῖται. καὶ τῆς αἰσθηταῖς ἐκροαῖς γινομένης ἡ μὲν ἑτέρα δι᾽ αὐτοῦ τοῦ φλεγμαίνοντος, ἡ δ᾽ ἑτέρα διὰ τῶν συναναστομουμένων αὐτῷ.
διὸ καὶ χρεία τῆς ἀνατομῆς ἐστιν εἰς διάγνωσιν τῆς
τοιαύτης κοινωνίας. αὗται μὲν οὖν ἀπὸ τῆς φύσεως
τῶν μορίων ἐνδείξεις προέρχονται, τὴν θεραπείαν ὑπ-
αλλάττουσαι τῶν φλεγμαινόντων· ἔτι τε πρὸς ταύταις
ἡ ἐκ τῆς θέσεώς τε καὶ διαπλάσεως, ἃς ὡς ὀργανικὸν
ἐνδείκνυται τὸ πεπονθός, οὐχ ὡς ὁμοιομερές.

10. Ἔτι δ᾽ ἄλλαι κοιναὶ τῶν ὀργανικῶν εἰσι καὶ τῶν
ὁμοιομερῶν, ὅταν ἐπισκεπτώμεθα τήν τ᾽ ἐνέργειαν
αὐτῶν, καὶ εἰ ἀραιὸν ἢ πυκνόν, ἢ ἀναίσθητον, ἢ
δυσαίσθητόν ἐστιν, ἢ εὐαίσθητον. εἰς ἃς ἁπάσας
ἀποβλέπειν χρὴ τὸν ἐπιχειροῦντα θεραπεύειν ὀρθῶς. |

901K   11. Ἐπεὶ δέ, ὡς ἀεὶ λέγομεν, οὐχ ἱκανόν ἐστιν αὐτὰ

inflamed to other parts, and the evacuation of the body externally. It is better to give attention to using both lest some contingent damage follows. Accordingly, since each evacuation of those things mentioned is twofold, there is the division of all the evacuating remedies into four. One component of the transfer of blood to other parts is getting rid of it from the inflamed parts and the other is drawing it from the nonaffected parts. On the evacuation of the body to the outside, one component is by perceptible discharges and the other is accomplished by those pathways that are "theoretical" (i.e. pores contemplated by reason).[18] And of the evacuation occurring by perceptible effluxes, one component is through the inflamed part itself and the other is through those things joined by an opening with it. On this account, anatomy is useful for the recognition of such associations. These indications proceed from the nature of the parts and change the treatment of parts that are inflamed. In addition to these, there are the indications from the position and conformation, which are indicative in respect to what is affected as an organic part and not as a *homoiomere*.

10. There are still other indications which are common to both organic parts and *homoiomeres* when we examine their function; whether they are loose in texture or dense, are without sensation, or with disturbed sensation, or with normal sensation. Anyone attempting to carry out the proper treatment must pay attention to all these factors.

11. Since, however, as I always say, it is not enough only

900K

901K

---

[18] That is, pores not directly perceptible.

μόνον γινώσκειν τὰ καθόλου χωρὶς τοῦ γεγυμνάσθαι
περὶ τὰ κατὰ μέρος, οὕτω καὶ νῦν πράξομεν οὐ διὰ
πάντων τῶν κατὰ μέρος ἰόντες, ἀλλ᾽ ὅσα περ ἂν ἡμῖν
ἱκανὰ δόξῃ τοῖς ἀναγνωσομένοις αὐτὰ γενήσεσθαι.
ὑποκείσθω τοίνυν ἧπαρ ἀρχόμενον φλεγμαίνειν καὶ
ζητείσθω τίς ἀρίστη θεραπεία γενήσεται τοῦ πάθους.
ἐπισκέπτου δὴ πρῶτον μὲν ἁπάντων, ἀφ᾽ ὧν εἶπον
ὁρμώμενος, εἰ δεῖται κενώσεως τὸ πᾶν σῶμα. κἂν
εὕροις δεόμενον, ἐφεξῆς σκέπτου τὴν ῥώμην τοῦ
κάμνοντος, εἰ δύναται κένωσιν ἀθρόαν ἐνεγκεῖν. ἔστω
δὴ πρότερον ἐρρῶσθαι τὴν δύναμιν· ἐφεξῆς σκέπτου
τὴν ἡλικίαν. εἰ γὰρ παιδίον εἴη, τὴν διὰ φλεβοτομίας
οὐκ οἴσει κένωσιν, ὡς ἔμπροσθεν ἐδείχθη· κατὰ μὲν
τὸν τῆς ἥβης καιρὸν οἱ παῖδες ἤδη φέρουσι τὴν διὰ
τῆς φλεβοτομίας κένωσιν.

ἀντισπαστέον οὖν ἅμα καὶ κενωτέον τὸ φερόμενον
ἐπὶ τὸ ἧπαρ αἷμα τῇ φλεβοτομίᾳ, κατὰ τὸν δεξιὸν
ἀγκῶνα τὴν ἔνδον φλέβα τέμνοντες, ἐπεὶ κατ᾽ εὐθὺ καὶ
δι᾽ εὐρείας ὁδοῦ τῇ κοίλῃ καλουμένῃ κοινωνεῖ· μὴ
902K φαινομένης δὲ ταύτης | τὴν μέσην τέμνειν· εἰ δὲ μήτ᾽
αὐτὴ φαίνοιτο, τὴν λοιπὴν καὶ τρίτην. τὸ δὲ ποσὸν τῆς
κενώσεως ἔκ τε τῆς κατὰ τὸ πλῆθος εὑρήσεις ποσότητος, ὅσα τ᾽ ἄλλα κατὰ τὸν ἔμπροσθεν εἴρηται λόγον,
ἡλικία τε καὶ φύσις ὥρα τε καὶ χώρα καὶ ἔθος, ἔτι τε
πρὸ τούτων καὶ ἡ δύναμις τοῦ κάμνοντος. ἐφ᾽ ἁπάντων
γὰρ ταῦτα κοινά. τὸ δ᾽ ἤτοι τὴν ἔνδον ἢ τὴν μέσην ἢ

---

[19] The veins mentioned are, respectively, the basilic, median

to know these things in general without becoming practiced in them individually, what I shall now do is not go through all of them one by one but only those that may seem to me to be sufficient for those who will read this. Therefore, let us assume that the liver is starting to become inflamed, and let us inquire into what the best treatment for the affection will be. Consider first, from all those things I spoke of when beginning, whether there is a need for evacuation of the whole body. And if you do find this is needed, consider next the strength of the patient and whether he is able to tolerate a complete evacuation. Let it be the case, first, that the capacity is strong. Next consider the age. If the patient is a child, he [or she] will not tolerate evacuation by phlebotomy, as was shown previously. However, children who have already reached puberty do tolerate evacuation by phlebotomy.

You must draw off and evacuate by phlebotomy the blood carried to the liver, incising the inner vein in the right antecubital fossa, since this is on a level and connects by a wide channel with the so-called [inferior] vena cava. If this [vein] is not visible, cut the medial vein, and if this is not visible, the remaining and third vein.[19] You will discover the amount of the evacuation from the amount of the abundance and those other things mentioned in the previous discussion—the age, the nature, the time of year, the place, the way of life, and more important than all these, the strength of the patient. These things are common in all cases. Which vein to cut, whether it should be the internal

902K

cubital, and cephalic; see *Gray's Anatomy*, 15th ed., p. 585, fig. 330. There is also a note by Tillaux on the relative merits of the veins described for venesection.

τὴν ὠμιαίαν φλέβα τέμνειν, ἢ τὴν παρὰ τὸ σφυρόν, ἢ
τὴν ἐν ἰγνύϊ, παρὰ τοῦ πεπονθότος μορίου τὴν
ἔνδειξιν ἔχει. καὶ διώρισται μὲν ἤδη κἀν τοῖς περὶ
φλεβοτομίας ὑπὲρ τῶν τοιούτων ἁπασῶν κενώσεων·

εἰρήσεται δὲ καὶ νῦν ὅσον εἰς τὰ παρόντα χρήσιμον, αὐτό τε τοῦτο πρῶτον, ὡς οὐκ ἀρκεῖ μόνον ὅτι
κενωτέον ἐστὶν ἐξευρεῖν, ὡς ἂν φαῖεν οἱ τὴν ἀμέθοδον
αἵρεσιν μετιόντες, οὐδὲν φροντίζοντες τῆς διαφορᾶς
τῶν πεπονθότων μορίων. οὐ γὰρ ὁ λόγος μόνον, ἀλλὰ
καὶ ἡ πεῖρα δείκνυσιν ἄλλην ἄλλῳ μορίῳ κένωσιν
ἁρμόττουσαν. ἐθεάσω γοῦν ἐνίους τῶν ἀρξαμένων
φλεγμαίνειν ὀφθαλμοὺς αὐτῷ μόνῳ τῷ καθαρθῆναι
διὰ τῆς κάτω γαστρὸς ἡμέρᾳ μιᾷ θεραπευθέντας. |
903K ὅπερ ἐὰν ἐφ᾽ ἥπατος ἀρχομένου φλεγμαίνειν ἐπιχει-
ρήσῃ τις πρᾶξαι, μεγίστην ἐργάσεται τὴν φλεγμονήν, ὥσπερ γε καὶ εἰ τῶν νεφρῶν ἢ τῆς κύστεως
φλεγμαίνειν ἀρξαμένων οὐρητικὰ ποτίζοι φάρμακα
καταμήνιά τε κινοῖ μήτρας φλεγμαινούσης· ἀντισπᾶν
γὰρ χρὴ τῶν ἀρχομένων ῥευματίζεσθαι πορρωτάτω
τὸ περιττόν, οὐχ ἕλκειν ἐπ᾽ αὐτά. κατὰ τοῦτον οὖν τὸν
λόγον οὐδὲ γαστρὸς οὐδ᾽ ἐντέρων ἀρξαμένων φλεγμαίνειν ὑπηλάτῳ χρῆσθαι προσήκει. τὴν δ᾽ αὐτὴν
ἔνδειξιν ἔχει τούτοις μὲν μήτρα, τοῖς δ᾽ οὐρητικοῖς

---

[20] These are the three works previously mentioned: *De venae
sectione adversus Erasistratum*, XI.147–86K, *De venae sectione
adversus Erasistrateos Romae degentes*, XI.187–249K, and *De*

or the medial vein, or the humero-cephalic vein, or the vein adjacent to the ankle, or that in the popliteal fossa, has its indication from the affected part. The distinction between all such evacuations has already been made in the writings on phlebotomy.[20]

I shall now speak about what is useful for our present purposes. The first thing is this: that it is not enough just to discover that you must evacuate, as those who attend to the amethodical school would say, not giving any thought to the difference between the affected parts. It is not reason alone but also experience which shows that evacuation is suitable for one part rather than another. At all events, you have seen some of those who are starting to develop inflammation in the eyes being treated in a single day by purging through the stomach downward. Should someone attempt to do this in the case of a liver starting to become inflamed, he will bring about a very severe inflammation, and the same applies with a developing inflammation of the kidneys or bladder, if someone were to drink diuretic medications, or if someone were to initiate menstruation when the uterus was developing inflammation. It is necessary to draw the superfluity in the opposite direction as far away as possible when parts are beginning to suffer from a flux, and not to draw to them. By the same argument, when there is an evolving inflammation in either the stomach or intestines, it is not appropriate to use purging below. The uterus has the same indication as these while the genitalia have the same indication as the urine-producing organs. In

903K

curandi ratione per venae sectionem, XI.250–316K. An English translation of all three is to be found in P. Brain (1986).

ὀργάνοις αἰδοῖα. τό γε μὴν ἐμέτοις χρῆσθαι τῶν
αἰδοίων πεπονθότων ἀντισπαστικόν ἐστι βοήθημα.

κατὰ δὲ τὸν αὐτὸν λόγον ἐπὶ μὲν τοῖς κατὰ τὴν
κεφαλὴν ἄπασιν ὑπήλατον φάρμακον, ὅσα δὲ κατὰ
τὸν φάρυγγα καὶ οὐρανίσκον, ἢ τὴν ὑπερῴαν, ἢ τὴν
γλῶτταν, ἢ ὅλως κατὰ τὸ στόμα φλεγμαίνειν ἄρχεται,
φυλακτέον ἐπὶ τούτων ἁπάντων τοὺς καλουμένους
ἀποφλεγματισμούς. ὅμοιον γὰρ τοῦτο τῷ καθαίρειν
904K κάτω τῶν ἐντέρων πεπονθότων καὶ τῷ | οὖρα κινεῖν
τῶν κατὰ τοὺς νεφροὺς ἢ κύστιν ἐχόντων κακῶς, ἢ
ἐμέτους τῶν κατὰ τὸν στόμαχον. ἄμεινον οὖν ἐπὶ τὴν
ῥῖνα παροχετεύειν, ἀρχομένων τῶν κατὰ τὸ στόμα
μορίων φλεγμαίνειν. οὕτω δὲ καὶ φλέβα τέμνειν, εἰ
μὲν ταῦτα πεπόνθοι, τὴν ὠμιαίαν ἐν χειρί, καὶ ταύτης
μὴ φαινομένης τὴν μέσην· εἰ δ' ἧπαρ ἢ θώραξ ἢ
πνεύμων ἢ καρδία, τὴν ἔνδον. ἐπὶ δὲ συνάγχης πρώτας μὲν τὰς ἐν χερσί, δευτέρας δὲ τὰς ὑπὸ τὴν γλῶτταν. τῶν δὲ κατ' ἰνίον πασχόντων καὶ τὴν ἐν ἀγκῶνι
μέν, οὐχ ἥκιστα δὲ καὶ τὴν ἐν τῷ μετώπῳ. ἐπὶ δὲ
νεφρῶν καὶ κύστεως αἰδοίου τε καὶ μήτρας τὰς ἐν τοῖς
σκέλεσι, μάλιστα μὲν τὰς κατὰ τὴν ἰγνύαν, εἰ δὲ μή,
τὰς παρὰ σφυρόν. ἀεὶ δ' ἐπὶ πάντων τὰς κατ' εὐθύ. τοῦ
μὲν ἥπατος ἀρχομένου φλεγμαίνειν, τὰς ἐν τῇ δεξιᾷ
χειρί, τοῦ δὲ σπληνός, ἔμπαλιν. ὥστ' εὐθὺς ἡ διαφορὰ

---

21 Medications that promote the discharge of mucus or
phlegm; on what these are and when they should be used, see *De
compositione medicamentorum secundum locos*, XIII.566K.

fact, the use of vomiting is a revulsive remedy for affections of the genital organs.

On the same grounds, in regard to all those things that involve the head, a purging medication [is appropriate] whereas, in the case of those parts in the pharynx, roof of the mouth, palate or tongue that are starting to become inflamed, you must, in all these cases, guard against the so-called "apophlegmatics."[21] This is similar to the purging downward of affected intestines, and to provoking urination in those in whom the kidneys or bladder are bad, or provoking vomiting in those in whom the stomach is bad. It is better, then, when the parts in the mouth are beginning to become inflamed, to redirect [the evacuation] via the nose. So too is it better to open a vein if these parts are affected; the (humero-)cephalic vein in the arm or, if this is not visible, the median cubital [vein], and if the liver, thorax, lungs or heart [are involved], the inner (basilic) vein. In cynanche,[22] first [cut the veins] in the arms and second those under the tongue. When parts in relation to the occiput [are affected], [cut the vein] in the elbow, and no less also that in the forehead. In respect to the kidneys, bladder, genitalia and uterus, [cut] those in the legs, especially those in relation to the popliteal fossa; if not, cut those beside the ankle. Always, in all these instances, cut straight. When the liver is starting to become inflamed, [cut] the veins in the right arm and, when the spleen is involved, the opposite (i.e. in the left arm). And so simply

904K

[22] The original Greek term is preserved in English although not in common use. Alternatives are the old English word "squinancy" (used by Peter English) or the current terms "quinsy" and "angina."

τῶν κενώσεων πρῶτον κατὰ τὴν τῶν μορίων διαφορὰν
ὑπαλλάττεται· καὶ δῆλον ὅτι τὸ κοινὸν τῆς ἐνδείξεως
οὐ μᾶλλον ὠφελείας ἢ βλάβης αἴτιον. ὅτι μὲν γὰρ
905K κενωτέον, ἔνδειξις | κοινή· τὸ δ᾽ ὅθεν ἢ ὅπως, οἱ
πεπονθότες τόποι διδάσκουσιν. οὕτω γοῦν καὶ οἱ τῶν
κώλων ἐπενοήθησαν δεσμοὶ θώρακος ἢ γαστρός, ἢ
τῶν κατὰ τὸν τράχηλον, ἢ τὴν κεφαλὴν μορίων φλεγμαινόντων. οὐ γὰρ δὴ τὸ φλεγμαῖνόν γε αὐτὸ δήσεις
κῶλον, ἀλλ᾽ ἐπὶ μὲν σκελῶν τὰς χεῖρας, ἐπὶ χειρῶν δὲ
τὰ σκέλη.

12. Καὶ μὴν καὶ τὸ ψύχειν καὶ στύφειν ἐν ἀρχῇ τὰ
φλεγμαίνοντα χωρὶς τῆς περὶ τῶν πεπονθότων ἐνδείξεως οὐ μᾶλλον ὠφελείας ἢ βλάβης αἴτιον. ἐπὶ μὲν
γὰρ τῶν κατὰ τὸ κῶλον μορίων ἀρκεῖ καὶ σπόγγον
ἐπιθεῖναι, βρέξαντας ἤτοι γε ὕδατι ψυχρῷ μικρὸν
ὄξους μιγνύντας ἢ ὕδατι μόνῳ· καθάπερ γε καὶ οἴνῳ
τινὶ τῶν αὐστηρῶν. ἥπατος δ᾽ ἀρξαμένου φλεγμαίνειν
οὐδεὶς ἂν χρήσαιτο τούτων οὐδενὶ νοῦν ἔχων ἄνθρω-
πος· ἀλλ᾽ οὐδ᾽ εἰ μήλινον ἐπιβρέξαις ἢ μύρσινον, ἢ
μαστίχινον, ἢ νάρδινον, ἢ σχίνινον, ἤ τι τῶν στυφόντων ἐλαίων ἢ καὶ τῶν ἄλλων ἐναφεψήσας ἀψίνθιον,
οὐδὲ τούτων οὐδὲν ἁρμόττει ψυχρόν, ὥσπερ οὐδὲ
κατάπλασμα ψυχρὸν οὐδέν. ἀλλὰ μῆλα μὲν ἐναφεψῶν
906K οἴνῳ, καὶ μάλιστα | κυδώνια, κατάπλασμα σκευάσοις
δι᾽ αὐτοῦ, φλεγμονῆς ἥπατος ἔτι ἀρχομένης, ψυχρὸν
δ᾽ οὐδὲ τοῦτο προσοίσεις· ὥσπερ οὐδὲ τὸ ἔλαιον
ἐσκευασμένον, ὡς εἴρηται, τοῖς ὀφθαλμοῖς ἢ τοῖς ἐν
τῷ στόματι μορίοις ἀρχομένοις φλεγμαίνειν. ὠτὶ δὲ

the difference of the evacuations varies primarily according to the difference of the parts; and it is clear that the general indication is not more useful than a cause of harm. For the general indication is that you must purge. The af-   905K
fected places teach you the where and the how. At all events, in this way too, bindings of the limbs are contrived when the chest, abdomen or the parts in the neck or head are inflamed. Definitely do not bind the inflamed limb itself but, in the case of the legs, bind the arms and, in the case of the arms, bind the legs.

12. Furthermore, to cool and contract what is inflamed at the outset, independent of the indication of the parts affected, is less a benefit than a cause of harm. In the case of parts in the limb, it is also sufficient to apply a sponge soaked in cold water, having mixed with it a little vinegar, or in water alone, just as it also is [to apply a sponge soaked] in one of the bitter wines. However, when the liver is starting to become inflamed, no man of sense would use any of these things, nor would you pour on apple extract, myrtle, mastich, nard, what is made from mastich or any of the astringent oils, nor also any other things which you have boiled in wormwood, nor is any of the cold things suitable just as no cold cataplasm is suitable. Nor would you apply something cold such as apples boiled down in wine or particularly quinces, if you should prepare a cataplasm from   906K
this when the liver has a still evolving inflammation, just as oil is not prepared, as was said, for the eyes or for those parts in the mouth which are starting to become inflamed.

κἂν ὄξος ἐγχέῃς μετὰ ῥοδίνου, βλάψεις οὐδέν. ἀλλ'
οὐκ ὀφθαλμοῖς γε φλεγμαίνουσιν ἀγαθὸν τοῦτο, καθάπερ οὐδὲ τὸ διὰ μόρων φάρμακον ἤ τι τῶν στομα-
τικῶν ὀνομαζομένων ἄλλο· πάντα γὰρ ἀνιαρὰ τὰ τοιαῦτα τοῖς ὀφθαλμοῖς, καίτοι κατὰ τὸ γένος ὄντα τῆς
ἐνδείξεως. ἀδένων δὲ φλεγμαίνειν ἀρχομένων ἤρκεσε
πολλάκις ἔλαιον μόνον θερμόν.

13. Ἀλλὰ καὶ ἡ ἄλλη δίαιτα τοῖς μὲν ἀδένας ἤ τι
τῶν κατὰ τὰ κῶλα μορίων ἔχουσι φλεγμαῖνον, ἕνα
μόνον λαμβάνει σκοπόν, ὡς τοσαῦτα καὶ τοιαῦτα
προσφέρεσθαι δεῖν, ὅσα δὴ καὶ οἷα πεφθήσεται
ῥᾷστα. διαφέρει δ' οὐδὲν ἢ χόνδρον, ἢ πτισάνην, ἢ
μελίκρατον, ἢ ῥόαν, ἢ μῆλον, ἤ τι τοιοῦτον προσενέγκασθαι. φλεγμαίνοντος δ' ἥπατος ἀκριβεστάτης διαί-
907K της ἐστὶ χρεία, καθάπερ γε καὶ γαστρός. τὸ | γὰρ
ἔργον αὐτῶν ἅπαντι τῷ ζῴῳ κοινόν ἐστι, καὶ μὴ
πεφθείσης καλῶς τῆς τροφῆς ἢ μὴ προσηκόντως
αἱματωθείσης, ἅπασι τοῖς τοῦ ζῴου μέλεσι μεγίστη
βλάβη προσγίνεται. τὰ δὲ τῶν κώλων μόρια τοσοῦτον
λαμβάνει τῆς τροφῆς, ὅσον τρέφεσθαι πέφυκε. διὸ
κἂν ἐκ μήλου καλῶς ἐν τῇ γαστρὶ πεφθέντος ἀφίκηται
πρὸς αὐτὰ τροφή, κἂν ἐκ χόνδρου, κἂν ἐκ πτισάνης,
οὐ μέγα διαφέρει. κατὰ δὲ τὸ ἧπαρ ὁπόσον ἡ διαφορὰ
τῶν τροφῶν δύναται καὶ ὡς μέγιστον ἐφ' ἑκάτερα,
πάρεστί σοι μανθάνειν.

14. Ἐκκενοῦσθαι μὲν δήπου χρὴ τοῦ ἥπατος ὅσον
ἀθροίζεται κατ' αὐτὸ τοῦ τε πικροχόλου χυμοῦ καὶ

In the ear, however, even if you pour in vinegar with oil of roses, you will do no harm. But this is not, in fact, good for the eyes when they are inflamed, just as the medication made from blackberries is not, nor any other of the so-called "stomatics."[23] All such things are distressing to the eyes and yet they are in the class of those things that are indicated. When glands are starting to become inflamed, warm oil alone is often satisfactory.

13. But also, the other diet for those who have inflammation involving the glands or one of the parts in the legs takes the one objective alone which is that, however many and varied the things are that need to be administered, they must be very easily concocted. It makes no difference whether gruel, barley gruel, melikraton, pomegranate, apple or some such thing is administered. However, when the liver is inflamed, there is need of the most precise diet, just as is also the case with the stomach. The action of these [organs] is common to every animal in that, if the nutriment is not digested properly or is not converted to blood in a suitable fashion, very great harm befalls all the parts of the animal. The parts of the limbs take as much of the nutriment as nourishes them naturally. Because of this, even if nourishment comes to them from apples properly digested in the stomach, or even from gruel or ptisan, there is no great difference. In the liver, how great the difference between the nutriments can be—that it is very great in each case—is there for you to learn.

14. Of course, with the liver it is necessary for however much of the picrocholic humor (bitter bile) and the ichors

907K

---

[23] For details on a number of such medications, see Galen's *De compositione medicamentorum secundum locos*, XII.919K ff., and also the pseudo-Galenic *Introductio sive medicus*, XIV.762K.

τῶν ἐκ τῆς φλεγμονῆς ἰχώρων. τοῦτο δ᾽ οὐκ ἂν γένοιτο
χωρὶς τοῦ διαρρύπτεσθαι μὲν τὰ κατὰ τὸ σπλάγχνον
ἀγγεῖα, τὸν δ᾽ εἰς τὴν νῆστιν καθήκοντα πόρον ἀνα-
στομοῦσθαι. χόνδρος οὖν ἐμπλάττων μὲν τοῦτον, ἐμπλάττων δὲ τὰ κατὰ τὸ ἧπαρ ἀγγεῖα καὶ μάλιστα τὰ
πέρατ᾽ αὐτῶν, κωλύει τήν τε χολὴν εἰς τὸ ἔντερον
ὑπιέναι καὶ τὴν τροφὴν εἰς ὅλον ἀναδίδοσθαι τὸ
908K σῶμα. χρεία τοίνυν ἐστὶ τῶν ἐκφραττόντων | ἐδεσμά-
των τε καὶ φαρμάκων αὐτῷ τε τῷ ἥπατι καὶ τῷ τοῦ
χοληδόχου πόρου στόματι. τὰ δὲ τοιαῦτα πάντα
γλίσχρα μὲν ἥκιστ᾽ ἐστί, λεπτὰ δὲ ταῖς συστάσεσι
καὶ δακνώδη ταῖς ποιότησιν.

ἀλλὰ πάλιν ὑπὸ τούτων δακνόμενα τὰ φλεγμαίνοντα παροξύνεται καὶ διὰ τοῦτο δεόμεθα τῶν ἄνευ
τοῦ δάκνειν αὐτὰ ῥυπτόντων, οἷόν πέρ ἐστι καὶ τὸ
μελίκρατον. ἀλλ᾽ ἴσμεν ὅτι τὰ γλυκέα πάντα καὶ
σπλῆνα καὶ ἧπαρ ἐπὶ πλεῖστον ἐξαίρει. λοιπὴ τοίνυν
ἄμεμπτος ὡς ἐν ἐδέσματι μὲν ἡ πτισάνη· χωρὶς γὰρ
τοῦ δάκνειν ῥύπτει· ὡς ἐν φαρμάκοις δ᾽ ὀξύμελι μεθ᾽
ὕδατος κεραννύμενον. καὶ γὰρ καὶ ἡ ῥόα καὶ τὸ μῆλον
ὅσα τ᾽ ἄλλα στύφει, συνάγοντα τοῦ χοληδόχου πόρου
τὸ στόμα, κωλύει τὴν χολὴν ἐκκρίνεσθαι· καὶ διὰ
τοῦτο βλάπτει φλεγμονὰς ἥπατος, καὶ μάλισθ᾽ ὅταν
ἐν τοῖς σιμοῖς ὦσι μέρεσι τοῦ σπλάγχνου· πρὸς γὰρ
τῇ διὰ φλεγμονὴν στενοχωρίᾳ καὶ ἡ ἐκ τῶν στυφόντων τε καὶ γλίσχρων ἐδεσμάτων προσέρχεται. καὶ
τὰ δάκνοντα δὲ μᾶλλον βλάπτει τὰς ἐν τοῖς σιμοῖς
φλεγμονάς· ὅσαι γὰρ ἐν τοῖς κυρτοῖς αὐτοῦ μέρεσι

from the inflammation collect in it, to be expelled. This would not occur without a thorough cleansing of the vessels in the viscus and the opening of the channel that comes down to the jejunum (i.e. the bile duct). Gruel, since it adheres to this and blocks up the vessels of the liver, and particularly the ends of these, prevents the bile from passing down to the intestine and the nutriment from being distributed to the whole body. Thus there is a need for foods and medications which relieve blockage in the 908K liver itself and in the opening of the bile duct. All such things are minimally viscid, thin in consistency, and stinging in quality.

But contrariwise, the parts that are inflamed, stung by these things, become irritated, and because of this we have need of things that cleanse them without stinging as, for example, melikraton does. But we know that all sweet things stir up the spleen and liver to a very great extent. Therefore, what is left irreproachable, as it is among food, is ptisan, for this cleanses without stinging, and oxymel among medications after it has been mixed with water. Furthermore, both pomegranate and apple, and those other things that are astringent, contract the opening of the bile duct and prevent the secretion of bile. And because of this, they harm the inflamed liver, and especially when they are in the concave parts of the organ, because the narrowing from astringent and viscid foods is added to the narrowing due to the inflammation. And things that are stinging are more harmful to inflammations in the concave [parts of the liver] because, with inflammations that occur

γίγνονται, φθάνει μεταβεβλημένα πρὸς αὐτὰς ἀφ-
909K ικνεῖσθαι τὰ ληφθέντα | καὶ μήτε τὸ στῦφον ἔτι
στύφειν ὁμοίως μήτε τὸ δάκνον δάκνειν μήτε τὸ διὰ
γλισχρότητα τοῖς στενοῖς ἀγγείοις ἐμπλαττόμενον
ἔτι μένειν ὁμοίως γλίσχρον. ἡ μεταβολὴ δ' αὐτοῖς
διττή, τῷ τε προπεπέφθαι καὶ διότι τῷ προϋπάρχοντι
κατὰ τὸ σπλάγχνον αἵματι μίγνυται. τοῦ δέ γε σιμοῦ
μέρους ἐν ἥπατι φλεγμαίνοντος εὐθὺς μὲν ἀναγκαῖόν
ἐστι καὶ τὰς ἐν τῷ μεσεντερίῳ συμφλεγμαίνειν φλέ-
βας· ἀπὸ γὰρ τῆς ἐπὶ πύλας ἅπασαι πεφύκασιν· εὐθὺς
δὲ καὶ τὰ προσπίπτοντα τοῖς στόμασιν αὐτῶν ἐπι-
δείκνυται τὴν ἑαυτῶν δύναμιν.

15. Ἆρά σοι δοκεῖ σμικρὰ διαφορὰ προσέρχεσθαι
τῇ κοινῇ τῶν φλεγμονῶν θεραπείᾳ παρὰ τῶν μορίων;
ἐμοὶ μὲν γὰρ μεγίστη φαίνεται, κἂν εἰ τὴν Θεσ-
σάλειον ἀναισθησίαν ζηλοῦντες οἴονται τὴν κοινὴν
ἔνδειξιν ἀρκεῖν μόνην. ἀναμνῆσαι δέ σε βούλομαι καὶ
τῆς καλῆς αὐτῶν θεραπείας, ἣν ἐπὶ Θεαγένους ἐποιή-
σαντο τοῦ Κυνικοῦ φιλοσόφου· ταύτην γὰρ ἔγνωσαν
οὐκ ὀλίγοι διὰ δόξαν τἀνθρώπου, δημοσίᾳ διαλεγο-
μένου κατὰ τὸ τοῦ Τραϊανοῦ γυμνάσιον ἑκάστης
910K ἡμέρας. | ὁ μὲν οὖν θεραπεύων αὐτὸν ἦν εἷς τῶν Σω-
ρανοῦ μαθητῶν, Ἄτταλος τοὔνομα. κατέπλαττε δὲ
ἑκάστης ἡμέρας τὸ ἧπαρ ἀρτομέλιτι, μὴ γινώσκων ὅτι
στύφεσθαι μετρίως δεῖται τὸ σπλάγχνον τοῦτο, διότι

---

24 On Attalus, Nutton (in F. Kudlien and R. J. Durling, 1991,

in its convex parts, the things taken which come to them are changed beforehand, so that what is astringent is not still similarly astringent, what is stinging does not sting, nor does what adheres to the narrowed vessels due to viscidity still remain viscid in the same way. The change in these things is twofold: that due to being previously concocted and that due to admixture with the blood previously present in the organ (liver). In fact, when there is inflammation of the concave part in the liver, it is immediately inevitable that the veins in the mesentery are also jointly inflamed as they all take their origin from the portal vein. Immediately, too, those things that come upon their openings reveal their own potency.

15. Does it seem to you that little difference accrues to the common treatment of inflammations from the parts? Because, to me, there seems to be a very significant difference, even if those who admire the obtuseness of a Thessalus think that the common indication is sufficient on its own. I would also like you to call to mind their wonderful treatment which they carried out in the case of Theagenes, the Cynic philosopher—many are aware of this because of the reputation of the man, who lectured in public each day at the gymnasium of Trajan. The man who treated him was one of the pupils of Soranus, Attalus by name.[24] Every day he applied a poultice of bread and honey to the liver, not realizing that this viscus needs to be

p. 14) writes: "The doctor, Attalus, is almost certainly to be identified with the royal physician, Statilius Attalus of Heraclea." (See also *EANS,* p. 179.) Theagenes was also a notable person of the time and a pupil of the Peregrinus about whom Lucian wrote an essay.

τῆς θρεπτικῆς δυνάμεως ἀρχὴ τοῖς ζῴοις ἐστὶ καὶ τὸ
φλεβῶδες γένος ἀπ᾽ αὐτοῦ πέφυκεν. οὕτως οὖν ἐθεράπευσε τὸ σπλάγχνον, ὡς τοὺς βουβῶνας ἀμίκτῳ καὶ
μόνῃ τῇ διὰ τῶν χαλώντων ἀγωγῇ, καταπλάττων μὲν
ἀρτομέλιτι, προκαταιονῶν δὲ ἐλαίῳ θερμῷ καὶ τρέφων
ἐκ χόνδρου ῥοφήματι. ταῦτα γὰρ ἀρκεῖ τὰ τρία σχεδὸν ἅπασι τοῖς νῦν ἀμεθόδοις Θεσσαλείοις εἰς τὴν
τῶν ὀξέων ἴασιν. ἔδοξε δέ μοι κατὰ μόνας εἰπεῖν τῷ
Ἀττάλῳ, Προσμιγνύναι τι τῶν στυφόντων καὶ μὴ
ψιλῇ χρῆσθαι τῇ διὰ τῶν χαλαστικῶν ἀγωγῇ. περὶ
μὲν οὖν τῆς τοῦ σπλάγχνου φύσεως οὐκ ἔμελλον ἐρεῖν
αὐτῷ· τοῦτο γὰρ ἦν ὄντως ὄνῳ μῦθον λέγειν· ὃ δ᾽ ᾤμην
εἰπὼν πείσειν αὐτόν, ᾧ καὶ πάντας ἀνθρώπους ὁρῶ
τάχιστα πειθομένους, τοῦτο διῆλθον μόνον· ὡς ἡ
μακρὰ πεῖρα ἐδίδαξε τοὺς ἰατροὺς θεραπεύειν ἧπαρ
ὕλῃ φαρμάκων μικτῇ· γεγραμμένην δ᾽ αὐτὴν εὑρήσεις
911K ἐν τοῖς θεραπευτικοῖς | γράμμασι τῶν ἰατρῶν.

Ἐὰν οὖν σοι δοκῇ, μῖξον, ἔφην, ἀψινθίου τι τῆς
κόμης μὲν ἀκριβῶς κεκομμένης τῷ καταπλάσματι,
τῆς πόας δ᾽ ὅλης τῷ ἐλαίῳ, καθάπερ ὁρᾷς ἄλλους
ἐναφέψοντας αὐτῷ μετρίως. τῷ καταπλάσματι δὲ
μυροβαλάνου πίεσμα καὶ ἴριν καὶ σχίνου τὸ ἄνθος ἢ
τῆς ναρδίτιδος βοτάνης τὴν ῥίζαν ἢ κυπέρου μῖξον·
οὐ χεῖρον δὲ καὶ δι᾽ οἴνου ποτ᾽ αὐτὰ κατασκευάσαι καὶ
μῖξαι ποτὲ τῆς ἰλύος αὐτοῦ, καί τι τῶν στυφόντων

---

25 Reference again to the three treatises listed in note 16
above.

drawn together moderately because it is the origin of the nutritive capacity in animals and because the class of veins is naturally derived from it. Therefore, he treated the viscus in the same way as he treated inguinal swellings, using relaxing agents unmixed and alone, applying a poultice of bread and honey moistened beforehand with warm oil, and fed [the patient] on gruel mixed with porridge, for these three things are almost enough to cure acute [diseases] as far as all the present-day amethodical Thessaleians are concerned. However, I decided to say this, at least, to Attalus: "Mix in something astringent and don't use the treatment by means of relaxing agents by themselves." I did not intend to tell him about the nature of the viscus, for truly this would be telling a tale to an ass, but what I did think to persuade him of by speaking (by which I see all men very swiftly persuaded) was to focus on this one thing alone: that long experience has taught doctors to treat the liver with a combination of medications. You will find that I have described this material in the therapeutic treatises for doctors.[25]

"Therefore," I said, "if it seems right to you, carefully mix some of the brayed leaf of wormwood with the poultice together with oil of the whole herb, just as you see the other ingredients being boiled down in it to a moderate degree. Mix in the poultice what is expressed from the myrobalan, iris, and the flower of mastich or the leaf and root of spikenard or cyperus.[26] And it is not a bad idea, on some occasions, to prepare these things with wine, and sometimes to mix the sediment of the latter, and to boil

911K

[26] Myrobalan is the cherry plum. On iris, see Dioscorides, I.1, on mastic, I.89, on spikenard, I.6–8 and on cyperus, I.4.

ἐναφεψῆσαι μήλων, ὁποῖα τὰ κυδώνιά τε καὶ στρούθια
καλούμενα καὶ ταῦτα δὴ τὰ πλεονάζοντα κατὰ τὴν
Ῥωμαίων πόλιν, ἃ προσαγορεύουσι κεστιανά. τὸ δ᾽
ἔλαιον, ὁρῶ γάρ σε καὶ τοῦτο μιγνύντα, μὴ τὸ τυχὸν
ἔστω, ἀλλ᾽ ἤτοι τὸ ἀπὸ τῆς Ἰσπανίας ἢ τὸ Ἰστρικόν,
ἢ τὸ ὀμφάκινον, ἢ σχίνινον, ἢ μύρτινον, ἢ μήλινον, ἢ
νάρδινον μύρον. πολλὴν δὲ καὶ ἄλλην ἔφην ὕλην
ἄφθονον εἶναι τῶν ἐναφεψεῖσθαι δυναμένων. καὶ γὰρ
σχίνου τοὺς ἁπαλοὺς κλῶνας καὶ μυρσίνης καὶ βάτου
καὶ ἀμπέλου καὶ μᾶλλον τῆς ἀγρίας, ἀφ᾽ ἧς καὶ τὴν
οἰνάνθην καλουμένην λαμβάνομεν. οὐ χεῖρον δ᾽ ἂν εἴη
912K καὶ τὸ Ἀττικὸν ὕσσωπον | τῷ τε καταπλάσματι καὶ
ταῖς κηρωταῖς μιγνύναι· καὶ γὰρ καὶ κηρωτάς τινας ἐξ
ὕλης τοιαύτης αὐτῷ συνεβούλευον ἐπιτιθέναι μετὰ τὸ
κατάπλασμα. καὶ συνάπτειν γ᾽ ἐπειρώμην ἐφεξῆς
αὐτῷ τὴν ὅλην ἀγωγήν, ἵνα καὶ τὰ καλούμενα πρὸς
τῶν ἰατρῶν ἐπιθέματα διὰ μικτῆς ὕλης σκευάζῃ.

Βέλτιον γάρ, ἔφην, ἐστὶν ἀρθέντος τοῦ καταπλάσματος ἐπικεῖσθαί τι τῷ σπλάγχνῳ.

καὶ ὁ Ἄτταλος ὑποτεμνόμενός μου τὸν λόγον, Εἰ
μὴ σφόδρα σ᾽ ἐτίμων, ἔφη, τούτων οὐδενὸς ἂν ἠνε-
σχόμην· ἐν οἷς γὰρ ἐναυάγησαν οἱ πρόσθεν ἰατροί,
πρὶν τὴν ὄντως ἰατρικὴν ὑπὸ τῶν ἡμετέρων εὑρεθῆναι,
ταῦτά μοι συμβουλεύεις ὥσπερ οὐκ εἰδότι. τρεῖς οὖν
ἡμέρας ἢ τέσσαρας, ἔφη, συγχώρησόν μοι προνοήσασθαι τοῦ Θεαγένους ὡς ἐγὼ βούλομαι, καὶ θεάσῃ
τελείως αὐτὸν ὑγιαίνοντα.

Τί οὖν, ἔφην, ἐὰν ἐξαίφνης ἱδρώτων ὀλίγων καὶ

down something from astringent apples, such as quinces, and those things which are abundant in the city of Rome, which they call *cestiana* apples.[27] Don't let the oil (for I see that you also mix this) be just what comes to hand; let it be from Spain or Istria, or grape, mastic, myrtle, or quince." And I spoke of much other material that is plentiful of those things that can be boiled down: the soft twigs of the mastich, myrtle and bramble, and of the grapevine, for preference the wild variety, from which we also take the so-called dropwort. It also might not be a bad idea to mix the Attic hyssop in the poultice and salves. I also recom- 912K mended to him salves from such a material to apply after the poultice. And I am accustomed to attempt to link the whole course of treatment to it sequentially so that you also prepare those things called "epithemata" by doctors from mixed material.

"What is better," I said, "is to lay something on the vis-cus when the poultice has been taken away."

Attalus, cutting my argument short, said: "If I did not have considerable respect for you, I would tolerate none of these things, for it was in these matters that earlier doctors came to grief, before the art of medicine was truly discov-ered by our [doctors], [and yet] you advise me on this as if I were ignorant. So give me three or four days," he said, "to take care of Theagenes as I want to, and you will see him completely restored to health."

"What then," I said, "if suddenly, after a few sweats that

27 On these see Pliny, *HN*, 15,11,10, #38.

GALEN

τούτων γλίσχρων ἐπιφανέντων ἀποθάνῃ, μνημονεύ-
σεις ὧν ὑπέσχου καὶ μεταθῇ τοῦ λοιποῦ;

καταγελῶν ἐπὶ τούτοις ὁ Ἄτταλος ἐχωρίσθη,
μηκέτ᾽ ἀποκρινόμενος μηδέν, ὥστ᾽ οὐδὲ περὶ τοῦ χόνδρου τι συμβουλεῦσαί μοι συνεχώρησεν, οὐδ᾽ ὅτι
913K   δεήσει τῶν οὐρητικῶν φαρμάκων | μιγνύναι τῷ ὕδατι
μικρὸν ὕστερον, ἐπειδὴ τὰ κυρτὰ τοῦ ἥπατος ἐπεπόνθει. καθάπερ γὰρ τὰ σιμὰ διὰ τῆς γαστρὸς ἐκκενω-
τέον ἐστίν, ὡς ὀλίγον ἔμπροσθεν εἶπον, οὕτω τὰ
κυρτὰ διὰ τῶν μετρίως οὐρητικῶν φαρμάκων, οἷόν
ἐστι τὸ σέλινον. ἐν δὲ τῷ χρόνῳ προϊόντι πεττομένης
ἤδη τῆς φλεγμονῆς καὶ τοῖς ἰσχυροτέροις ἐγχωρεῖ
χρήσασθαι, τῷ ἀσάρῳ καὶ τῇ Κελτικῇ νάρδῳ καὶ τῷ
καλουμένῳ φοῦ καὶ πετροσελίνῳ καὶ σμυρνίῳ καὶ
μήῳ· καθάπερ γε καὶ διὰ τῆς γαστρὸς κενοῦν, εἰ τὰ
σιμὰ πεπόνθασι, κνίκον μιγνύντα τοῖς ἐδέσμασι καὶ
ἀκαλήφην καὶ λινόζωστιν, ἐπίθυμόν τε καὶ πολυπό-
διον καὶ πάνθ᾽ ὅσα μετρίως ὑπάγει. ἔτι δὲ μᾶλλον ἐν
ταῖς παρακμαῖς αὐτοῖς τε τούτοις χρῆσθαι θαρσαλεώ-
τερον ἢ πρόσθεν, ὅσα τε τούτων ἐστὶ σφοδρότερα, τὰ
μὲν ἐναφέψοντας τῇ πτισάνῃ, τὰ δὲ κόψαντας, ὡς
χνοώδη γενέσθαι· διδόναι δὲ καὶ ταῦτα διὰ πτισάνης
ἢ μεθ᾽ ὕδατος. ἐγὼ γοῦν καὶ πολυποδίου τι ποτὲ συν-
έψησα τῇ πτισάνῃ καὶ μέλανος ἐλλεβόρου[7] φλοιόν.

καὶ διὰ τῶν κλυσμάτων δὲ κενοῦν αὐτοὺς προσῆκεν, ἐν ἀρχῇ μὲν ἀρκουμένους ἁλσὶν ἢ νίτρῳ ἢ

7 ἐλλεβόρου K (ἐλε- B); ἐλλ- fort. (dial. Att.; ἑλλ- dial. Ion.)

376

are viscid appear, he dies—will you remember what you promised and change in the future?"

After these exchanges Attalus went off laughing derisively but giving no further response; so he did not agree to consult with me about the spelt grits at all, nor that he would need to mix diuretic medications with water a little later because the convexity of the liver had been affected. For just as the concavity (of the liver) must be evacuated via the stomach, as I stated a little earlier, so too must the convexity be evacuated by moderately diuretic medications, an example being celery.[28] When the inflammation has already "ripened" (i.e. softened) with the passage of time, it is also possible to use stronger [medications]—hazelwort, Celtic spikenard, the so-called wild spikenard, parsley, Cretan alexanders, and spignel, just as it is possible to evacuate via the stomach, if the concave surface is affected, [using] safflower mixed with foods, nettle, mercury, epithymum, polypody, and all those things that empty moderately. And particularly in those inflammations that are abating, use these things more boldly than before, and those that are stronger than them, boiling down some in ptisan and pulverizing others so they become a fine powder. Give these things by means of ptisan or with water. At all events, I have also sometimes boiled some polypody in psitan and the bark of black hellebore.

And it is also appropriate to purge these [patients] by means of a clyster, being satisfied in the beginning with salts of either niter or aphronitrum that have been mixed

913K

---

[28] For the long list of medications to follow, see the list of medications in the Introduction, section 10.

914K ἀφρονίτρῳ μεμιγμένοις | τῷ μελικράτῳ· κατὰ δὲ τὰς
παρακμὰς καὶ μάλιστα ἐὰν σκιρρῶδές τι καταλείπη-
ται τῆς φλεγμονῆς, ἰσχυρότερα μιγνύντας φάρμακα·
τὸ γοῦν ὕσσωπον ἐναφεψόμενον τῷ ὕδατι τηνικαῦτα
καὶ τὴν ὀρίγανον καὶ τὴν κολοκυνθίδα καὶ τὸ λεπτὸν
κενταύριον. ἐπιτηδειότατα γάρ ἐστι σκιρρωθῆναι τὰ
δύο σπλάγχνα, τό θ᾽ ἧπαρ καὶ ὁ σπλήν, ἐὰν ἀμελήσῃ
τις αὐτῶν ἢ τοῖς γλίσχροις ἐδέσμασι χρήσηται, καθάπερ καὶ ὁ Ἄτταλος ἐπὶ τοῦ Θεαγένους ἑκάστης
ἡμέρας χόνδρον προσφέρων καὶ μηδὲν διδοὺς τῶν
ἐκφραττόντων τε καὶ ῥυπτόντων.

ἀλλὰ τό γε συμβὰν τῷ Θεαγένει, μᾶλλον δὲ τῷ
Ἀττάλῳ, καιρὸς εἰπεῖν. ὡς γὰρ ὑπέσχετό μοι μετὰ
τρεῖς ἡμέρας ἐπιδείξειν τὸν ἄνδρα τῆς φλεγμονῆς τοῦ
ἥπατος ἀπηλλαγμένον, ὁ μὲν ἔτι δὴ καὶ μᾶλλον ἐπὶ
πλεῖστόν τε κατήντλει τὸ σπλάγχνον ἐλαίῳ θερμῷ·
κατέπλαττέ τε συνεχέστερον ἐκ τῆς ἐπιμελείας ἐλπίζων αὐτῷ[8] προχωρήσειν τὰ τῆς θεραπείας ἄμεινον,
ἀπεκρίνατό τε πυνθανομένοις γαυριῶν ὑπὲρ τοῦ Θεαγένους τὰ βελτίω. ἀλλὰ συνέβη γε καθ᾽ ὃν ἐγὼ
τρόπον εἶπον, ἐξαίφνης ἀποθανεῖν αὐτόν. καὶ τὸ πάν-
915K των γελοιότατον, ὁ μὲν | Ἄτταλος ἦγέ τινας τῶν ἠρω-
τηκότων φίλων ὅπως διάγοι, δεῖξαι βουλόμενος αὐτὸν
οὕτως ἔχοντα καλῶς ὡς λούεσθαι μέλλειν, ἀγαλλό-
μενός τε μετὰ πολλῶν εἰσῆλθεν εἰς τὸν οἶκον ἐν ᾧ
κατέκειτο· τὸν Θεαγένη δὲ τεθνεῶτα λούειν ἐνεχείρουν
ἔνιοι τῶν φίλων, ταῦτα δὴ τὰ νενομισμένα, Κυνικοί τέ
τινες ὄντες καὶ ἄλλως φιλόσοφοι. διὸ καὶ μέχρι τοῦ

378

with melikraton. In the inflammations that are abating,   914K
and particularly if some hardness of the inflammation remains, mix stronger medications. In fact, in these circumstances, hyssop boiled down in water, and oregano, colycyn-
thida and centaury. The two viscera, the liver and spleen,
are the most prone to becoming scirrhous, if someone has
neglected them or used viscid foods, just like Attalus, who,
in the case of Theagenes, administered gruel every day
and gave nothing to clear the obstruction or cleanse.

But it is now time to say what happened to Theagenes,
or rather Attalus. For as he promised to show me the man
free of inflammation of the liver after three days, he
poured warm oil over the viscus even more than before,
and he applied poultices more frequently, hoping that by
this care he would provide better treatment for him. He
took pride in answering those who inquired about
Theagenes that things were better. But what actually happened to him was as I said—the man suddenly died. And,
the most ridiculous thing of all, Attalus brought in some of   915K
those friends who asked him how he (the patient) was doing, since he wished to show him as being so well that he
was about to take a bath and, in an exalted state, with many
people he went into the house in which Theagenes was lying. Some of Theagenes' friends, who were Cynics and
philosophers of other persuasions, were taking on the task
of washing Theagenes who was dead—this was in fact
the customary practice. On which account, Attalus hap-

---

8 B; αὐτῷ K

νεκροῦ παραγενέσθαι συνέβη τῷ Ἀττάλῳ μετὰ τοῦ
χοροῦ τῶν θεατῶν, ἅτε μηδενὸς ἔνδον οἰμώζοντος.
οὔτε γὰρ οἰκέτης οὔτε παιδίον οὔτε γυνὴ τῷ Θεαγένει
ἦν, ἀλλ' οἱ φιλοσοφοῦντες μόνοι παρῆσαν αὐτῷ φίλοι,
τὰ μὲν ἐπὶ τοῖς τεθνεῶσι νομιζόμενα πράττοντες, οὐ
μὴν οἰμώζειν γε μέλλοντες. οὕτω μὲν ὁ Θεσσάλειος
ὄνος εὐδοκίμησεν, ἐπὶ πολλῶν θεατῶν ἐπιδείξας
ἀπηλλαγμένον τῆς φλεγμονῆς ἐντὸς τῶν τεττάρων
ἡμερῶν, ὡς ὑπέσχετο, τὸν ἄνθρωπον. οἱ δ' ἄλλοι
Μεθοδικοὶ μυρίους ἀποκτείνοντες ὁσημέραι τὴν ἀγωγὴν τῆς θεραπείας οὐδέπω καὶ νῦν ὑπαλλάξαι τολ-
μῶσιν, οὐδὲ πειραθῆναί ποτε κἂν ἅπαξ τῆς τοῖς ἄλλοις ἰατροῖς, οἷς ὄντως ἐσπουδάσθη τὰ τῆς τέχνης
ἔργα, γεγραμμένης· οὕτω δευσοποιόν τι πρᾶγμά
916K ἐστιν ἀμαθία | σφοδρά, καὶ μᾶλλον ὅταν ἀλαζονίᾳ
μιχθῇ. τοιοῦτοι μὲν οὖν ἐν ἅπασιν οἱ Θεσσάλειοι.

16. Χρὴ δ' ἡμᾶς φεύγοντας τὰ τοιαῦτα ἁμαρτή-
ματα κἂν εἰ μηδὲν ἄλλο, ἀλλ' οὖν τῇ ἐμπειρίᾳ πιστεύ-
ειν. ὅπερ ἀεὶ παραινῶ τοῖς ἀγυμνάστοις περὶ τὸν
λόγον. ἄμεινον γὰρ αὐτοῖς ἐστι μηδ' ὅλως ἐξ ἀναλο-
γισμῶν τι λαμβάνειν, ὅταν ἀμαθεῖς τε ἅμα καὶ ἀγύ-
μναστοι τῶν λογικῶν ὦσιν μεθόδων, ἃς νῦν ἡμεῖς
γράφομεν. ἡ γάρ τοι τοῦ ἥπατος οὐσία ῥᾷστα σκίρ-
ροις ἁλίσκεσθαι πέφυκεν, ἔχουσά τι καὶ φύσει πηλῶδες, ὡς παίζων τις ἔλεγεν ἰατρὸς τῶν καθ' ἡμᾶς. ἡ δὲ
τοῦ σπληνὸς ἀραιοτέρα μέν ἐστι τοῦ ἥπατος, ἁλίσκε-

pened to be in attendance right up to Theagenes' death, along with the chorus of spectators, inasmuch as nobody within the house was lamenting because Theagenes had no house slave, child or wife—those who practiced philosophy were the only friends present for him and were the ones carrying out the things customary for the dead, although they were not intending to lament. And so the Thessaleian ass was held in good repute when he displayed to the many spectators the man delivered from the inflammation within four days, as he promised. However, the other Methodics, although they killed countless people every day, did not even once have the courage to change the basis of treatment, nor did they dare, at any time, to put to the test what had been written about by other doctors whose practice in the craft is highly respected, so deeply ingrained is their ignorance, and particularly when it is mixed with quackery. In all cases, the Thessaleians are like this.

916K

16. It is, however, necessary for us to avoid such errors, and even if nothing else, to put our trust in experience. This is what I always advise those unpracticed in reasoning. It is better for them not to totally accept something on the basis of reasoning when they are ignorant of, and, at the same time, unpracticed in the logical methods which I am now writing about. For the substance of the liver is, by its nature, very easily susceptible to scirrhosity, being also naturally "muddy,"[29] as one of the doctors among us was wont to say in jest. However, the [substance] of the spleen is thinner than that of the liver but is more frequently

[29] For this term, in reference to water, see Aristotle, *History of Animals,* 549b14.

381

ται δὲ συνεχέστερον ἐκείνου τῷ σκιρρώδει παθήματι
διὰ τὴν τῆς τροφῆς ἰδέαν ᾗ χρῆται. δέδεικται γὰρ ὑπὸ
τοῦ παχέος αἵματος τρεφόμενος, ὃ καθάπερ τις ἰλύς
ἐστι τοῦ καθαρωτέρου καὶ ῥᾷστα γίγνεται μέλαινα
χολή· διὸ καὶ μελαγχολικὸν αὐτὸ ἢ μέλαν καλοῦμεν
περίττωμα. ταῦτα μὲν οὖν τὰ δύο σπλάγχνα καὶ ἡ
917K   ἐμπειρία δείκνυσιν ἐναργῶς ἁλισκόμενα | τοῖς σκιρ-
ρώδεσιν ὄγκοις.

οἱ νεφροὶ δὲ τῷ κατακεκρύφθαι λανθάνουσι τὴν
ἁφήν· ὅ γε μὴν λόγος ἡμᾶς διδάσκει καὶ τούτους
ἑτοίμως ἁλίσκεσθαι σκίρροις· καὶ διὰ τοῦτο τὰς μὲν
ἀλύτους τὸ πάμπαν εἶναι νεφρίτιδας, τὰς δὲ δυσλύ-
τους· ἄμφω γὰρ ἔχουσιν οἱ νεφροὶ τὰ λελεγμένα τῶν
προειρημένων σπλάγχνων ὑπάρχειν ἑκατέρῳ, τήν τε
τῆς οὐσίας ποιότητα καὶ τὴν τῶν διερχομένων ἐν
αὐτοῖς περιττωμάτων φαυλότητα. διὸ καὶ οἱ τὰ παχύχυμα τῶν ἐδεσμάτων ἐσθίοντες ἁλίσκονται τῷ τῆς
λιθιάσεως πάθει. προορᾶσθαι τοιγαροῦν χρὴ τοσοῦτον μᾶλλον, ὅσον δυσιατότεραι τῶν τριῶν τούτων
εἰσὶ σπλάγχνων αἱ σκιρρώδεις διαθέσεις, ὅπως μή τις
αὐταῖς περιπέσῃ. μάλιστα δ᾿, ὡς εἴρηται, περιπίπτουσιν οἱ φλεγμηνάντων αὐτῶν ἐδέσμασι χρώμενοι παχεῖς ἢ γλίσχρους χυμοὺς γεννῶσι. καὶ μέντοι καὶ τὰ
ἰάματα τῶν τοιούτων παθῶν ὡμολόγηται πᾶσιν εἶναι
τὰ τέμνοντα καὶ διαλύοντα καὶ θρύπτοντα, τοῦ γένους
μὲν ὄντα δηλονότι τοῦ τῶν ἐκφραττόντων καὶ ῥυπτόντων, ἰσχυρότερά γε μὴν ταῖς δυνάμεσι. μεμάθηκας δ᾿

afflicted by a scirrhous affection than the latter because of the kind of nourishment which it uses. For it has been shown that when it (i.e. the spleen) is nourished by thick blood, which is like some slime of more pure blood, this very readily becomes black bile, because of which we call this a melancholic or black superfluity. And experience clearly shows that these two viscera are subject to scirrhous swellings. 917K

The kidneys, by virtue of the fact that they are concealed, elude the touch, but reason, at least, teaches us that they too are readily subject to scirrhosity. Because of this, some of the renal affections are altogether irremediable while some are difficult to remedy, since the kidneys have both things that have been spoken of that are in each of the previously mentioned viscera: the quality of the substance and the poor quality of the superfluities passing through them. And because of this too, those who eat foods that are viscid are seized by the affection of lithiasis (renal calculi). Accordingly, it is necessary to make greater provision, to the extent that the scirrhous conditions of these three organs are harder to cure, so that someone is not subject to them. As has been stated, those people particularly come to grief who use foods that generate thick and viscid humors when they are suffering from inflammation. Indeed, the cures of such affections are agreed by all to be those things that cut, dissolve, and break up, which are obviously of the class of agents that open up obstructions and cleanse, but are stronger in their potencies. You have learned the potency of these and their material in the trea-

ἐν τῇ Περὶ τῶν ἁπλῶν φαρμάκων πραγματείᾳ τήν τε
δύναμιν αὐτῶν καὶ τὴν ὕλην.

918K   διὸ καὶ νῦν ὁ | λόγος μοι γενήσεται σύντομος
ἀρκουμένῳ ταῖς καθόλου δυνάμεσι καὶ μόνῃ τῇ μεθόδῳ μετὰ παραδειγμάτων ὀλίγων. ὅπου γὰρ ἥ θ᾽ ὕλη
τῶν δυνάμεων ἤδη προπαρεσκεύασταί σοι καὶ τὰ
συνενδεικνύμενα τὴν θεραπείαν ἔμπροσθεν εἴρηται,
καταλείπεται νῦν οὐδὲν ἄλλο, πλὴν τῶν οἰκείων ἑκάστου νοσήματος ἐνδείξεων ἐπιμνησθῆναι. τὰ συνενδεικνύμενα δὴ λέγω δηλονότι δύναμιν καὶ φύσιν καὶ
ἡλικίαν καὶ ὥραν καὶ χώραν καὶ ἔθος, ὅσα τ᾽ ἄλλα
τοιαῦτα. καὶ τοίνυν περὶ τῆς ἀπὸ τῶν μορίων ἐνδείξεως
ἐφεξῆς ἐρῶ, τὴν ἀρχὴν ἀπὸ τῶν κατὰ τὸ ἧπαρ ποι-
ησάμενος. ἐνδείκνυται γὰρ τοῦτο τὰ μὲν ἔξωθεν ἐπιτι-
θέμενα κατὰ τὰς φλεγμονὰς αὐτοῦ μικτῆς εἶναι χρῆναι δυνάμεως οὐ μόνον ἐν γενέσει τῆς φλεγμονῆς
οὔσης, τοῦτο μὲν γὰρ κοινὸν ἁπασῶν φλεγμονῶν,
ὁπότε γε καὶ μόνοις τοῖς ἀποκρουστικοῖς βοηθήμασι
κατὰ τὸν χρόνον ἐκεῖνον οὐκ ἄν τις ἁμάρτοι χρώμενος, ἀλλὰ κἀπειδὰν μήτ᾽ ἐπιρρέῃ μηδὲν ἔτι, μήτε
ἀπώσασθαι δυνατὸν ᾖ τὸ ἐν τῷ φλεγμαίνοντι μορίῳ
περιεχόμενον.

    γίγνεται γὰρ καὶ τοῦτο διὰ πλείους αἰτίας. ἐν ἀρχῇ
μὲν οὖν ὀλίγον τε τὸ ἐπιρρέον ἐστὶ καὶ λεπτότερον ὡς |
919K   τὸ πολύ· ἐνίοτε δὲ καὶ ἡ κατ᾽ αὐτὸ τὸ δεχόμενον μόριον
ἰσχυροτέρα δύναμις, ὡς ἂν μηδέπω κεκμηκυῖα καὶ
τὸ περιεχόμενον αὐτὸ κατὰ τὸ φλεγμαῖνον οὐδέπω
βιαίως ἐσφηνωμένον. ἀκμαζούσης δὲ τῆς φλεγμονῆς

tise *On the Mixtures and Potencies of Simple Medica-*
tions.[30]

Because of this, my discussion now will be brief in that I   918K
am satisfied with the potencies in general and with the
method alone along with a few examples. For where the
material of the potencies has already been prepared for
you and those things jointly indicating the treatment were
spoken of before, nothing else remains now apart from
making mention of the specific indications of each disease.
Obviously, the joint indications I speak of are capacity,
nature, age, time of year, place, custom and other such
things. Therefore, I shall speak in order about the indication from the parts, making a start from those pertaining to
the liver. For this [viscus] indicates that those remedies applied externally to inflammations of it need to be of mixed
potency, not only when the inflammation is at the beginning, for this is common to all inflammations (when, in
fact, someone using the revulsive remedies on their own
would not fail at that time), but also when nothing is still
flowing and it is not possible to drive away what is contained in the inflamed part.

This occurs for many reasons. At the start, what flows
[from the part] is small in amount and thinner than usual.
Sometimes, however, the capacity in the receiving part   919K
itself is stronger, so that, as it is not yet consumed, what
is contained in the inflamed part is not yet severely impacted. When the inflammation is at its peak and the con-

[30] For example, see *De simplicium medicamentorum tempe-*
ramentis et facultatibus, XII.6, 15, and 102K.

τό τε περιεχόμενον αἷμα πολὺ καὶ πολλάκις παχύτερον, ἐσφηνωμένον τε σφοδρῶς, ἥ τε τοῦ μορίου δύναμις ἀσθενεστέρα· δεόμεθα δὲ καὶ ταύτης ἰσχυούσης, ὡς ὠθεῖν δύνασθαι τὸ περιττὸν ἀφ᾽ ἑαυτῆς· ἐπειδὴ τῶν στυφόντων φαρμάκων ἡ δύναμις οὐχ ἱκανὴ τηνικαῦτα τὸ πᾶν ἐργάσασθαι μόνη. συνάγουσα μὲν γὰρ καὶ σφίγγουσα καὶ οἱονεὶ πιλοῦσα καὶ θλίβουσα τὰ σώματα δύναται τὰ λεπτότερα ταῖς συστάσεσιν ὑγρὰ πρὸς τοὺς περικειμένους ἀποπέμπειν τόπους, οὐ μὴν ἄνευ γε τοῦ συνεπισχεῖν τι καὶ τὴν ἐν τῷ πάσχοντι μορίῳ δύναμιν, ἀξιόλογον αὐτοῖς γίνεται τὸ ἔργον. τηνικαῦτα γοῦν ἡ μὲν ἀπὸ τῆς φλεγμονῆς ἔνδειξις τῶν ποιητέων ἐστὶ μία· καλοῦσι δ᾽ αὐτὴν διαφόρησιν, ἐκκενοῦσαν λόγῳ θεωρητοῖς πόροις τὸν ἐν τῷ φλεγμαίνοντι μορίῳ χυμόν. ἡ δ᾽ ἀπ᾽ αὐτοῦ τοῦ μορίου πρὸς τοὐναντίον ἀντισπᾷ, κελεύουσα φυλάττειν αὐτοῦ

920K τὸν τόνον. | ἐπιπλεκομένων οὖν ἀλλήλαις ἐναντίων ἐνδείξεων ἐπιπεπλέχθαι χρὴ καὶ τὸ φάρμακον. μηροῦ δὲ φλεγμαίνοντος, ἢ κνήμης, ἢ πήχεος, ἢ βραχίονος, ἢ καὶ ἐν αὐτοῖς ἀδένων, οὐ δεόμεθα φυλάττειν τὸν τόνον. οὗτος ὁ σκοπὸς ἔστω σοι κοινὸς ἐπὶ πάντων τῶν μορίων, ὧν ἔργον τι τοιοῦτόν ἐστιν, ὡς ὅλῳ τῷ σώματι χρήσιμον ὑπάρχειν.

17. Οὐκοῦν οὐδὲ τὸν σπλῆνα παντάπασι χαλᾶν χρή, καὶ γὰρ καὶ οὗτος ἐκκαθαίρει τοῦ ἥπατος ὅσον ἰλυῶδές τέ ἐστι καὶ μελαγχολικόν, ἐπεὶ δ᾽ ὑπὸ τοιούτου τρέφεται. καὶ διὰ τοῦθ᾽, ὅταν ἔμφραξίς τις ἢ φλεγμονὴ κατ᾽ αὐτὸν γένηται, καὶ μάλιστα ὅταν ᾖ

tained blood is great in amount and often thicker and strongly impacted, and the capacity of the part is weaker, we also need this to be stronger so as to be able to drive out the superfluity from it when the potency of the astringent medications is not sufficient, under these circumstances, to do the whole job alone. For it is able, by drawing together and binding, and as if compacting and compressing the bodies, to send away the fluids that are thinner in consistency to the surrounding places. However, without some contribution from the capacity in the affected part the action does not become worthy of note among them. Anyway, under these circumstances, the indication from the inflammation is one of the things that must be acted upon, and they call this dispersal, since it evacuates the humor in the inflamed part by the "theoretical" pores. However, the indication from the part itself is to drawing toward what is opposite, urging preservation of its strength. Therefore, when opposite indications are interwoven with one another, it is necessary to use a medication that is compound. If the thigh, shank, forearm, upper arm, or also the glands in these structures are inflamed, we do not need to preserve the strength. This must be your general indicator in the case of all the parts whose action is of the kind that is useful to the whole body.

920K

17. It is then necessary not to relax the spleen altogether, for this [viscus] also purifies the liver of whatever is muddy and melancholic because it is nourished by such [a humor]. Also because of this, whenever some blockage or inflammation occurs in it, and particularly whenever it is

ἐσκιρρωμένος, τῶν ἐκφραττόντων καὶ τεμνόντων
ἰσχυροτέρων δεῖται. οἷον γάρ ἐστι φάρμακον ἥπατι τὸ
ἀψίνθιον, τοιοῦτον τῷ σπληνὶ καππάρεως φλοιός·
ὁποῖον δ' ἥπατι τὸ καλούμενον εὐπατόριον, τοιοῦτον
τῷ σπληνὶ τὸ σκολοπένδριον. ὁμοίων μὲν γὰρ δεῖται
φαρμάκων κατὰ τὸ γένος ἀμφότερα τὰ σπλάγχνα·
τοσούτῳ δ' ἰσχυροτέρων ὁ σπλήν, ὅσῳ παχυτέρα
921K χρῆται | τροφῇ. σκιρρουμένοις οὖν αὐτοῖς αἱ προσ-
ήκουσαι τροφαὶ κοιναὶ μὲν τῷ γένει, διαφέρουσαι δὲ
τὸ μᾶλλόν τε καὶ ἧττόν εἰσιν. ὅθεν εἰ καὶ δι' ὀξυμέλιτος ἡ κάππαρις ἐσθίοιτο, χρησίμη μὲν ἀμφοτέροις
ἐστὶ τοῖς σπλάγχνοις, ἀλλ' οὔτ' ἴση τὸ πλῆθος οὔθ'
ὁμοίως κεκραμένον ἔχουσα τὸ ὀξύμελι· πλείων τε γὰρ
αὕτη καὶ δι' ἀκρατεστέρου τοῦ ὀξυμέλιτος ἐπὶ τοῦ
σπληνὸς ὠφελιμωτέρα γίγνοιτ' ἄν. αὕτη μὲν οὖν ἡ
διαφορὰ κατὰ τὸ μᾶλλόν τε καὶ ἧττον. τῶν δὲ ὁμοίων
κατὰ γένος ἢ εἶδος ἢ ὡς ἄν τις ἐθέλοι ὀνομάζειν
βοηθημάτων, ἔκ τε τῆς ἐνεργείας αὐτῶν εἴληπται καὶ
διαπλάσεως.

ἀπὸ δὲ τῆς εἰς[9] τὰ παρακείμενα κοινωνίας, ὅπερ
ἐστὶ ταὐτὸν τῇ θέσει, τὸ τὰ μὲν κυρτὰ τοῦ ἥπατος
ἐκκαθαίρεσθαι διὰ νεφρῶν, τὰ δὲ σιμὰ διὰ τῆς κάτω
γαστρός. ἐπὶ δὲ σπληνὸς τὴν ἑτέραν μόνην εἶναι
κένωσιν τῶν περιττῶν, ἡ γὰρ ἐπὶ τοὺς νεφροὺς οὐκ
ἔστι τούτῳ τῷ σπλάγχνῳ. διὰ τοῦτ' οὖν ὅταν φλεγμαίνῃ, τοῖς κατωτερικοῖς ὀνομαζομένοις φαρμάκοις

9 K; πρός B, *recte fort.*

scirrhous, it requires stronger [medications] that clear the blockage and are cutting. Wormwood is such a medication for the liver; the bark of the caper plant is such a medication for the spleen. Likewise, as so-called agrimony (eupatorium) is to the liver, so scolopendrium is to the spleen.[31] Both viscera require similar medications in terms of class although, to the extent that the spleen is stronger, so it uses thicker nutriment. Therefore, for viscera that have become scirrhous, the appropriate nutriments are common in terms of class but different in terms of quantity. Consequently, if [the patient] eats oxymel or capparis, this is useful to both viscera, but not to an equal extent. The same applies to oxymel when it has been mixed, for if it is greater in amount and made from purer oxymel, the more beneficial it would be for the spleen. The actual difference relates to quantity. For similar remedies in terms of class or kind (or whatever anyone might wish to term it), the choice is made from the function and conformation of the viscera.

921K

Based on the association with those structures adjacent (which is the same as position), the purging of the convexity of the liver [takes place] through the kidneys, whereas the purging of the concavity takes place through the stomach downward. In the case of the spleen, the purging of superfluities is via the stomach alone, since purging by the kidneys does not apply to this viscus. Because of this, whenever the spleen is inflamed, we provoke it with the so-

---

[31] For these four medications in Dioscorides see III.26 for absinthum (wormwood), II.204 for capparis, IV.41 for eupatorion, and III.152 for scolopendrium.

ἐρεθίζομέν τε καὶ διαρρύπτομεν αὐτόν, ὅπως μεθιῇ καὶ
922K χαλάσῃ τὰ περιττά. | διττὸς δ' ὁ τρόπος ἐστὶ τῆς τῶν
τοιούτων φαρμάκων χρήσεως. ἐπὶ μὲν τῶν ἀνωτέρω
κειμένων διὰ τῶν μὲν ἐσθιομένων καὶ πινομένων, ἐπὶ
δὲ τῶν κατωτέρω διὰ κλυστῆρος ἐνιεμένου, ἐπειδὴ τῶν
μὲν ἐσθιομένων καὶ πινομένων ἡ δύναμις ἐκλύεται
πρὶν ἐπὶ τὰ κάτω μέρη προχωρῆσαι, τὰ δ' ἐνιέμενα
τὴν ἀρχὴν οὐδ' ἐπαναβῆναι δύνανται πρὸς τὴν
νῆστιν· ἀλλ' εἰ καὶ πάνυ σφόδρα βιάζοιο, τάχ' ἂν
ἅψαιο τῶν λεπτῶν ἐντέρων μόνων. καὶ τοῦτ' οὖν αὐτὸ
παρὰ τῆς τῶν μορίων θέσεως ἐδιδάχθημεν, ἐνιέναι
μέν τι τοῖς κατωτέρω κειμένοις ἐντέροις, ἄνωθεν δὲ
διδόναι τοῖς τ' ἀνωτέρω καὶ αὐτῇ τῇ γαστρὶ καὶ
σπληνὶ καὶ στομάχῳ. λέγω δὲ νῦν στόμαχον, ὅπερ
δὴ καὶ κυρίως ὀνομάζουσιν· ἐνίοτε γὰρ οὕτω καλοῦσι
καὶ τὸ στόμα τῆς γαστρός· ὥσπερ ὅταν εἴπωσι συγ-
κόπτεσθαί τινας στομαχικῶς. ἀλλ' ἐπί γε τοῦ κυρίως
ὀνομαζομένου στομάχου καὶ τὰ καταπλάσματα κατὰ
τῆς ῥάχεως ἐπιτίθεμεν, οὐκ ἔμπροσθεν, ὥσπερ ὅταν
τὸ στόμα τῆς γαστρὸς φλεγμαίνῃ· κατὰ γάρ τοι τῆς
923K ῥάχεως ὁ στόμαχος ἐπίκειται διά | τε τοῦ τραχήλου
καὶ τοῦ θώρακος φερόμενος κάτω μέχρι τῆς γαστρός.

18. Οὔτ' οὖν ταῦτα γινώσκουσιν οἱ Θεσσάλειοι, καὶ

---

32 As previously mentioned (and as Galen himself recognizes
here), there is some confusion surrounding the terminology of
the stomach/esophagus/abdomen generally. The meaning is clear
here. For "gastric syncope," see Galen's *In Hippocratis librum de
acutorum victu*, XV.609K, and Siegel (1973), pp. 251–53. Sted-

called purging medications and thoroughly cleanse it in such a way that it lets go of and releases the superfluities. The mode of use of such medications is twofold: in the case of those parts situated above, it is through things eaten or drunk while, in the case of those parts situated below, it is through the insertion of clysters *per rectum*, since the potency of things that are eaten and drunk is dissipated before they advance to the parts below, whereas things that are first inserted *per rectum* are not able to rise upwards to the jejunum, although if they are very strongly impelled, they might perhaps reach as far as the small intestines only. This is also what we were taught from the position of the parts: to insert something *per rectum* for the intestines situated below and to give something from above (*per os*) for those structures above—the stomach itself, the spleen, and the esophagus. I speak now of esophagus (*stomachus*) which is surely also the proper sense of this term. For sometimes people also name the opening of the stomach in this way, as when they say some people suffer "gastric (*stomachical*) syncope."[32] But in the case of what is properly called the esophagus, we also place poultices down along the spine and not in front, as we do whenever the opening of the stomach is inflamed, for in truth the *stomachus* (esophagus) lies down along the spine and runs through the neck and chest extending down as far as the stomach below.

18. The Thessaleians, then, do not know these things

man lists a deglutition (swallow) syncope, defined as follows: "faintness or unconsciousness upon swallowing. This is nearly always due to excessive vagal effect on the heart that may already have bradycardia or atrioventricular block."

922K

923K

διὰ τοῦτο πάντας ὁμοίως θεραπεύουσιν, οὔθ᾽ ὅτι πᾶν
τὸ φλεγμαῖνον μέλος, ἐὰν μὴ στεγνὸν ἔχῃ τὸ περι-
κείμενον ἑαυτῷ δέρμα, χαλᾷ τι καὶ μεθίησιν ἔξω τῶν
λεπτομερῶν ἰχώρων. καὶ διὰ τοῦτ᾽ ἔκ τε τῶν κατὰ τὸ
στόμα καὶ τὴν ῥῖνα καὶ τὴν φάρυγγα καὶ στόμαχον,
ἔντερά τε καὶ γαστέρα καὶ τὰ σπλάγχνα πάντα, ῥεῖ τι
πρὸς τοὐκτός. οὕτω δὲ καὶ κατὰ τὴν ἔνδον ἐπιφάνειαν
τοῦ θώρακος ἐκκρίνεταί τις ἰχώρ, ὅταν φλεγμαίνῃ.
καθάπερ οὖν ὁ μὲν σπλὴν καὶ τὰ σιμὰ τοῦ ἥπατος
ἐκκαθαίρεται διὰ τῶν ἐντέρων, οἱ νεφροὶ δὲ καὶ κυρτὰ
τοῦ ἥπατος διὰ τῶν οὔρων, οὕτως ὁ θώραξ, ὅταν γε τὰ
ἔνδον αὐτοῦ φλεγμαίνῃ, κατὰ τὸν ὑπεζωκότα μεθίησί
τι πρὸς τὴν μεταξὺ χώραν ἑαυτοῦ καὶ τοῦ πνεύμονος·
ἐκκαθαρθῆναι δὲ τοῦτο δεήσεται διὰ τῶν αὐτῶν ὁδῶν
τῷ πνεύμονι. δώσομεν οὖν τοῖς οὕτω κάμνουσι φάρμακα τῆς λεπτυνούσης δυνάμεως, ὅπως ἀναστομῶνται
924K   μὲν αἱ ὁδοί, τέμνωνται δὲ τὰ | δι᾽ αὐτῶν ὁδοιπορήσειν
μέλλοντα, καὶ μάλισθ᾽ ὅταν ᾖ παχέα ταῖς συστάσεσιν
ἢ γλίσχρα, καθάπερ ἐπὶ τῶν ἐμπύων ἐστίν.

ἐκλεξώμεθα δὲ καὶ τούτων ὅσα μὲν μέτρια, φλεγμαινόντων ἔτι τῶν πεπονθότων μελῶν, ὅσα δὲ ἰσχυρότερα, κατὰ τὰς ἀκριβεῖς παρακμὰς τῶν φλεγμονῶν
ἢ καὶ τελέως μὲν αὐτῶν πεπαυμένων, ἐκκριθῆναι δὲ
τῶν περιττῶν δεομένων. μέτριον μὲν οὖν ἐν τοῖς τοιούτοις ἐστὶν ὅ τε τῆς πτισάνης χυλὸς καὶ τὸ μελίκρατον, ἰσχυρότερον δὲ τὸ τῆς ἀκαλήφης σπέρμα, καὶ
ὅταν ἐμβληθῇ βραχύ τι τῷ μελικράτῳ τῶν δριμέων
βοτανῶν, οἷον ὀριγάνου καὶ ὑσσώπου καὶ καλαμίνθης

and, because of this, they treat everyone in the same way. Nor do they know that every structure that is inflamed, if the skin surrounding it is not obstructed, releases and lets go of some of the fine-particled ichors externally. And it is due to this that something from those things in the mouth, nose, pharynx, esophagus, intestines, stomach and all the viscera, flows to the outside. In this way, too, some ichor in the internal aspect of the chest [wall] is separated whenever this part is inflamed. In just the same way as the spleen and the concave part of the liver are purged via the intestines, while the kidneys and the convex part of the liver are purged via the urine, so the chest [wall], when the internal parts of it are inflamed in relation to the pleura and something passes into the space between itself and the lung, will need to be purged via the same channels as the lung. Therefore, we will give those who are suffering in this way, medications of a thinning potency so that the channels are opened up, and medications that are cutting 924K for the things that are going to find a passage through these channels, especially when these things are thick or viscid in consistency, as they are in the case of the empyemas.

However, we must select those that are moderate when the affected members are still inflamed, but those that are stronger in the abatement of the inflammation or when it has ceased completely, since the superfluities need to be evacuated. Moderate among such things are the juice of ptisan and melikraton; stronger is the seed of the stinging nettle, and whenever a little of one of the sharp herbs is put into the melikraton. Examples of the former are oregano,

καὶ γλήχωνος, ἴρεώς τε τῆς Ἰλλυρίδος ἡ ῥίζα. πλέον
δ᾽ εἰ μίξαις τούτων ἢ καὶ τὴν ἴριν οὕτω κόψαις καὶ
σήσαις ὡς χνοώδη ποιῆσαι, κἄπειτα ἐπιβάλλοις τῷ
μελικράτῳ, τμητικώτατον ἕξεις φάρμακον. οὕτω δὲ
καὶ τὸ διὰ πρασίου σκευαζόμενον ὀξύμελί τε καὶ ἄλλα
τοιαῦτα, τέμνειν ἱκανῶς πέφυκε τὰ παχύτερα τῶν ἐν
θώρακι καὶ πνεύμονι περιττῶν. καὶ πάντων αὐτῶν τὴν
εὐπορίαν ἔχεις ἐν ταῖς περὶ τῶν φαρμάκων πραγμα-
τείαις. |

925K   19. Οὐ μόνον δὲ τὴν εἰρημένην τῶν βοηθημάτων
διαφορὰν ἐκ τῶν πεπονθότων ἐμάθομεν τόπων, ἀλλὰ
καὶ τὸ τὰ μὲν ἐπιπολῆς φλεγμαίνοντα τοῖς τῆς
φλεγμονῆς ἰδίοις βοηθήμασιν ἰάσασθαι, τὰ δ᾽ ἐν τῷ
βάθει μετὰ τοῦ μιγνῦναι τι καὶ τῶν δριμυτέρων· ἐκλύεται γὰρ ἡ δύναμις εἰς τὸ βάθος αὐτῶν διαδιδομένων.
οὕτω δὲ καὶ ἡ σικύα βοήθημα γενναῖον εὕρηται τῆς τ᾽
ἔξω φορᾶς τῶν ἐν τῷ βάθει καὶ τῆς οἱονεὶ μοχλείας
τῶν ἤδη σκιρρουμένων. ἀλλ᾽ οὐ χρηστέον ἐστὶ σικύᾳ
κατ᾽ ἀρχὰς ἐπὶ μορίου φλεγμαίνοντος, ἀλλ᾽ ἐπειδὰν
ὅλον τὸ σῶμα κενώσῃς καὶ χρεία σοι γένηται κενῶσαί
τι καὶ ἐκμοχλεῦσαι τῶν κατὰ τὸ φλεγμαῖνον ἢ πρὸς
τοὐκτὸς ἀποσπάσασθαι. γινομένων δ᾽ ἔτι τῶν παθῶν
οὐκ αὐτοῖς τοῖς ἀρχομένοις κάμνειν μέλεσιν, ἀλλὰ
τοῖς συνεχέσιν αὐτῶν ἐπιβάλλειν τὴν σικύαν ἀντι-
σπάσεως ἕνεκεν. οὕτω γοῦν καὶ μήτρας αἱμορραγού-
σης πρὸς τοὺς τιτθοὺς ἐπιβάλλομεν σικύαν, ἐπ᾽ αὐτῶν
ἐρείδοντες μάλιστα τῶν κοινῶν ἀγγείων, θώρακός τε
926K   καὶ | μήτρας, τὸ τῆς σικύας στόμα. κατὰ δὲ τὸν αὐτὸν

hyssop, catmint, pennyroyal, and the root of the Illyrian iris. However, if you mix more of these, or if you also beat and sift the iris in such a way as to make a fine powder and then you put it into the melikraton, you will have a medication that is very cutting. In this way too, the medication made from horehound prepared with oxymel, and other such things are, by nature, sufficient to cut the thicker superfluities in the chest wall and lung. And you have an abundance of all these things in the treatises on medications.[33]

19. Not only did we learn the stated differentiation of the remedies on the basis of the affected places, but also that inflamed parts on the surface are cured by the specific remedies of the inflammation, whereas those that are deep are cured by admixture of one of the more pungent [remedies] as well, for the potency of remedies is dissipated when they pass down to the depths. Likewise, the cupping glass is found to be an excellent remedy for the outward passsage of those things in the depths and for the dislodgement, as it were, of those things that are already scirrhous. An exception is that the cupping glass must not be used on the inflamed part at the start but only when you evacuate the body, and there is a need for you to evacuate and dislodge any of the things in the inflammation, or to drive them away to the outside. When affections are still evolving, do not apply the cupping glass to the actual structures that are beginning to suffer but to those parts adjacent to them for the purpose of revulsion. And in this way, we place a cupping glass on the breasts when there is uterine bleeding, fixing the mouth of the cupping glass particularly

925K

926K

---

[33] The three treatises listed in note 16 above.

τρόπον αἱμορραγίας διὰ ῥινῶν γιγνομένης ἐπιβάλλομεν τοῖς ὑποχονδρίοις μεγίστας σικύας. οὕτω δὲ καὶ
πᾶσαν ἄλλην αἱμορραγίαν ἀντισπῶμεν ἐπὶ τἀναντία
διὰ τῶν κοινῶν φλεβῶν, ὥσπερ αὖ πάλιν ἕλκομεν, εἰ
τούτου δεοίμεθα· κατὰ γοῦν ἐφηβαίου τε καὶ βουβῶνος ἐπιτίθεμεν σικύαν, ἔμμηνα κινῆσαι βουλόμενοι.
καὶ κατ᾽ ἰνίου δὲ σικύα τιθεμένη γενναῖόν ἐστι βοήθημα ῥεύματος ὀφθαλμῶν. χρὴ δὲ προκεκενῶσθαι τὸ
σύμπαν σῶμα· πληθωρικοῦ γὰρ ὄντος αὐτοῦ, καθ᾽ ὅ τι
περ ἂν ἐρείσεις μέρος τῆς κεφαλῆς τὴν σικύαν, ὅλην
αὐτὴν πληρώσεις.

20. Οὗτος οὖν ὁ κοινὸς σκοπὸς ἁπάσης φλεγμονῆς
οὐχ ὡσαύτως ἐφ᾽ ἑκάστου τῶν μορίων ἐπιτελεῖται.
προσέρχεται δ᾽ ἅπασι τοῖς εἰρημένοις οὐ σμικρὰ
μοῖρα καὶ ἡ τοῦ προσενεχθησομένου φαρμάκου φύσις. οὐ γὰρ ἁπλῶς εἰ στῦψαι δεοίμεθα τὴν ἀρχομένην
φλεγμονήν, ἅπαν τὸ στῦφον προσοίσομεν ἐπὶ τῶν
καταπίνεσθαι μελλόντων, ἀλλ᾽ ὅσοις ἂν αὐτῶν οὐδε-
927K μία μέμικται δύναμις φθαρτική. χάλκανθος | γοῦν ἐν
τοῖς μάλιστα στύφει, καθάπερ γε καὶ τὸ μῖσυ καὶ
σῶρυ καὶ χαλκῖτης καὶ διφρυγές, ὅ τε κεκαυμένος
χαλκὸς ἥ τε λεπὶς αὐτοῦ καὶ τὸ ἄνθος· ἀλλ᾽ ἔστι
βλαβερὰ τὰ φάρμακα ταῦτα καταπινόμενα· διόπερ
οὐδὲ τοῖς στοματικοῖς ἀσφαλῶς μίγνυνται· παραρρεῖ

---

[34] The singular is used for the Greek plural here in line with
modern usage.   [35] See the list of medications (Introduction,
section 10) for these substances, and also Dioscorides, V.114, 117,
119, 115, and 143 respectively.

to those vessels that are common to both the chest and the uterus. In the same way, when hemorrhage occurs through the nose, we place a very large cupping glass on the hypochondrium.[34] The same also applies to every other hemorrhage: we draw it away to the opposite parts through the common veins, just as we also draw it back again should we need to do this. And if we wish to set in motion the menstrual flow, we place the cupping glass on the pubes and the inguinal glands. Also, a cupping glass placed over the inion (external occipital protuberance) is an excellent remedy for a flux of the eyes. It is, however, necessary to evacuate the whole body beforehand because, if it is plethoric, regardless of what part of the head you might place the cupping glass on, you will fill the whole head [with blood].

20. Therefore, this common indicator of every inflammation is not produced in the same way in each of the parts. Parts also make no little contribution to all those things mentioned and the nature of the medication that is going to be applied. For if we need to draw together the incipient inflammation, we will not simply employ any astringent in the case of the things that are going to be swallowed, but those of them in which no destructive potency is mixed. Chalcanthum, at any rate, is among those things   927K
that draw together particularly just as, in fact, are also misu, sory, chalcitis, pyrites, and copper that has been burned, and the flake and "flower" of this.[35] But these medications are harmful when swallowed and so are not safe to mix with the stomatics, for sometimes something

γὰρ ἐνίοτε αὐτῶν τι μέχρι τῆς γαστρός. οὐ μὴν οὐδ'
ἀλόη καλῶς ἂν μιχθείη τοῖς καταπίνεσθαι μέλλουσι
φαρμάκοις, ἕνεκα φλεγμονῆς τῶν ἔνδον· ἐπειδὴ καὶ
ταύτῃ μέμικταί τις δύναμις καθαρτική. παρηκμακυίας
μέντοι τελέως τῆς φλεγμονῆς, εἴ τις μικρὸν ἀλόης
μίξειεν ἕνεκα τοῦ τὴν γαστέρα κινῆσαι μὴ διακεχωρη-
κυῖαν τελέως, οὐδὲν βλάψει. βέλτιον δὲ διὰ λινοζώστι-
δος ἢ ἀκαλήφης ἢ κνίκου ἤ τινος τῶν τοιούτων ὑπάγειν τὴν γαστέρα τῶν οὕτω καμνόντων.

ὅλως δέ, ἄν τις ἀφέλῃ τὴν ἀπὸ τῶν μορίων ἔνδειξιν, οὐδὲν κωλύει τὴν ἰατρικὴν οὐχ ἐξ μησίν, ἀλλ'
ἐξ ἡμέραις ὅλην ἐκμαθεῖν. οὐ μὴν οὐδὲ προσθέντες
τὴν ἀπὸ τῶν μορίων ἔνδειξιν ἔχοιμεν ἂν ἤδη τὸ πᾶν
εἰς τὴν θεραπείαν ἄνευ τοῦ τὰς περὶ τῶν φαρμάκων
ἐκμαθεῖν μεθόδους. ἐπ' ἐκείνας οὖν ἰτέον ἐστὶ τῷ
928K   μέλλοντι τελέως ἰάσασθαι τὰ νοσήματα. | νυνὶ γάρ,
ὡς πολλάκις εἶπον, εἰ καί τινος ἐμνημόνευσα φαρμάκου, παραδείγματος ἕνεκα τοῦτ' ἔπραξα.

21. Προσθῶμεν οὖν ἔτι τῆς ἀπὸ τῶν μορίων ἐνδείξεως ὅσα μήπω λέλεκται, καταβάλλοντες προφανέ-
στατα τὴν τῶν Θεσσαλείων αἵρεσιν· οἳ μήτε ἀνατομῆς ἁπτόμενοι μήτ' ἐνεργείας ἢ χρείας εἰδότες, ὅταν
ἴδωσί τινα κροκιδίζοντα καὶ καρφολογοῦντα, τολμῶ-
σιν ὀξυροδίνῳ καταβρέχειν τὴν κεφαλὴν ἡμῖν ἑπόμε-
νοι. διὰ τί γὰρ οὐ τὸν θώρακα μᾶλλον; εἴπερ ἐνδεικτε-

of these flows through until it reaches the stomach. Nor would it be good for aloes to be mixed with medications that are going to be swallowed for the sake of inflammation of those structures that are internal, since there is also some cathartic potency mixed with it. Nevertheless, when the inflammation has completely abated, if someone were to mix a small amount of aloes for the purpose of activating the stomach when it is passing absolutely nothing through, it will do no harm. It is better, however, to purge the stomach of those suffering in this way by means of mercury, stinging nettle, safflower or something of this sort.

On the whole, then, should someone take the indication from the parts, there is nothing to stop him thoroughly learning the entire healing art not in six months but in six days. If we don't add the indication from the parts, we already have everything pertaining to treatment apart from thoroughly learning the methods relating to the medications. It is to those that someone who intends to cure diseases completely must go. For the present, as I often stated, if I made mention of some medication, I did this by way of an example. 928K

21. What I must still add is the indication from those parts I have not yet spoken about, since I reject quite explicitly the sect of Thessaleians who, neither grasping anatomy nor understanding functions or uses, dare to drench the head with oxyrrhodinum[36] following me, whenever they see someone picking at the blankets or tugging at bits of hair. Why is the chest not better if they discover the remedies indicatively? However, it is also possible for a person

---

36 Oil of roses mixed with vinegar; see Galen, *De simplicium medicamentorum temperamentis et facultatibus*, XI.559K.

κῶς μὲν εὑρίσκουσι τὰ βοηθήματα, δυνατὸν δ' ἐστὶ
καὶ τῆς καρδίας πασχούσης φρενιτικὸν γίνεσθαι τὸν
ἄνθρωπον. ὁ μὲν γὰρ Ἐμπειρικὸς ἐκ τῆς πείρας φησὶ
τὴν τῶν τοιούτων βοηθημάτων εὕρεσιν ἐσχηκέναι, τῷ
δὲ καὶ ταύτην ἀτιμάσαντι καὶ τὴν τῶν ἐνεργειῶν
ζήτησιν φυγόντι πόθεν ἐπῆλθεν ἀντὶ τοῦ θώρακος
ἑλέσθαι τὴν κεφαλὴν ἐπὶ τῶν φρενιτικῶν ἐπιβρέχειν·
ἀλλὰ τοῦτό γε τὸ ὀξυρρόδινον ὃ τῇ κεφαλῇ προσ-
φέρομεν ἐπὶ τῶν φρενιτικῶν, ὥσπερ τις ἔλεγε τῶν
ἑταίρων, οὐ μόνον τοὺς ἀμεθόδους Θεσσαλείους, ἀλλὰ
929K καὶ τοὺς ἄλλους ἅπαντας ἐξελέγχει φανερῶς, | ὅσοι
κατὰ τὴν καρδίαν ἡγοῦνται τὸ ψυχῆς ἡγεμονικὸν
ὑπάρχειν. ἰδὼν γοῦν ποτε τῶν ἀπ' Ἀθηναίου τινὰ τὴν
κεφαλὴν αἰονῶντα ῥοδίνῳ καὶ ὄξει μεμιγμένοις ἐκώ-
λυον ἀξιῶν ἐπιφέρειν τῷ θώρακι τὸ βοήθημα· βε-
βλάφθαι μὲν γὰρ τῷ παραφρονοῦντι τὸ ἡγεμονικόν,
εἶναι δ' ἐν καρδίᾳ τοῦτο κατὰ τὸν Ἀθήναιον, οὔκουν
ὀρθῶς αὐτὸν ποιεῖν ἀποστάντα τοῦ θώρακος ἐνοχλεῖν
τῇ κεφαλῇ καὶ πράγματα παρέχειν ἀπαθεῖ μορίῳ,
νυνὶ μὲν ὀξυρροδίνῳ καταντλοῦντα, νυνὶ δὲ ἀποκεί-
ροντα καὶ σπονδύλιον ἢ ἔρπυλλον ἤ τι τοιοῦτον προσ-
φέροντα· καὶ εἰ χρονίζοι τὸ πάθημα καὶ τὸ καστόριον
ἢ καὶ νὴ Δία σικύαν· ὅμοιον γὰρ εἶναι τοῦτο τῷ
φλεγμαίνοντος μηροῦ τῇ περόνῃ προσάγειν τὸ βοήθημα.

καὶ μὲν δὴ κἀπὶ τῶν ληθαργικῶν οὐδείς ἐστιν ὃς οὐ
προσφέρει τῇ κεφαλῇ τὰ βοηθήματα· καὶ τοῦτο γὰρ
τὸ πάθος ἐναντίον μέν πώς ἐστι κατὰ τὴν ἰδέαν τῇ

to become phrenitic when the heart is affected. The Empiric says that he has come upon the discovery of such remedies by experience, but why does someone who disdains experience and shuns the search for functions choose to pour water on the head rather than the chest in those with phrenitis? But this oxyrrhodinum, which we apply to the head in those with phrenitis, clearly refutes not only the amethodical Thessaleians, as one of their own adherents used to say, but also all the others who think the 929K *hegemonikon* (authoritative part) of the soul is in the heart. Anyway, sometimes when I see one of the students of Athenaeus[37] moistening the head with oil of roses and vinegar mixed together (i.e. oxyrrhodinum), I stop [him], thinking it proper for him to apply the remedy to the chest, since it is the *hegemonikon* which has been injured in someone with delirium, and this is in the heart according to Athenaeus. He is not, therefore, acting correctly if he rejects the chest to trouble the head, thereby creating a difficulty for an unaffected part by now pouring oxyrrhodinum over it and cutting off the hair, and applying spondylium, or tufted thyme, or some other such thing, or if the affection becomes chronic, also castor, or even, by Zeus, a cupping glass, for this is like applying the remedy to the shank when the thigh is inflamed.

And even in those with lethargy, there is nobody who does not apply the remedies to the head, for this affection is also, in a way, opposite in terms of kind to phrenitis. It

---

[37] Athenaeus of Attalea was a first-century AD doctor who practiced in Rome. He is generally regarded as a Pneumatist; see *EANS*, pp. 166–67.

φρενίτιδι. γίνεται δ' ἐγκεφάλου πάσχοντος, ἐν ᾧ τῆς ψυχῆς ἐστι τὸ ἡγεμονικόν. ὅταν μὲν οὖν ὁ πλεονάζων ἐν ἐγκεφάλῳ ψυχρὸς ᾖ χυμός, ἀναισθησία τε καὶ ἀκινησία καταλαμβάνει τὸν ἄνθρωπον· ὅταν δὲ θερ-

930K μός, | εὐκινησία μᾶλλον, ὡς ἂν εἴποι, τις ἅμα τῇ τοῦ λογισμοῦ βλάβῃ. συμβαίνει γάρ, ὡς ἐν τοῖς περὶ τούτων δέδεικται λόγοις, διὰ μὲν τὴν ψύξιν ἡ ἀργία, διὰ δὲ τὴν θερμασίαν ἡ ἄμετρος κίνησις, ἐκ δὲ τῆς χυμῶν μοχθηρίας ἡ ἄνοια. φλεβοτομητέον οὖν ἐστιν ἐπὶ τῶν τοιούτων παθῶν κατ' ἀρχὰς εὐθὺς ἰσχυρᾶς μὲν οὔσης εἰς τοσοῦτον τῆς δυνάμεως ὡς ἐνεγκεῖν ἀλύπως τὴν φλεβοτομίαν, ἑτέρου δὲ μηδενὸς κωλύοντος ὧν ἐν τοῖς περὶ φλεβοτομίας εἴπομεν, οἷον ἤτοι πλήθους ὠμῶν χυμῶν ἢ παιδικῆς ἡλικίας ἢ ὥρας ἢ χώρας ἐσχάτως θερμῆς ἢ ψυχρᾶς.

τοῦτο μὲν οὖν κοινὸν ἀμφοτέροις τοῖς νοσήμασιν, ὅσα τε μετὰ καταφορᾶς καὶ ὅσα μετ' ἀγρυπνίας γίνεται. κοινὸν δὲ καὶ τὸ κατὰ τὴν ἀρχὴν ὀξυρόδινον προσφέρειν· ἀπώσασθαι γὰρ χρὴ τῆς κεφαλῆς τὸν χυμόν, ὁποῖος ἂν εἴη. τὰ δ' ἐφεξῆς ἐναντία, πραΰνειν μὲν γὰρ προσήκει τὰ μετὰ τῶν ἀγρυπνιῶν, ἐπεγείρειν δὲ τὰ μετὰ τῆς ἀκινησίας. εἰκότως οὖν ἀκμαζόντων αὐτῶν τοῖς μὲν ἀγρυπνιτικοῖς καὶ περικοπτικοῖς νοσήμασι τὰς διὰ μήκωνος κωδειῶν ἐπιβροχὰς προσοίσο-

931K μεν, ὀσφρανούμέν τε καὶ διαχρίζομεν, ἤτοι | τὰ πτερύγια τῆς ῥινὸς ἐκ τῶν ἔνδοθεν μερῶν ἢ τὸ μέτωπον ὁμοίοις φαρμάκοις. καρῶσαι γὰρ χρὴ καὶ ναρκῶσαι ποιῆσαι τὸ ἡγεμονικόν, ἐμψύχοντα δηλονότι τὸν ὑπερ-

occurs when the brain, in which the *hegemonikon* (author-itative part) of the soul lies, is affected. Therefore, when-ever the humor predominating in the brain is cold, anes-thesia and akinesia befall the person. However, when it is hot, there is more normal movement (*eukinesia*), as one 930K might say, along with damage to reasoning. For inertia (in-activity) is what happens due to cold, as has been shown in the discussions of these matters, whereas excessive move-ment happens due to heat, and folly due to the vitiosity of humors. You must carry out phlebotomy right at the outset in such affections, if the strength of the capacity is such as to bear the phlebotomy without harm and there is none of the other things I spoke about in the writings on phlebot-omy to contraindicate it: for example, an excess of crude humors, young age, the season, the place, or extreme heat or cold.[38]

This, then, is common to both diseases; [that is], both those that occur with lethargy and those that occur with wakefulness. Common also [to both] is the application of oxyrrhodinum at the outset, for it is necessary to repel the humor from the head, whatever sort it might be. Those things which are opposites follow, for it is appropriate to calm in the diseases with wakefulness and to rouse in the diseases with inertia. It is, then, reasonable, when things are reaching their peak in diseases with wakefulness and raving, that we apply the washings from poppy heads, and that we introduce the smells to, and smear over with, ei-ther the alae of the nostrils internally or the forehead with 931K similar medications. For it is necessary to make the *hege-monikon* sleepy and numb, cooling, obviously, the overly

[38] See *De curandi ratione per venae sectionem*, XI.269K ff.

τεθερμασμένον ἐγκέφαλον. ἐπὶ δὲ τῶν ἐναντίων παθῶν
ἐπεγεῖραι καὶ τεμνεῖν καὶ θερμῆναι προσήκει τὸ πάχος τοῦ λυποῦντος χυμοῦ, ὅστις ἄνευ μὲν τοῦ σήπεσθαι καταφορὰς βαθείας ἐργάζεται, χωρὶς πυρετῶν,
ἃς ὀνομάζουσιν ἀποπληξίας καὶ κάρους καὶ κατοχάς.
εἰ δὲ καὶ σήποιτό ποτε, μετὰ πυρετοῦ γίνεται τὰ
τοιαῦτα καὶ καλεῖται τὸ τοιοῦτο νόσημα λήθαργος.

ἐναφεψοῦντες οὖν ὄξει θύμον καὶ γλήχωνα καὶ
ὀρίγανον, ὅσα τ' ἄλλα τοιαῦτα, τῇ ῥινὶ τῶν οὕτω
διακειμένων προσοίσομεν, ὅπως ὁ ἀτμὸς ἐπὶ τὸν ἐγκέφαλον ἀναφερόμενος τέμνῃ τὸ πάχος τοῦ χυμοῦ.
μετὰ δὲ ταῦτα καὶ τὸν οὐρανίσκον ἰσχυροῖς καὶ δρι-
μέσι φαρμάκοις χρίσομεν. ἑξῆς δὲ τούτων καὶ πταρ-
μικοῖς χρησόμεθα. καὶ κατὰ τῆς κεφαλῆς ἐπιθήσομεν
ὁμοίας δυνάμεις φαρμάκων, ἄχρι καὶ τοῦ νάπυος,
ἐπιτείνοντες ἀεὶ τὸ σφοδρὸν αὐτῶν, εἰ χρονίζει τὸ
πάθος. ἀλλὰ καὶ ταῖς σικύαις ἐπ' ἀμφοτέρων χρονι-

932K ζόντων χρησόμεθα καὶ τῷ καστορίῳ. | πέττει γὰρ αὐτὸ
καλῶς ἐπὶ προήκοντι τῷ χρόνῳ παραληφθέν, ὥστε
κἀνταῦθα πάλιν εἰς κοινὴν θεραπείαν ἄγεσθαι λήθαρ-
γόν τε καὶ φρενίτιδα ἐπὶ τῆς παρακμῆς. ἐν μὲν δὴ τοῖς
τοιούτοις πάθεσιν ἐξελέγχονται προφανῶς οἵ τε ἀπὸ
τοῦ Θεσσάλου πάντες οἵ τ' ἐν τῇ καρδίᾳ τὸ ἡγεμονικὸν τῆς ψυχῆς μέρος εἰπόντες ἰατροί. μὴ γὰρ ὅτι
τῶν εἰρημένων τινὸς εὐπορῆσαι βοηθημάτων αὐτοῖς
ἐστι δυνατόν, ἀλλὰ μηδὲ ὅτῳ μορίῳ προσφέρειν δεῖ
τὰ βοηθήματα. οὐ γὰρ δὴ ὥσπερ ἐπ' ὀφθαλμίας ἢ
πλευρίτιδος ἢ συνάγχης αὐτός τε ὁ κάμνων αἰσθάνε-

heated brain. However, in the opposite affections, it is appropriate to rouse and to cut and heat the thickness of the distressing humor which, without putrefaction, creates deep somnolence quite apart from fever. People call these [conditions] apoplexies, torpors, and catalepsies. However, if at some time there is putrefaction, such things occur with a fever, and this disease is called lethargy.

We shall apply Cretan thyme boiled down in vinegar, pennyroyal, oregano and other such things to the nose of those in such a state, so that the air being carried up to the brain might cut the thickness of the humor. After this, we shall also rub the palate with potent and acrid medications. Next after these, we shall also use ptarmic medications. And we shall apply medications of similar potencies to the head, even going as far as mustard, always increasing the strength of these if the affection is chronic. But also, in both cases, if they are chronic, we shall use the cupping glass and castor. For the latter concocts well when taken   932K
at the proper time. As a result, here again lethargy and phrenitis lead to a common treatment at the time of their abatement. Certainly, in such affections, all those who follow Thessalus, doctors who say the authoritative part of the soul is in the heart, are quite clearly refuted. For it is not only that the abundance of the remedies spoken of is not available to them, but they also [don't know] to which part they need to apply these remedies. It is certainly not like ophthalmia, phrenitis or cynanche (quinsy, sore

ται τοῦ πεπονθότος μορίου, ἡμῖν τε διὰ τῆς ἁφῆς καὶ
τῆς ὄψεως εἰς γνῶσιν ἥκει, κατὰ τὸν αὐτὸν τρόπον ἐπί
τε ληθάργου καὶ φρενίτιδος ἐπιληψίας τε καὶ παρα-
πληξίας καὶ σπασμῶν καὶ τετάνων, ἔτι τε τῆς καλουμένης ἰδίως κατοχῆς. ἐφ' ὧν ἁπάντων ἡ μὲν ἰδέα τῶν
βοηθημάτων ἐκ τῆς τοῦ πάθους φύσεως εὑρίσκεται, τὸ
δὲ χωρίον ᾧ μάλιστα χρὴ προσφέρειν αὐτὰ διὰ τοῦ
προεγνῶσθαι τὰς ἐνεργείας τε καὶ χρείας τῶν μορίων.

22. Ὅτι δ' ἀεὶ μεμνῆσθαι χρὴ τῶν συνενδεικνυ-
933K μένων | ἁπάντων, κἂν παραλειφθῇ ποτε ἐπὶ τῷ λόγῳ,
πολλάκις εἴρηται πρόσθεν. ἀλλὰ νῦν γε τὴν ἀπὸ τῶν
μορίων ἔνδειξιν μόνην πρόκειται διελθεῖν, οἷον εὐθὺς
ἐπὶ τῶν κατὰ τὴν κεφαλήν, εἰ καὶ μηδὲν ἄλλο, τοῦτο
γοῦν ἔνεστιν ἅπαντι νοῆσαι προχείρως, ὅτι πρόκειται
τοῦ ἐγκεφάλου πρῶτον μὲν ἡ παχεῖα μῆνιγξ, ἐοικυῖα
ταῖς ἐκτὸς ταύταις βύρσαις· ἐπ' αὐτῇ δ' ἐστὶ τὸ
κρανίον. ἀναγκαῖον οὖν ἐκλύεσθαι τῶν ἐπιτιθεμένων
φαρμάκων τὴν δύναμιν ἐν τοῖς προβλήμασι πυκνοῖς
καὶ σκληροῖς οὖσι· καὶ εἰ μὴ ῥαφαὶ κατὰ τὸ τῆς
κεφαλῆς ὀστοῦν ὑπὸ τῆς φύσεως ἐγεγόνεισαν, οὐδὲν
ἂν μέγα τῶν φαρμάκων οὐδὲν ἤνυσεν ἔξωθεν ἐπιτιθέ-
μενον. ἐπεὶ δὲ καὶ αἱ ῥαφαί, καὶ μάλιστα ἡ στεφα-
νιαία, παρίησιν ἔσω ῥαδίως οὐ τὰς ποιότητας μόνας
τῶν ἐπιτιθεμένων φαρμάκων, ἀλλὰ καὶ τὰς οὐσίας,
ὅταν γε ᾖ λεπτομερής, εἰκότως πολλὰ τῶν κατὰ τὸν
ἐγκέφαλον ὠφελεῖται παθῶν ὑπὸ τῆς τῶν ἔξωθεν ἐπιτιθεμένων φαρμάκων δυνάμεως. ἐγὼ γοῦν ἀπ' ἐμαυτοῦ
πειραθεὶς οἶδα, καταντληθέντος ῥοδίνῳ ψυχρῷ, ταχί-

throat) where the sufferer himself is aware of the affected part; rather, it comes to recognition through our tactile and visual examination, and the same applies in lethargy, phrenitis, epilepsy, hemiplegia, convulsion and *tetanos*, as well as what is called catalepsy specifically. In all such cases, the kind of remedy is discovered from the nature of the affection, while the place to which it is particularly necessary to apply the remedy is discovered through a prior knowledge of the functions and uses of the parts.

22. I have often said before that it is always necessary to call to mind all the joint indications even though they may sometimes be neglected in the argument. But now what I propose to go over is the indication from the parts alone, the immediate example being the parts pertaining to the head. If nothing else, it is at least readily possible for everyone to know this—that the thick membrane (dura mater), which is similar to those hides that are external, is placed first over the brain, and after it there is the cranium. It is necessary, then, to release the potency of the applied medications onto obstacles that are thick and hard, and if the sutures in the bones of the head had not been created by Nature, none of the medications applied externally would achieve anything significant. Since, however, there are also the sutures, and in particular the coronal suture, they allow easy inward passage, not only of the qualities of the applied medications but also of their substances, at least whenever this is fine-particled. So in all likelihood, there is benefit for many of the affections in the brain through the potency of externally applied medications. Anyway, I know this from my own experience, having bathed [the head] with

933K

στης τε καὶ σαφεστάτης εἴσω διαδόσεως αἰσθανό-
934K μενος ἐν τῷ | κατὰ τὸ βρέγμα τόπῳ. καὶ μέντοι καὶ
διαφορὰ παμπόλλη τοῖς ἀνθρώποις ὁρᾶται πρὸς ἀλλήλους ἐναργῶς ἐπὶ τῆς ῥαφῆς τῆσδε καὶ πρὸ τῆς
ἀνατομῆς. ἐθεασάμεθα γοῦν τινων ἐξυρημένων ἔτι ἐν
τῷ μασσᾶσθαι σαφεστάτην κίνησιν τῆς συναρθρώ-
σεως τῶν κατὰ τὴν στεφανιαίαν ῥαφὴν ὀστῶν, ὡς
εἶναι πρόδηλον ὅτι χαλαρὰ τοῖς ἀνθρώποις ἐκείνοις ἡ
σύνθεσις ἦν τῶν ὀστῶν τῆς κεφαλῆς.

εἰκότως οὖν κατὰ τοῦτο μάλιστα τὸ χωρίον ἐπι-
φέρουσιν ἅπαντες ἰατροὶ τὰς ἐπιβροχὰς τῇ κεφαλῇ,
κατὰ διαδοχὴν μὲν ἀπὸ τῶν πρώτως εὑρόντων ἐπὶ
τοὔργον ἐρχόμενοι, θεώμενοι δὲ καὶ αὐτοί, ὅσοι γε
προσέχουσι τὸν νοῦν οὐκ ἀργῶς τοῖς γινομένοις,
ὅπως μὲν ἡ ῥαφὴ φαίνηται κινουμένη σαφῶς, ὅπως
δ᾽ αἰσθάνωνται κατὰ τοῦτο τὸ μέρος οἱ ἄνθρωποι,
τάχιστα θερμαινόμενοί τε καὶ ψυχόμενοι διὰ τῶν
ἔξωθεν αὐτοῖς ὁμιλούντων· πρὸς γὰρ αὖ τοῖς ἄλλοις
καὶ λεπτότατόν ἐστιν ἐν τούτῳ τῷ μέρει τὸ κρανίον καὶ
ἀραιότατον. ὅταν οὖν τινος εἴσω διϊκνεῖσθαι φαρμάκου τὴν δύναμιν ἰσχυρῶς ἐθελήσῃς, κατὰ τοῦτο μάλιστα τὸ χωρίον ἐπιτίθει. κάλλιον δὲ καὶ μετὰ ἀνα-
935K τρίψεως αὐτὸ πράττειν, | ἀποκείραντα τῶν τριχῶν ἢ
ξυρῶντα τελέως· εἰ δ᾽ ὑγρὸν εἴη τὸ προσφερόμενον, ἐξ
ὑψηλοτέρου βάλλοντα καὶ οἷον κατακρουνίζοντα· δι-
ϊκνεῖται γὰρ εἴσω μᾶλλον ὑπὸ τῆς βολῆς ὠθούμενον.

ὥσπερ δ᾽ ἐνταῦθα τὸ σφοδρότερον τῆς βολῆς συμ-
φέρον ἐστίν, οὕτως ἐπ᾽ ὀφθαλμῶν ἀλυσιτελές· ἐν κε-

cold oil of roses and observed its very swift and safe distribution inward at the site of the bregma. Nevertheless, a 934K
considerable difference is clearly seen in respect to this suture, [comparing] one [person] to another, even before
dissection. At all events, in the palpation of those who have
been shaved, I have seen very distinct movement at the
junction of the bones at the coronal suture, so it is clear
that in those people the junction of the bones of the head is
loose.[39]

It is, therefore, reasonable for all doctors to apply irri-
gations to the head to this part (i.e. the coronal suture) in
particular, following on those who first discovered it when
they happened upon the action, since those men, at least
those who were not remiss in directing their attention to
what was occurring, saw how the suture is obviously mov-
able, and how people are aware that when things are brought
into contact with this part externally they are very quickly
heated or cooled. For again, in comparison to other [parts],
the cranium is both very thin and very loose in texture in
this part. Therefore, whenever you want the potency of
some medication to penetrate inwardly strongly, you place
it especially in relation to this part. It is also better to do
this with rubbing, after cutting the hair or shaving it off 935K
completely. If what is to be applied is moist, put it on from
a height as if pouring it down, for it penetrates inwardly
better when impelled by a force.

But just as here the greater strength of the force is ad-

[39] The Greek term used here—*synarthrosis*—is now used to
describe a junction of bones that is "immovable or nearly immovable" (S), which would apply to the cranial sutures once fusion has
occurred.

φαλῇ μὲν γὰρ ὀστοῦν ἐστὶ τὸ πληττόμενον, ἐπὶ
ὀφθαλμοῦ δὲ ὑμενώδη τινὰ καὶ ἀσθενῆ σώματα. καὶ
κατὰ μὲν τὴν κεφαλὴν ἕτερον μέν ἐστι τὸ πληττόμενον, ἄλλο δὲ τὸ θεραπευόμενον, ὃ τῆς μὲν πληγῆς
οὐκ αἰσθάνεται, τῆς δὲ διὰ τὴν βολὴν ἀφικνουμένης
εἰς αὐτὸ δυνάμεως ἀπολαύσει· κατὰ δὲ τὸν ὀφθαλμὸν
οὐκ ἄλλο μέν τι τὸ πληττόμενον, ἄλλο δὲ τὸ θερα-
πευόμενόν ἐστιν· ἀλλ' ὅπερ χρὴ θεραπευθῆναι, τοῦτο
καὶ πλήττεται σφοδρῶς. ἔτι τε πρὸς τούτοις τὸ μὲν
ὀστοῦν τῆς κεφαλῆς ἀναίσθητόν ἐστιν, αἰσθητικώ-
τερον δὲ μόριον ὁ ὀφθαλμός. ἐγχεῖν οὖν αὐτῷ τὰ
φάρμακα, πρῶτον μὲν ἐπαίροντας τὸ ἄνω βλέφαρον,
ὡς ὅτι μαλακώτατα, δεύτερον δὲ μὴ καταράσσοντας,
καθάπερ ἐπὶ τῆς κεφαλῆς. ἐξευρίσκειν δὲ καὶ αὐτὰ
τὰ ἐγχεόμενα μετὰ τῶν φαρμάκων ὑγρά, φύσεως ἀδη-
κτοτάτης. |

936K   καί μοι δοκοῦσιν οἱ παλαιοὶ μετὰ πολλῆς περισκέψεως ἐπὶ τὸ τῶν ᾠῶν ὑγρὸν ἀφικέσθαι, τό τε
ἀδηκτότατον αὐτοῦ καὶ τὸ γλίσχρον ἑλόμενοι. διὰ μὲν
γὰρ τοῦ μηδ' ὅλως δάκνειν[10] ὁ προειρημένος αὐτοῖς
ἐπληροῦτο σκοπός· διὰ δὲ τοῦ γλίσχρου συντέλειά
τις εἰς ἀνωδυνίαν ἐγίγνετο· λεαίνειν γὰρ πέφυκε τὰ
τοιαῦτα τῶν ὑγρῶν ἁπάσας τὰς τραχύτητας, ὅσαι
διὰ ῥεῦμα γίγνονται δριμύ. καὶ προσέτι μονιμώτερα
τῶν ὑδατωδῶν τε καὶ λεπτῶν ὑγρῶν ἐστι τὰ παχέα
μετρίως καὶ γλίσχρα. ὅτι μὲν οὖν τὸ χωρὶς τοῦ
δάκνειν γλίσχρον, ὅταν καὶ μετρίως ᾖ θερμόν, ἀνω-
δυνώτατόν ἐστιν, ἔμαθες δή που κἀπὶ τῶν κατὰ τὴν

410

vantageous, so in the eyes it is disadvantageous because, in the head, it is the bone that is struck, whereas in the eye it is certain membranes and weak bodies. And in the head, it is one thing which is struck but another thing which is treated, something which has no perception of the force, but has the benefit of the potency which comes to it through the force. In the eye, what is struck is not something different but what is being treated, so what must be treated is also struck strongly. In addition to these things, the bone of the head is insensitive whereas the eye is a very sensitive part. Therefore, in pouring medications on the latter, first lift the eyelid upward, as it is very soft; second, do not let the medication fall down [onto the eye] as in the head. And, in discovering those moist things to pour on along with the medications, [choose those] which are least stinging in nature.

And the ancients seem to me, after careful consideration, to have come to the liquid [part—i.e. the white] of the egg, [thereby] choosing the least stinging and viscid part of it. For by it being not stinging at all, the previously mentioned objective was fulfilled for them, while by it being viscid, there was some contribution toward it being painless, insofar as such liquids, by their nature, soften all the harshest things which occur due to a sharp flux. And besides, the more stable of the liquids that are watery and thin are those that are moderately thick and viscid. You learned, certainly, that to some degree what is viscid without being stinging, whenever it is also moderately warm, is

936K

---

10 μηδ' ὅλως δάκνειν K; μὴ δάκνειν B

γαστέρα δακνωδῶν διαχωρημάτων, ἐφ' ὧν ἐνιέμενον
στέαρ εὐθέως πραΰνει τὴν ὀδύνην. ὅτι δὲ καὶ διαμένειν
ἄμεινόν ἐστι τὸ τοιοῦτον ὑγρὸν ἐν τῷ πεπονθότι μορίῳ
πρόδηλον.

ἐπὶ μὲν γὰρ τῶν κατὰ τὴν γαστέρα τὸ κλύζειν
συνεχῶς ἀνιαρόν· ἐπὶ δὲ τῶν κατὰ τὸν ὀφθαλμὸν
ἀνατείνειν τὸ βλέφαρον. ἥ γε μὴν εὐαισθησία τοῦ
μορίου καὶ τὸ λεῖον αἱρεῖσθαι πάντως τὸ μέλλον
ἐνίεσθαι καὶ μηδὲν ἐν αὐτῷ ἔχειν τραχὺ καὶ ψαμμῶδες
937K ἐνδείκνυται. διὰ τοῦτ' οὖν | ἐπενοήθη καλῶς ἥ τ' ἐκ τῶν
ᾠῶν ὑγρότης καὶ τὸ λελειῶσθαι χρῆναι σφόδρα ἀκριβῶς, ὅσα γε τῶν γεωδῶν σωμάτων ἀναμίγνυται τοῖς
ὀφθαλμικοῖς φαρμάκοις. ὅταν γε μὴν ὀδύναι γίγνων-
ται σφοδραὶ κατ' αὐτούς, ἀναμνησθεὶς ὅσα περὶ γενέσεως ἁπασῶν ὀδυνῶν ἔμαθες, ἐπισκέπτου κατὰ τίνα
διάθεσιν ἐξ αὐτῶν ὀδυνᾶσθαι συμβαίνει τὸν ὀφθαλμὸν ἐν ταῖς φλεγμοναῖς, ὑπὲρ ὧν νῦν ὁ λόγος ἐστίν.
ἤτοι γὰρ ἐπὶ τὸ δάκνεσθαι σφοδρῶς ἐκ τῆς τῶν
ἐπιρρεόντων δριμύτητος, ἢ διὰ τὸ τείνεσθαι πεπληρω-
μένους τοὺς χιτῶνας αὐτῶν, ἢ δι' ἔντασίν τινα παχέων
ὑγρῶν ἢ πνευμάτων φυσωδῶν, ὀδύναι γίνονται σφοδραὶ κατ' αὐτούς. τὰς μὲν οὖν δήξεις διά τε τῶν
καθαιρόντων φαρμάκων ἀντισπῶντας κάτω καὶ κε-
νοῦντας θεραπεύειν προσήκει καὶ αὐτῷ τῷ μορίῳ τοῦ
ᾠοῦ τὸ ὑγρὸν ἐγχέοντας, ὅπως ἀλύπως ἐκκλύζηται
σὺν αὐτῷ τὸ δριμὺ ῥεῦμα. προπεπεμμένης δὲ τῆς
φλεγμονῆς ἤδη καὶ κενοῦ τοῦ σώματος ὄντος, ἐπιτηδειότατα τούτοις ἐστὶ λουτρά· καὶ γὰρ ἀνώδυνοι

most pain-relieving in the case of stinging things passing through the stomach, in which case grease, when inserted immediately, soothes the pain. It is also clear that it is better for such liquid to remain in the affected part.

In the case of things befalling the stomach, it is distressing to purge continually, whereas in the case of those befalling the eye, lift up the eyelid. In fact, the normal sensation of the part also shows that what is chosen to be put in should be completely smooth and have nothing rough and sandlike in it. Because of this, then, what is considered good [for the eyes] is the liquid from eggs and those earthlike bodies that are mixed with the eye medications, which must be very painstakingly triturated. Whenever severe pains occur in the eyes, having recalled those things you learned about the genesis of all pains, you must give consideration to some condition from among those in which it happens that the eye feels pain in the inflammations, which is what the discussion is now about. Severe pains occur in the eyes due either to the severe stinging from the pungency of things flowing in, or to the stretching of its membrane when it has been filled full [with humors], or to a distension of thick humors or flatulent *pneuma*. It is appropriate to treat the stingings with purging medications, effecting revulsion downward and evacuation, or by pouring the white of an egg on the part itself, whereby the stinging flux is painlessly washed away with it. However, when the inflammation has already "ripened" and the body has been evacuated, bathing is very suitable for these [cases];

937K

παραχρῆμα γίνονται καὶ παύεται τὸ ἐπιρρέον ὑγρὸν
τοῖς ὀφθαλμοῖς, ἐκκριθέντος μὲν τοῦ πλείστου δι᾿
ὅλου τοῦ σώματος ἐν τοῖς λουτροῖς, ἐπικερασθέντος
δὲ τοῦ λοιποῦ. |

938K   τὰς δ᾿ ἐπὶ τῇ πληρώσει τάσεις διά τε κενώσεως
αἵματος καὶ γαστρὸς ὑπαγωγῆς καὶ τρίψεως τῶν κάτω
μορίων ἰάσασθαι προσήκει· εἰ δ᾿ ἀναγκαζοίμεθά ποτε,
καὶ δεσμοῖς τῶν κώλων, ἔπειτα πυριάσεσιν αὐτοῦ τοῦ
φλεγμαίνοντος μορίου, δι᾿ ὕδατος ποτίμου θερμοῦ
συμμέτρως· τὰς δ᾿ ἐντάσεις προκενώσαντα κἀπὶ τούτων τὸ πᾶν σῶμα καὶ μέντοι καὶ ἀντισπάσαντα κάτω
τὴν ῥοπὴν τῶν χυμῶν. ἑξῆς αὐτοῖς τοῖς τοπικοῖς
ὀνομαζομένοις βοηθήμασι θεραπεύειν, οὐ τοῖς ἀποκρουομένοις καὶ ἀναστέλλουσι φαρμάκοις χρώμενον,
ἀλλὰ τοῖς διαφοροῦσι. πυριατέον οὖν αὐτούς, ὡς ἀρτίως εἴρηται, καὶ τὸ τῆς τήλεως ἐγχυτέον ἀφέψημα,
προπλύναντας ἐπιμελῶς τὴν τῆλιν, ὅπως μηδὲν αὐτῇ
προσιζηκὸς ἤτοι κόνεως ἢ ψάμμου λάθῃ. διαφορητικὸν γὰρ ἀλύπως ἐστὶ τὸ φάρμακον τοῦτο, πάντων
μάλιστα τῶν ὀφθαλμοῖς προσφερομένων. μεμνῆσθαι
δὲ χρὴ τῶν κοινῶν παραγγελμάτων ἐπὶ πάντων τῶν
κατὰ μέρος. ὧν ἓν καὶ τόδ᾿ ἐστίν, ὡς τὰ διαφορητικὰ
φάρμακα πλήθους ὄντος ἐν ὅλῳ σώματι μορίοις τισὶ
939K προσφερόμενα | πληροῖ μᾶλλον ἢ κενοῖ. ταῦτά τε οὖν
καὶ τὰ ἄλλα θεραπεύων νοσήματα καὶ τὰς φλεγμονάς,
ὑπὲρ ὧν νῦν ὁ λόγος ἐστί, μηδενὶ θαρρήσεις τῶν
διαφορητικῶν ὀνομαζομένων βοηθημάτων, πρὶν τῇ
τοῦ σώματος ὅλου χρήσασθαι κενώσει. πρόσεχε δὲ

not only do they immediately become pain-free but also the moisture flowing to the eyes ceases, since the greater part has been expelled through the whole body by the baths and the remainder has become diluted.

It is appropriate to cure the tensions due to repletion by 938K evacuation of blood, downward purging of the stomach, rubbing of the lower parts and sometimes, if we are compelled [to do so], also by binding of the limbs, and afterward with fomentations of the inflamed part itself by water that is fresh and moderately hot. It is appropriate to cure the distensions by prior evacuation of the whole body with respect to the humors and by effecting revulsion of their movement downward. Next, treat with the so-called topical remedies themselves; do not use the medications that are repelling or repercussive but those that are discutient. You must warm them, as I said just now, and you must pour on the decoction of fenugreek, after carefully washing the fenugreek[40] beforehand so that nothing clinging to it, either dust or sand, escapes notice. This medication is the most painlessly dispersing, especially of all those applied to the eyes. It is, however, necessary to remember the common precepts regarding all these things individually, one of which is this: with discutient medications, if they are applied to certain parts when there is abundance in the whole body, they fill rather than empty. Therefore, when 939K treating these and other diseases, as well as the inflammations, which is what the present discussion is about, do not be over-zealous in using any of the remedies that are termed discutient before purging the whole body. Direct

40 On the use of fenugreek (*Trigonella foenum-graecum*), see Dioscorides, II.124.

κἀκείνῳ τὸν νοῦν ἐπὶ πάντων παθῶν, οὐ μόνον φλεγμονῶν, ὡς ἐνίοτε τὸ μὲν ὅλον σῶμα μετρίως διάκειται
κατά τε ποιότητα καὶ συμμετρίαν χυμῶν· ἐν δέ τι
τῶν ὑπερκειμένων ἢ δύο τῷ κάμνοντι μορίῳ τὴν ἑαυτοῦ περιουσίαν ἐκπέμπει· καθάπερ ἀμέλει κἀπὶ τῶν
ὀφθαλμῶν οὐ σπανιάκις, ἀλλὰ καὶ πάνυ πολλάκις
ἰδεῖν ἐστι γιγνόμενον, ἐπιπεμπούσης αὐτοῖς τῆς κεφαλῆς τὸ ῥεῦμα.

πρόδηλον οὖν, οἶμαι, κἀπὶ τούτων ἐστὶν ὡς χρὴ
τὴν κεφαλὴν ἰάσασθαι προτέραν, εὑρόντα τὴν διάθεσιν αὐτῆς, ᾗ τῶν περιττωμάτων ἡ γένεσις ἕπεται. καὶ
τά γε χρονίζοντα τῶν ὀφθαλμῶν ῥεύματα θεραπεύομεν ἀφιστάμενοι μὲν αὐτῶν τῶν ὀφθαλμῶν, ἐπὶ δὲ
τὴν τῆς κεφαλῆς ἀφικνούμενοι πρόνοιαν· ἐκ μὲν τοῦ
γένους τῶν δυσκρασιῶν οὖσαν, ἐνδεικνυμένην δὲ
θεραπείαν ἐναντίαν ἑαυτῇ, καθὸ δέδεικται πρόσθεν.
ὡς τὰ πολλὰ μὲν οὖν ἤτοι ψυχρὰ δυσκρασία γίγνεται
940K βλαβερὰ κατὰ τὴν κεφαλὴν | ἢ ὑγρά· καὶ δῆλον ὅτι
καὶ ἀμφότεραι συνέρχονται. σπανιώτεραι δέ εἰσιν αἱ
διὰ θερμότητα δριμὺ ῥεῦμα τοῖς ὀφθαλμοῖς ἐπιπέμ-
πουσαι, καθ᾽ ἃς οὐ προσήκει τοῖς διὰ θαψίας καὶ τοῦ
νάπυος χρῆσθαι φαρμάκοις, ἀλλὰ τοὐναντίον ἅπαν,
ἐλαίῳ μὲν ὀμφακίνῳ τε καὶ Ἱσπανῷ καὶ ῥοδίνῳ, λουτροῖς δὲ ποτίμων ὑδάτων πλείοσιν. ἐνίοτε μὲν οὖν ὁ
ἐγκέφαλος ἐπιπέμπει τὸ ῥεῦμα· καὶ χρὴ τούτου μὲν
τὴν κρᾶσιν ἐπανορθοῦσθαι τοῖς ὅλης τῆς κεφαλῆς
ἐπιθέμασιν.

ἐνίοτε δὲ τῶν ἀγγείων ἐστὶ τὸ πάθος ἤτοι τῶν

your attention to that in all affections, not only inflammations, as sometimes the whole body is in a moderate state with regard to the quality and balance of humors, and it is one or two of the parts situated above that send their own surplus to the suffering part. This is, of course, not rare in the case of the eyes, but is also very frequently seen to occur when the head sends forth the flux to them.

I think it is clear, then, that even in these cases, it is necessary to cure the head first, discovering the condition in it which the generation of superfluities follows. And in fact, we treat the chronic fluxes of the eyes by disregarding the eyes themselves and coming to a prior consideration of the head. When [the condition] is derived from the class of the *dyskrasias*, it indicates treatment that is opposite to itself, as has been shown before. So then, the many cold or moist *dyskrasias* are harmful to the head, and it is also clear that 940K both occur together. More rare are the *dyskrasias* which due to heat send an acrid flux to the eyes. In these it is not appropriate to use the medications made from the thapsia or sinapi (white mustard)[41] but the complete opposite, i.e. oil from unripe olives, Spanish oil, or oil of roses, and frequent baths of fresh water. Sometimes, the brain sends the flux and it is necessary to correct the *krasis* of this with epithemata applied to the whole head.

Sometimes, however, the affection is of the vessels,

---

[41] On the two medications, see Dioscorides, IV.157 and II.184 respectively.

φλεβῶν ἢ τῶν ἀρτηριῶν ἀτονωτέρων ὑπαρχουσῶν, ὡς δέχεσθαι τὴν τῶν ἄλλων ἀγγείων περιουσίαν. ἡνίκα ἐκτέμνοντές τι μέρος αὐτῶν ἢ καὶ διατέμνοντες ὅλα μέχρι πολλοῦ βάθους, διαλαμβάνομεν οὐλῇ σκληρᾷ τὰ μεταξὺ διορίζοντες μόρια τοῦ τμηθέντος, ὡς μηκέτ᾽ εἶναι συνεχῆ, μηδ᾽ ἐπιρρεῖν ἐκ τοῦ ἑτέρου πρὸς τὸ ἕτερον. ἀλλ᾽ ὅταν γε τῶν ἐν τῷ βάθει κειμένων ἀγγείων τῶν ἄνωθεν ἡκόντων ἅμα τοῖς νεύροις ἐπὶ τοὺς ὀφθαλμοὺς ἢ τὸ πάθημα, τούτων οὐδὲν οἷόν τ᾽ ἐστὶ πρᾶξαι· διὸ καὶ δυσίατα πάντα τὰ τοιαῦτα ῥεύματα
941K γίγνεται. τὰ δ᾽ ἔξωθεν ἀγγεῖα | καὶ χωρὶς χειρουργίας ἔνεστι ῥῶσαι, φαρμάκοις καταχρίοντα τονωτικοῖς.

ἐνίοτε δὲ καὶ θερμὸν αἷμα καὶ ἀτμῶν μεστὸν ἐπὶ τὴν κεφαλὴν ἀναφέρεται καὶ πληθύει μάλιστα κατὰ τὰς ἀρτηρίας. ἐφ᾽ οὗ χρησιμώτατον εὕρηται βοήθημα τοῖς ἰατροῖς ἡ ἀρτηριοτομία. χρὴ δὲ ξυροῦντα τὴν κεφαλὴν ἐπιμελῶς ἅπτεσθαι τῶν τ᾽ ὀπίσω καὶ καθ᾽ ἑκάτερον οὖς ἀρτηριῶν καὶ τῶν ἐν τῷ μετώπῳ τε καὶ τοὺς κροτάφους. ὅσοι δ᾽ αὐτῶν θερμότεραί σοι φαίνονται τῶν ἄλλων εἶναι καὶ μᾶλλον σφύζειν, ἐκείνας τέμνειν· ὅσαι δὲ μικραί τέ εἰσι καὶ ὑπὸ τῷ δέρματι, κἂν μέρος αὐτῶν ἐκτέμνῃς, ὥσπερ ἐν τοῖς σκέλεσιν ἐπὶ τῶν κιρσῶν εἰώθαμεν πράττειν, ἄμεινον ἐργάσῃ. καὶ τῶν καθ᾽ ἡμᾶς γέ τις ὀφθαλμικῶν οὐχ ὁ φαυλότατος ἐξέκοπτε τῶν ἐποχουμένων τοῖς κροταφίταις μυσὶν ἀρτηριῶν οὐκ ὀλιγίστην μοῖραν.

when either the veins or arteries are very weak, so that they receive the surplus of the other vessels. At which time, if we cut out some part of these or cut through them completely to a great depth, we divide by means of hard scar tissue, effecting a separation between the parts of what has been cut, so there is no longer continuity and nothing flows from one part to the other. But whenever there is an affection of the deep vessels, [that is] of those [vessels] that come upward with the nerves to the eyes, it is not possible to do anything for these. Because of this, all such flows are difficult to cure. However, it is possible to strengthen those vessels which are superficial without surgery by pouring   941K on strengthening medications.

Sometimes, too, blood that is hot and full of air is carried upward to the head and is particularly abundant in the arteries. For this reason, arteriotomy is found to be the most useful remedy by doctors. After carefully shaving the head, you should palpate the arteries behind each ear, and those on the forehead and at the temples.[42] Cut any of these that seem to you hotter than the others and more pulsatile. Those that are small and under the skin, even if you cut a part of them, as I am accustomed to do in the legs in the case of varicose veins, it will be better. In fact, one of the eye doctors among our contemporaries, and not the worst one, used to cut out a by no means small part of the artery lying on the temporal muscles.

[42] Presumably the posterior auricular, frontal, and superficial temporal arteries.

ὡς τὰ πολλὰ μὲν οὖν ἀνασπᾶται τῆς ἐκμηθείσης ἀρτηρίας τὰ καταλειπόμενα μόρια πρὸς τὸ συνεχὲς ἀμφοτέρων, καὶ μᾶλλον γίνεται τοῦτο ἐπί τε τῶν μικρῶν ἀγγείων καὶ ἧττον σφυζόντων. εἰ δ' ἐν τῷ

942K γυμνοῦν φαίνοιτό | σοι μέγα τὸ ἀγγεῖον ἢ μεγάλως σφύζοι, ἀσφαλέστερον αὐτῷ βρόχον περιβάλλοντα πρότερον, οὕτως ἐκκόπτειν τὸ μεταξύ. γιγνέσθωσαν δ' οἱ τοιοῦτοι τῶν βρόχων ἐξ ὕλης δυσσήπτου· τοιαύτη δ' ἐστὶν ἐν Ῥώμῃ μὲν ἡ τῶν Γαϊτανῶν[11] ὀνομαζομένων, ἐκ μὲν τῆς τῶν Κελτῶν χώρας κομιζομένων, πιπρασκομένων δὲ μάλιστα κατὰ τὴν Ἱερὰν Ὁδόν, ἥτις ἐκ τοῦ τῆς Ῥώμης ἱεροῦ κατάγει πρὸς τὰς ἀγοράς. τούτων μὲν οὖν ἐν Ῥώμῃ ῥᾷστον εὐπορῆσαι· καὶ γὰρ εὐωνότατα πιπράσκεται.

κατ' ἄλλην δὲ πόλιν ἰατρεύοντί σοι παρασκευάσθω τῶν νημάτων τι τῶν σηρικῶν ὀνομαζομένων. ἔχουσι γὰρ αἱ πλούσιαι γυναῖκες αὐτὰ πολλαχόθι τῆς ὑπὸ Ῥωμαίων ἀρχῆς, καὶ μάλιστα ἐν μεγάλαις πόλεσιν, ἐν αἷς εἰσι πολλαὶ τῶν τοιούτων γυναικῶν. εἰ δὲ μὴ παρείη τοῦτο, τῶν κατ' ἐκείνην τὴν χώραν ἐν ᾗπερ ἂν ὢν τυγχάνῃς ἐκλέγου τὴν ἀσηπτοτέραν ὕλην, οἷα πέρ ἐστιν ἡ τῶν ἰσχνῶν χορδῶν· αἱ μὲν γὰρ εὔσηπτοι ταχέως ἀποπίπτουσι τῶν ἀγγείων. ἡμεῖς δὲ βουλόμεθα περισαρκωθέντων αὐτῶν ἀποπίπτειν τὸν βρόχον. ἡ γὰρ ἐπιτρεφομένη τοῖς ἀποτετμημένοις μέρεσι

943K τῶν ἀγγείων σὰρξ | ἐπίθεμα γίγνεται καὶ μύει τὸ στόμιον αὐτῶν. ἐπειδὰν δὲ φθάσῃ γενέσθαι, καιρὸς ἤδη τοῖς βρόχοις ἀκινδύνως ἀπορρυῆναι. τὰς μέντοι

So in many instances, the remaining parts of the cut artery retract in continuity with themselves, and this occurs particularly in the case of small vessels and those that are less pulsatile. If the exposed vessel seems to you large, or to pulsate strongly, it is safer to place ligature[s] around it beforehand and then cut what is between them. Such ligatures should be of a material which does not rot easily— the kind that in Rome is called Gallic, which is brought down from the country of the Celts and is sold particularly along the Sacred Way that leads down from the temple of Roma to the markets. It is very easy to provide these in Rome and they are sold very cheaply. 942K

For you who are practicing medicine in another city, there must be provision of one of the suture materials spun from the so-called silks. Rich women in the Roman Empire very often have these, especially in large cities in which there are many such women. If this is not available, choose the material most resistant to decay in that place in which you happen to be, like for example the material of dried gut, for those materials that readily decay quickly fall off the vessels. We do, however, want the ligature to fall off after the vessel has been covered by flesh, for the flesh that grows between the severed parts of the vessel creates a covering for, and closes over the opening of, these parts. Whenever this, which you anticipate, occurs, it is now time 943K

---

11 B; Γαϊετανῶν K

φλέβας, ὅτ᾽ ἄν ποτε ἐκτέμνῃς τι μόριον αὐτῶν, οὐκ ἀναγκαῖον οὕτως ἀσήπτοις ὕλαις διαδεῖν, ἀλλ᾽ ἀρκεῖ καὶ τῶν ἄλλων τις. ἐπὶ μὲν γὰρ τῶν ἀρτηριῶν ἡ διηνεκὴς κίνησις ἀνοίγνυσι τὰ στόματα τῶν τετμημέ- νων ἀγγείων· ἐπὶ δὲ τῶν φλεβῶν, ὅταν ἅπαξ μύσῃ καθ᾽ ὁντιναοῦν τρόπον ἤτοι πιληθέντα δι᾽ ἐπιδέσεως ἢ στυφθέντα διὰ φαρμάκων, ἐπιτρέπει τῇ πέριξ σαρκὶ περιφύεσθαι, καὶ μάλισθ᾽ ὅταν ἀκίνητον ἔχῃ τὸ μέρος ὁ χειρουργηθεὶς ἄνθρωπος, ἔτι δὲ μᾶλλον, ἐὰν καὶ ἀνάρροπον ἔχῃ ἐπὶ κενῷ τῷ σύμπαντι σώματι. καὶ γάρ τοι καὶ τοὺς κιρσοὺς οὕτω θεραπεύομεν. ὀνομάζε- ται δὲ κιρσὸς ἡ ἀνευρυσμένη φλέψ. ἀνευρύνεται δ᾽ ἐν ὄρχεσί τε καὶ σκέλεσι τοὐπίπαν.

ἐπεὶ δὲ γραφομένων ἔτι τῶνδε τῶν ὑπομνημάτων ἠξίωσαν οὐκ ὀλίγοι τῶν ἑταίρων ἐπὶ τῇ τελευτῇ τῆς ὅλης πραγματείας ἁπάντων ἐφεξῆς με τῶν κατὰ χει- ρουργίαν μνημονεῦσαι, διὰ τοῦτο καὶ νῦν ὁ περὶ τῶν κιρσῶν ἀναβεβλήσθω λόγος. ὄντων δ᾽ οὐκ ὀλίγων

944K κατὰ | μέρος ἐν ὀφθαλμοῖς παθῶν εἰδικωτέρας θερα- πείας δεομένων, οὐδ᾽ ὑπὲρ ἐκείνων ἔτι λέγειν ἐνταυθοῖ προσήκει. τῷ μὲν γὰρ ἐπιμελῶς ἀνεγνωκότι τὰ πρόσθεν εἰρημένα καὶ φύσει συνετῷ ῥᾷστόν ἐστι κατὰ τὴν ἀκολουθίαν ἐξευρίσκειν ἅπαντα· τοῖς δὲ μὴ τοιούτοις ἄμεινον ἰδίᾳ γράψαι θεραπευτικὴν πραγμα- τείαν ἁπάντων τῶν ἐν ὀφθαλμοῖς παθῶν, ἐπεὶ καὶ πολλοὶ τῶν ἑταίρων οὕτως ἀξιοῦσιν.

for the ligatures to be taken off without danger. However, with the veins, if at some time you cut some part of these, it is not necessary to bind them thoroughly with nondecaying materials in the same way; one of the other things is sufficient. In the case of the arteries, the continuous movement opens the mouths of the cut vessels, whereas with the veins, whenever they have been completely closed by whatever means—either compressed by a bandage, or contracted by medications—rely on surrounding flesh growing around them, especially whenever the person carrying out the procedure holds the part immobilized, and still more, if he holds it elevated after the whole body has been evacuated. Certainly, this is also how we treat varices. A dilated vein is termed a varix. Moreover, they are dilated, in general, in the testicles and the legs.

However, since these treatises are still to be written, many of my colleagues thought it right that in the final part of this whole work, I mention in order all those things that pertain to surgery, and because of this, let the discussion about varices be deferred for the present. And although there are many separate affections in the eyes which need 944K more specific treatments, it is no longer appropriate to speak about those here. For to someone who has assiduously read what has been said hitherto and is naturally perceptive, it is a very easy matter to discover all the things in the following treatment. For those who are not like this, it is better to write specifically regarding the therapeutic approach to all the affections in the eyes, since there are also many of my colleagues who think this worthwhile.

ΒΙΒΛΙΟΝ Ξ

1. Περὶ μὲν τῶν παρὰ φύσιν ὄγκων ὁπόσοι μέν εἰσι καὶ ὁποῖοι γέγραπται πρόσθεν ἰδίᾳ καθ' ἓν βιβλίον. ὡς δ' ἄν τις αὐτοὺς θεραπεύοι μεθόδῳ, τῆς προκειμένης πραγματείας ἴδιον ὄν, ἐν τῷ τρισκαιδεκάτῳ τῶνδε τῶν ὑπομνημάτων ἠρξάμεθα λέγειν. ἐπεὶ δ' ἐν τοῖς ἔμπροσθεν ὑπὲρ ἁπάντων πυρετῶν ὁ λόγος ἐγεγόνει, βέλτιον ἔδοξέ μοι περὶ πρώτης φλεγμονῆς διελθεῖν ὡς ἂν συνεχέστατά τε γινομένης καὶ πυρετοὺς ἐπιφερούσης πολλάκις. εἴρηται μὲν οὖν τι κἀν τῇ τῶν πυρετῶν θεραπείᾳ περὶ τῆς φλεγμονῆς ἅμα ταῖς |
ἄλλαις αὐτῶν αἰτίαις· ἀλλ' ὁ τέλειός τε καὶ ἴδιος αὐτῆς λόγος ἐν τῷ πρὸ τούτου βιβλίῳ γέγραπται, τὴν μέθοδον τῆς θεραπείας ὁποία τίς ἐστι διερχομένων ἡμῶν, οὐ τὴν τῶν βοηθημάτων ὕλην, ὅ τι μὴ παραδείγματος ἕνεκεν, ὡς κἀπὶ τῶν ἔμπροσθεν ἐποιήσαμεν.

οὐ πόρρω δὲ τῆς φλεγμονῆς ἕτερον νόσημά ἐστιν ἐρυσίπελας ὀνομαζόμενον, ἐπὶ χολώδει χυμῷ συνιστάμενον, ὡς ἐδείχθη. βέλτιον δ' ἴσως αὐτὸ μακροτέρῳ λόγῳ διορίσαι τῆς φλεγμονῆς. κοινὰ μὲν οὖν

BOOK XIV

1. I have previously written about the swellings contrary to   945K
nature—how many there are and of what sort—separately
in one book.[1] How someone might treat them methodi-
cally, which is the particular task of the work before us, I
started to speak about in the thirteenth [book] of these
treatises. Because in those [books] that preceded [the thir-
teenth], discussion had taken place about all fevers, it
seemed to me better to go over inflammation first, as it
occurs very frequently and often brings fever with it.
Therefore, I also said something, even in the treatment of
fevers, about inflammation together with the other causes   946K
of fevers, but the complete and specific discussion of inflammation has been given in the book prior to this one
where I went over the method of treatment in terms of
kind, but not the material of the remedies, other than for
the sake of an example, as I did in the earlier books.

Not far removed from inflammation, there is another
disease called erysipelas which arises from a bilious humor, as was shown. It is, perhaps, better to differentiate
this from inflammation by means of a lengthier discus-

[1] Again, "swelling" rather than "tumor" is used to translate
onkos because of the more specialized meaning the latter term has
acquired. The book referred to is *De tumoribus praeter naturam*,
VII.705–32K.

425

ἀμφοῖν ὅ τε παρὰ φύσιν ὄγκος ἐστὶ καὶ ἡ θερμασία,
διαφέρει δὲ πρῶτα μὲν καὶ μάλιστα τῇ χροιᾷ. ἐρυθρᾶς
μὲν γὰρ οὔσης αὐτῆς φλεγμονὴν τὸ πάθος ὀνομάζουσιν, ὠχρᾶς δ' ἢ ξανθῆς ἢ ὥσπερ ἐξ ὠχροῦ καὶ ξανθοῦ
χρώματος μικτῆς, ἐρυσίπελας. ἀτὰρ οὖν καὶ ὁ σφυγμὸς ἴδιον σύμπτωμά ἐστι τῆς μεγάλης φλεγμονῆς·
καὶ γὰρ καὶ διὰ βάθους γίνεται μᾶλλον, ὥσπερ γε
ἐρυσίπελας ἐν τῷ δέρματι μᾶλλον ἢ ἐν τῷ βάθει.
λεπτὸς γὰρ κατὰ τὴν σύστασιν ὁ τῆς ὠχρᾶς χολῆς
χυμός, ὥστε διαρρεῖ ῥᾳδίως ἐπὶ τὸ δέρμα, τὰ σαρ-
947K κώδη καὶ ἀραιὰ μόρια διερχόμενος. | ἡ δὲ τοῦ δέρματος
πυκνότης οὐκέθ' ὁμοίως εὔπορος τῇδε τῇ χολῇ, πλὴν
εἰ πάνυ λεπτὴ καὶ ὑδατώδης εἴη, τοιαύτη γὰρ μάλιστα
καὶ ἡ καθ' ἑκάστην ἡμέραν ἐστὶ συναπερχομένη τοῖς
ἱδρῶσι.

κα πολλῶν ὄψει κατὰ τὰ βαλανεῖα ταῖς στλεγ-
γίσιν ἀποξεόντων τὸν ἱδρῶτα τοιοῦτον τὴν χρόαν,
οἷόν περ καὶ τὸ οὖρόν ἐστι τοῖς ἐπὶ πλέον ἀσιτήσασιν.
οἶσθα γὰρ ὡς χρονιζόντων ἡμῶν ἐν ταῖς ἀσιτίαις
ὠχρότερον ἐξ ὑδατώδους γίγνεται τὸ οὖρον· ὕστερον
δέ ποτε καὶ ξανθόν, εἰ μὴ φθάσειέ τις ἐπιτέγξαι τὸν
αὐχμὸν τοῦ σώματος, ὑγραινούσῃ τε τροφῇ καὶ ποτῷ.
κατὰ φύσιν μὲν οὖν διοικουμένου τοῦ σώματος ὁ
πικρόχολος χυμὸς ἀδήλως διαπνεῖται· παρὰ φύσιν δὲ
κατ' ἄλλα τε πάθη πλεονάζων φαίνεται, περὶ ὧν αὖθις
εἰρήσεται, καὶ μέντοι κατὰ τουτὶ τὸ νῦν ἡμῖν προκείμενον, ὃ καλοῦσιν ἐρυσίπελας. ὅταν γὰρ ἤτοι πολὺ
πλέον τοῦ κατὰ φύσιν ἢ παχύτερος γενόμενος ἀθρόως

sion.[2] Common to both are the abnormal swelling and heat; the difference lies primarily and particularly in the color. When the color is red [doctors] call the affection inflammation whereas, when it is pale, or yellow, or a color like a mixture of pale and yellow, [they call it] erysipelas. But throbbing too is a specific symptom of a major inflammation, and it also occurs more in deep locations, just as erysipelas occurs more in the skin than in deep locations because the humor of yellow bile is thin in consistency, so that it flows readily to the skin, passing through the fleshy and loose-textured parts. However, the dense texture of the skin is no longer similarly easy for this bile to pass through, unless it is particularly thin and watery, for this especially is what escapes each day along with the sweat.

947K

And among the many people who scrape themselves with strigils in the bathhouses, you will see sweat of this color. It is also like the urine in those who fast to excess. You know that when we spend a long time fasting, the urine becomes rather pale from wateriness, and subsequently sometimes also yellow in someone who does not moisten the dryness of the body beforehand with moistening food and drink. Therefore, when the body is ordered in accord with nature, the picrocholic humor (bitter bile) is dissipated insensibly whereas, when the body is ordered contrary to nature, it appears abundant in other affections about which more will be said later, and in relation to the subject now before us, which they call erysipelas. For whenever there is either very much more [picrocholic humor] than accords with nature, or it is quite thick, it will be

[2] Erysipelas, in the current sense, is an inflammation. For the ancient use of the term, see the list of diseases and symptoms in the Introduction (section 9).

ἐνεχθῇ πρὸς τὸ δέρμα, διακαίει τε τοῦτο καὶ εἰς ὄγκον
αἴρει.

2. Κρεῖττον δ' ἐστίν, ὥσπερ ἀεὶ λέγομέν τε καὶ
πράττομεν, οὕτω καὶ νῦν ἡμᾶς ἀπὸ τῶν πραγμάτων, |
948K οὐ τῶν ὀνομάτων ἄρξασθαι, δευτέραν ἀμείνω τῆς
πρόσθεν ἀρχῆς τῷ λόγῳ τήνδε θεμένους. ὅταν αἵματος πολλοῦ κατασκήψαντος εἴς τι μόριον, ὡς μὴ
στέγεσθαι πρὸς τῶν ἀγγείων τῶν κατ' αὐτό, διεκπίπτῃ
τι δροσοειδῶς ἐκ τῶν ἀγγείων εἰς τὰς τῶν μυῶν
χώρας, ἃς ἔχουσι μεταξὺ τῶν συντιθέντων αὐτοὺς
ὁμοιομερῶν σωμάτων, ὄγκος μὲν γίγνεται διὰ τὸ πλῆ-
θος· ἕπεται δὲ αὐτῷ τάσις μὲν τοῦ δέρματος, ὀδύνη δ'
ἐν τῷ βάθει καὶ πόνος μετὰ σφυγμοῦ καὶ ἁπτομένοις
ἀντιτυπία τις, ἔρευθός τε καὶ θερμότης, ὡς ἂν καὶ τοῦ
δέρματος ἀπολαύοντος ὧν πάσχουσιν αἱ ὑπ' αὐτὸ
σάρκες. ἀνάλογόν τε τοῖς νῦν εἰρημένοις καὶ περὶ τὰ
σπλάγχνα γίνεται διάθεσις. ἔστι γὰρ δὴ κἂν τούτοις
ἰδία σάρξ, ἣν ἔνιοι παρέγχυμα προσαγορεύουσιν, εἰς
ἣν ἀτμοειδῶς ἐκ τῶν πεπληρωμένων ἀγγείων ἐκκρινό-
μενον τὸ αἷμα τὰ προειρημένα συμπτώματα ἐργάσεται. μία μὲν οὖν ἥδε διάθεσις αἱματώδους ἔκγονος
ῥεύματος ἐν σαρκοειδέσι σώμασι μάλιστα γινομένη.

δευτέρα δ' ἑτέρα χολώδους περὶ τὸ δέρμα συνιστα-
μένη μάλιστα τό τ' ἐκτὸς τοῦτο, τὸ κοινὸν ἁπάντων
949K σκέπασμα τῶν μορίων, | καὶ τὸ καθ' ἕκαστον τῶν
ἐντὸς περιτεταμένων ὑμενῶδές τε καὶ λεπτόν. ὥσπερ
οὖν ἡ προτέρα συνεπιλαμβάνει τι καὶ τοῦ δέρματος,
οὕτω καὶ ἥδε τῆς ὑποκειμένης αὐτῷ σαρκός. εἰ δὲ καὶ

carried all together to the skin, burning this and raising it to a swelling.

2. It is better, as I always say and do, and as I am now doing, to begin from the matters [themselves] and not 948K from the names, and it would be better to lay this down as secondary to the aforesaid starting point of the discussion. Whenever a lot of blood has rushed down into some part such that it is not contained by the vessels in that part, then something dewlike escapes from the vessels into the spaces of the muscles that exist between the *homoiomerous* bodies which constitute them, so a swelling arises due to the excess. Tension of the skin follows this, and deep pain and distress, along with throbbing and some resistance to those palpating, and redness and heat, as if the skin is also affected by the things the flesh under it suffers. A condition analogous to those now being discussed also occurs involving the viscera. For assuredly, there is a specific flesh in these too, which some call "parenchyma," in which the blood, evacuated vaporously from the overfilled vessels, will bring about the aforementioned symptoms. This, then, is one condition springing from a bloody flux which occurs particularly in fleshy bodies.

Another, and second, arising from a bilious flux, exists around the skin, particularly that which is external and is the common covering of all parts, but also that which is 949K membranous and thin, enclosing each of the parts within. Therefore, just as the prior [condition] also involves some of the skin, in the same way too, this condition involves the flesh underlying it. If the humor is thicker and more acrid,

παχύτερος ὁ χυμὸς εἴη καὶ δριμύτερος, ἀποδέρει τὴν
ἐπιδερμίδα καί ποτ᾽ ἐν τῷ χρόνῳ πρὸς τὸ βάθος
ἐξικνεῖται τοῦ δέρματος ἡ ἕλκωσις. αὕτη μὲν οὖν ἡ
διάθεσις ἐρυσίπελας ὀνομαζέσθω, διττὴν δὲ ἔχον, ὡς
εἴρηται, διαφοράν, ἤτοι χωρὶς ἑλκώσεως, ἢ σὺν ταύτῃ
γιγνόμενον. ἡ προτέρα δὲ μονοειδής ἐστι καὶ καλείσθω φλεγμονή. ὅταν οὖν μήτε ἀκριβῶς ᾖ χολῶδες τὸ
ῥεῦμα μήθ᾽ αἱματῶδες, ἀλλ᾽ ἐξ ἀμφοῖν μικτόν, ἀπὸ
μὲν τοῦ κρατοῦντος ἐν τῇ μίξει τοὔνομα αὐτῷ τιθέσθω,
κατηγορείσθω δὲ τοῦτο τὸ κρατούμενον, ὡς ἤτοι
φλεγμονὴν ἐρυσιπελατώδη καλεῖν ἡμᾶς ἢ ἐρυσίπελας
φλεγμονῶδες. εἰ δὲ μηδέτερον ἐπικρατοίη, μέσον ἐρυ-
σιπέλατός τε καὶ φλεγμονῆς ὀνομαζέσθω τὸ πάθος.

3. Ἡ δὲ τῆς θεραπείας μέθοδος ὡς ἐπὶ τῶν ἄλλων
συνθέτων, οὕτω καὶ νῦν γιγνέσθω τὴν ἀρχὴν ἀπὸ |
950K τῶν ἁπλῶν ποιησαμένοις. κοινὸς μὲν οὖν σκοπὸς
ἅπασι τοῖς οὕτω παρὰ φύσιν ὄγκοις ἡ κένωσις. οὕτω
δ᾽ εἶπον ἀναμιμνήσκων τοῦ πλήθους τῶν ἐργαζομένων
αὐτοὺς χυμῶν. εἰ γάρ τις τούτους κενώσειε, τὴν κατὰ
φύσιν ἕξιν ἀναλήψεται τὸ μόριον. ἀλλὰ καὶ ἡ κένωσις
ὁμοίως ἅπασι διττή, μία μὲν οὖν ἀπωθουμένων ἡμῶν
αὐτοὺς εἰς ἕτερα μόρια, δευτέρα δὲ διαφορούντων ἔξω
κατὰ τὴν ἄδηλον αἰσθήσει διαπνοήν. ἐπεὶ δ᾽ οὐ τῷ
ποσῷ μόνον ἀνιᾷ τὸ ἐρυσίπελας, ἀλλὰ καὶ τῷ ποιῷ,
σφοδρὰν ἔχον τὴν φλόγωσιν, ἐμψύξεως δεήσεται
περιττοτέρας ἢ κατὰ τὴν φλεγμονήν. οὐ μὴν ἀκίν-

it excoriates the epidermis and, on some occasions, ulceration of the skin goes deep over time. Therefore, let this condition be called erysipelas, being of two types, occurring either without or with ulceration, as was said. The former is of one form—let it be called inflammation. Whenever the flux is neither entirely bilious nor entirely bloody but is a mixture of both, let the name for this be taken from what is predominant in the mixture, and let this signify the predominant [component], so that we call it either an erysipelitic inflammation or an inflammatory erysipelas. And if neither predominates, let the affection be termed intermediate between erysipelas and inflammation.

3. As is the case in other composite [conditions], so also now the initial method of treatment must be by doing those things derived from the simple conditions. Therefore, evacuation is, in this way, the common aim for all the swellings contrary to nature. However, I speak thus bearing in mind the great number of humors which produce them. For if someone were to evacuate these, the part would regain the state of accord with nature. But also the evacuation is, in a similar manner, twofold for all instances: the first component is when we drive the humors back to other parts and the second is when we dissipate them externally in imperceptible transpiration. Since erysipelas causes distress not only by quantity but also by quality, and is severe in respect to the burning heat (*phlogosis*), it will require cooling to a greater extent than a phlegmon.[3] Nor

950K

3 The point here is taken to be the distinction between the severity of the heat in a generalized inflammation of the skin (as seen in what is currently termed erysipelas) and that in a localized purulent infection indicated by the somewhat archaic term "phlegmon."

δυνός γε ἡ τοιαύτη θεραπεία τῷ παντὶ σώματι, διὰ τὸ
φέρεσθαι τὴν χολὴν ἐνίοτε πρός τι τῶν ἐπικαίρων
μορίων, ὅπου γε οὐδ' ὅταν αἷμα ψύχηται πλεονάζον,
ἀκίνδυνον ἐκ τῶν ἀκύρων μερῶν ἀπωθεῖσθαι τὸ ῥεῦμα.

καθάπερ οὖν ἐπ' ἐκείνου μετὰ τῆς τοῦ παντὸς σώματος κενώσεως τοῖς ἀποκρουστικοῖς ὀνομαζομένοις
ἐχρώμεθα βοηθήμασιν, οὕτω καὶ νῦν πράξομεν· ἀντὶ
μὲν φλεβοτομίας χολαγωγῷ φαρμάκῳ καθαίροντες,
αὐτὸ δὲ τὸ πεπονθὸς μέρος ἐμψύχοντες. ὅρος δ' ἔστω
951K τοῦ ψύχειν ἡ | τῆς χρόας μεταβολή. καὶ γὰρ τό γε
ἀκριβὲς ἐρυσίπελας εὐθὺς ἅμα ταύτῃ παύεται, τὸ δ'
οὐκ ἀκριβές, ἀλλ' ἤδη πως φλεγμονῶδες πελιδνὸν
ἀποφαίνει τὸ δέρμα, ψυχόντων ἐπὶ πλέον. εἰ δὲ μηδ'
οὕτως τις παύοιτο, μελαίνεται, καὶ μάλιστα ἐπὶ τῶν
πρεσβυτικῶν σωμάτων· ὥστ' ἔνια τῶν οὕτω ψυχθέν-
των οὐδὲ τοῖς διαφορητικοῖς φαρμάκοις ἐκθεραπεύε-
ται τελέως, ἀλλ' ὑπολείπει τινὰ περὶ τὸ μόριον ὄγκον
σκιρρώδη. μεταβαίνειν οὖν ἄμεινον ἀπὸ τῶν ψυχόντων τε καὶ στυφόντων ἐπὶ τἀναντία καθ' ὃν ἂν καιρὸν
ἴδῃς ἠλλοιωμένον τὸ χρῶμα τοῦ πάσχοντος μορίου,
πρὶν ἤτοι πελιδνὸν ἢ καὶ παντάπασι μέλαν γενέσθαι.

λέλεκται δ' ἐν ταῖς περὶ φαρμάκων πραγματείαις ἡ
τῶν ψυχόντων ὕλη, τὸ στρύχνον καὶ τὸ ἀείζωον, ἥ τ'
ἀνδράχνη καὶ ἡ κοτυληδὼν καὶ τὸ ψύλλιον, ὅ θ'
ὑοσκύαμος καὶ ἡ θριδακίνη καὶ ἡ σέρις, ὅ τε ἀπὸ τῶν

---

[4] The term "cholagogic" in modern usage has a more re-

is such a treatment free of danger to the whole body because sometimes the bile is carried to one of the important parts. But at least whenever the excess blood becomes cooled, there is no danger in driving back the flux from the unimportant parts.

Therefore, just as in that case, along with the evacuation of the whole body, we make use of the so-called repulsive remedies, so too will we do this now. But instead of phlebotomy and purging with a cholagogic medication,[4] we cool the affected part itself. Let the end point of the cooling be the change of color. For, in fact, pure erysipelas 951K also immediately ceases along with the change of color whereas, if it is not pure (erysipelas) but is already to some extent inflammatory, the skin appears livid when you cool to excess. Unless someone stops this in some way it becomes black, particularly in the case of elderly bodies, so that some of those cooled like this are not completely cured with the discutient medications but some induration remains around the swollen part. Therefore, it is better to change from the cooling and astringent [medications] to their opposites at the time when you see the color of the affected part change and before it becomes either livid or altogether black.

I have spoken about the material of things that are cooling in the treatises on medications[5]—sleepy nightshade, houseleek, purslane, navelwort, fleawort, henbane, wild lettuce, endive, the lentils from marshes, the salves [made]

stricted sense, indicating promotion of the flow of bile from the gallbladder into the intestine only.

[5] See particularly *De simplicium medicamentorum temperamentis et facultatibus*, XI.740K ff.

τελμάτων φακός, αἵ τε δι' ὕδατος πάνυ ψυχροῦ κηρω-
ταὶ καὶ τἆλλα ὅσα τοιαῦτα. μεταπεσούσης δὲ τῆς
φλογώσεως τοῦ πεπονθότος μορίου, καταπλάττειν
αὐτό, πρὶν πελιδνὸν γενέσθαι δι' ὠμῆς λύσεως, ὀνο-
μάζω δ' οὕτω τὸ κρίθινον ἄλευρον. εἰ δὲ καὶ πελιδνὸν |
952K γενέσθαι φθάσειεν, ἀποσχάζοντα καταπλάττειν αἰο-
νᾶν τε τὰ μὲν πλεῖστα δι' ὕδατος θερμοῦ· καὶ θάλασ-
σα δὲ καὶ ἅλμη ποτὲ συνοίσει. καὶ αὐτῷ δὲ τῷ καταπλάσματι μίγνυται τοιοῦτον ὕδωρ ἢ ὄξος ἢ ὀξάλμη.
ἐν τούτῳ δὲ τῷ καιρῷ καὶ κοριανοῦ μετ' ἀλφίτων ἔνιοι
πειραθέντες ἔγραψαν ὡς ἀγαθὸν εἴη φάρμακον ἐρυσιπελάτων.

εἶτ' αὖθις ἕτεροι κατ' ἀρχὰς χρησάμενοι μεγάλης
βλάβης τῷ κάμνοντι κατέστησαν αἴτιοι. καὶ ἡ διὰ τοῦ
ῥοδίνου δὲ κηρωτὴ λαμβάνουσα τῆς τιτάνου κατὰ τὸν
αὐτὸν τρόπον ἐνίοις γέγραπται βοηθεῖν ἐρυσιπέλασι
καὶ ἄλλα τινὰ τοιαῦτα τῶν θερμαινόντων ἱκανῶς, ὧν
οὐδέν ἐστιν ἐρυσιπέλατος ἴαμα πρὶν μεταπεσὸν αὐτὸ
παύσασθαι μὲν ὅπερ ἦν ἐξ ἀρχῆς, ἕτερον δ' ἐναντίον
ἐκείνῳ γενέσθαι. πῶς γὰρ οὐκ ἐναντίον ἔσται τῷ[1]
θερμῷ πάθει τὸ ψυχρόν, ἢ τῷ ξανθῷ τὴν χρόαν ἢ
ὠχρῷ τὸ πελιδνὸν ἢ μέλαν; ὥσπερ δὲ πολλάκις ἐπι-
μίγνυται φλεγμονῇ τὸ ἐρυσίπελας, οὕτως ἐνίοτε τῷ
οἰδήματι· καὶ καλείσθω τηνικαῦτα τὸ μικτὸν ἐξ ἀμφοῖν ἐρυσίπελας οἰδηματῶδες, ὥσπερ γε κἀπειδὰν
953K ψυχόμενον | σκληρόν τε καὶ δύσλυτον γένηται, κλη-
θήσεται σκιρρῶδες ἐρυσίπελας. καὶ τοίνυν καὶ ἡ θερα-

---

[1] B; τό K

from very cold water and other such things. When the burning heat of the affected part undergoes a change, apply a poultice of uncooked wheat (this is what I call barley meal) to the part before it becomes livid. If it has already become livid, apply a poultice moistened copiously with 952K warm water to those areas that have been scarified; both seawater and brine are sometimes useful. And mix some water of this kind with the poultice itself, or vinegar, or a vinegar/brine mixture.[6] Some men, speaking from experience, wrote that coriander with barley groats is a good medication for erysipelas at this time.

Then again others, when they used this from the beginning, were responsible for great harm to the patient. Also, the salve made from oil of roses, which partakes of white earth, has in the same way been described by some to be a remedy for erysipelas, [as have] others among those agents that are sufficiently heating. None of these is a cure for erysipelas before it has undergone a change, ceasing to be what it was from the start and becoming something else opposite to that. For how will cold not be the opposite of a hot affection, or what is livid or black [the opposite] of yellow and pale in terms of color? Just as erysipelas is often mixed with inflammation, so too is it sometimes mixed with edema. Under such circumstances, the mixture of both should be called "edematous erysipelas," just as whenever, by being cooled, it becomes hard and difficult to resolve, it 953K will be called "scirrhous erysipelas." Therefore, your treat-

---

6 The Greek term "oxalme" is used by both Linacre and Peter English; see also *De alimentorum facultatibus*, VI.616K. The same term is given to a natural product from Sicily used medicinally; see Aristotle, *Meteorologica*, 359b15–16.

πεία καθάπερ ἐν ἅπασι τοῖς συνθέτοις, οὕτω κἀπὶ
τούτων σοι γιγνέσθω· καὶ μάλιστα μὲν ἀνθισταμένῳ
πρὸς τὸ κρατοῦν, οὐκ ἐπιλελησμένῳ δ᾽ οὐδὲ τῆς ἀπὸ
τοῦ μιχθέντος ἐνδείξεως.

4. Ὥσπερ γὰρ ἐπὶ χολώδει ῥεύματι τὸ ἐρυσίπελας,
οὕτως ἐπὶ φλέγματι τὸ οἴδημα γίγνεται χαῦνός τις
ὄγκος ἀνώδυνος. ἴσμεν δὲ δήπου καὶ ἄλλως οἰδήματα
γινόμενα περὶ τοὺς πόδας ἐν ὑδερικαῖς διαθέσεσι καὶ
φθόαις καὶ καχεξίαις ἑτέραις ἰσχυραῖς. ἐπ᾽ ἐκείνων
μὲν οὖν σύμπτωμά ἐστι τὸ οἴδημα τοῦ κατέχοντος
πλήθους τὸν ἄνθρωπον, οὐδὲ μιᾶς ἰδίας ἐξαιρέτου
θεραπείας δεόμενον. ἀρκεῖ γάρ, εἴπερ ἄρα, τηνικαῦτα
ἀνατρίβειν τὰ σκέλη ποτὲ μὲν δι᾽ ὀξυροδίνου, ποτὲ δὲ
δι᾽ ἐλαίου καὶ ἁλῶν, ἢ καὶ αὐτῷ τῷ ὀξυροδίνῳ τῶν
ἁλῶν ἐπεμβάλλοντα. διὰ δὲ τὸν φλεγματώδη χυμὸν
ἐπιρρυέντα μορίῳ συνιστάντος οἰδήματος, ἱκανὸς ἐνίοτε καὶ σπόγγος μόνος ὕδατι βεβρεγμένος ὀλίγον
ὄξους ἔχοντι. γενέσθω δ᾽ ἡ κρᾶσις ἤτοι γ᾽ ὡς ἂν καὶ
954K   πίοι τις ἢ οὐ πολλῷ γε τούτου | πλέον ἔχουσα τὸ ὄξος.
ἐπιδεῖν δὲ χρὴ τὸν σπόγγον ἐκ τῶν κάτω μερῶν
ἀρχόμενον καὶ ἄνω τελευτῶντα. καινὸς δ᾽ ἔστω πάντως, εἰ μέλλοι τι χρηστὸν ἐργάσασθαι· μὴ παρόντος
δέ σοι καινοῦ, διαρρύπτειν χρὴ καὶ καθαίρειν τὸν
ἐπιτυχόντα νίτρῳ τε καὶ ἀφρονίτρῳ καὶ τῇ καλουμένῃ
κονίᾳ στακτῇ. μὴ καταστάντος δὲ ἐπὶ τοῖσδε τοῦ

ment too must be like this in these cases, as in all those things that are composite—that is, particularly directed against what prevails without overlooking the indication from what is mixed.

4. Just as erysipelas arises due to a bilious flux, so too does edema arise due to phlegmatic flux and is a spongy, painless swelling. However, we also recognize, of course, edemas that arise in other ways, involving the feet in the dropsical and wasting conditions, and other severe *kachexias*. In those, edema is a symptom of the *plethora* afflicting the person and does not require one specific and special treatment. If it does indeed need treatment, it is sufficient under these circumstances to rub the legs with oxyrrhodinum on some occasions and with oil and salts on others, or also with oxyrrhodinum after mixing in salts. However, when edema arises due to a phlegmatic humor flowing into the part, sometimes a sponge moistened with water having a little vinegar [in it] is sufficient alone. Make the mixture such that either someone might drink it, or at least that it has not much more vinegar [than this]. It is, however, necessary to bind the sponge on, beginning from the parts below and ending at the parts above. And [the sponge] must be absolutely fresh if it is going to be beneficial. If a fresh [sponge] is not available to you, it is necessary to wash thoroughly and purify whatever sponge there happens to be with niter, saltpeter and so-called "purified lye."[7] If [the edema] does not subside with these things, when you are

954K

---

[7] It is not clear what these substances are. LSJ has "native sodium carbonate." Linacre has *colatum lixivium* (purified lye). L&S have "efflorescence of salt-petre" (potassium nitrate). Peter English has "nitre," "salt-petre," and "boiled water."

οἰδήματος, ἐπειδὰν αὖθις ἐπιδέῃς, ἐπεμβάλλειν καὶ
βραχὺ τῆς στυπτηρίας καὶ σπόγγον καινὸν προσοι-
στέον· εἰ δ' οὐκ ἔχοις καινόν, ἄμεινον χρῆσθαι τούτῳ
δὴ τῷ καλουμένῳ πρὸς τῶν πολλῶν ἐλλυχνίῳ. μαλακὸν δ' ἔστω πρὸ πάντων, ὁποῖον τὸ Ταρσικόν· εἰ δ'
ἐκείνου τις εὐποροίη, θαρρῶν χρήσθω· βέλτιον γὰρ
ἐνεργήσει τοῦ σπόγγου. δενέσθω δὲ δηλονότι τῷ τὴν
στυπτηρίαν ἔχοντι ὀξυκράτῳ καὶ ἐπιδείσθω, καθότι
προείρηται, κάτωθεν ἄνω. σφίγγειν δ' οὕτω συμμέτρως, ὡς ἐν κατάγματι, καὶ μᾶλλον μὲν τὰ κάτω
πρῶτα· κατὰ βραχὺ δ' ἐκλύειν τὴν σφίγξιν· οὐ μὴν εἰς
τοσοῦτόν γε, ὡς χαλαρὸν γενέσθαι τι μέρος τῆς
ἐπιδέσεως. ἀγαθὸν δὲ φάρμακον εἰς ταῦτα καὶ τὸ
γλαύκιον αὐτό τε καθ' ἑαυτὸ δι' ὀξυκράτου λυθέν, ἔτι
955K τε μᾶλλον τὸ δι' | αὐτοῦ συντιθέμενον ἡμέτερον φάρ-
μακον, οὗ τὴν σύνθεσιν ἔχεις ἐν τῇ περὶ τῶν φαρμάκων πραγματείᾳ λεγομένῃ.

ἦσαν δέ μοι μέχρι δεῦρο τρεῖς, ἥ τε περὶ τῶν
ἁπλῶν, καὶ μετὰ ταύτην ἡ περὶ τῆς συνθέσεως αὐτῶν,
εἶθ' ἡ περὶ τῶν εὐπορίστων. ἔοικα δὲ καὶ τετάρτην
ἄλλην ποιήσειν, ἐπειδὴ καὶ πολλοὶ τῶν ἑταίρων οὕτως
ἀξιοῦσιν· ἐν ᾗ περὶ τῶν κοινῶν καὶ ἰδίων ἑκάστου
μορίου καὶ πάθους φαρμάκων ὁ λόγος ἐστί μοι. ἀλλὰ

---

8 The primary meaning listed by LSJ is "a lamp wick," but
ellychnium is also listed as a surgical dressing, citing this passage
and Soranus, 2.11.

9 See Dioscorides, II.105, and Galen, XI.439K.

438

binding it again, throw in a little alum, and apply a fresh sponge. If you don't have a fresh sponge, it is better to use that which is called "ellychnium"[8] by many people. Above all, this must be soft, like that from Tarsus. If some of that is available, use it boldly because it works better than a sponge. Obviously it must be steeped in the oxykratos[9] having astringency, and must be bound in the manner previously described—i.e. upward from below. Bind moderately tightly in this way, as with a fracture, and particularly bind the things below first. Gradually loosen the binding, but not to such a degree that some part of it becomes loose. Also glaucium[10] is a good medication for these things, either by itself or dissolved in oxykratos; even better is my own medication compounded from this. You have its composition in my treatise on medications.[11]

955K

Up to this point there have been three [such treatises] by me: one about the simples, and after this, one about the combination of these, and one about their ease of procurement. It seems likely I shall also do another, fourth treatise, since many of my colleagues think it right that I do so. In this, my discussion will be about the medications that are common and specific for each part and affection.[12] But

---

[10] LSJ has "the juice of the horned poppy (Glaucium corniculatum)"; see also Dioscorides, III.86, I.64, and Galen, XI.857K. "Glaucium" is Linacre's term. Peter English translates this as "Glaucian Medicine."

[11] See *De simplicium medicamentorum temperamentis et facultatibus*, XI.857K.

[12] The four treatises referred to are Galen's three surviving major pharmacological treatises (see Book 13, note 16) and presumably an additional treatise either never completed or now lost.

τοῦτό γε τὸ διὰ τοῦ γλαυκίου φάρμακον οὐκ οἰδήματα
μόνον, ἀλλὰ καὶ πολὺ δὴ μᾶλλον αὐτῶν ἐρυσιπέλατά
τε καὶ φλεγμονὰς² ἀρχομένας ἰᾶται, καὶ μάλιστα
τὰς θερμάς. εὔδηλον δὲ ὅτι καὶ τὰς ἐρυσιπελατώδεις
φλεγμονὰς ἐρυσιπέλατά τε τὰ φλεγμονώδη θεραπεύσεις ταὐτῷ τούτῳ τῷ φαρμάκῳ. οὐ μὴν τάς τε σκιρ-
ρουμένας ἤδη τῶν φλεγμονῶν ἢ τὰ ψυχθέντα τῶν
ἐρυσιπελάτων, οὔθ' ὅλως οὐδεμίαν διάθεσιν σκιρ-
ρώδη· περὶ ὧν εἴρηταί μέν τι κἀν τῷ πέμπτῳ Περὶ τῆς
τῶν ἁπλῶν φαρμάκων δυνάμεως, εἰρήσεται δὲ καὶ νῦν.
ὁ μὲν γὰρ χυμὸς ὁ τὸ τοιοῦτον πάθος ἐργαζόμενος |

956K ἤτοι γλίσχρος ἐστὶν ἢ παχύς, ἢ ἀμφότερα. τῆς θεραπείας δὲ αὐτοῦ κοινὸς ὁ σκοπὸς ἐκκενῶσαι τὸ περιεχόμενον ἐν τῷ μορίῳ παρὰ φύσιν ἅπαν. ἴδιος δ' ὁ
τρόπος τῆς κενώσεως· ἀπορρῦψαι γὰρ αὐτὸν χρὴ
δυσλύτως ἐμπεπλασμένον.

ἐὰν οὖν ἀθρόως ἕλκουσί τε καὶ διαφοροῦσι φαρμάκοις ἐγχειρῇ τις κενοῦν, ἄνευ τοῦ μαλάττειν τε καὶ
χεῖν τοῖς ὑγραίνουσι καὶ θερμαίνουσιν, ἐν ὀλίγαις μὲν
ἡμέραις ταῖς πρώταις ἀξιόλογον ἐπίδοσιν δόξει λαμβάνειν ἡ θεραπεία, τό γε μὴν ὑπόλοιπον τῆς διαθέσεως ἀνίατον ἔσται· διαφορηθέντος γὰρ ἅπαντος τοῦ
λεπτομεροῦς κατ' αὐτήν, ὅμοιόν τι πήξει λιθώδει τὸ
λοιπὸν λήψεται. καὶ γὰρ οὖν καὶ τῶν ἐν ταῖς ἀρθρίτισι
πώρων ἡ γένεσις ἐξ ὑγροῦ παχέος ἐστὶ καὶ γλίσχρου
μὴ κατὰ βραχὺ διαφορηθέντος, ἀλλ' ἀθρόως ξηραν-
θέντος ὑπὸ βιαίων φαρμάκων. οὕτω δὲ κἂν τοῖς νεφροῖς οἱ λίθοι γεννῶνται, κατοπτηθέντος ἐν αὐτοῖς

this medication [made] from glaucium cures not only edemas but also, much more than them, erysipelas and incipient inflammations, especially the hot ones. However, it is also clear that you will treat erysipelitic inflammation and inflammatory erysipelas with this same medication, but not inflammations that are already scirrhous, nor erysipelas that is cold, nor, in general, any scirrhous condition. Something was said about these in the fifth [book] of *On the Mixtures and Potencies of Simple Medications*,[13] and will also be said now. The humor which brings about such an affection is either viscous, or thick, or both. The indicator of the treatment of this, which is general, is to evacuate everything contrary to nature contained in the part. The manner of evacuation is, however, specific, for it is necessary to cleanse thoroughly what has indissolubly filled up.

956K

Therefore, if someone should attempt to evacuate suddenly and with medications that draw and disperse without softening and melting with medications that moisten and heat, the treatment will seem to make notable progress in the first few days, but what remains of the condition will be incurable because, when everything in it that is fine has been dispersed, what is left will take on something like a stony solidity. Thus, the genesis of gout is from a thick and viscous humor when this is not dispersed gradually but is dried out all at once by strong medications. In the same way, too, stones are generated in the kidneys when a thick

[13] See *De simplicium medicamentorum temperamentis et facultatibus*, XI.704–88 K.

---

[2] τε καὶ φλεγμονάς B; τε τὰς φλεγμονάς K

χυμοῦ παχέος καὶ γλίσχρου. διὰ τοῦτ᾽ οὖν ἐπὶ τῶν σκιρρωδῶν διαθέσεων οὐδὲν τῶν ἰσχυρῶς θερμαινόντων ἢ ξηραινόντων φαρμάκων ἁρμόττει, μόνα δὲ ὅσα

957K μετὰ τοῦ μαλάττειν ἱκανὰ | διαφορεῖν ἐστιν, οἷον ὁ ἐλάφειος μυελὸς καὶ μόσχειος καὶ στέαρ αἴγειόν τε καὶ ταύρειον καὶ λεόντειον, ἔτι τε πρὸς τούτοις ἀμμωνιακὸν θυμίαμα καὶ βδέλλιον ἑκάτερον, καὶ μᾶλλον τὸ Σκυθικὸν ὅσῳ περ ὑγρότερόν ἐστιν. ὡσαύτως δὲ καὶ ὁ στύραξ ὁ ὑγρότερος ἀμείνων τοῦ ξηροῦ. τούτοις οὖν προσέχων τοῖς σκοποῖς καὶ τὰς ἐπιπεπλεγμένας διαθέσεις ἰᾶσθαι δυνήσῃ, κατὰ τὴν εἰρημένην ἤδη πολλάκις ἐπὶ πάντων τῶν συνθέτων παθῶν μέθοδον.

5. Ἐμοὶ δὲ καιρὸς ἂν εἴη περὶ τῆς τῶν πασχόντων μορίων ἀναμνῆσαι διαφορᾶς, ἧς ἀεὶ μνημονεύειν ἠξίουν ἐπὶ πάντων τῶν νοσημάτων τε καὶ συμπτωμάτων. εἴρηται δ᾽ οὐκ ὀλίγα καὶ κατὰ τὸ πρὸ τούτου γράμμα περὶ τῆς ἀπ᾽ αὐτῶν ἐνδείξεως. ὁ μὲν γὰρ πρῶτος σκοπὸς τῆς ἰάσεως ἁπάντων τῶν παρὰ φύσιν ὄγκων ἐν οἷς οὐδέπω γεγόνασι πῶροι κένωσίς ἐστιν. ἡ δὲ τῶν σκιρρουμένων μορίων κένωσις ὑπὸ τῶν προειρημένων γίγνεται φαρμάκων, ἃ καλεῖν ἔθος ἐστὶ τοῖς ἰατροῖς μαλακτικά. τῷ δ᾽ εἶναι τῶν μορίων τὰ μὲν

958K ἀραιότερα φύσει, | τὰ δὲ πυκνότερα, καὶ τὴν κένωσιν ἑτέρων δεῖσθαι κατ᾽ εἶδος βοηθημάτων ἀναγκαῖόν ἐστι. διὰ τοῦτ᾽ οὖν ἐπὶ τενόντων τε καὶ συνδέσμων ἔδοξέ μοι βέλτιον εἶναι παραμιγνύναι τι τῇ διὰ τῶν

---

14 Bdellium is the gum from trees of the genus *Commiphora*.

and viscous humor is "baked" in them. Because of this, then, in the case of the scirrhous conditions, none of the medications that are strongly heating or drying is suitable, but only those that are sufficient to disperse as well as soften; for example, the marrow from deer or calves, or fat   957K
from goats, bulls or lions and, besides these, ammoniacal incense and both forms of bdellium, but more so the Scythian by virtue of its being more moist, and similarly storax, the more moist [form being] better than the dry.[14] Therefore, paying attention to these indicators, you will be able to cure the offending conditions according to the method mentioned often already in the case of all composite affections.

5. It is now time for me to make mention of the variations in the affected parts which, in all the diseases and symptoms, I always consider worthwhile calling to mind. A lot was said in the book prior to this one about the indication from these.[15] The primary indicator of the cure of all the swellings contrary to nature in which stones have not yet formed is evacuation. The evacuation of the scirrhous parts is by the previously mentioned medications which it is customary for doctors to call softening agents (emollients). Because some of the parts are more loose-textured by nature and some more dense, the evacuation necessar-   958K
ily requires different kinds of remedies. For this reason, in the case of tendons and ligaments [which have become scirrhous],[16] it seemed to me better, in the treatment by

Storax is the resin from any of the trees or shrubs from the genus *Styrax*. See Dioscorides I.80 and I.79 respectively. Scythia was a region north and east of the Black Sea.   [15] See Book 13, chapter 8ff.   [16] Added following Linacre, p. 706.

μαλαττόντων φαρμάκων ἀγωγῇ τῆς τμητικῆς ὀνομαζομένης, ὧν τοῖς μάλιστα ὄξος ἐστὶ τοιοῦτον. ἐνίοτε
μὲν οὖν αὐτῷ χρώμεθα κἀπὶ τῶν ἄλλων μορίων σκιρ-
ρωθέντων, ὡς ἐρῶ μικρὸν ὕστερον.

ἐπὶ δὲ τενόντων τε καὶ συνδέσμων ὧδέ πως. ὄξει
δριμυτάτῳ σβέννυμι λίθον διάπυρον, εἰ μὲν οὖν οἷόν
τε, τὸν πυρίτην λίθον καλούμενον, ὃς οὐδ᾽ αὐτός ἐστι
σπάνιος ἐν ταῖς μεγάλαις πόλεσι, μὴ παρόντος δὲ
τούτου, τὸν μυλίτην· ὀνομάζουσι δὲ οὕτως ἐξ οὗ τὰς
μύλας ἐφ᾽ ὧν ἀλοῦσι κατασκευάζουσιν. εἶτα ἀναφερο-
μένου τινὸς ἀτμοῦ θερμοῦ μετὰ τὸ καταχυθῆναι τοῦ
λίθου τὸ ὄξος, ἐν ἐκείνῳ τὸν ἐσκιρρωμένον σύνδεσμον
ἢ τένοντα διακινεῖν ἀναγκάζω· καὶ μετὰ τοῦτο πάλιν
ἐπιτίθημι τὸ μαλακτικὸν φάρμακον. ἐλαίῳ γε μὴν ἀπ᾽
ἀρχῆς τῆς θεραπείας οὐχ ὕδατι καταντλῶ τὸ πεπονθὸς μόριον, ἄχρι παντὸς ἑκάστης ἡμέρας. ἐχέτω δὲ
τοῦτο μηδεμίαν στύψιν, ἀλλ᾽ ἀκριβῶς ἔστω λεπτομε-
959K   ρές, οἷόν | πέρ ἐστι τὸ Σαβῖνον. ἐναφεψῶ δ᾽ ἐνίοτε τῷ
ἐλαίῳ καὶ τῆς ἀλθαίας τὴν ῥίζαν, ἀγρίου τε σικύου
καὶ εἴ τις ἄλλη τοιαύτη. καὶ τούτῳ χρῶμαι καθ᾽
ἑκάστην ἡμέραν, ὡς εἶπον. ἡ δὲ δι᾽ ὄξους θεραπεία
χρήσιμός ἐστιν ἐπὶ προσήκοντι τῷ πάθει, προπαρε-
σκευασμένου τοῦ μέρους ὑπὸ τῶν μαλακτικῶν.

ἐπενόησα δέ τινα καὶ σύνθετα φάρμακα δι᾽ ὄξους,
ἃ μεταξὺ τῶν μαλακτικῶν ἐπιτίθημι πρὸς μίαν ἡμέραν. ἡ γάρ τοι τοῦ ὄξους δύναμις, ἐάν τις αὐτῇ
μετρίως τε καὶ κατὰ τὸν προσήκοντα καιρὸν χρήσηται, ὠφελεῖ τὰς τοιαύτας διαθέσεις τέμνουσα καὶ

the emollient medications, to mix in one of the so-called cutting [medications], of which vinegar particularly is an example. Therefore, we sometimes use this, even in the case of other scirrhous parts, as I shall say a little later.

In the case of tendons and ligaments [the treatment] is somewhat like this: I quench a red hot stone with the sharpest vinegar, or if this is not possible, a so-called "firestone" which is not scarce in the big cities, or if this is not available, a millstone, for so people call what they prepare for grinding grain. Then, while a certain hot vapor is being carried up after the vinegar has been poured on the stone, I force the ligament or tendon that has become scirrhous to move in that. After this, I again apply the emollient medication. In fact, from the beginning of the treatment, I bathe the affected part with oil and not water continually each day. This oil must have no astringency but must be strictly fine-particled like the Sabine oil.[17] Some- 959K
times I also boil down the root of marshmallow, wild cucumber, or something else of this kind, in the oil. And as I said, I use this every day. The treatment with vinegar is useful for the appropriate affection, if the part has been previously prepared by the emollient [medications].

I also devised some composite medications with vinegar which I apply for one day between the emollients. Certainly, the potency of the vinegar, if someone will use this potency moderately and at the appropriate time, helps such conditions by cutting and dissolving the thick hu-

[17] This is oil made from overripe olives, i.e., the opposite to omotribes; see *De simplicium medicamentorum temperamentis et facultatibus*, XI.872K.

διαλύουσα τοὺς παχεῖς χυμούς· εἰ δὲ ἀμετρότερον ἢ
οὐκ ἐν καιρῷ τῷ προσήκοντι, τὸ λεπτότερον ἐξαρ-
πάζουσα τὸ καταλεῖπον ἐᾷ λιθοῦσθαι. καὶ μέντοι καὶ
μέχρι πλείονος, εἴ τις αὐτῷ χρῷτο, τῆς οὐσίας ἅπτεται
τῶν νεύρων. διὰ τοῦτ᾽ οὖν οὔτε πολλάκις οὔτε κατ᾽
ἀρχὰς οὔτ᾽ ἐν χρόνῳ πολλῷ χρηστέον ἐστὶ τοῖς δι᾽
ὄξους φαρμάκοις ἐπὶ συνδέσμων τε καὶ τενόντων. ἐπὶ
μέντοι σπληνὸς ἢ τῶν σαρκωδῶν μερῶν τοῦ μυὸς
ἐσκιρρωμένων ἀκίνδυνος ἡ χρῆσις· ἀραιά τε γὰρ
ταῦτα φύσει καὶ φόβος οὐδείς ἐστι πληγῆναί τι νεῦ-
960K ρον ὑπὸ τῆς | δυνάμεως αὐτοῦ. τῷ μὲν οὖν ἀμμωνιακῷ
θυμιάματι μετ᾽ ὄξους πολλοὶ καὶ ἄλλοι χρῶνται κατὰ
τοῦ σπληνὸς ἐπιτιθέντες πηλῶδες τῷ πάχει, τὸ μικτὸν
ἐξ ἀμφοῖν ἐργαζόμενοι· καὶ πολλάκις γε τοῦτο μόνον
ἤρκεσεν εἰς τὴν θεραπείαν αὐτοῦ.

ἐπὶ δὲ τῶν μυῶν ἄλλον μὲν οὐκ εἶδον, ἐγὼ δ᾽
ἐχρησάμην πολλάκις ἐν τῷ μεταξὺ τῶν μαλακτικῶν.
ἐπ᾽ ἐκείνοις μὲν οὖν οὐδεμία σαφὴς ὠφέλεια γίγνεται·
προμαλαχθέντος δ᾽ ὑπ᾽ αὐτῶν τοῦ σκιρρώδους ὄγκου
μεγίστην ὠφέλειαν ἐργάζεται τὸ δι᾽ ὄξους λυθὲν ἀμ-
μωνιακόν. ἀπόχρη δὲ καὶ τούτῳ κατὰ μίαν ἢ καὶ
δευτέραν ἡμέραν χρησαμένους ἐπανελθεῖν αὖθις ἐπὶ
τὰ μαλακτικά· πάλιν δ᾽ ἐκείνοις χρησαμένους ἡμέραις
πλείοσιν ἀφικνεῖσθαι πάλιν ἐπὶ τὸ δι᾽ ὄξους φάρμακον, εἴτ᾽ οὖν ἀμμωνιακὸν εἴτε καὶ τῶν ἄλλων τι τῶν
μαλακτικῶν ὅσα μικρὸν ἔμπροσθεν εἶπον. οὐ γὰρ δὴ
ἄλλο γέ τι δεῖ προσφέρειν, οἷα τὰ πολλὰ τῶν ξηρῶν
ὀνομαζομένων ἐστὶ φαρμάκων· εὐδοκιμήσει μὲν γὰρ

mors. If, however, [the use] is too immoderate or not at the appropriate time, it forcibly draws out what is thinner and allows what is left to become stony. Furthermore, if someone uses it for too long, it attacks the substance of the nerves. Because of this, then, you must not use the medications made with vinegar frequently, or at the beginning, or for a long time in the case of ligaments and tendons. However, in the case of the spleen or the fleshy parts of muscles when they have become scirrhous, the use [of vinegar] is without danger because these structures are loose-textured in nature and there is no fear of any nerve being damaged by its potency. Many others use the ammonia- 960K cal fumigations along with vinegar, applying these to the spleen when it is claylike in thickness, making a mixture from both ingredients. And often this alone is sufficient for its treatment.

In the case of the muscles, I have not seen anyone else [use this], but I do often use it in between the emollients. With those, nothing clearly beneficial occurs, but when the scirrhous swelling has been softened by them beforehand, the ammoniacal mixture dissolved with vinegar produces a very considerable benefit. It is enough to use this on the first or even the second day and then go back to the emollients. Then, when I have used those again for many days, I come back once more to the medication made with vinegar, either the ammoniacal or one of the other emollients that I spoke about a little earlier. Nor is there need to apply anything else, such as the many so-called drying medications, for the medication is highly regarded in the begin-

ἐν ἀρχῇ τὸ φάρμακον, ἀνίατον δ᾽ ἐργάζεται τὸ λείψανον τῆς διαθέσεως. διὰ τοῦτο γοῦν ἐγὼ πολλάκις ἐν
961K τῷ μεταξὺ καὶ καταπλάσματι τῷ δι᾽ ἀλθαίας | ἐχρη-
σάμην· ὀνομάζουσι δὲ αὐτὴν ἀναδενδρομαλάχην οἱ
πολλοί· ταύτης οὖν ἡ ῥίζα λυομένη μετὰ στέατος
ἀγαθὸν εἰς τὰ τοιαῦτα φάρμακα. ἔστω δὲ τὸ στέαρ, εἰ
μὲν οἷόν τ᾽ εἴη, χήνειον· εἰ δὲ μὴ παρείη τοῦτο,
ὀρνίθειον· εἰ δὲ μηδὲ τοῦτο, τῷ τῶν ὑῶν χρηστέον.
ἀλλὰ καὶ τῆς ἀγρίας μαλάχης, ἥτις πανταχόθι φύ-
εται, τὰ φύλλα λειωθέντα μετά τινος τῶν εἰρημένων
ὀνήσει. ἀμείνω δὲ τῶν ὠμῶν ἐστι τὰ προαφηψημένα
μετρίως.

τὰ μὲν δὴ τοιαῦτα βοηθήματα καὶ τὸ ποικίλλειν,
ὡς εἴρηται, τὴν θεραπείαν ἁπάντων μορίων ἐστὶ κοινά, ὅσα περ ἂν ἁλῷ πάθει σκιρρώδει, καθάπερ γε καὶ
ἡ τῶν οἰδημάτων ἴασις, ἣν ὀλίγον ἔμπροσθεν εἶπον,
ἐξαλλάττεται κατὰ τὰ μόρια ταῖς εἰρημέναις διαφο-
ραῖς. αὐτίκα γέ τοι τοῖς καθ᾽ ὑποχόνδριον οἰδήμασιν
οὐκ ἄν τις ψυχρὸν ἐπιθείη σπόγγον ἐξ ὀξυκράτου·
καθάπερ οὐδὲ τοῖς ἄλλοις ὄγκοις τοῖς κατ᾽ αὐτό. τίς
δ᾽ ἀψίνθιον ἐναποζέσας ἐλαίῳ κατήντλησέ ποτε τὸ
γόνυ; τίς δὲ ὀφθαλμὸν ἤ τι τῶν ἔνδον τοῦ στόματος
ὁπωσοῦν πάσχον; ἀλλ᾽ ἥπατί τε κακοπραγοῦντι καὶ
σπληνὶ προσφέρεται πολλάκις ὠφελιμώτατα.

962K μαρτυρεῖ δ᾽ αὐτοῖς καὶ | ἡ πεῖρα δεικνῦσα τὴν
δύναμιν ἐναργῶς οὐδὲν ἧττον τοῦ λόγου. νυνὶ δ᾽ ἡμεῖς
μόνην ἐν τῇδε τῇ πραγματείᾳ τὴν κατὰ μέθοδον εὕρεσιν τῶν ἰαμάτων ὁποία τίς ἐστι διερχόμεθα. κατὰ μέν-

ning but makes what is left of the condition incurable. Anyway, because of this, I have often used in between times a poultice made with marshmallow, although many people call this hollyhock. A good thing for such medications is the root of this dissolved with fat. If possible, it must be goose fat, or if this is not available, chicken fat, or if this is not, you must use the fat of pigs. But the leaves of the wild mallow, which grows everywhere, when triturated with one of those things spoken of, will also be beneficial. It is better if [the leaves] are moderately decocted beforehand rather than being uncooked.

961K

Now such remedies and the variability in treatment are, as I said, common for all parts that are troubled by a scirrhous affection. In the same way too, the treatment of the edemas, which I mentioned a little earlier, changes in relation to the parts through the differences mentioned. For example, surely nobody would apply a sponge cooled by oxykraton for edema involving the hypochondrium, just as they would not for other swellings involving this. And who would ever pour wormwood decocted in oil on the knee? Who would do so in respect to the eye or any of those things within the mouth that are affected in any way? But in a liver that is failing or a spleen, it is often applied with the greatest benefit.

962K

It is also evident from these things that experience shows the potency clearly, no less than theory. For the present, I am only going through the discovery of cures by method in this work; [that is], what kind of discovery [is in-

GALEN

τοι τὰς περὶ φαρμάκων ἀμφοτέρας ἐμίξαμεν, ἅπαντα
γινώσκειν ἀξιοῦντες τὸν ἰατρόν, ὅσα τ᾽ ἐκ πείρας
εὕρηται μόνης, ὅσα τ᾽ ἐκ λόγου μόνου· καὶ τρίτα γε
πρὸς αὐτοῖς ὅσα συντελούντων ἀμφοτέρων εἰς τὴν
εὕρεσιν. ἅπασι δ᾽ οὖν αὐτοῖς ὀρθῶς χρήσεται μόνος ὁ
γεγυμνασμένος ἐν τῇδε τῇ μεθόδῳ.

6. Περὶ μὲν οὖν φλεγμονῆς καὶ σκίρρου καὶ οἰδήματος ἀρκείτω τὰ εἰρημένα μεμνημένων ἡμῶν ὡς
σκίρρον ὀνομάζομεν ὄγκον σκληρὸν ἀνώδυνον, οὐ
μὴν ἀναίσθητόν γε πάντως· ὁ γὰρ τοιοῦτος ἀνίατος· οἱ
δ᾽ ἄλλοι πάντες ὥσπερ ἀναισθητότερόν τε καὶ δυσαι-
σθητότερον ἀποφαίνουσι τὸ πάσχον μόριον, οὕτως οὐ
παντελῶς ἀναίσθητον, ὅταν γ᾽ αἰσθητικὸν ᾖ φύσει·
τοὺς γὰρ συνδέσμους ἴσμεν ἀναισθήτους ὄντας. εἰ δέ
τις ἐκείνους μόνους τῶν παρὰ φύσιν ὄγκων ὀνομάζειν
963K ἀξιοῖ σκίρρους, ὅσοι παντάπασίν εἰσιν | ἀναίσθητοι,
τοὺς δὲ ἄλλους οὐ σκίρρους, ἀλλ᾽ ὄγκους σκιρρώδεις,
ὑπὲρ ὀνόματος ἴστω ζυγομαχῶν καὶ καλέσομεν οὕτω
καὶ ἡμεῖς τὸ πάθος, ὅταν ἐκείνῳ διαλεγώμεθα· καὶ γὰρ
καὶ ἔθος ἡμῖν ἐστιν οἷς ἄν τις ὀνόμασι χαίρῃ, τούτοις
αὐτῷ διαλέγεσθαι.

7. Καιρὸς οὖν ἤδη περὶ τῶν ἐμφυσημάτων διελθεῖν,
οὐ τὴν αὐτὴν ἐχόντων θεραπείαν τοῖς οἰδήμασιν. ἐκεῖνα μὲν γάρ, ὡς ἔφην, ὑπὸ φλεγματώδους γίνεται
χυμοῦ· καὶ διὰ τοῦτο θλιβόντων εἴκει μέχρι βάθους
ἱκανοῦ τῶν δακτύλων ἐγκαταβαινόντων εἰς αὐτά. τὰ
δ᾽ ἐμφυσήματα φυσώδους πνεύματος ἀθροιζομένου
γίγνεται ποτὲ μὲν ὑπὸ τῷ δέρματι, ποτὲ δὲ ὑπὸ τοῖς

450

volved]. However, I mixed both [experience and method in the works] on medications, thinking it worthwhile for the doctor to know all these—those that are discovered by experience alone and those that are discovered by theory alone. And third, in addition to these, there are those that come to discovery from both [experience and theory] jointly. It is only the person who has had thorough training in this particular method who will use all these [medications] correctly.

6. Therefore, let what I have said be enough about inflammation, both scirrhous and edematous, bearing in mind that I call scirrhous a hard, painless swelling, but not one that is altogether without sensation, for such a thing is incurable. All the others show the affected part more anesthetic or dysesthetic, but not altogether anesthetic, at least whenever it is capable of sensation by nature, for we know that ligaments are without sensation. If, however, someone thinks it right to call those alone, of the swellings contrary to nature, scirrhosities—[that is] those that are altogether anesthetic—but the others not scirrhosities but scirrhous swellings, he must realize that he is contending about terminology, and that I too shall name the affection in this way whenever I dispute with him. Furthermore, it is my custom to use the same terms that please him when debating with him.

7. It is time now to go over the inflations which do not have the same treatment as the edemas. For the latter arise due to a phegmatic humor, as I said, and because of this yield when palpating fingers are pressed into them to a sufficient depth. The inflations, on the other hand, arise when a flatulent *pneuma* collects, sometimes under the skin, sometimes under the periosteal membranes, or the mem-

963K

451

περιοστέοις ὑμέσιν, ἢ τοῖς τοὺς μῦς περιέχουσιν ἤ τι
τῶν σπλάγχνων. ἀθροίζεται δ᾽ οὐκ ὀλίγον ἐνίοτε καὶ
κατὰ τὴν γαστέρα καὶ τὰ ἔντερα κἂν τῷ μεταξὺ
τούτων τε καὶ τοῦ περιτοναίου. καὶ διαφέρει γε τῶν
οἰδημάτων τῷ μὴ βοθροῦσθαι πιεζόμενα καὶ ψοφεῖν
ὥσπερ τύμπανον· ἔτι τε καὶ τῷ περιέχεσθαι πολλάκις
ἐν αἰσθητῇ κοιλότητι καὶ ταύτῃ γ᾽ ἔστιν ὅτε μεγίστῃ.

964K   σκοπὸς | δ᾽ ἔστω σοὶ κἀπὶ τούτων ὁ μὲν ἀπάντων
αὐτῶν κοινός, ἐκκενῶσαι τὸ παρὰ φύσιν, ἐν ὅτῳ περ
ἂν εἴη περιεχόμενον· ὁ δ᾽ ἴδιος ἐπὶ τῷ κοινῷ, τὸ διὰ
τῶν λεπτομερεστάτων τε καὶ θερμοτέρων ταῖς δυνάμεσι φαρμάκων διαφορῆσαι. ἐπὶ μὲν οὖν τῶν κατὰ τὴν
γαστέρα τε καὶ σπλάγχνα λεπτομερὲς ἔλαιον ἐργάσεται τοῦτο, πήγανον ἐναπεζεσμένον ἔχον ἤ τι τῶν
θερμαινόντων σπερμάτων, οἷόν πέρ ἐστι τό τε τοῦ
κυμίνου καὶ τὸ τοῦ σελίνου τε καὶ πετροσελίνου. καὶ
ποτε καὶ σικύα μεγάλη χωρὶς ἀμυχῶν δὶς ἢ τρὶς
ἐπιβαλλομένη κατὰ μέσην τὴν γαστέρα. περιλήψεται
δὲ δηλονότι τὸν ὀμφαλὸν ἡ τηλικαύτη τε καὶ οὕτω
τιθεμένη. κατὰ δὲ τὰ κῶλα καὶ τοὺς ὑπὸ τῷ δέρματι
μῦς ἤ τινας τῶν περιοστέων ὑμένων ἐμφυσήματος
γενομένου, χωρὶς μὲν ὀδυνῶν ὑγρόν τι τῶν λεπτομερεστάτων ἱκανόν, οἷόν πέρ ἐστι τὸ τῆς στακτῆς ὀνομαζομένης κονίας, ἀναλαμβανομένης σπόγγῳ καινῷ·
συνούσης δὲ ὀδύνης ὑπαλείφειν τὸ μόριον ἐλαίῳ χαλαστικῷ.

    γίνονται δὲ ἐκ πληγῶν αἱ τοιαῦται διαθέσεις· ἐν αἷς

branes surrounding the muscles, or one of the viscera. It also gathers sometimes, and to no small extent, in the stomach and intestines, and in the space between these and the peritoneum. And this differs from the edemas by not pitting when pressed, and by making a sound like a drum, and often by being encompassed within a cavity capable of sensation, which is sometimes large.

In these cases, your aim must be that which is common to all these things—namely, to evacuate what is contrary to nature in whatever it might be contained. The specific aim in addition to the general aim is the dispersal by medications that are very fine-particled and quite heating in their potencies. Therefore, in the case of the inflations involving the stomach and viscera, a fine-particled oil which has rue or one of the warming seeds (for example, cumin, common celery or parsley) boiled down in it, will do this; and sometimes a large cupping glass is placed over the middle of the abdomen, without scarification, two or three times. Clearly, under these circumstances, it is placed in such a way that it will surround the umbilicus. When inflation occurs without pain in the limbs, or the muscles under the skin, or some of the periosteal membranes, one of the very fine-particled fluids is sufficient, as for example, so-called lye ashes[18] is, when it is taken up by a fresh sponge. However, when pain is present, smear the part with a relaxing oil.

Such conditions arise from blows in which either a

964K

[18] Linacre has lixivium (or lye ashes); see Pliny, 28,18,75 #244; Caelus Aurelianus, tard 2.3.70; and Galen, XIII.569K. Peter English has "Lee made of Ashes administered with the root of Sparage" (p. 314).

ἤτοι μῦς τις ἢ περιόστεος ὑμὴν θλᾶται. κατὰ μὲν οὖν
965K τοῦ περιοστέου τὸν εἰρημένον ἐπιτιθέναι | σπόγγον·
ἐπὶ δὲ τῶν μυῶν, ὀδυνῶνται γὰρ ἐνίοτε, παρηγορικω-
τέρου χρεία φαρμάκου. διόπερ οὐ μόνῃ τῇ κονίᾳ
χρώμεθα κατὰ τούτους, ἀλλὰ μιγνύντες αὐτῇ τὸ καλούμενον ἕψημα μετ' ἐλαίου βραχέος. ἄμεινον δὲ μηδ'
ὅλως κατ' ἀρχὰς τότε μιγνύναι τὴν κονίαν, ἀλλὰ τῷ
ἑψήματι χρῆσθαι μετὰ οἴνου καὶ ὄξους βραχέος, ἔλαιον ἐπιχέοντα συμμέτρως· ἐπειδὰν δὲ μίξῃς αὐτὰ θερ-
μήνας συμμέτρως, ἐπιτίθει βρέχων ἔριον ἄπλυτον, ὃ
καλοῦσιν οἰσυπηρόν. εἰ δὲ μὴ τοῦτο ἔχεις, ἀλλὰ τὴν
οἴσυπον ἐκείνου ἐπεμβάλλειν τῷ μιχθέντι, διὰ τῶν
εἰρημένων· ὅτι δ' ἀμείνων ὁ Ἀττικὸς οἴσυπος ἅπαντος
ἄλλου, κἂν ἐγὼ μὴ λέγω, γινώσκεις. καὶ τοίνυν καὶ ἡ
δι' αὐτοῦ κηρωτὴ τῶν ἅπασι γινωσκομένων ἐστὶ φαρ-
μάκων· καὶ χρῶνταί γε πάμπολλοι κατὰ τῶν ἐν ὑποχονδρίῳ φλεγμονῶν ταύτῃ. καὶ ταύτης οὖν ἐμβάλλων,
ὅταν οἴσυπος μὴ παρῇ, τὸ δέον ἐργάσῃ. χρὴ γὰρ
τοὺς τεθλασμένους μῦς παρηγορῆσαι, διὰ φαρμάκου
μικτὴν ἔχοντος δύναμιν, ὡς καὶ πέττειν ἅμα καὶ διαφορεῖν καὶ στύφειν μετρίως· ὡς ὅταν γε μηδὲν ἔχῃ
966K στύψεως, αὐξάνει | τὰς φλεγμονὰς ἐνίοτε καὶ μάλιστα
ἐπὶ τῶν πληθωρικῶν σωμάτων.

μεμνημένος οὖν τῶν εἰρημένων τριῶν σκοπῶν ἐπὶ
τῶν ἐμπεφυσημένων διὰ πληγήν τινα μυῶν, εἰ μὲν
ὀδύνη μείζων εἴη, τῷ παρηγορικωτέρῳ χρῶ τρόπῳ· μὴ

---

19 The two terms—oesyperon and oesypus—are applied to

muscle or periosteal membrane is bruised. Therefore, place the previously mentioned sponge over the periosteum. In the case of the muscles, for they are sometimes 965K painful, there is need for a more soothing medication. This is why we do not use lye ashes alone in these (i.e. the muscles), but mix with it what is called "hepsema" with a little oil. It is better not to mix lye ashes at all at the beginning but to use the hepsema with wine or a little vinegar, pouring the oil on moderately. Whenever you mix these, having heated them moderately, place them on unwashed wool soaked in water, which they call "oesypum" (wool grease).[19] If, however, you don't have this, but do have the grease from it, apply it with the mixture of those things spoken of. You know that the Attic wool grease is better than any other, even if I do not say so. Therefore, the salve from this is also better than all the known medications, and the great majority use this for inflammations involving the hypochondrium. If you also put this [salve] on whenever wool grease is not available, you will achieve what is needed. It is necessary to soothe the bruised muscles with a medication which has a mixed potency such as to be digestive and discutient, and at the same time, moderately astringent because, whenever it has no astringency, it sometimes increases the inflammation, particularly in the 966K case of plethoric bodies.

Having called to mind the three previously stated indicators in the case of muscles inflated due to a blow, if pain is a major factor, use the more soothing method; if pain is

the greasy sweat and dirt of unwashed wool used for cosmetic purposes by Roman women; see Dioscorides, II.74, and Pliny, 29.2.10 #35.

παρούσης δὲ ταύτης, ἀγωνιστικωτέρῳ. καλεῖν δὲ οὕτως εἴωθα τοὺς διὰ συντόμων ὁδῶν ἐπὶ τὸ τέλος
ἰόντας. αἱ σύντομοι δ᾽ ὁδοὶ διὰ τῶν ἰσχυρὰν δύναμιν
ἐχόντων γίνονται φαρμάκων· ἰσχυρὰ δὲ δύναμίς ἐστι
μάλιστα μὲν ἐν τῇ κονίᾳ τε καὶ τῷ ὄξει, δεύτερον δ᾽ ἐν
οἴνῳ. τούτων οὖν πλέον μίξεις, ὅταν τοῦ παρηγορεῖ-
σθαι καταφρονήσῃς. αὐτῶν δὲ τούτων πάλιν ἀποκρού-
σασθαι μὲν βουλόμενος οἶνον ἐμβαλεῖς πλείονα. κάλλιστος δ᾽ εἰς ταῦτα μέλας αὐστηρός. διαφορῆσαι δὲ
προαιρούμενος, τὴν κονίαν. ὄξος δ᾽ αὐτῇ μιγνύμενον
εἰς ἄμφω συνεργεῖ· διότι καὶ μικτὴν δύναμιν ἔχειν
ἐδείχθη. ἀνωδύνου δὲ γινομένου τοῦ μυὸς ἔξεστί σοι
μὴ παρούσης τῆς κονίας ἀντ᾽ αὐτῆς ἐμβάλλειν ἀφρό-
νιτρον· ἔστω δὲ τοῦτο μὴ λιθῶδες, ἀλλ᾽ ἀφρῶδες
μᾶλλον. ἔστι δὲ τὸ μὲν λιθῶδες σκληρὸν καὶ πυκνὸν
967K καὶ δυσκόλως διαχεόμενον ἐν τῷ μίγνυσθαι | τοῖς
εἰρημένοις ὑγροῖς· τὸ δὲ ἀφρῶδες μαλακὸν καὶ χαῦνον, ἔτι τε λευκότερον τοῦ λιθώδους. τοῦτ᾽ οὖν καὶ
λύεται τάχιστα πρὸς τῆς ὑγρότητος, ὀνίνησί τε μάλιστα λεπτομερὲς ὑπάρχον.

ὅσα δὲ τῶν τοιούτων ἐμφυσημάτων ἀμεληθέντα
χρονίζει, πρῶτον μὲν ἐπ᾽ αὐτῶν τῷ διὰ τῆς κονίας, ὡς
εἴρηται, χρήσαιο· δεύτερον δὲ τῶν ἐμπλαστωδῶν τινι.
παράδειγμα δ᾽ ἐρῶ καὶ τῶνδε. γλοιὸν ἀναζέσας διήθη-
σον πρῶτον, ὡς γενέσθαι καθαρόν· εἶτ᾽ αὖθις ἐμβαλὼν τῇ κακκάβῃ τίτανον ἄσβεστον λείαν ὡς ἄλευρον

not present, use the more potent. I am accustomed to speak of the latter as approaching the goal via the shortest path. The shortest path is via the medications having a strong potency and a strong potency is greatest in lye ashes and vinegar, and next in wine. You will, therefore, mix a greater amount of these whenever you are not concerned about soothing. When you wish to drive away these same things again, you will put in more wine. Best for this purpose is black, bitter [wine]. When you have made a prior decision to disperse, [then use] lye ashes. If vinegar has been mixed with it, it works jointly to both ends because it was shown to have a mixed potency. When the muscle is not painful, it is possible, if lye ashes is not available to you, to put in aphronitrum in its place. However, this must not be stony but more frothy. What is stony is hard, dense and dispersed with difficulty when it is mixed with the previ-   967K
ously mentioned fluids. What is frothy is soft and spongy, and, in addition, is whiter than the stony. This also dissolves very quickly in the presence of the liquid and is particularly advantageous because it is fine-particled.

In the case of inflations that have been neglected and are chronic, you use, in the first place, the medication made with lye ashes, as I said, and in the second place, one of the plasters. I shall also mention an example of these: *gloios*[20] which you have boiled down and filtered first so that it becomes pure, and after that, when you have thrown it into a three-legged pot, stuff in unslaked lime as smooth

[20] The *gloios* referred to here is perhaps that described in Dioscorides, I.36 as "Rupos gymnasion," or grime from the gymnasium walls. The sweat scraped off wrestler's bodies is another possibility; see Dioscorides, I.35.

ἔμπλαττε μέχρι πηλώδους συστάσεως. ἀγαθὸν δ' ἐπὶ
τούτοις ἐστὶ καὶ τὸ διὰ τοῦ συκομόρου φάρμακον, ὅσα
τ' ἄλλα τοιαῦτα. νυνὶ γάρ, ὡς εἴρηται πολλάκις ἤδη,
παραδείγματα μόνα γράφω τῶν φαρμάκων, ὧν ἡ
καθόλου δύναμις ὑπὸ τῆς θεραπευτικῆς εὑρίσκεται
μεθόδου. ὥσπερ δὲ τῶν φαρμάκων τῆς χρήσεως ἐν-
ταυθοῖ παραδείγματα γράφεται χάριν τοῦ νοῆσαί τε
σαφέστερον ὅλην τὴν μέθοδον, εὐπορώτερόν τε γίνεσθαι περὶ τὴν εὕρεσιν τῆς ὕλης, οὕτω καὶ τῶν πεπονθότων μορίων τοῦ σώματος.

αὐτίκα γέ τοι πάθος ἐστὶ τὸ καλούμενον ὑπὸ τῶν |
968K νεωτέρων πριαπισμός, ἐπειδὴ τὸ αἰδοῖον ἀκουσίως
ἐξαίρεται τῶν οὕτω διακειμένων· ὃ θεασάμενός τις τῶν
ἐν τοῖσδε τοῖς ὑπομνήμασι προγεγυμνασμένων ἑτοίμως γνωριεῖ τοῦ τῶν ἐμφυσημάτων ὑπάρχον γένους.
ἀναμνησθεὶς τά τε κατὰ τὴν ἀνατομὴν φαινόμενα τοῦ
μορίου καὶ τὰ κατὰ τοὺς φυσικοὺς ὑπὲρ τῆς ἐνεργείας
αὐτοῦ καὶ τῆς χρείας λόγους, οὐ χαλεπῶς ἐννοήσει ὅτι
τὸ πληρούμενον ἀτμώδους πνεύματος, τὸ σηραγγῶδες
νεῦρον, ὃ τὴν ἰδίαν οὐσίαν συνίστησι τοῦ αἰδοίου, τὸ
πάθος ἐργάζεται τοῦτο. πνεῦμα δ' ἀτμῶδες ἐν τοῖς τῶν
ζῴων σώμασιν ἐμάθομεν ἐκ χυμῶν θερμαινομένων
ἠρέμα γίγνεσθαι. ἔνθα μὲν γὰρ ἰσχυρόν τ' ἐστὶ τὸ
ἔμφυτον θερμόν, ᾗ θ' ὑγρότης τοῦ μορίου κατειρ-
γασμένη τελέως, εἰς ἀτμοὺς λεπτομερεῖς λυομένη,
κατὰ τὴν ἄδηλον αἰσθήσει διαπνοὴν εἰς τὸ περιέχον

as wheat meal until it becomes claylike in consistency. Good in addition to these is also the medication made from sycamore, and other such things. Now, as I already said often, I shall write down a single example of the medications, the potency of which is discovered, in general, by the therapeutic method. Just as here examples of the use of the medications are written for the sake of knowing the method as a whole more clearly and being better provided in regard to the discovery of the material, so too [are examples] of the affected parts of the body.

To begin with, let me tell you, there is an affection called priapism in young men when the penis of those in 968K such a state is erect against the person's will. Someone who is well versed in these books, when he sees this, will readily recognize it to be of the class of inflations. Having called to mind the appearances of the part from anatomy, and those things pertaining to its function and use in [the books] on the physical theories,[21] he will realize without any difficulty that when the "spongy nerve" is filled with vaporous *pneuma* which coexists with the specific substance of the penis, it brings about this affection. And we learned that vaporous *pneuma* in the bodies of animals arises from humors that are slowly heated. When the innate heat is strong, or the moisture of the part is overcome completely and is dissolved to a thin vapor, it flows off to the surround-

[21] See Book 15 of *De usu partium*. M. T. May (1968) has the following translation of the sentence relevant to the present discussion: "For that it does happen when the hollow nerve is filled with *pneuma* is appropriate to the business now in hand; how it happens belongs to a work on natural philosophy" (vol. 2, p. 659). The "spongy nerve" is presumably the corpus cavernosa.

ἀπορρεῖ. ἔνθα δ᾽ ἤτοι τὸ σύμφυτον θερμὸν ἀσθενέστε-
ρόν ἐστιν ἢ τὸ κατὰ φύσιν ὑγρὸν ἡμίπεπτόν τε καὶ
παχὺ καὶ γλίσχρον, ἐνταῦθα παχύτερος ὁ ἀτμὸς ἢ ὡς
διαπνεῖσθαι γεννᾶται, καὶ μάλισθ᾽ ὅταν καὶ τὸ μόριον
αὐτὸ πυκνωθῇ ποτε. πολλάκις δὲ τὸ μὲν ἐν τῷ μορίῳ
969K περιεχόμενον | ὑγρὸν ὑπόψυχρόν τ᾽ ἐστὶ καὶ παχὺ καὶ
γλίσχρον· αὐξηθεῖσα δὲ ἡ θερμότης εἰς ἀτμοὺς αὐτὸ
διαλύει παχεῖς. ὅπερ καὶ μάλιστά σοι σκεπτέον ἐστὶ
καὶ διοριστέον, ὡς πρὸς τὴν θεραπείαν.

ἡ μὲν γὰρ ἀρχὴ κοινὴ τῶν διαθέσεων ἀμφοτέρων
προκενῶσαι τὸν ὅλον ὄγκον, ἣν ἂν οἷόν τ᾽ εἴη δέξα-
σθαι κένωσιν. εἴρηται δὲ πολλάκις ὑπὲρ τῆς ἐν τοῖς
κενωτικοῖς βοηθήμασι δυνάμεως, ἅπερ ἐστὶ φλεβοτομία καὶ κάθαρσις, ἥ τε διὰ τῶν ὑπηλάτων φαρμάκων καὶ ἡ διὰ τῶν ἀνωτερικῶν ἢ ἐμετικῶν ὀνομα-
ζομένων· ἔτι τε τρίψις πολλὴ καὶ κίνησις πᾶσα καὶ
λουτρὰ καὶ μάλισθ᾽ ὅσα διαφορητικῶν ὑδάτων ἐστί.
οὕτω δὲ καὶ τὰ δριμέα φάρμακα χριόμενα διαφορεῖ
καὶ πάνθ᾽ ἁπλῶς ὅσα θερμαίνουσί τε καὶ ξηραίνουσι.
κατὰ συμβεβηκὸς δέ τι καὶ ἡ ἀσιτία δέδεικται κενοῦν
καὶ μάλισθ᾽ ὅταν ᾖ τὸ περιέχον θερμόν. ὅπερ ἂν οὖν
ἐκ τούτων τῶν βοηθημάτων ὁ κάμνων ἐπιτηδειότατος
ᾖ προσίεσθαι, τούτῳ κενωτέον αὐτὸν ἐπιτιθέντα τῷ
μορίῳ φάρμακον, εἰ μὲν θερμότερον εἴη γεγονός, τῶν
ψυχόντων, ἀνάλογον τῇ πλεονεξίᾳ τῆς θερμασίας· εἰ
970K δὲ μή, | κατ᾽ ἀρχὰς μὲν πάντως μετρίως ψῦχον, ὕστε-
ρον δ᾽ οὐκ ἀναγκαῖον. οὕτω δὲ καὶ τὰ κατὰ τὴν ὀσφὺν
ἅπαντα μόρια παραληπτέον ἐστὶ φαρμάκῳ τὴν αὐτὴν

ings through insensible transpiration. However, when either the innate heat or natural fluid is semidigested, thick and viscid, then the vapor is generated too thickly to allow transpiration, and particularly when the part itself is also made dense on some occasions. And oftentimes, if the fluid contained in the part is somewhat cold, thick and viscid, when the heat is increased, it breaks this up into thick vapors. You must particularly consider and distinguish this because it bears on the treatment. 969K

For the common principle of both conditions is first to evacuate the whole mass [of the body] if it can tolerate evacuation. I have often spoken about the potency in the evacuating remedies, which are phlebotomy and evacuation. The latter [is effected] both through medications which carry off downward (purgatives) and through those acting upward, namely the emetics. Besides these, there is considerable rubbing and all movement and bathing, particularly bathing in discutient fluids. In like manner too, the pungent medications disperse when rubbed on, as all those that are warming and drying generally do. By chance, it also happens that fasting has been shown to purge, particularly whenever the ambient air is hot. Therefore, you must evacuate the patient with whichever of these remedies he is fit to tolerate, applying this medication to the part. If the part has become too hot, [apply] one of the cooling [agents] in proportion to the excess of heat. However, if it has not, being altogether moderately cold 970K from the beginning, later this is not necessary. Similarly you must also make provision for all the parts in the loins

δύναμιν ἔχοντι. καὶ τὴν ἄλλην δίαιταν ἄφυσόν τε καὶ
ξηραντικὴν παραληπτέον. γίνεται δὲ οὐ πολλοῖς μὲν
τὸ πάθος τοῦτο, νεανίαις γε μὴν μᾶλλον ἢ κατ᾽ ἄλλην
ἡλικίαν· ὥστε καὶ ἡ φλεβοτομία μάλιστα αὐτοὺς
ὀνίνησιν, ὡς ἂν καὶ τῆς ἡλικίας οὐκ ἀρνουμένης
αὐτῶν.

οἶδα γοῦν ἐγώ τινα τῶν ὑπ᾽ ἐμοῦ θεραπευθέντων ἐν
τρισὶν ἡμέραις εἰς τὸ κατὰ φύσιν ἐπανελθόντα διά τε
φλεβοτομίας καὶ φαρμάκου τοιοῦδε· κηρωτὴν διὰ
ῥοδίνου τοῦ ἁπλοῦ ποιήσας ὑγρὰν οὕτως, ὡς ἐπὶ τῶν
καταγμάτων χρώμεθα μεθ᾽ ὕδατος ψυχροῦ, καὶ γὰρ
θέρους ἦν ἀρχή, δεύσας τε καὶ ἀναφυράσας ἐπέθηκα
τῷ αἰδοίῳ καὶ ταῖς ψόαις ὀνομαζομέναις. οὗτος μὲν
οὕτως ἐθεραπεύθη. τούτου δ᾽ οὐχ ἧττον ἄλλος ἐπὶ τῇ
φλεβοτομίᾳ τῷ διὰ τοῦ χαμαιμήλου χρησάμενος
ὑγρῷ φαρμάκῳ. δίδωμι δ᾽ αὐτοῖς καὶ τῆς νυμφαίας
πίνειν τό γε κατ᾽ ἀρχάς, ἐφεξῆς δὲ τῆς ἄγνου τὸ
σπέρμα· καὶ εἰ ἐπιχρονίζοι, πηγάνου δαψιλὲς ἐσθίειν.
971K ἔστι γὰρ καὶ τοῦτο | τὸ παράγγελμα κοινὸν ἐπὶ
πάντων σχεδὸν τῶν διὰ μοχθηροὺς χυμοὺς συστάν-
των νοσημάτων, ὡς ἐπὶ τῇ τελευτῇ τοῖς θερμαίνουσι
καὶ ξηραίνουσι χρῆσθαι· τελέως γὰρ ἐκκόπτει ταῦτα
τὸ καταλειπόμενον τοῦ χυμοῦ.

8. Γλῶττάν γε μὴν οὕτως ἐξαρθεῖσαν εἴδομεν, ὡς
μὴ χωρεῖσθαι πρὸς τοῦ στόματος τοῦ ἀνθρώπου, μήτε
πεφλεβοτομημένου ποτὲ καὶ τὴν ἡλικίαν ἐξηκοντού-
του· δεκάτη δέ που σχεδὸν ὥρα τῆς ἡμέρας ἦν, ἡνίκα
τὸ πρῶτον εἶδον αὐτόν. καί μοι καθαρτέος ἔδοξεν

with a medication having the same potency. And a different regimen, which is without flatulence and which causes drying, must be provided. This affection does not occur in many people, but more in youths than in any other age group, so that phlebotomy is especially beneficial, as their age presents no contraindication.

Anyway, I know one of those I treated who returned to an accord with nature in three days with phlebotomy and a medication such as the following: a salve made from oil of roses alone as a liquid, of the kind we use with cold water in the case of fractures. And because it was the beginning of summer, having wet and mixed it up well, I applied it to the penis and what are called the psoas muscles. This is how he was treated. No less than this person, there was another [in whom], after phlebotomy, I used a moist medication made with chamomile. I also give these [patients] drinks made from the waterlily, at least at the beginning, and after that, the seed of the chaste tree.[22] If [the problem] is chronic, I give them an abundance of rue to eat, because this is also the common precept in nearly all the diseases that are as-   971K
sociated with bad humors, i.e. to use, at the end, those things that heat and dry. For finally, these things eradicate what remains of the humor.

8. I have in fact seen a tongue so swollen that it could not be contained in the mouth of the person—someone aged sixty who had never been phlebotomized. It was almost the tenth hour of the day when I first saw him, and he seemed to me to be someone who must be purged with the

[22] For a description of the widespread use of this medication, see Dioscorides, I.135.

εἶναι τοῖς συνήθεσι καταποτίοις, ἃ διὰ τῆς ἀλόης καὶ
σκαμμωνίας καὶ κολοκυνθίδος συντίθεμεν, εἰς ἑσπέραν δοθέντος τοῦ φαρμάκου· ἐπ᾽ αὐτὸ μέντοι τὸ πεπονθὸς μέρος ἐπιθεῖναί τι συνεβούλευσα τῶν ψυχόντων
τήν γε πρώτην· ὕστερον γάρ, ἔφην, ἁρμοσόμεθα πρὸς
τὸ ἀποβαῖνον. ἀλλ᾽ ἑνί γέ τινι τῶν ἰατρῶν οὐκ ἐδόκει·
καὶ διὰ τοῦτο τῶν μὲν καταποτίων ἔλαβεν· ἀνεβλήθη
δ᾽ ἡ περὶ τοῦ τοπικοῦ φαρμάκου σκέψις εἰς τὴν ὑστεραίαν, ἡνίκα καὶ μᾶλλον ἀνύσειν τι τὸ δοκιμασθὲν
ἠλπίζετο, προκεκενωμένου δὲ τοῦ παντὸς σώματος,
972K ἀντισπάσεώς τε πρὸς τὰ κάτω γεγενημένης. | ἀλλὰ
διά γε τῆς νυκτὸς ἐναργέστατον ὄναρ αὐτῷ γενόμενον
ἐπήνεσέ τε τὴν ἐμὴν συμβουλήν, ὥρισέ τε τοῦ φαρμάκου τὴν ὕλην, θριδακίνης χυλῷ διακλύζεσθαι κε-
λεῦσαν· ᾧ δὴ καὶ μόνῳ χρησάμενος ὁ ἄνθρωπος
ὤνητο τελέως, ὡς μηκέτ᾽ ἄλλου δεηθῆναι. τό γε μὴν
ἐπὶ τοῦ πριαπισμοῦ μᾶλλον ἐμετικοῖς φαρμάκοις χρῆσθαι τῶν ὑπηλάτων, ἐπὶ δὲ τῆς γλώττης ἔμπαλιν,
εὔδηλον ὡς ἐκ τῆς τοῦ μορίου θέσεως ἔχει τὴν ἔνδειξιν. ἡ γὰρ ὑφ᾽ Ἱπποκράτους ἀντίσπασις ὀνομαζο-
μένη τὴν εὕρεσιν οὐκ ἀπὸ τῆς οὐσίας, ἀλλ᾽ ἀπὸ τῆς
θέσεως τοῦ θεραπευομένου λαμβάνει μορίου.

9. Καιρὸς οὖν ἤδη μεταβαίνειν ἐφ᾽ ἕτερον ὄγκου
γένος, ἀπὸ τοῦ πράγματος ἀρξαμένους μᾶλλον ἢ τῆς
προσηγορίας· αὕτη γὰρ ἀναμφισβήτητός τ᾽ ἐστὶ καὶ
ὄντως ἐπιστημονικὴ διδασκαλία. τὸ μὲν οὖν ἐπιρρεῖν
τινα χυμὸν ἅπασι τοῖς τοιούτοις ὄγκοις ἐν τῷ Περὶ
τῶν παρὰ φύσιν ὄγκων ἐπιδέδεικται γράμματι. τὸ δὲ

customary little pills which I compound from aloes, scammony and colocynth, the medication being given toward
evening. However, I advised that one of the cooling
[agents] be placed on the affected part itself as the first
measure. Later, I said, we will adapt according to what
happens. But to one of the doctors this did not seem good
and because of this, the patient took some of the little pills.
Consideration of the topical medication was put off to the
next day when he hoped something which was tried might
be effective after prior evacuation of the whole body and a
revulsion downward had occurred. However, during the   972K
night, a very clear dream appeared to him, which approved
my advice and determined the material of the medication,
ordering a thorough washing with the juice of lettuce. And
certainly, when he used this alone, the man benefited completely so as to no longer need anything else. On the one
hand, in priapism, it is better to use emetic medications
rather than those that purge downward. Contrariwise, in
the case of the tongue, it clearly takes the indication from
the position of the part. For what is termed revulsion by
Hippocrates takes its discovery not from the substance but
from the position of the part being treated.

9. It is, then, already time to pass on to another class of
swelling, starting from the matter rather than the name,
for this is, beyond dispute, the truly scientific teaching.
That there is the flow of some humor to all such swellings has been demonstrated in the work *On Abnormal*

μὴ τὸν αὐτὸν ἐν ἅπασιν εἶναι διὰ τῆς αἰσθήσεως
ἐναργῶς φαίνεται, διαφερόντων γε τῶν ὄγκων οὐ τῇ
973K χρόᾳ μόνον, ἀλλὰ καὶ ταῖς κατὰ θερμότητα | καὶ
ψυχρότητα καὶ σκληρότητα καὶ μαλακότητα διαφο-
ραῖς. ὁ μὲν οὖν ἐρυθρὸς ὄγκος ἐναργῶς ἐνδείκνυται
τὸν χυμὸν ὑπάρχειν αἷμα, καθάπερ γε καὶ ὁ ξανθὸς
καὶ ὠχρὸς τὴν τοιαύτην χολήν· ὁ δ' ὑπόλευκός τε καὶ
χαῦνος τὸ φλέγμα. γίνονται δέ τινες ὄγκοι παρὰ
τούσδε, τῇ μὲν χρόᾳ μεταξὺ τῶν ἐρυθρῶν τε καὶ
μελάνων, οἷόν περ τὸ φαιόν ἐστι χρῶμα. καλοῦσι δ'
αὐτὸ πολλάκις ἐν τοῖς τοῦ σώματος μέρεσι γενόμενον
οἱ πλεῖστοι τῶν ἰατρῶν πελιδνόν. ἡ δ' ἀντιτυπία καὶ
τούτοις τοῖς ὄγκοις ἱκανή· καὶ εἰ φλέβας ἀξιολόγους
ἔχοι τὸ μόριον, ἐξαιρουμένας ἰδεῖν ἔστιν αὐτὰς ὑπὸ
παχέος τε καὶ μελαντέρου πως αἵματος, οἷόν περ
ἐνίοτε καὶ κατὰ γαστέρα πολλοῖς τῶν ἡπατικῶν ἐκκρί-
νεται. καί τινες τῶν ἰατρῶν οὐ κακῶς μοι δοκοῦσιν
εἰκάζειν αὐτὸ τῇ τοῖς οἴνοις ὑφιζανούσῃ τρυγί.

ἐπὶ πλέον δ' οὖν ὁ χυμὸς οὗτος ἐκθερμανθεὶς ἤτοι
διὰ σῆψιν ἢ πυρετὸν φλεγμονώδη τὴν μέλαιναν ἐργάζεται χολήν, ἧς οὔτε ζῷόν τι γεύεται μέχρι καὶ τῶν
μυῶν ἥ τε γῆ ξύεται πρὸς αὐτῆς, ἕκαστόν τε τῶν
ἀποξυθέντων μερῶν εἰς ὕψος αἴρεται· καὶ καλεῖται τὸ
974K γιγνόμενον, | ὥς που καὶ Πλάτων ἔφη, ζέσις τε καὶ
ζύμωσις. ἔστι γὰρ ὁ τοιοῦτος χυμός, οἷόν περ τὸ ὄξος
ἐδείξαμεν ὑπάρχειν, ὑφ' οὗ καὶ αὐτοῦ κατὰ γῆς ἐκχυ-
θέντος ταὐτὸ γίνεται σύμπτωμα. διὸ καὶ καλοῦσιν οἱ
παλαιοὶ τὸν χυμὸν τὸν τοιοῦτον ὀξύν, καθάπερ τὸν

Swellings.[23] However, that [the humor] is not the same in all swellings is clearly apparent through perception, since they differ not only in color but also in the differences in 973K heat and coldness, and in hardness and softness. Thus, a red swelling clearly shows the humor to be blood just as a yellow or pale swelling shows it be this sort of bile (i.e. yellow bile), while a whitish or spongy swelling shows it to be phlegm. Some swellings occur besides these, which are between red and black in color, like the color gray. The majority of doctors call this livid when it occurs often in the parts of the body. The resistance in these swellings is also considerable, and if the part has major veins, it is possible to see them lifted up somehow by the thick and rather black blood of the sort that is sometimes separated through many of the hepatic veins in the abdomen. Certain doctors seem to me, not inappropriately, to liken this to the settling sediment in wines.

This humor, when it has been heated still more, either through putrefaction or an inflammatory fever, creates the black bile which no animal, not even mice, would taste. Earth is scratched by it, each of the parts scratched being raised to a height and this occurrence is called, as Plato 974K also said somewhere, "seething and fermentation."[24] For there is such a humor—the kind we showed vinegar to be—due to which this same occurrence happens when it is poured out onto earth. Accordingly, the ancients also call such a humor "acidic" just as they call pale bile, "bitter."

[23] See *De tumoribus praeter naturam*, VII.707K.
[24] The translation follows Linacre's punctuation of the Latin; see his p. 717. The reference to Plato is *Timaeus* 66B.

τῆς ὠχρᾶς χολῆς πικρόν. οὐχ ἥκιστα δὲ καὶ κατὰ τοὺς ἐμέτους φαίνεται τοιοῦτος. ὥσπερ δ᾽ ὀλίγον ἔμπροσθεν ἔλεγον, ἐν μέν τι πάθος ἀκριβῆ σκίρρον ὑπάρχειν, ᾧ πρὸς τοῖς ἄλλοις ἀναισθησία σύνεστι, τοὺς δ᾽ ἄλλους ὄγκους, ὅσοι μηδέπω παντάπασίν εἰσιν ἀναίσθητοι, διχῶς ὀνομάζεσθαι πρὸς τῶν ἰατρῶν, ἤτοι σκίρρους, ἐπειδὴ τοῦ γένους εἰσὶ τῶν σκίρρων, ἢ σκιρρώδεις ὄγκους· οὕτω κἀπὶ τῶν χυμῶν ἡ μὲν ἀναμφισβητήτως μέλαινα χολὴ τοιαύτη τίς ἐστιν οἵαν ἄρτι διῆλθον, ὀξεῖα καὶ ζυμοῦσα τὴν γῆν, ἀηδής τε πᾶσι τοῖς ζῴοις. ἡ δ᾽ ἐπιτηδεία γενέσθαι τοιαύτη καλεῖται διχῶς, ἤτοι μελαγχολικὸς χυμὸς ἢ μέλαινα χολή, τῶν οὕτως ὀνομαζόντων αὐτὴν ἐρούντων ἂν ἑτέραν μὲν εἶναι μέλαιναν ἐν τῷ κατὰ φύσιν ἔχειν τῷ ζῴῳ ἑκάστης ἡμέρας γινομένην, ἑτέραν δὲ τὴν ἐκ συγκαύσεως καὶ οἷον κατοπτήσεως ἀποτελουμένην. |

975K   ὅπερ δ᾽ ἀεὶ παρακελεύομαι, καταφρονεῖν μὲν ὀνομάτων, ἐπιστήμην δ᾽ ἀσκεῖν ἀκριβῆ τῆς τῶν πραγμάτων φύσεως, οὕτω καὶ νῦν ποιητέον, ὀνομάζοντας μὲν ὡς ἂν ἐπέλθῃ, λόγῳ δ᾽ ἑρμηνεύοντας ὡς γίνονταί τινες ὄγκοι παρὰ φύσιν ὑπὸ τοῦ τοιούτου χυμοῦ τὴν φύσιν οἷά πέρ ἐστιν ἐν οἴνῳ μὲν ἡ τρύξ, ἐν ἐλαίῳ δ᾽ ἀμόργη· καὶ ὡς οὗτοι τῷ χρόνῳ προϊόντι σηπομένου τοῦ χυμοῦ διὰ τὴν ἐν τοῖς ἀγγείοις σφήνωσιν ἕλκονται. ὥσπερ οὖν τἄλλα πάθη πάντα παμπόλλην ἔχει διαφορὰν κατὰ τὸ μέρος ἐν τῷ μᾶλλόν τε καὶ ἧττον, οὕτω καὶ τοῦτο. τῆς γάρ τοι φλεγμονῆς ἡ μὲν ἐξέρυθρός ἐστιν,

No less does it also appear in such vomitings. As I said a little earlier, one such affection is, in a strict sense, a scirrhus which, besides its other [features], is without sensation; the other swellings that are not yet entirely without sensation are named in two ways by doctors—either a scirrhosity because it is of a class with the scirrhus, or a scirrhous swelling.[25] In this way, in the humors what is indubitably black bile is of such a character which I went over just now, and is acidic and ferments earth and is unpleasant to all animals. It is suitable to name such a thing in two ways: either a black bilious (melancholic) humor or black bile, since those who name it in this way would say that the one is black and arises each day in the animal in accord with nature while the other is produced from burning and, as it were, roasting.

What I always advise is to pay little regard to names but   975K
to cultivate a precise knowledge of the nature of matters, which is what we must also do now, naming [things] as they come up but interpreting by theory those swellings contrary to nature as they occur subject to the nature of such a humor like, for example, the lees in wine or the watery part in oil, and that in these things, with the passage of time, when the humor putrefies, there is ulceration due to obstruction in the vessels. Therefore, just as all the other affections have a very considerable difference individually in terms of more or less, so too does this. Certainly, on the one hand there is the marked redness of inflammation, while on the other hand there is redness that is only slightly

[25] It is not entirely clear what distinction is being made here in modern terms. Stedman defines scirrhus as an "obsolete term for any fibrous indurated area, especially an indurated carcinoma."

ἢ δ' ὀλίγῳ τινὶ τοῦ κατὰ φύσιν ἐρυθροτέρα. τὸ δ' οὖν
εἶδος ἢ γένος ἢ ὅπως ἂν ἐθέλῃς ὀνομάζειν, ἀμφοτέραις ταὐτόν· ἐρυθρότεραί τε γάρ εἰσι τοῦ κατὰ φύσιν,
ὀδύνη τε πάντως αὐταῖς σύνεστι, παμπόλλην ἔχουσα
καὶ ἥδε διαφορὰν ἐν τῷ μᾶλλόν τε καὶ ἧττον· οὕτω δὲ
καὶ ἡ ἀντιτυπία καὶ ἡ τάσις τοῦ δέρματος οὐκ ἴση
πάσαις· ἀλλὰ κοινόν γε ἐπ' αὐτῶν ἀντιτυπώτερον
εἶναι τὸ μέρος ἢ πρόσθεν ἦν ὅτ' εἶχε κατὰ φύσιν,
976K ἐξῆρθαί τε εἰς ὄγκον | τινὰ συνεκτεινομένου τε καὶ
παρατεινομένου τοῦ δέρματος, εἰς ὅσον ἂν ὁ ὄγκος
αἴρηται.

οὕτως οὖν καὶ τὸ νῦν ἡμῖν ἑρμηνευόμενον πάθος
ἐνίοτε μὲν ἀμυδρὰ καὶ σμικρὰ τὰ συμπτώματα καὶ
λαθεῖν πως δυνάμενα τοὺς πολλοὺς ἐπιφέρει, πολλάκις δ' οὕτως ἰσχυρὰ καὶ μεγάλα καὶ σαφῆ πᾶσιν,
ὡς μηδὲ παῖδα λαθεῖν. ἀλλὰ τό γε κοινὸν ἐν ἅπασι
τοῖς κατὰ μέρος ἐφ' ἑαυτὸ καλοῦν τὴν νόησιν ἔν γε
νόσημα τὸ τοιοῦτον ἐνδείκνυται καὶ προσηγορίαν
μίαν ἐπ' αὐτῷ ἀναγκάζει τίθεσθαι. μεγάλων μὲν οὖν
ἁπάντων ὄντων οὐδεὶς ἀμφισβητεῖ τῆς προσηγορίας,
ἀλλ' ὀνομάζουσι συμφώνως τὸ τοιοῦτον πάθημα καρ-
κίνον. ἀρχόμενον δ' ἔτι λανθάνειν εἰκός ἐστι τοὺς
πολλούς, ὥσπερ ἀμέλει καὶ τὰ τῆς γῆς ἀνίσχοντα
φυτά· καὶ γὰρ καὶ ταῦτα μόνοις τοῖς ἀγαθοῖς γεωρ-
γοῖς διαγινώσκεται. τίς οὖν ἥ τε κοινὴ καὶ ἰδία τῆς
θεραπείας ἔνδειξις ἐπὶ καρκίνου, καιρὸς ἤδη λέγειν.

ἡ μὲν κοινὴ κενῶσαι μὲν ἐν τῷ παραχρῆμα τὸν γεν-
νῶντα τὸ πάθος χυμόν, ὁμοίῳ γένει κενώσεως τῇ τῶν

470

more than normal. Therefore, the kind or class, or whatever else you might wish to call it, is the same for both, for they are redder than normal and pain is present in all of them, this too having a very great difference in terms of more or less. In the same way, the resistance and the tension of the skin are not equal in them all, but what is common is that the part is more resistant than it was before when it was normal, and is raised to a swelling when the 976K skin is extended together [with it] and stretched out to whatever degree the swelling might raise it.

Therefore, in this way too, the affection I am now explaining sometimes brings symptoms that are faint and slight, and are somehow able to escape the notice of many. However, they are often so severe, major and clear to all as not even to escape the notice of a child. But what is common in all of the individual instances, what draws the understanding to itself, shows it to be one particular disease and compels us to apply one name to it. In all those instances that are major, nobody disputes the name, but by common consent, they term such an affection "cancer." However, when it is still incipient, it is likely to escape the notice of many, just as of course the emerging plants of the earth are also overlooked, for these are detected by good farmers only. It is now time to state what the common and the specific indications of treatment are in cancer.

The common [indication] is to immediately evacuate the humor which generates the affection in it with a similar

977K ἄλλων ὄγκων ἔμπροσθεν εἰρημένη· | κωλῦσαι δὲ τοῦ
λοιποῦ μάλιστα μέν, εἰ οἷόν τε, μηδ' ἀθροίζεσθαι
κατὰ τὰς φλέβας τοιοῦτον χυμόν· εἰ δὲ μή, ἀλλὰ
κενοῦν τε πάντως αὐτὸν ἐκ διαλειμμάτων ἅμα καὶ τῷ
ῥωννύειν τὸ μόριον, ἵνα μηδὲν φέρηται πρὸς αὐτὸ τῆς
τῶν χυμῶν περιουσίας. ὥσπερ οὖν τὸν πικρόχολον
χυμὸν ἐκκενοῦμεν, καθαίροντες φαρμάκῳ τοιοῦτον ἕλκειν ἐπιτηδείῳ χυμόν, οὕτω καὶ τὸν μελαγχολικὸν
ἐκκενώσομεν ἢ διὰ τῶν ἁπλῶν τινος, οἷόν ἐστι τὸ
ἐπίθυμον, οὗ πλῆθος δραχμῶν τεττάρων ἐν ὀρρῷ γάλακτος ἢ μελικράτῳ δίδομεν· ἤ τινος τῶν συνθέτων,
οἷόν πέρ ἐστι καὶ τὸ ἡμέτερον συγκείμενον ἐκ δυοῖν
καὶ τριάκοντα ἁπλῶν φαρμάκων. ἀλλὰ τὴν μὲν τούτων
ὕλην ἐν ἑτέροις ἔχεις γεγραμμένην·

ἐνταῦθα δὲ τὰ τῆς μεθόδου λεγέσθω, μετὰ γάρ τοι
τὴν κάθαρσιν ἐπὶ πάντων τούτων ἐρρέθη πρόσθεν ὡς
ἤτοι γ' ὀπίσω χρὴ διώσασθαι τὸν κατασκήψαντα
χυμὸν εἴς τι μόριον ἢ διαφορῆσαι· καὶ ὡς κατ' ἀρχὰς
μὲν ἔν τε τῷ τῆς καθάρσεως καιρῷ καὶ πρὸ αὐτῆς
ἀπωθῆσαι, διαφορεῖν δὲ προκαθάραντα ἀκριβῶς ὅλον
978K τὸ σῶμα. μετρίας μέν τοί γε | τῆς καθάρσεως γενο-
μένης μικτὸν εἶναι χρὴ τὸ προσφερόμενον φάρμακον
ἐκ διαφορούσης τε καὶ ἀποκρουομένης δυνάμεως. ἑκά-
τεραι δ' εἰσὶν ἄπρακτοι περὶ τὸν παχὺν χυμόν· αἱ μὲν
γὰρ ἀσθενεῖς αὐτῷ τῷ μηδὲν ἐργάζεσθαι μέγα, αἱ δ'
ἰσχυραὶ τῷ σφοδρῶς μὲν ἤτοι διαφορεῖν ἢ ἀπωθεῖσθαι τὸ λεπτότερον ἐν τῷ κατὰ τὰς φλέβας αἵματι· τὸ
δὲ παχὺ καὶ μελαγχολικόν, ὃ τῇ τρυγὶ προσεικάζομεν,
οὔτ' ἐκκενοῦν οὔτ' ἀποκρούεσθαι.

472

class of evacuation to that spoken of previously for the other swellings. Thereafter, particularly if it is possible, 977K prevent such a humor collecting in the veins. If this is not possible, then evacuate it completely from the interstices along with also strengthening the part so that it does not bring upon itself any excess of the humors. Therefore, just as we evacuate the picrocholic humor (bitter bile), purging such a humor with a medication suitable for drawing, so too do we purge the melancholic humor, either with one of the simples (epithymum is an example, of which we give an amount of four drachms in a serum of milk or in melikraton), or by one of the compound medications (for example, my own compound [made] from thirty-two of the simple medications). But you have the material of these recorded in other [treatises].[26]

Here, however, let the matters of method be spoken of, for certainly, after the purging in all these cases, it was said before that it is necessary to either drive back the humor which has fallen down to some part, or to disperse it, and that at the outset, both at the time of the purging and prior to it, to repel and disperse it, having previously purged the whole body thoroughly. In fact, when there has been a moderate purging, the medication must be applied after a 978K mixture of discutient and revulsive potencies. Each of the two singly is ineffectual in respect of the thick humor, for those potencies that are weak in this effect nothing significant while those that are strong in this either disperse violently or drive out what is thinner in the blood in the veins. What is thick and melancholic, which I compare to the sediment [of wine], is neither purged nor repelled.

[26] See particularly Galen, *De simplicium medicamentorum temperamentis et facultatibus*, XII.208K.

τούτοις οὖν χρωμένῳ κατ᾽ ἀρχὰς μὲν ἧττον ὁ ὄγκος ἐπίδηλος ἔσται, τὸ λείψανον δ᾽ αὐτοῦ δύσλυτον ἀπεργασθήσεται. διὸ τῶν συμμέτρων ταῖς δυνάμεσι φαρμάκων ἐστὶ χρεία, μήτε νικωμένων διὰ τὴν ἀσθένειαν μήτε παχυνόντων ἰσχυρῶς τὸ αἷμα διὰ τὸ σφοδρὸν τῆς ἐνεργείας, ἔτι δὲ πρὸς τούτοις ἀδήκτων παντάπασιν· ἡ γὰρ κακοήθεια τοῦ πάθους ὑπὸ τῶν δακνόντων παροξύνεται καὶ ὥσπερ εἰώθασι λέγειν ἀγριοῦται. διὰ τοῦτ᾽ οὖν ὅσα σύμμετρα μέν ἐστι ταῖς δυνάμεσιν, ἄδηκτα δὲ ταῖς ποιότησιν, ἁρμόττει τοῖς τοιούτοις πάθεσιν. εὐπορία δὲ τῆς ὕλης αὐτῶν, ὡς ἐν τοῖς περὶ φαρμάκων ὑπομνήμασι ἐδείχθη, διὰ τῶν κεκαυμένων

979K καὶ πεπλυμένων | μεταλλικῶν ἐστι. τὰ μὲν γὰρ διὰ τούτων συγκείμενα φάρμακα μεγάλως τοὺς ἀρχομένους καρκίνους ἅμα ταῖς καθάρσεσιν ἰᾶσθαι δύναται· τοὺς μείζονας δ᾽ ἱκανὸν αὐτοῖς ἐστι κωλύειν αὐξηθῆναι. τούς γε μὴν ἰαθέντας, ὅπως μηκέτι γεννηθῶσι προφυλάξασθαι, τῆς ὑγιεινῆς ἐστι πραγματείας ἔργον, ἧς μόριόν ἐστι καὶ ἡ περὶ τῶν ἐδεσμάτων. εἴ γε μὴν ἐγχειρήσεις ποτὲ διὰ χειρουργίας ἰᾶσθαι καρκῖνον, ἄρξαι μὲν κενοῦν ἀπὸ καθάρσεως τοῦ μελαγχολικοῦ χυμοῦ. περικόψας δὲ πᾶν ἀκριβῶς τὸ πεπονθός, ὡς μηδεμίαν ἀπολείπεσθαι ῥίζαν, ἔασον ἐκχυθῆναι τὸ αἷμα καὶ μὴ ταχέως ἐπίσχῃς, ἀλλὰ καὶ θλῖβε τὰς πέριξ φλέβας, ἐκπιέζων αὐτῶν τὸ παχὺ τοῦ αἵματος· εἶτα θεράπευε τοῖς ἄλλοις ἕλκεσι παραπλησίως.

474

Therefore, by using these from the beginning, the swelling will be less obvious but the remnant of it will be made difficult to disperse. Accordingly, there is need of medications moderate in their capacities which can neither be overcome due to weakness nor thicken the blood strongly due to the strength of their action and, in addition to these things, are altogether nonbiting, for the malignancy of the affection is provoked by those that are biting and, as people are accustomed to say, it becomes "wild." Because of this, then, those that are moderate in their potencies and nonbiting in their qualities are suitable for such affections. An abundance of the material of these, as was shown in the treatises on medications, comes from metallics[27] that have been burned and washed. For the 979K medications compounded with these have a significant capacity to cure incipient cancers [when used] along with the purgatives, while in the more advanced [cancers], they are sufficient to prevent them from being increased. In fact, in respect of those who are cured, how to guard against further generation (recurrence) is a function of the business of health, a part of which concerns foods. At any rate, if you will sometimes undertake to cure the cancer through surgery, you will start by evacuating the melancholic humor by purging. When you have accurately cut off the whole affected part so as to leave no root behind, allow the blood to flow out and do not quickly control it, but also compress the surrounding veins, squeezing out of these the blood that is thick. Then proceed to treat it similarly to other wounds.

[27] On these, see Dioscorides, V.84ff., and Galen, XII.208K and XIII.478K.

10. Ἔστι γε μὴν καὶ ἄλλο πάθος ὑπὸ χυμοῦ παχέος τε καὶ ζέοντος γινόμενον. ἄρχεται δὲ τὰ πολλὰ
μὲν ἀπὸ φλυκταίνης, ἐνίοτε δὲ καὶ χωρὶς ταύτης. ἀλλὰ
κνᾶταί γε πάντως ἐν ἀρχῇ τὸ μόριον· εἶτ᾿ ἀνίσταται
980K φλύκταινά τις, | ἧς ῥηγνυμένης ἕλκος ἐσχαρῶδες
γίνεται. πολλάκις δὲ οὐ μία φλύκταινα γεννᾶται κνη-
σαμένων, ἀλλὰ πολλαὶ μικραὶ καθάπερ τινὲς κέγχροι
καταπυκνοῦσαι τὸ μέρος· ὧν ἐκρηγνυμένων ὁμοίως
ἐσχαρῶδες ἕλκος γεννᾶται. κατὰ δὲ τοὺς ἐπιδημήσαν-
τας ἄνθρακας ἐν Ἀσίᾳ καὶ χωρὶς φλυκταινῶν ἐνίοις
εὐθέως ἀπεδάρη τὸ δέρμα.

ἅπασιν οὖν, ὡς ἔφην, ἕλκος ἐσχαρῶδες γίνεται,
ποτὲ μὲν τεφρώδους ἐσχάρας, ποτὲ δὲ μελαίνης. ἥ τε
πέριξ ἅπασα αὐτοῖς σὰρξ εἰς ἐσχάτην ἀφικνεῖται
φλόγωσιν· οὐ μὴν τῇ χρόᾳ γ᾿ ἔοικεν ἐρυσιπέλατι, ἀλλ᾿
ἔτι καὶ τῆς φλεγμονῆς ἐπ᾿ αὐτῷ γίνεται τὸ χρῶμα
μελάντερον, ὡς εἰ καὶ μίξαις ἐρυθρῷ πλέονι τοῦ μέλανος ἔλαττον. ὅτι δὲ πυρέττουσιν οἱ οὕτως ἔχοντες ἐξ
ἀνάγκης οὐδὲν ἧττον, ἀλλ᾿ ἔτι καὶ μᾶλλον ἐκείνων οἷς
ἐρυσιπελατώδης ἐστὶν ἡ φλεγμονή, πρόδηλον παντί.
καὶ μὲν δὴ καὶ ὡς ἀπὸ φλεβοτομίας ἀρκτέον ἐστὶ τῆς
ἰάσεως εὔδηλον εἶναι νομίζω τοῖς μεμνημένοις ὅσα
περὶ φλεβοτομίας εἴρηται κατὰ τὴν τῶν πυρετῶν
θεραπείαν. οὐκ ἄδηλον δ᾿ οὐδ᾿ ὅτι μέχρι λειποθυμίας
981K ἐπ᾿ αὐτῶν ἡ κένωσις γινομένη | μειζόνως ὠφελήσειε,
πλὴν εἰ μὴ τῶν ἄλλων τι παρείη τῶν κωλυόντων

10. There is another affection that also arises due to a thick and seething humor. However, it begins in many instances from a pustule (*phlyktaina*), although sometimes also without this. But certainly at the start, a pustule springs up, and when this bursts, a crusted wound arises. Often, however, it is not a single, itchy pustule that is produced, but many small ones like grains of millet thickly studded over the part and when these break out, a crusted wound is generated in the same way. In the blisters (*anthrakes*) which are prevalent in Asia and without the pustules [described above], the skin in some is immediately excoriated.[28]

In all these, a crusted wound arises, as I said, the crust sometimes being ashlike in color and sometimes black. All the flesh around these develops a severe burning heat. It does not, in fact, resemble erysipelas in color but the still darker color of inflammation arises in it, as if you were to mix less of the black with more red. It is clear to everyone that those affected in this way are inevitably not less but more febrile than those in whom the inflammation is erysipelitic. Moreover, that you must start the treatment with phlebotomy is, I think, quite clear to those who recall what was said about phlebotomy in the treatment of fevers. What is not obscure is that the evacuation [of blood] which occurs to the point of fainting will be of greater benefit in these cases, unless one of the other things which contrain-

980K

981K

[28] It is not clear what precisely the distinction is here. *Phlyktainai* (translated as "blisters" by Jones) are described under empyema in Hippocrates, *Prognostic*, xvii, while *anthrakes* (translated by Jones as "pustules") are described in Hippocrates, *Epidemics*, iii.7, in what is clearly a generalized disease.

φλεβοτομεῖν. ἐπὶ δὲ τοῦ πεπονθότος μέρους, ὅσον μὲν ἐπὶ τῇ φλογώσει τῶν χυμῶν, τῶν ψυχόντων ἐστὶ χρεία· διὰ δὲ τὸ πάχος τοῦ χυμοῦ καὶ μέντοι καὶ τὴν κακοήθειαν οὔτ᾽ ἀποτρέψεις ποτὲ τὸ ῥεῦμα· καὶ εἰ τούτου ποτὲ τύχοις, ἕτερόν τι ἐν τῷ βάθει βλάψεις. οὐ μὴν οὐδ᾽ ἐπιρρεῖν αὐτῷ συγχωρητέον ἐστίν, ἀλλ᾽ εὑρίσκειν φάρμακα μετὰ τοῦ μετρίως ἀποκρούεσθαι καὶ διαφορεῖν δυνάμενα. τοιοῦτον δ᾽ ἐστὶ τό τε δι᾽ ἀρνογλώσσου κατάπλασμα καὶ τὸ δι᾽ ἑφθῆς φακῆς, ἄρτου κλιβανίτου μιγνύντων ἡμῶν αὐτοῖς ὅσον ἁπαλόν. ἔστω δὲ μὴ πάνυ τι καθαρός, ὥσπερ οὐδὲ ῥυπαρός γε ἄγαν. ὁ μὲν γὰρ ἀκριβῶς καθαρὸς ἐμπλαστικωτέρας οὐσίας ἐστίν, ὁ δὲ πιτυρίας ἁδρομερεστέρας.

κατ᾽ αὐτοῦ δὲ τοῦ ἕλκους ἐπιτιθεμένου τινὸς τῶν ἰσχυρῶν φαρμάκων, οἷόν ἐστι τὸ Ἄνδρωνος ἢ Πασίωνος ἢ Πολυείδου, λύοντα μετά τινος τῶν γλυκέων οἴνων αὐτά, μέχρι γλοιώδους συστάσεως. κάλλιστοι δὲ οἴνων εἰς τοῦτο Θηραῖός τε καὶ Σκυβελίτης ἐστίν· ὧν μὴ παρόντων σιραίῳ χρηστέον, ὃ καλεῖται παρ᾽ ἡμῖν ἕψημα. τὰ δὲ τοῖς ἄλλοις ἕλκεσι προσφερόμενα φάρμακα πέττοντά τε καὶ διαπυΐσκοντα, ταῦτα οὐ χρὴ νῦν προσφέρειν· αὐξήσεις γὰρ τὴν σηπεδόνα τοῦ μορίου. καὶ μὴν καὶ ἀποσχάζειν τοὺς τοιούτους ὄγκους, ὅταν γε προφλεβοτομήσῃς, οὐκ ἀνεπιτήδειον· ἔστωσαν δὲ βαθύτεραι τῶν συμμέτρων αἱ ἀμυχαὶ διὰ τὸ πάχος τοῦ λυποῦντος χυμοῦ. παυσαμένης δὲ τῆς φλογώσεως ὁμοίως τοῖς ἄλλοις ἕλκεσιν εἰς οὐλὴν

982K

dicates phlebotomy is present. In the affected part, there is a need of those things that are cooling to the extent that there are the humors due to inflammation. But because of the thickness of the humor and, of course, the malignancy, you will never turn back the flux. And if, at some time, you should achieve this, you will harm something else in a deep location. You must certainly not allow it to flow to the [part] itself, but find medications that are able to repel and disperse it in a moderate way. Such a medication is the poultice made with plantain, or with boiled lentils, after we have mixed with these baked wheat bread that is soft. This must not be something very refined, just as it must not in fact be very unrefined. For what is completely unrefined is of a more adherent substance, whereas the bread made with bran is more coarse-grained.

On the wound itself, place one of the strong medications, examples of which are those of Andron, Pasion or Polyeides, dissolving them in one of the sweet wines until there is a glutinous consistency.[29] The best of the wines for 982K this [purpose] are those from Thera or Scybela. If neither of these is available, you must use new wine boiled down, which I call "hepsema." You must not now apply the medications that are applied to other wounds, which are digesting and pus-producing, because you will increase the putrefaction of the part. And further, to scarify such swellings, at least when you have carried out phlebotomy, is not unsuitable. But the scarifications must be deeper than just moderate due to the thickness of the distressing humor. However, when the inflammation has ceased, you will

[29] For the eponymous troches devised by these men, see the list of medications in the Introduction (section 10).

ἄξεις τὸ ἕλκος. ἀρκεῖν ἡγοῦμαι καὶ περὶ τῶν ἀνθράκων εἰρῆσθαι τοσαῦτα.

11. Τῶν δ' ἄλλων ὄγκων ἐφεξῆς μνημονεύσω καὶ πρῶτόν γε τῶν καλουμένων χοιράδων. γίνονται δ' αὗται σκιρρουμένων ἀδένων. καὶ ἡ θεραπεία γ' αὐτῶν, ὅσον μὲν ἐπὶ τῷ πάθει, κοινὴ τοῖς ἐν ἄλλῳ τινὶ μέρει γινομένοις σκίρροις· ὅσον δ' ἐπὶ τῇ φύσει τοῦ μορίου, κατά τινας ἀδένας ἑτέρους προσλαμβάνει σκοποὺς διττούς. ἄμεινον δ' ἴσως κἀνταῦθα διαστείλασθαί τι

983K περὶ τῶν ὀνομάτων ἕνεκα σαφοῦς | διδασκαλίας. ὅσοι μὲν γὰρ ἀδένες ἀγγείων σχιζομένων ἀναπληροῦσι τὸ ἐν μέσῳ, στήριγμα γινόμενοι τῆς σχίσεως αὐτῶν, οὐ μεγάλη τούτων ἡ χρεία τοῖς ζῴοις ἐστίν, ἀλλ' ἐκ περιττῆς μὲν προνοίας ὥσπερ ἄλλ' ἄττα τοὺς τοιούτους ἀδένας ἡ φύσις ἐδημιούργησε. τῶν δ' ἤτοι σίελον ἢ γάλα παρασκευαζόντων ἢ σπέρμα, καὶ μέντοι καὶ ὅσοι φλεγματώδη τινὰ γεννῶσιν ὑγρότητα κατὰ μεσεντέριον ἢ φάρυγγα καὶ λάρυγγα, μείζων ἡ χρεία. καί τινες οὐδὲ ἀδένας ὀνομάζουσι τοὺς τοιούτους, ἀλλ' ἀδενώδη σώματα πολὺ τῶν ἄλλων ἀδένων ἀραιότερά τε καὶ σπογγοειδέστερα τὴν οὐσίαν ὄντα· καὶ μέντοι καὶ καθήκουσιν εἰς τοὺς τοιούτους ἀδένας ἀρτηρίαι τε καὶ φλέβες αἰσθηταί. σκιρρωθέντας τε θεραπευτέον αὐτοὺς ὡσαύτως τοῖς ἄλλοις ἅπασι μορίοις. ὅσοι δ' ἐν ταῖς μεταξὺ χώραις τῶν ἀγγείων εἰσίν, ἕτερος ἐν τούτοις προσέρχεται σκοπὸς τῆς ἰάσεως· ἐν ᾧ συναιρεῖται τῷ πάθει τὸ μέρος. ἔστι δὲ καὶ αὐτὸς

bring the wound to a scar in a similar way to other wounds. Such things are, I think, enough to say about the blisters.

11. Next, I shall mention the other swellings, and first, the so-called scrofulous swellings;[30] these are scirrhous (indurated) glands. Their treatment, to the extent that it depends on the affection, is common to the scirrhosities occurring in any other part. To the extent that it depends on the nature of the part, it takes in addition two indicators in relation to certain other glands. However, it is perhaps better here to expand somewhat on the terms for the sake of clear instruction. Some glands fill the space between 983K vessels that have divided, being the basis of their division. Their use to animals is not great, but Nature, out of her abundant forethought, crafted such glands just as it crafted others. Of those adapted for saliva, milk or sperm, and of course those [adapted for] the generation of a certain phlegmatic fluid in the mesentery, pharynx and larynx, the use is greater.[31] And there are some people who do not term such things glands but call them glandular bodies, since they are very much less dense and more spongy than other glands in terms of substance, and perceptible arteries and veins pass down to such glands. You must treat these, when they are indurated, in a similar way to all other parts. For those that lie in the spaces between vessels, another aim of cure is added in those in which the part is removed with the affection. And this removal itself is also

30 These are glandular swellings possibly attributable to tuberculosis. Linacre calls them "choerada." Peter English uses the later, popular term "King's Evil."

31 This represents something of an anatomical potpourri—salivary glands, mammary tissue, testes and lymph nodes.

οὗτος διττός, ἤτοι γ᾽ ἐκκοπτόντων ἡμῶν σμίλῃ τὸ
984K πεπονθὸς ὅλον, ὡς | ἐπὶ τῶν καρκίνων, ἢ σηπόντων
φαρμάκοις. ἥτις δ᾽ ἐστὶν ἡ τῶν τοιούτων φαρμάκων
ὕλη κατὰ τὰς περὶ τῶν φαρμάκων πραγματείας ἔχεις.

12. Ἐμοὶ δ᾽ ἤδη καιρὸς ὑπὲρ τῶν ἄλλων ὄγκων
εἰπεῖν, ὧν πρῶτός ἐστι τὸ καλούμενον ἀπόστημα.
διττὸν δὲ καὶ τούτου τὸ γένος, ἐν μὲν ὅταν ἐκπυη-
σάσης φλεγμονῆς ἀθροισθῇ τὸ πῦον, οἷον ἐν κόλπῳ
τινί· τὸ δ᾽ ἕτερον ἄνευ φλεγμονῆς προηγησαμένης
ὑγροῦ τινος εὐθὺς ἐξ ἀρχῆς ἄλλοτε μὲν ἄλλου κατ᾽
εἶδος, ἀλλὰ πάντως γε μὴν δριμέος ἀθροιζομένου
κατά τι μόριον. ὑποδέρει δὲ τοῦτο τὰ περικείμενα
σώματα· χώραν δὲ αὐτῷ παρασκευάζων ἤτοι μεταξὺ
δυοῖν χιτώνων ἢ ὑπό τισιν ὑμέσιν. ὑποδέρει δὲ πάντως
μὲν τῷ πλήθει διατείνων, ἔστι δ᾽ ὅτε ἐν τῷ χρόνῳ
δριμύτητά τινα σηπεδονώδη προσλαμβάνων. εὑρί-
σκονται δὲ διαιρουμένων σμίλῃ τῶν τοιούτων ἀποστημάτων ἰδιότητες οὐχ ὑγρῶν μόνων, ἀλλὰ καὶ στερεῶν τινων σωμάτων οὐκ ὀλίγαι. καὶ γὰρ ὀνύχων καὶ
τριχῶν καὶ ὀστῶν καὶ ὀστράκων καὶ λίθων καὶ πώρων
985K θραύσμασιν εὑρέθη τινὰ σώματα παραπλήσια. | καὶ
μέντοι καὶ τῶν ὑγρῶν αὐτῷ τὸ μὲν οἷον βόρβορος ἢ
πηλός, ἢ ἐλαίου τις ἰλύς, ἢ οἴνου τρύξ· τὸ δ᾽ οὕτω
δυσῶδες ὡς δυσχεραίνειν ἅπαντας. ἀλλὰ ταῦτα μέν
ἐστι σπανιώτερα.

συνηθέστατα δὲ γιγνόμενα τοῦ γένους τοῦδε τῶν
νοσημάτων εἴδη ἐστὶ τρία, προσηγορίας ἕκαστον αὐτῶν ἰδίας τετυχηκός, ἀθήρωμα καὶ μελικηρὶς καὶ στε-

twofold: either we cut out what is affected in its entirety with a scalpel, as in the case of cancers, or we putrefy with medications. What the material of such medications is you have in the treatises on medications.[32]

984K

12. It is now time for me to speak about other swellings, the first of which is what is called an abscess. There are also two classes of this: one whenever the inflammation has undergone suppuration and the pus is collected as if in some cavity, and the other when there is a fluid of one kind or another right from the start without preceding inflammation, which is, in general, pungent, collected in some part. This excoriates the surrounding bodies when it prepares a space for itself either between two coverings or under certain membranes. At all events it excoriates, distending by its magnitude, while it is possible, over time, for it to take on a certain acrid putrefaction. When such abscesses are divided by a scalpel, peculiar things are discovered, and not only fluids but also quite a number of solid bodies. For nails, hairs, bones, hard shells, stones, fragments of stones and certain other similar bodies are found. Furthermore, among the fluids in them, there is that which is like mire or earth, or slime of oil, or sediment of wine, foul-smelling in such a way as to disgust everyone.[33] But these are rarer.

985K

The most commonly occurring of this class of diseases are three in kind—atheroma, meliceris and steatoma, each

[32] Taken as a general reference to Galen's three major pharmacological treatises; see Book 13, note 16.

[33] From the description, teratomas are included under abscesses.

ἄτωμα, ἀπὸ τῆς ὁμοιότητος τῶν περιεχομένων οὐσιῶν κατὰ τοὺς ὄγκους. ἔστι γὰρ αὐτῶν ἡ μέν τις οἷόν περ τὸ στέαρ, ἡ δὲ οἷον μέλι, καί τις ἀθήρᾳ παραπλήσιος. οἱ σκοποὶ δὲ τῆς θεραπείας κοινοὶ διαφορῆσαι τὸ περιεχόμενον ἢ σῆψαι πᾶν ἢ ἐκτεμεῖν. ἔνιοι μὲν οὖν ὄγκοι τοῖς τρισὶν ὑποπίπτουσι σκοποῖς, ὅσοι λεπτότερον ὑγρὸν ἔχουσιν, ὡς ἡ μελικηρίς· ἔνιοι δὲ τοῖς δύο μόνοις, ὥσπερ τὸ ἀθέρωμα· καὶ γὰρ ἐκτεμεῖν καὶ σῆψαι οἷόν τε τοῦτο· τὸ δὲ στεάτωμα διὰ χειρουργίας μόνης θεραπεύεται μήτε σαπῆναι μήτε διαφορηθῆναι δυνάμενον.

ἐπὶ δὲ τῶν ἐν βάθει συνισταμένων ἀποστημάτων καὶ μάλιστα κατὰ τὰ σπλάγχνα τὰ διὰ τῶν ἀρωμάτων
986K φάρμακα λυσιτελέστατά | εἰσιν, ὧν ἡ δύναμις εἰς ἀτμούς τε λῦσαι καὶ διαφορῆσαι τὸ συνιστάμενον ὑγρόν. ἔστι δὲ καὶ ἄλλα πολλὰ μὲν τοιαῦτα· μάλιστα δὲ αὐτῶν εὐδόκιμα τό τε διὰ τῶν ἐχιδνῶν, ὅπερ ὀνομάζουσι θηριακὴν ἀντίδοτον, ἥ τ' ἀθανασία καλουμένη καὶ ἀμβροσία· ταῦτα μὲν οὖν πολυτελῆ· τῶν δ' εὐτελῶν ἄριστόν ἐστιν τὸ ἡμέτερον, ὃ διὰ τῆς Κρητικῆς καλαμίνθης σκευάζομεν. ἅπαντα δὲ τὰ τοιαῦτα κατὰ τὴν περὶ τῶν φαρμάκων πραγματείαν ἀθροίσομεν,[3] ἣν, ὡς ὀλίγον ἔμπροσθεν ἔφην, ἐπὶ ταῖς τρισὶ ταῖς ἔμπροσθεν ἄμεινον εἶναι νομίζω προσθεῖναι, χάριν τοῦ λείπεσθαι μηδέν.

[3] B (cf. colligemus KLat); ἀθροιζόμενα K

of them acquiring its name from the likeness of the substance contained in the swelling.[34] For one of these is like fat, one like honey, and one like gruel. The indicators (aims) of treatment are common [to all three]: to disperse what is contained, to putrefy the whole thing or to excise it. Some swellings fall under three indicators: [that is] those that have a thinner fluid like the meliceris. Some fall under two only, like the atheroma, for this can either be excised or putrefied. The steatoma is treated by surgery alone, neither putrefaction nor dispersion being possible.

In the case of all the abscesses situated in deep locations, and particularly those in relation to the viscera, the medications made from the aromatic herbs are most useful; the potency of these either dissolves to vapor or disperses the associated fluid. And there are many other such things; particularly highly regarded among these is that made from vipers which they call the theriac antidote, or what are called *athanasia* and *ambrosia*.[35] These are very expensive, while of those that are cheap, the best is my own medication that I prepare with Cretan catmint. All such things are gathered together in the work on medications in respect to which, as I said a little earlier, I think it is better for these three things to be added so nothing is left out.

986K

[34] It is not entirely clear what the modern counterparts of these three swellings or tumors are, hence the retention of the Greek names (which are also found in Linacre and Peter English). Possible candidates are, respectively, lipomas, fluid-filled cysts, and sebaceous cysts. It is also unclear what the process of "digestion" is. [35] According to LSJ, these are alternative names for the same substance, although Galen's subsequent comments suggest three separate items are being listed.

13. Ἐπεὶ δὲ τῶν κατὰ τὴν χειρουργίαν πραττομένων οἱ σκοποὶ τὸ μέν τι κοινὸν ἔχουσι, τὸ δ' ἴδιον, ἄμεινον εἶναί μοι δοκεῖ μὴ διασπᾶν αὐτούς, ἀλλ' ἀθρόως ἅπαντας ἐν τοῖς τελευταίοις τῆσδε τῆς πραγματείας εἰπεῖν. νυνὶ δὲ τοσοῦτον ἔτι περὶ τῶν παρὰ φύσιν ὄγκων ῥητέον ἐστίν, ὡς ὅσοι μὲν αὐτῶν ὅλῳ τῷ γένει παρὰ φύσιν εἰσί, ἐνδείκνυνται τὴν ἄρσιν, ὑπαγόμενοι κοινοτέρῳ σκοπῷ τῷ κατὰ πάντων ἐκτεταμένῳ τῶν τοιούτων, ὅσα ταῖς οὐσίαις | ὅλαις ἐξέστηκε τοῦ κατὰ φύσιν, ὥσπερ ἐπὶ τῶν στεατωμάτων καὶ ἀθερωμάτων ἔχει. τούτου δὲ γένους ἐστὶ καὶ ἡ καλουμένη μυρμηκία καὶ ἡ ἀκροχορδών, ὅ τ' ἐν τῇ κύστει λίθος ὑπόχυμά τε καὶ ἡ τῆς μύλης κύησις, ἐπὶ γυναικῶν, ὀνομάζουσι δ' οὕτω τὴν ἀδιάπλαστον σάρκα· πάντα γὰρ τὰ τοιαῦτα τελέως ἐκκόψαι σπεύδομεν.

ὧν δὲ καὶ ὁ πεπονθὼς τόπος ἕν τι τῶν κατὰ φύσιν ἐστὶ μορίων, ὁ μὲν πρῶτος σκοπὸς ἰᾶσθαι τὸ πάθος, ὁ δ' ἐπ' αὐτὸ δεύτερος, ὅταν ἀνίατον ᾖ, συνεκκόψαι τῷ πάθει τὸ μέρος, ὡς ἐπὶ καρκίνου τε καὶ τῶν ἀθεραπεύτων ἁπάντων ἑλκῶν. ἔμπαλιν δ' ὡς ἐπὶ τῶν ὑποχυμάτων ἀποπίπτοντες τοῦ πρώτου σκοποῦ πρὸς ἕτερον ἄγομεν αὐτὰ τόπον ἀκυρώτερον. ἔνιοι δὲ καὶ ταῦτα κενοῦν ἐπεχείρησαν, ὡς ἐν τοῖς χειρουργουμένοις ἐρῶ. νυνὶ δ' ἀρκέσει τοσοῦτον εἰπεῖν, ὡς τὸ κατὰ τὰς ὑδροκήλας ὑγρὸν ἀλλότριόν ἐστι τῆς τοῦ σώματος οὐσίας ὅλῃ τῇ φύσει· καὶ τὸ κατὰ τοὺς ἀσκίτας ὑδέρους ὕδωρ. ὧν ἡ κένωσις ἤτοι διὰ φαρμάκων γίγνεται διαφορητικῶν ἢ διὰ | χειρουργίας· ἐπὶ μὲν

987K

988K

13. Since the indicators of those matters pertaining to surgery have something common and something specific, it seems to me better not to separate them, but to speak of them all collectively in the final [sections] of this particular treatise. Now what must still be said about the unnatural swellings is that those of them that are in the whole class "contrary to nature" indicate removal, since they are subsumed under the more common indicator of excision in relation to all such things that in their whole substance depart from an accord with nature, just as obtains in the cases of steatomas and atheromas. Of this class also are the so-called sessile warts and the thin-necked warts. Further, there are stone in the bladder, cataract, and the hard formation of the womb in women which they call the "as-yet-unformed flesh."[36] For all such things we strive for complete excision.

987K

On the other hand, of those where the affected place is one of the parts in accord with nature, the first indicator is to cure the affection, but second after this, whenever the affection is incurable, it is to cut out the part together with the affection, as in the case of a cancer and all untreatable ulcers. Contrariwise, having abandoned the first indicator, as in the case of cataracts, we lead these things to another, less important place. Some [doctors], however, also attempt to evacuate these things, as I shall speak of in the [writings] on surgery. For the moment it will suffice to say this: that the fluid in hydroceles is foreign to the substance of the body in its whole nature, as is the fluid in ascites also. The evacuation of these things is by discutient medications

36 Presumably fibroids.

τῆς ὑδροκήλης διὰ καθέσεως σίφωνος, ἐπὶ δὲ τῶν
ὑδέρων διὰ παρακεντήσεως.

συνεκτέμνεται δὲ τῷ πάθει τὸ πεπονθὸς μόριον,
ὥσπερ ἐπὶ τῶν ἔμπροσθεν εἰρημένων, οὕτω κἀπὶ τῶν
κηλητῶν τοῦ περιτοναίου τι μέρος. ὡσαύτως δὲ καὶ ὁ
γαργαρεὼν ἐνίοτε τῷ πάθει συνεκτέμνεται κατά τε τὰ
σκέλη καὶ τοὺς ὄρχεις αἱ φλέβες τοῖς κιρσοῖς, ὅ τ᾽ ἐν
τῇ ῥινὶ χιτὼν τῷ πωλύπῳ καὶ ὁ τετρημένος ὀδοὺς τῷ
τρήματι. ἀλλὰ τούτων μὲν οὐδὲν δυνατόν ἐστιν ἐς τὸ
κατὰ φύσιν ἀγαγεῖν· ἐπὶ δὲ τοῦ γαργαρεῶνος ἐργάζεσθαι χρὴ τοῦτο παντὶ τρόπῳ καὶ μὴ σπεύδειν
ἐκτεμεῖν· ὅταν δὲ ἰσχνὸς καὶ ἱμαντώδης γένηταί ποτε,
τηνικαῦτα ἀφαιρεῖν. τοιοῦτον μὲν οὖν αὐτὸν ἐργάσεται χρόνος μακρότερος· οἷον δ᾽ Ἱπποκράτης ἔγραψε
κατὰ τὸ Προγνωστικὸν ὀλίγων ἡμερῶν ἀριθμός.

οὕτω δὲ καὶ τἆλλα τὰ κατὰ μέγεθος ἐξιστάμενα τοῦ
κατὰ φύσιν, ἐν οἷς ἐστι καὶ τὰ ὑπερσαρκοῦντα πάντα
καὶ ἐγκανθίδες, οἵ τε κατὰ τὴν ἕδραν ὀνομαζόμενοι
θύμοι. τινὰ δὲ τῶν τοιούτων ἑλκῶν ἐπαμφοτερίζει ταῖς
ἰδέαις, ὥσπερ αἵ τε πολὺ τοῦ πέριξ δέρματος ἐξέχου-
989K σαι τῶν | οὐλῶν καὶ τὰ κατὰ τοὺς ὀφθαλμοὺς πτερύ-
για. τῆς γε μὴν ἰάσεως ἐπ᾽ αὐτῶν ὁ σκοπὸς πρόδηλος.
ἐκκόπτεσθαι[4] γὰρ δεῖ τὰ τοιαῦτα πάντα· καὶ χρὴ
σκοπούμενον ἀεὶ τοὺς τρόπους τῆς ἀναιρέσεως ἐπὶ τὸν
ἄριστον ἐξ αὐτῶν ἰέναι. σκοποὶ δ᾽ εἰσὶ τῆς κρίσεως

---

[4] B; ἐγκόπτεσθαι K

or by surgery: in the case of hydrocele it is by insertion of a   988K
siphon and in the case of hydrops it is by paracentesis.

On the other hand, the affected part is excised along
with the affection in ruptures of some part of the peritoneum, as in the case of those things previously spoken of.
In like manner too, the uvula is sometimes excised together with its affection, as are the veins that are varicose
in the legs and the testicles, and the membrane of the nose
with a polyp, and a tooth bored through with a hole. But it
is not possible to bring any of these things to an accord with
nature whereas, in the case of the uvula, one must try to do
this in every way and not hasten to excise it, although
sometimes, when it becomes thin and fibrous, remove it
under these circumstances. A longer period of time will effect this; for example, Hippocrates wrote in the *Prognostic*
that the number was a few days.[37]

In the same way too, other things depart from an accord
with nature in relation to magnitude; among them are all
the excesses of flesh, the encanthides (tumors of the inner
angle of the eye) and the so-called warty excrescences in
relation to the anus. Some wounds such as these partake of
both kinds, as do the scars that stand out very prominently
from the surrounding skin[38] and pterygia in the eyes. The   989K
indicator of the cure in these things is clear; it is necessary
to eradicate all such things and always to consider which of
the ways of excision is the best to proceed with. There are
three indicators relating to the judgment of what is best:

[37] This exact statement was not located in the *Prognostic*, although avoidance of early excision in relation to the uvula is spoken of in *Prognostic*, xxiii.26–38.

[38] Presumably keloid.

τῶν ἀρίστων τρεῖς, ἥ τε τοῦ χρόνου τῆς θεραπείας
βραχύτης, τό τ᾽ ἀνωδύνως αὐτὴν ἐργάσασθαι, καὶ
τρίτος ἐπὶ τούτοις ἡ ἀσφάλεια. καὶ αὐτῆς τῆς ἀσφα-
λείας ἴδιοι σκοποὶ τρεῖς· εἷς μὲν καὶ πρῶτος ὡς τυχεῖν
τοῦ τέλους πάντως· ἕτερος δὲ τὸ κἂν ἀποτύχωμέν ποτε
τοῦ τέλους, ἀλλὰ μηδέν γε βλάψαι τὸν κάμνοντα· καὶ
τρίτος, ὡς μὴ ῥᾳδίως ὑποτροπιάσαι τὸ νόσημα.

κατὰ ταὐτά σοι κρίνοντι τὴν ἀρίστην ὁδὸν τῆς
ἰάσεως ἐπὶ πάντων τῶν νῦν ἡμῖν προκειμένων εὑρε-
θήσεται ποτὲ μὲν ἡ διὰ τῆς χειρουργίας αἱρετωτέρα,
ποτὲ δὲ ἡ διὰ τῶν φαρμάκων. ἡ μὲν οὖν διὰ τῆς
χειρουργίας ἔν γε τοῖς νῦν ἡμῖν προκειμένοις ἐπὶ τὴν
ἀναίρεσιν αὐτῶν σπεύδει τελείως ἐκκόψαι τοῦ ζῴου τὸ
παρὰ φύσιν, ὅλῳ τῷ γένει προαιρουμένη· τούτου δ᾽
990K ἀποτυγχάνουσα τοῦ σκοποῦ δεύτερον | ἔχει τὸν τῆς
μεταθέσεως ἐπὶ τῶν ὑποχυμάτων. ἡ δὲ διὰ τῶν φαρμάκων πρώτῳ μὲν χρῆται σκοπῷ κενῶσαί τε καὶ
διαφορῆσαι τὸ παρὰ φύσιν· εἰ δ᾽ οὗτος ἀδύνατος εἴη,
διὰ τὴν τοῦ μορίου φύσιν ἢ καὶ τὸ τοῦ πάθους ἀνίατον
ἐκπυῆσαί τε καὶ διασῆψαι· δεύτερος δ᾽ ἐπ᾽ αὐτῶν
σκοπὸς οὗτος. οὕτω γοῦν κἀπὶ τοῦ γαργαρεῶνος ποι-
οῦμεν, εἰς τὸ κατὰ φύσιν μὲν πρῶτον ἐπανάγοντες
αὐτόν· εἰ δ᾽ ἀποτύχοιμεν τούτου, τελέως ἐκκόπτοντες
ἤτοι διὰ χειρουργίας ἢ διὰ φαρμάκων καυστικῶν.
ἀλλὰ τὰ μὲν φάρμακα κάλλιον εἰς τὴν περὶ τῶν
φαρμάκων ἀναβάλλεσθαι πραγματείαν, ἐπειδὴ τετάρτην ἄλλην ἄμεινον ἔδοξεν ἐπὶ τρισὶ πραγματεύεσθαι·

the shortness of time of the treatment, that it is done painlessly, and third, in addition to these, safety. And there are three indicators specific to safety itself. One, and the first, is to reach the goal completely. Another is that even if we fail sometimes to achieve the goal, at least we do nothing that will harm the patient, and the third is that the disease does not readily recur.

In these considerations, the best path of cure for you who are making the judgment in all the matters now lying before us will be found sometimes to be by means of surgery as the most preferred, and sometimes to be through medications. The path through surgery, at least in those matters now before us, which hastens to the destruction of these things by excising what is contrary to nature completely from the organism, is preferable in its entire class. However, if this fails to reach its objective, there is a second option which is the goal of change, as in cataracts. The path through medications uses evacuation and dispersion of what is contrary to nature as the first objective. However, if this is not possible because of the nature of the part or because of the incurability of the affection, [the next step is] to bring about suppuration and putrefaction. This is the second objective in the case of medications. Anyway, this is what we do in the case of the uvula, restoring it first to an accord with nature. If we fail in this, finally we eradicate it, either by surgery or by caustic medications. But the medications are better deferred to the treatise about medications, since it seemed preferable to provide another,

990K

491

τὴν δὲ χειρουργίαν ἐπὶ τῇ τελευτῇ τάξαι τῆσδε τῆς
πραγματείας.

14. Οὔκουν ἔτι χρὴ διατρίβειν ἐν τοῖσδε· μετα-
βῆναι δὲ πρὸς τὰ παραπλήσια ἰάσεως δεόμενα τοῖς
εἰρημένοις. ἔστι δὲ ταῦτα τά θ' ὑπερβάλλοντα τοῦ
προσήκοντος ἐν ἀριθμῷ τε καὶ πηλικότητι καὶ τὰ κατ'
ἄμφω ταῦτα ἐνδέοντα. νυνὶ μὲν οὖν ὠνόμασα τοῦ
991K προσήκοντος· εἰ δὲ | καὶ τοῦ συμφέροντος ἢ κατὰ
φύσιν ἢ χρησίμου ποτ' εἴποιμι, τὴν αὐτὴν ἕξει δύναμιν ὁ λόγος. εἴρηται μὲν ἑτέρωθι περὶ τῶνδε μακρό-
τερον· ἀναμνῆσαι δὲ καὶ νῦν ἐπὶ κεφαλαίων ἅπαντα
ἄμεινον εἶναί μοι δοκεῖ. πρῶτον τῶν κατὰ τὸ σῶμα
πάντων ἐστὶν οὗ μάλιστα χρῄζομεν, ἡ τῶν μορίων
ἀόχλητος ἐνέργεια. ταύτην τε ἔχοντες φύσει διὰ τοῦτο
συνήθως λέγομεν ὀρέγεσθαι τοῦ κατὰ φύσιν ἐνεργεῖν.
ἐπεὶ δὲ τοῦτο αὐτὸ συνῆπται τῷ κατὰ φύσιν ἔχειν,
εἰκότως οὐδὲν ἡγούμεθα διαφέρειν ἢ κατὰ φύσιν ἔχειν
εἰπεῖν, ἢ κατὰ φύσιν ἐνεργεῖν. ἀλλὰ καὶ χωρὶς προσ-
θήκης ἐνίοτε λέγομεν ἐφίεσθαι τοῦ κατὰ φύσιν ὑπ-
ακουομένου τῇ λέξει τοῦ ἔχειν ἢ ἐνεργεῖν, ἢ ἀμφοτέρων. αὕτη μὲν ἡ αἰτία τοῦ καλῶς εἰθίσθαι τοῖς
ἰατροῖς τῇ τοῦ κατὰ φύσιν χρῆσθαι φωνῇ περὶ πάντων ὧν αἱρούμεθα κατὰ τὸ σῶμα.

μαθεῖν δ' ἐστὶν ὅτι μὴ πρῶτον μηδὲ δι' αὐτὸ τὸ
κατὰ φύσιν ἡμῖν ἐστιν αἱρετέον, ἀλλὰ δευτέρως τε καὶ
κατὰ συμβεβηκός, ἐν οἷς ἡ φύσις ἀποτυγχάνει. καὶ
992K γὰρ ἕκτος δάκτυλος εὐθὺς ἐξ ἀρχῆς συνεγενήθη | τισὶ
καὶ λείπων πέμπτος ἐγένετο καί τινα τοιαῦτα ἕτερα, τὰ

fourth part in addition to the three, and to assign the surgical [method] to the last part of this treatise.

14. Therefore, we ought not waste time further on these things but rather go on to those things needing similar kinds of cures to those spoken of. These are the things that exceed what is proper in number and magnitude, and those that are lacking in both these respects. Just now I used the term "what is proper," but if I were to say at some 991K time "what is of benefit," or "in accord with nature," or "of use," the argument will have the same force. More has been said about these matters elsewhere. It seems to me better to mention them all now under the chief points. The first is that of all the things in the body, what we most have need of is the undisturbed function of the parts. And because we have this naturally, on this account we customarily say we strive for function in accord with nature. Since this itself is linked to being "in accord with nature," I think it is unreasonable to make a distinction between saying "to be in accord with nature" and "to function in accord with nature." But also, without addition, we sometimes say "aim at an accord with nature," understanding in the phrase either "to be" or "to function" or both. This is the actual reason for it being customary for doctors rightly to use the term "accord with nature" about all the things which we desire in relation to the body.

It is possible to learn that it is not up to us to choose primarily or for its own sake what is in accord with nature but secondarily and contingently from among the things in which Nature fails. For in some people a sixth digit is present right from birth and in some a fifth is missing, and 992K

493

μὲν ἀριθμῷ, τὰ δὲ μεγέθει τοῦ προσήκοντος ἐσφαλμένα. καὶ εἴπερ συνεχῶς μὲν ταῦτα, σπάνια δ' ἐγίνετο τὰ κατορθώματα, τοὐναντίον ἂν ἐπὶ τῶν τῆς φύσεως ἔργων ἐπράττομεν, οὐ φυλάττοντες ὥσπερ νῦν, ἀλλ' ἀναιροῦντες αὐτά. οὗτος ὁ λόγος ἀεί σοι μνημονευέσθω, διαφέρων εἰς τὴν τῶν ὀνομάτων χρῆσιν, ἐξαπατῆσαι δυναμένων τοὺς ἀσκέπτους ἐμποιῆσαί τε διαφωνίας φαντασίαν, ἐὰν ὁ μέν τις λέγῃ τὸ παρὰ φύσιν ἅπαν ἐκκόπτειν δεῖν· ὁ δὲ τὸ βλάπτον ἢ ἀσύμφορον ἢ λυμαινόμενον ταῖς ἐνεργείαις.

15. Αὖθις οὖν ἀναλαβόντες ὑπὲρ τῶν πραγμάτων λέγωμεν, ἐπειδὴ τὰ τῶν ὀνομάτων ἡμῖν διώρισται, τὴν ἀρχὴν τῷ λόγῳ τήνδε ποιησάμενοι. παράκειται τοῖς εἰρημένοις νοσήμασι τὰ κατ' ἀριθμὸν ἢ μέγεθος ἐξεστῶτα τοῦ προσήκοντος. ἐφ' ὧν ἐκκόπτειν μὲν χρὴ τὸ περιττόν, ἤτοι κατὰ μέγεθος ἢ κατ' ἀριθμόν· ἀνατρέφειν τε καὶ κατασκευάζειν τὸ λεῖπον, ὅταν γε δυνατὸν ᾖ τοῦτο πρᾶξαι. τὸν γάρ τοι πέμπτον δάκτυλον ἢ τοιοῦτόν τι μόριον ἕτερον οὐχ οἷόν τε | γεννῆσαι τοῖς ἰατροῖς, ἀλλ' ἔστι μόνης τῆς φύσεως ἔργα τὰ τοιαῦτα πάντα. τὸ μέντοι τελέως ἀφαιρεῖν τὸ κατ' ἀριθμὸν ὑπερβάλλον ἢ ἀποκόψαι τι τοῦ κατὰ τὸ μέγεθος ὑπεραυξηθέντος οὐδ' ἡμῖν ἀδύνατον. ἐν γοῦν ἔστι τῶν ἰατρῶν ἔργων οὐ τὸ φαυλότατον, ὅταν εἰς πολυσαρκίαν ἐκτραπῇ τὸ σῶμα τοσαύτην ὥστε μηδὲ βαδίζειν ἀλύπως δύνασθαι μηδ' ἅψασθαι τῆς ἕδρας διὰ τὸν ὄγκον τῆς γαστρός, ἀλλὰ μηδ' ἀναπνεῖν ἀκωλύτως, ἐκτήκειν αὐτὸ καὶ καθαίρειν· ὥσπερ γε

993K

there are other such variants where things have fallen short of "what is proper," either in number or magnitude. And if these things were frequent while those things that are correct were rare, we would be doing the opposite to the actions of Nature if we did not preserve them as they are but took them away. You must always be mindful of this argument, which bears on the use of names, since names can deceive those who are unreflecting and create the illusion of discord, if someone were to speak of the need to excise everything contrary to nature while another were to speak of harm, inconvenience or injury to the functions.

15. Therefore, to reiterate, let me speak again about these matters, since I made the definition of terms the beginning of this discussion. Present among the diseases mentioned are those which depart from what is proper in terms of number and magnitude, in which it is necessary to excise the excess in either magnitude or number, and nurture and prepare what remains, at least whenever it is possible to do this. For certainly, with respect to a fifth finger or some other such part, it is not possible for doctors to create [this], but all such actions are for Nature alone. However, it is not impossible for us to completely remove what is in excess in number, or to excise what is increased unduly. At all events, it is one of the occupations of doctors, and by no means the most trivial, that whenever the body turns to obesity to such a degree that the person is neither able to walk without distress nor to sit on account of the mass of the belly, and is not able to breathe freely, to melt this away and get rid of it. It is just the same when some-

993K

495

κἀπειδὰν ἐν ἀτροφίᾳ γένηται παραπλησίᾳ τοῖς ἐχομέ-
νοις φθόῃ, τῆς ἀναθρέψεως αὐτοῦ προνοεῖσθαι. πολλάκις δ' οὐχ ὅλον, ἀλλ' ἕν τι μέρος ἐν ἀτροφίᾳ γίνεται
προηγησαμένης ἤτοι παραλύσεως ἢ δυσκρασίας μορίου. καί σοι καιρὸς ἤδη περὶ τῆς τούτων θεραπείας
ἐπισκέπτεσθαι τὴν ἀρχὴν ἀπὸ τῶν εἰς τὴν πολυσαρκίαν ἄμετρον ἐκπεσόντων ποιησαμένῳ.

δέδεικται δ' ἐν τοῖς Περὶ κράσεων ἡ θερμοτέρα τε
καὶ ξηροτέρα κρᾶσις ἰσχνὸν ἐργαζομένη τὸ σῶμα.
τοιαύτην οὖν σοι ποιητέον ἐστὶ τὴν τῶν παχέων σωμάτων, εἰ μέλλοι γενήσεσθαι σύμμετρος. μεμάθηκας
994K   δ' ἐν τῷ κατ' ἐκείνην μὲν τὴν | πραγματείαν, ἀλλὰ καὶ
κατὰ τὴν Τῶν ὑγιεινῶν οὐδὲν ἧττον, ὀξέα γυμνάσια
καὶ ἡ λεπτύνουσα δίαιτα καὶ φάρμακα τοιαῦτα καὶ
τῆς ψυχῆς αἱ φροντίδες ἀποφαίνουσι τήν τε κρᾶσιν
ὅλην θερμοτέραν καὶ ξηροτέραν καὶ διὰ ταύτην τὸ
σῶμα λεπτότερον. ἐν μὲν δὴ τοῖς γυμνασίοις οἱ ὀξύτα-
τοι δρόμοι μάλιστα ἁρμόζουσιν. ἡ δ' ὕλη τῆς λεπτυνούσης διαίτης ἰδίᾳ γέγραπται καθ' ἓν ὅλον βιβλίον.
εἰ δὲ καὶ φαρμάκων δέοιντο τῶν λεπτυνόντων, εἴρηται
μὲν καὶ ταῦτα κατὰ τὰς περὶ τῶν φαρμάκων πραγμα-
τείας·

εἰρήσεται δὲ καὶ νῦν ὅσα δραστικώτατα τῶν τοιούτων ἐστίν, οἷς ἀξιῶ σε χρῆσθαι καθαίρειν ἐπιχειροῦν-
τα πολυσαρκίαν ἄμετρον. οἷς οὖν εἰώθασιν ἕνεκα

---

39 See, particularly, *De temperamentis*, Book 2, chapter 4.

thing like atrophy occurs in those with consumption, and
we provide for the nourishing of the body. Often, however,
it is not the whole body but some single part that becomes
atrophic when either paralysis or *dyskrasia* of the part
leads the way. It is timely for you now to give consideration
to the treatment of these things, making a start from those
[bodies] that have degenerated into excessive obesity.

It has been shown in the work *On Krasias (Mixtures)*
that a hotter and colder *krasis* makes the body lean.[39] Such
a *krasis* is, then, what you must bring about for thickened
bodies if they are to become normal. You have learned this
matter in that treatise, but also no less in *On the Preserva-* 994K
tion of Health,[40] *viz.* that brisk exercise, a thinning diet,
such medications, and anxieties of the soul represent a
krasis that is, as a whole, hotter and drier, and because of
this *krasis*, the body is thinner. Among exercises, very swift
running is particularly suitable. The material of the thinning diet has been written about separately in one complete book.[41] If people also require medications that are
thinning, these were spoken about in the treatises on medications.[42]

However, what I shall now speak about is which of
these medications is most effective and which ones I think
it worthwhile for you to use in attempting to get rid of
excessive obesity. Some [doctors] are accustomed to use

[40] See *De sanitate tuenda*, Book 6, chapter 8.

[41] Presumably *De victu attenuante*, not included in Kühn but
found in K. Kalbfleisch (1898), CMG V.4,2, and in an Italian translation by N. Marinone (1973).

[42] See, particularly, *De simplicium medicamentorum tempera-*
mentis et faculatatibus, XI.594K and XI.627K.

παθῶν ἀρθριτικῶν ἔνιοι χρῆσθαι τμητικοῖς τὴν δύναμιν ἱκανῶς οὖσι, τούτοις καὶ σὺ χρῶ, θεραπεύειν
ὑπερβάλλουσαν εὐσαρκίαν πειρώμενος. ἔστι δὲ τοιαῦτα πηγάνου τὸ σπέρμα καὶ μᾶλλον τοῦ ἀγρίου σὺν
αὐτοῖς τοῖς κορύμβοις, ἀριστολοχία θ' ἡ στρογγύλη
καὶ τὸ λεπτὸν κενταύριον, ἥ τε γεντιανὴ καὶ τὸ πόλιον·
ὅσα τε τῶν οὐρητικῶν ὀνομαζομένων ἰσχυρά, καθάπερ τὸ πετροσέλινον. ἕκαστον γὰρ τῶν τοιούτων καὶ
995K αὐτὸ | καθ' αὑτὸ καὶ σὺν ἄλλοις λεπτύναί τε τοὺς
χυμοὺς ἱκανὸν καὶ κενῶσαι, τὰ μέν τοι δι' οὔρων
αἰσθητῶς, τὰ δέ τοι καὶ κατὰ τὴν ἄδηλον αἰσθήσει
διαπνοήν· ἀλλὰ καὶ οἱ διὰ τῶν κεκαυμένων ἐχιδνῶν
ἅλες ἱκανῶς λεπτύνουσι. καὶ πολλοὶ τῶν ἰσχνοτέρων
ἢ μέσως εὐσάρκων ὑπὸ τῶν τοιούτων φαρμάκων πό-
σεως ἀπώλοντο, κατοπτηθέντος αὐτοῖς τοῦ αἵματος.
ὥρμησαν δ' ἐπ' αὐτὰ θεασάμενοί τινας ἀπηλλαγμέ-
νους ἀρθριτικῶν παθῶν, οὐκ ἐπιλογισάμενοι τὴν κρᾶσιν τῶν ὠφεληθέντων, ὑγροτέραν τε καὶ φλεγματι-
κωτέραν οὖσαν, οἷά πέρ ἐστι καὶ ἡ τῶν παχέων, ἐφ' ὧν
ἀκίνδυνος ἡ τῶν τοιούτων χρῆσις φαρμάκων.

ἐγὼ γοῦν ἐθεράπευσα τινὰ νεανίσκον ἐτῶν ἐγγὺς
τεσσαράκοντα, παχὺν ἱκανῶς γεγονότα, τῇ τε πρὸς
τοὺς ἀρθριτικοὺς ἀντιδότῳ καὶ τοῖς ἁλσὶ τῆς θηριακῆς, αὐτῇ τε τῇ θηριακῇ μετὰ τοῦ καὶ τῇ ἄλλῃ διαίτῃ
τῇ λεπτυνούσῃ χρῆσθαι καὶ γυμνασίων δρόμοις ὠκέ-
σι. παρεσκεύαζον δ' αὐτὸν ἐπὶ τὸν δρόμον, ἀνατρίβων
μὲν πρῶτον ὠμολίνοις τραχέσιν, ἄχρι τοῦ φοινίξαι τὸ
δέρμα, τρίβων ἐφεξῆς τρίψει δι' ἐλαίου τῶν διαφορη-

498

those that are extremely thinning in terms of capacity for affections of the joints, and you must also use these when trying to treat an excess of normal flesh. Such medications are the seeds of rue, and especially that which is wild, with the actual clusters of ivy fruit, aristolochia which is round, thin centaury, gentian, hulwort and those of the so-called urine-producing medications that are strong, like parsley. For each of these, either by itself or with other 995K medications, is sufficient to thin the humors and to evacuate [them], some perceptibly through the urine and others through imperceptible transpiration. But the salt [made] from burned vipers is also adequately thinning. And many of those who are quite thin or moderately plump have perished due to the drinking of such medications because their blood becomes overheated. However, [doctors] rush to these medications when they see some are freed from affections of the joints, without considering the *krasis* of those who were helped—whether it was more moist and more phlegmatic, as it is in those who are obese and in whom the use of such medications is without danger.

Anyway, I treated a young man, close to forty, who was excessively fat, with an antidote against the arthritides and with the salt of theriac, and with theriac itself, along with the thinning diet, and exercised him with fast running. However, I did prepare him for the running by rubbing him first with rough cloths until the skin was reddened, and next rubbing him with oil containing some of the

996K τικῶν τι φαρμάκων ἔχοντος, | ᾧ καὶ μετὰ δρόμον
ἐχρώμην αὖθις. ἔστι δὲ ταῦτα σικύου ῥίζα τοῦ ἀγρίου
καὶ ἡ ἀλθαία καὶ ἡ γεντιανὴ καὶ ἡ ἀριστολοχία καὶ ἡ
τοῦ πάνακος ῥίζα καὶ τὸ πόλιον καὶ τὸ κενταύριον. ἐν
δὲ τῷ χειμῶνι καὶ μετὰ τὸ λουτρὸν ἐπαλείφειν συμφέρει τῷ εἰρημένῳ ἐλαίῳ. οὐκ εὐθὺς δὲ τὴν τροφὴν
ἐπὶ τοῖς λουτροῖς διδόναι προσῆκεν, ἀλλὰ κοιμᾶσθαι
πρότερον, ἐπιτρέπειν δ', εἰ βούλοιντο καὶ αὖθις λού-
σασθαι πρὶν ἢ τραφῆναι. κάλλιον δ' ἐστὶ καὶ τὸ ὕδωρ
τῶν διαφορητικῶν.

εἰ μὲν οὖν αὐτοφυὲς ἔχοιμεν, ἐκείνῳ χρωμένους
ὁποῖον ἐστι καὶ τὸ κατὰ τὴν Λέσβον ἀπὸ τεσσαράκοντα σταδίων τῆς Μιτυλήνης· εἰ δὲ μὴ κατασκευ-
άζοντας αὐτοὺς παραπλήσιον. ἔστι δὲ τὸ κατὰ τὴν
Μιτυλήνην καὶ χρόᾳ καὶ δυνάμει τοιοῦτον, ὁποῖον ἂν
γένοιτο μιχθέντος ἁλὸς ἄνθους ὕδατι θαλάσσης.
τουτὶ γὰρ τὸ ὕδωρ καὶ τοῖς ὑδεριῶσι καὶ τοῖς ἄλλοις
οἰδαλέοις ἐπιτήδειόν ἐστιν, ἰσχυρῶς ξηραῖνον· ὡσαύτως δὲ δὴ καὶ τοῖς πολυσάρκοις καὶ μάλιστα ὅταν
αὐτοὺς ἀναγκάζῃ τις ἐν αὐτῷ κολυμβᾶν ὀξύτατα καὶ
λουσαμένους πλέον μὴ παραχρῆμα πίνειν ἢ ἐσθίειν,
997K ἀλλ' ἤτοι κοιμᾶσθαι πρότερον | ἢ πάντως γε ἡσυ-
χάζειν. εἰδέναι δὲ χρὴ καὶ προλέγειν τῷ θεραπευομένῳ τὴν πολυσαρκίαν ὡς ἔσθ' ὅτε διὰ κίνησιν
ἀθροωτέραν εἰκός ἐστι καὶ πυρέξαι αὐτόν· ὅτι τε μήθ'
ὁ πυρετὸς ἀνάρμοστος εἰς τὰ παρόντα γενήσεται, τοῦ
ἰατροῦ καλῶς ἅπαντα πράττοντος. εὔδηλον γὰρ ὡς
ὅταν ἐπὶ κόπων πυρέξωσιν καὶ οἱ οὕτω θεραπευόμενοι,

discutient medications, which I used again after the run- 996K
ning. These [medications] are the root of the wild cucumber, marshmallow, gentian, aristolochia, the root of panax,
hulwort and centaury. In the winter, it is also of use to
smear [the patient] with the aforementioned oil after a
bath. It is, however, not appropriate to give [patients] food
immediately after bathing, but [for them] to sleep first,
and to allow them, if they wish, to bathe again before being
nourished. It is better, too, if the water has discutient properties.

If we are to depend on natural [waters], they should use
the kind that is in Lesbos, forty stadia from Mytilene. If
not, let them prepare what is similar. The water at
Mytilene is of such a kind, in terms of color and potency,
as would be the case if lustrous salts were mixed with
seawater. For this water is suitable both for those with
hydrops and for the other swellings, being strongly drying.
Similarly, it is also suitable for those who are obese, and
particularly when someone also compels them to swim
more in it quickly, and after bathing more, not to drink or
eat immediately but either to sleep beforehand, or at least 997K
rest completely. It is, however, also necessary to realize
you should say beforehand to the one being treated for
obesity that there are times when, due to more concentrated movement, he is likely to also become febrile, and
that the fever will not be inappropriate to the prevailing
circumstances, if the doctor does everything properly. It is
clear that whenever those who are treated in this way be-

καταστήσαντες αὐτῶν τὸν πυρετόν, αὖθις ἐπὶ τὴν
αὐτὴν ἰδέαν τῆς ὅλης θεραπείας ἀφιξόμεθα. φεύγειν δ᾽
ἐπ᾽ αὐτῶν χρὴ καὶ τοὺς τροφίμους οἴνους, οἷοί πέρ
εἰσιν οἱ παχεῖς· τοῖς δὲ ὑδατώδεσι χρῆσθαι, τουτέστι
τοῖς λευκοῖς μὲν τῇ χρόᾳ, λεπτοῖς δὲ κατὰ τὴν σύστασιν, ἢ τοῖς τεθαλαττωμένοις.

16. Ὅσους δ᾽ ἀνατρέφειν βουλόμεθα καταλελεπτυ-
σμένους, οἶνον μὲν δώσομεν τὸν παχύν, ἐδέσματα δὲ
τὰ παχύχυμα καὶ γυμνάσια τὰ βραχέα καὶ τρίψιν τὴν
μετρίαν· καὶ ἁπλῶς εἰπεῖν, ἅπαντα τοῖς εἰρημένοις
ἐναντία πράξομεν. ἐπιτήδειον δ᾽ αὐτοῖς ἐστι καὶ τὸ
πιττοῦσθαι δι᾽ ἡμερῶν ἤτοι τριῶν ἢ τεττάρων· κάλλιστον γὰρ τοῦτο φάρμακον εἰς σάρκωσιν, ὥστ᾽ εἰ καί
998K   τι μόριον ἔν ποτε πάθῃ, διὰ τούτου | τοῦ βοηθήματος
ἀνατρέφειν αὐτό. καὶ ἡμῖν ἤρκεσεν ἐπὶ πάντων σχεδὸν μόνον τοῦτο· καὶ γὰρ καὶ θερμαίνει καὶ ὑγραίνει
πλῆθος αἵματος ἐπισπώμενον. οὔτ᾽ οὖν συνεχῶς χρὴ
προσφέρειν αὐτὸ τοῖς κάμνουσι σώμασιν, οὔθ᾽ ὅτε
χρὴ καταχρίειν πολλάκις, ἀλλ᾽ ἐν μὲν χειμῶνι δίς, ἐν
θέρει δ᾽ ἅπαξ ἀρκεῖ. τοῖς δ᾽ ἐκ γενετῆς ἔχουσιν ἰσχνό-
τερά τινα μόρια καὶ οἱ ἀνδραποδοκάπηλοι βοηθοῦσι
διὰ τοῦ βοηθήματος τοῦδε μετὰ τῆς καλουμένης ἐπι-
κρούσεως, ἔστι δὲ κἀκείνοις συμμετρία τις, ὡς μὴ
μᾶλλον τοῦ δέοντος γίγνοιτο μήτ᾽ ἔλαττον, οὔσης
τοιᾶσδε. ναρθήκια λεῖα μετρίως ἀληλιμμένα κατὰ τῶν
ἰσχνῶν μορίων ἐπαράσσουσιν, ἄχρι περ ἂν ἐξαρθῇ
μετρίως· ἐν τούτῳ γὰρ τὸ ὅλον ἐστίν, ὥσπερ καὶ
Ἱπποκράτης ἔλεγεν ἐπὶ καταντλήσεως ὕδατος θερμοῦ,

come febrile through fatigue when we establish their fever, we will come again to the same form of the whole treatment. It is also necessary, in their case, to avoid wines that are nourishing, [that is] those that are thick, whereas we do use those that are watery, that is to say, those that are white in color and thin in consistency, or have been mixed with seawater.

16. However, when we wish to feed up those who have been made very thin, we will give wine which is thick, foods which are thick-humored, little exercise and moderate rubbing—in short, we will do all the things opposite to those mentioned [for obesity]. The application of pitch over three or four days is also suitable for them, for this is the best medication for the creation of flesh, so that even if one single part is at some time affected, build it up with 998K this remedy. And pitch alone is almost sufficient for us in the case of all [these], for when much blood is drawn, it heats and moistens. Therefore, it is not necessary to apply this continuously to the patients' bodies nor is it necessary to smear them often; twice in winter and once only in summer is enough. In those who have certain parts thinner from birth, slave traders also help with this remedy along with the so-called "percussion,"[43] so there is some symmetry in those parts such that they are not more or less than necessary. This is how [it is done]: they strike rods against the thin parts until they are moderately raised, for the whole matter is in this, just as Hippocrates said in the case of pouring on warm water that "first [the part] is raised up

[43] For this method, see also Paulus Aeginata, 3.69 and 6.51.

τὸ μὲν πρῶτον ἀείρεται, ἔπειτα δ᾽ ἰσχναίνεται. πάντ᾽ οὖν ἃ εὐσαρκῶσαι βουλόμεθα μόρια καὶ τρίβειν χρὴ καὶ καταντλεῖν καὶ παίειν καὶ πιττοῦν ἄχρι περ ἂν ἐξαρθῇ· γενομένου δὲ τούτου, παραχρῆμα παύεσθαι χρὴ πρὶν ἄρξασθαι διαφορεῖσθαι. τὰ γάρ τοι θερ-
999K μαίνοντα πάντα | καθάπερ ἕλκειν πέφυκεν, οὕτω καὶ διαφορεῖν. ἐὰν οὖν ἀναμείνῃς διαφορηθῆναι τὸ ἑλ-χθέν, οὐδὲν ἕξεις πλέον.

οὕτω καὶ πυγάς τις ἀνδραποδοκάπηλος ἔναγχος ηὔξησεν ἐν ὀλίγῳ χρόνῳ παιδὸς ὑπολέπτου, συμμέ-τρως μὲν τῇ κατακρούσει χρώμενος ἑκάστης ἡμέρας ἢ παρὰ μίαν, συμμέτρως δὲ πιττῶν. ἀλλὰ τοῖς γε τὸ σύμπαν σῶμα λεπτοῖς καὶ λούεσθαι μετὰ τροφὴν ἐπιτήδειον. ὥσπερ δὲ τοῖς λεπτύνουσι βοηθήμασι κίν-δυνος ἦν ἀκολουθῆσαι πυρετὸν ὑπερθερμανθέντος ἀμέτρως τοῦ σώματος, οὕτω καὶ τοῖς λουομένοις ἐπὶ τροφαῖς κίνδυνός ἐστι κατὰ τὸ ἧπαρ ἔμφραξιν γενέσθαι, καὶ μάλιστα διὰ τὸ τῶν ἐδεσμάτων εἶδος· ἐμφράττει γὰρ καὶ ἄλλως τὰ παχύχυμα, χρωμένων ἐπὶ πλέον αὐτοῖς. ὅπου δὲ καὶ ἄλλως τοῦτο δρᾶν πέφυκε, πολὺ δὴ μᾶλλον ἐπὶ βαλανείοις ἅμα τροφῇ τοῦτο δράσει.

γίνεταί γε μὴν καὶ λίθων ἐν νεφροῖς σύστασις ἐπὶ τῇ τοιαύτῃ διαίτῃ χρονιζούσῃ· διὰ τί δέ, οὐ πᾶσι γίνεται πρόδηλον. ἐνίους μὲν γὰρ εἰκός ἐστιν ἤτοι πυκνοὺς ἔχειν τοὺς νεφροὺς ἢ στενὰς τὰς ἀναστο-μώσεις τῶν ἐν ἥπατι φλεβῶν, ἐνίους δὲ τἀναντία. καὶ
1000K τούτων | διάγνωσις οὐδεμία σαφής ἐστιν· ἀλλ᾽ ἐπερω-τᾶν χρὴ τὸν διαιτώμενον, ὡς εἴρηται, συνεχῶς εἴ τις

and then reduced."[44] All those parts that we wish to be "well-fleshed," we must rub and pour [water] on, strike, and apply pitch to until they become raised. When this has happened, we must immediately stop before they start to be dispersed. Certainly, all things that are heated disperse, just as they naturally attract. If you wait for what has been attracted to be dispersed, you will have nothing more. 999K

In this manner, a slave trader recently increased the buttocks of a rather thin child in a short time by using a moderately narrow incision either every day or on alternate days and applying pitch moderately. But for those who are thin in the whole body, bathing after food is also useful. However, just as with the thinning remedies, there is always the danger of a fever following when the body is immoderately overheated, so too, in those bathing after food, there is a danger of blockage occurring in the liver, and this particularly depends on the kinds of foods, for those that are otherwise thick-humored cause blockage if we use them to excess. Where people are accustomed by nature to do this under other circumstances, they will do it much more by bathing at the same time as taking nourishment.

In fact, the formation of stones in the kidneys also occurs due to the prolonged use of such a diet. It is, however, clear why this does not occur to everyone in that it is likely for some to have thickened kidneys or narrowed openings in the veins of the liver, and some to have the opposite. And 1000K there is no clear way of recognizing these variants. Instead, it is necessary to inquire of the person following the diet, as

[44] Hippocrates, *Office*, 13.

αὐτῷ βάρους αἴσθησις ἐν ὑποχονδρίῳ δεξιῷ καὶ κατὰ
τοὺς ψόας γίγνεται. κἂν αἴσθηταί ποτε τοιούτου τινός,
αὐτίκα δι᾿ ὀξυμέλιτος διδόναι κάππαριν ἐν ἀρχῇ τῆς
τροφῆς, ἄχρι περ ἂν καταστῇ τὸ βάρος. ἐπὶ δὲ τῶν
δυσκόλως ἀνατρεφομένων μορίων καὶ πλέον ἤδη κατ-
εψυγμένων ἐχρησάμην ἐνίοτε καὶ θαψίᾳ, ποτὲ μὲν
μετὰ μέλιτος ἐπιχρίων τὸ μόριον, ἔστι δ᾿ ὅτε καὶ
κηρωτῆς· ἐπισπᾶται γὰρ καὶ αὕτη τοῖς μορίοις οἷς ἂν
ἐπιτεθῇ πλῆθος αἵματος.

ἐφ᾿ ὧν δὲ ὀλίγον ἐνδεῖ τῷ δέρματι τοῦ αἰδοίου πρὸς
τὸ κατὰ φύσιν, ἐπὶ τούτων ἄνευ θαψίας πολλάκις μόνῃ
τῇ τάσει τὸ δέον εἰργασάμην, ἵνα χάρτου μαλακὴν
καὶ εὔτονον ἐν κύκλῳ περιελίττων ὑποκεχρισμένῳ τῷ
δέρματι κόμμεως. εὔδηλον δὲ δήπουθεν ὅτι καὶ τὸ τῆς
ἰνὸς πέρας ἐπικολλᾶν χρὴ διὰ κόμμεως τῷ ὑποβεβλη-
μένῳ ἄνω μέρει τῆς ἰνός· ἐν τάχει τε γὰρ ξηραίνεται
καὶ ἀλύπως σφίγγει. προϋποτιθέναι δὲ χρὴ τοῦ δέρ-
1001K ματος τῆς ποσθῆς ἐκ τῶν ἔνδον | μερῶν στρογγύλον
τι σύμμετρον, ὃ καὶ μετὰ τὸ κολλῆσαι τὴν ἶνα ῥᾳδίως
ἐξαιρήσεις. ἔνιοι δὲ τῶν διὰ θαψίας ἐπαγόντων τὴν
ποσθὴν τὸ στρογγύλον τοῦτο μολύβδινον ἐποίησαν
ὥσπερ τι σωληνάριον· εἶτ᾿ ἔξωθεν αὐτὸ περιτείνοντες
τὸ δέρμα τῆς ποσθῆς καταδοῦσιν ἱμάντι μαλακῷ. καὶ
γένοιτ᾿ ἄν ποτε καὶ τοῦτο χρήσιμον ἐφ᾿ ὧν ἐνδεῖ πολὺ
τοῦ δέρματος· εἰ δ᾿ ὀλίγον εἴη τὸ λεῖπον, ἀρκεῖ μόνον,
ὡς εἴρηται, τὸ χαρτίον ἐν κύκλῳ περιελιττόμενον· ἐν
αὐτῷ δὲ τῷ περιβάλλειν τε τῷ δέρματι καὶ κατα-

I said, if he has any sense of heaviness continuously in the right hypochondrium and in the loin muscles. If he does at some time have the sense of such a thing, then immediately give oxymel with capers at the beginning of food until the heaviness settles down. In those parts that are restored with difficulty, and some when they are already more chilled, I am accustomed to use thapsia[45] also, sometimes smearing the part with honey and at other times with a salve also, for this draws out the excess blood from the parts to which it is applied.

In those who have a minor lack of the skin of the penis in terms of what is normal, I have often brought about what is needed by stretching alone without thapsia, wrapping a soft and firm strip of papyrus in a circle around the skin smeared with gum.[46] It is, of course, clear that it is also necessary to smear the upper edge of the strip with gum by placing it underneath, for then it dries quickly and binds painlessly. It is necessary to set something moderately round under the skin of the penis from the parts within—something which you will also remove easily after 1001K gluing the strip [of papyrus]. However, some of those who lay this round object on the penis with thapsia make it of lead, like a small pipe, and then, stretching the skin of the penis, bind it with a soft cord externally. This is also useful sometimes in the case of those who have a major lack of skin. If there is a little remaining, it is enough just to wind the papyrus around in a circle, as was said. However, in placing this around the skin and gluing it, we must put the

[45] *Thapsia garganica*; see Dioscorides, IV.157.
[46] This is presumably a treatment of *leipodermos*; see Celsus, VII.25.1, LCL, vol. 3, pp. 420–23, and the pseudo-Galenic *Definitiones medicae,* XIX.445K; for a modern account see J. P. Rubin (1980), *Urology*, XVI, pp. 121–24.

κολλᾶν αὐτὸ προϋποκεῖσθαι χρὴ τὸ σωληνάριον. ἐγὼ
δὲ εἴωθα, εἰ καὶ μηδὲν τούτων παρῇ, τοῦ χάρτου
σύμμετρον ἑλίττων ἐνθεῖναι στήριγμα τοῦ περιβλή-
ματος, ἵν᾽ ὕστερον, ὅταν ἀκριβῶς παγῇ τὸ περιελιττό-
μενον ἔξωθεν ἐξαρθέντος τοῦ στηρίγματος, εὐκόλως
οὐρεῖν ὑπάρχῃ τῷ θεραπευομένῳ. πρόδηλον δ᾽ ὅτι καὶ
τοῦτο τὸ πάθημα τοῦ γένους τῶν νοσημάτων ἐστίν, ὃ
κατὰ πηλικότητα τοῦ κατὰ φύσιν ἐξέστηκεν, ἐνίοτε
μὲν ἀποσαπείσης τῆς ποσθῆς γινόμενον, ἐνίοτε δὲ ἐξ
ἀρχῆς ἔλαττον συγγενόμενον. ὑπάγεται δὲ καὶ χει-
1002K ρουργίας τρόπῳ διττῷ· ποτὲ μὲν ἄνω κατὰ τὴν | ἀρχὴν
τοῦ αἰδοίου τὸ δέρμα τεμνόντων κυκλοτερῶς, ἕνεκα
λυθείσης αὐτοῦ τῆς συνεχείας ἕλκεσθαι κάτω μέχρι
τοῦ σκεπάσαι τὴν καλουμένην βάλανον ὅλην· ἐνίοτε
δὲ ὑποδερόντων σμίλῃ κατὰ τὰ ἔνδον ἀπὸ τῆς κατὰ
τὴν βάλανον ῥίζης, εἶθ᾽ ἑλκόντων κάτω, κἄπειτα
δεσμευόντων, ὡς εἴρηται, μαλακῷ τινι.

λεχθήσεται δὲ περὶ τῶν τοιούτων τρόπων ἐπὶ προήκοντι λόγῳ, καθάπερ γε καὶ περὶ κολοβωμάτων·
οὕτως γὰρ ὀνομάζουσι τὰ κατὰ χεῖλος ἢ πτερύγιον
ῥινὸς ἢ οὖς ἐλλείποντα. μεθοδεύεται γάρ πως καὶ
ταῦτα· πρῶτον μὲν ὑποδερόντων ἑκατέρωθεν τὸ δέρμα,
μετὰ δὲ τοῦτο ἐπαγόντων καὶ συναγόντων ἀλλήλοις
τὰ χείλη τῶν δερμάτων, ἀφαιρούντων τε τὸ τετυλω-
μένον ἑκατέρου, κἄπειτα ῥαπτόντων τε καὶ κολλών-
των. ἐκ ταὐτοῦ δὲ τοῦ γένους εἰσὶ καὶ αἱ κατὰ τὸν

---

[47] Coloboma here appears to refer to both congenital and

small pipe under beforehand. If none of these things is available, I am accustomed, after winding the papyrus around moderately, to put in a support of what is placed around, so that later, after what surrounds is fixed in place accurately on the exterior, it is possible, when the support is removed, for the man being treated to pass urine easily. It is clear that this affection is also of the class of diseases in which there has been a departure in terms of magnitude from what accords with nature, sometimes occurring when there is putrefaction of the penis and sometimes when there is a congenital deficiency. Surgery is introduced in a twofold manner; sometimes above in relation to the origin 1002K of the penis, when doctors cut the skin in a circular fashion so as to draw it down after its connection is released until it covers the so-called glans in its entirety, and sometimes, after stripping off the skin with a scalpel internally from the root of the glans when this is drawn downward, to bind, as I said, with something soft.

I shall speak about such ways as the discussion proceeds, just as I shall also speak about the colobomata, for so they term deficiencies in the lip, the alae of the nostrils and the ears.[47] For these are also treated in some way by first stripping off the skin below on each side and, after this, leading on and drawing together the margins of the skin on each side, having removed what of each has become hard, and then suturing and conglutinating them. Also from this same class are the rhyades[48] involving the

aquired deficiencies involving different structures. The transliterated term has been retained, although it is now used primarily in connection with the eye.

[48] Rhyades, which is a term not now in use, appears to be an acquired tissue defect following eye surgery; see Celsus, VII.7.4.

μέγαν κανθὸν ῥυάδες, ἢ μειωθέντος ἐπὶ πλέον ἢ τελέως ἀπολλυμένου τοῦ κανθοῦ. τελέως μὲν οὖν ἀπολλυμένου παντάπασιν ἀνίατον γίνεται τὸ νόσημα, μειωθέντος δὲ διὰ τῶν μετρίως στυφόντων θεραπεύεται
μετὰ τοῦ προκαθᾶραι πρῶτον μὲν ὅλον τὸ σῶμα,
δεύτερον δὲ τὴν κεφαλήν. ἔστι δὲ μετρίως στύφοντα
1003K φάρμακα τά τε διὰ γλαυκίου καὶ κρόκου | καὶ τὰ
νάρδινα καλούμενα, καὶ μάλισθ᾽ ὅσα δι᾽ οἴνου σκευ-
άζεται. συνελόντι δὲ εἰπεῖν, ἐπὶ πάντων ἐν οἷς ἀπώ-
λετό τις οὐσία μάλιστα μὲν αὐτὴν ὁμοιοτάτην πειρᾶσθαι χρὴ κατασκευάζειν· εἰ δ᾽ ἀδύνατον εἴη τοῦτο,
τῆς γε αὐτῆς χρείας ἐστοχασμένων ἡμῶν· ἔσται δὲ
καὶ αὐτὴ κατὰ τοῦθ᾽ ὁμοία.

τοῦ γοῦν τῆς κνήμης ὀστοῦ πολλάκις ἀναγκασθέν-
τες ἐκκόψαι συχνόν, εἰς τὴν χώραν αὐτοῦ φῦσαί τινα
ἑτέραν οὐσίαν τὴν φύσιν προκαλούμεθα διὰ τῶν σαρ-
κωτικῶν φαρμάκων, ἥτις ἐν ἀρχῇ μὲν οἷά περ σκληρὰ
σάρξ ἐστιν, ὕστερον δὲ πώρου σκληροτέρου λαμβάνει
σύστασιν, καὶ τῷ χρόνῳ κρατυνθεῖσα πρὸς τὰς βαδί-
σεις ἀντ᾽ ὀστοῦ⁵ γίνεται ἐπιτηδεία. καὶ σκυταλίδας δὲ
δακτύλων ἐκκόπτοντες ὁρῶμεν ἐν τῇ χώρᾳ τῶν ἐκκο-
πεισῶν ἑτέραν οὐσίαν, οἵαν περ εἴρηκα γεννωμένην.
ὅτι δὲ καὶ φλέβας αἰσθητὰς ἐνίοτε γεννωμένας εἴδομεν εἴρηται πρόσθεν. αὗται μὲν οὖν οὐχ ὅμοιαι ταῖς
ἀπολωλυίαις, ἀλλ᾽ αἱ αὐταὶ γίγνεσθαι λέγοιντ᾽ ἄν,
ὥσπερ γε καὶ ἡ ἐν τοῖς κοίλοις ἕλκεσι σάρξ. ἡ δ᾽ ἐν τῷ
τυλοῦσθαι ταύτην οὐλὴ γεννωμένη δέρματι μέν ἐστιν
1004K ὁμοιοτάτη, δέρμα δ᾽ οὐκ ἔστι· | πυκνοτέρα γοῦν αὐτοῦ

greater canthus [of the eye] when the canthus is either reduced still more or destroyed altogether. When it is destroyed altogether, the disease becomes completely incurable; when it is reduced, it is treated by moderate
astringents along with prior purging, first of the whole
body and second of the head. The medications that are
moderately astringent are those made from the glaucium,
crocus, and the so-called nard, and particularly those that   1003K
are prepared with wine. In short, in all those cases in which
some substance is destroyed, we must especially attempt
to provide that which is most like [what is destroyed] while,
if this is not possible, we aim at the actual use and this will
be in relation to these same things.

At all events, often when we are forced to cut out a large
part of the bone of the shank, we call forth some other substance (in terms of nature) to grow up in its place by way of
flesh-producing medications. This is, in the beginning, a
kind of hard flesh and later takes on the consistency of a
harder callus, and over time is strengthened and becomes
serviceable for walking in place of bone. And when the
phalanges of the fingers have been cut out, we see in the
space of what is cut out, another substance generated of
the kind I have described. Also, I said before that I sometimes saw the perceptible veins regenerate. These are not
the same as those that have been destroyed, although they
might be said to become so, like the flesh in hollow
wounds. In being hardened, this scar that is generated is
very similar to skin but is not actually skin and seems to be   1004K

---

5 B (cf. ossis loco KLat); αὐτοῦ

511

φαίνεται διά τε τῆς ὄψεως καὶ τῆς ἁφῆς, καὶ μέντοι
καὶ τῷ λογισμῷ τεκμαιρομένοις ἐκ τοῦ μὴ φύειν τρίχας. ταῦτ᾽ οὖν ἔχων ἀεὶ πρόχειρα πρὸς τὰς θεραπείας
εὐπορήσεις ὧν σε χρὴ πράττειν.

17. Ἐμοὶ δ᾽ ἤδη καιρὸς ἐπ᾽ ἄλλα προϊέναι νοσημάτων εἴδη, κοινωνοῦντα τοῖς προειρημένοις. ὁ μὲν
οὖν ἕρπης ὀνομαζόμενος ἐκ τοῦ αὐτοῦ γένους ἐστὶ τοῖς
ἡλκωμένοις ἐρυσιπέλασιν, ἡ σαρκοκήλη δὲ τοῖς σκίρ-
ροις. ὀφίασις δὲ καὶ ἀλωπεκία καὶ ἡ πτίλωσις ἐκ τοῦ
γένους ἐστὶ τῶν νοσημάτων ἐν οἷς ἀπόλωλέ τι τῶν
κατὰ φύσιν, ὥσπερ γε καὶ ἡ μυρμηκία τῷ ὅλῳ γένει
παρὰ φύσιν ἐστί. τρίτη δ᾽ ἀπάντων αὐτῶν ἡ διαφορά.
τινὰ μὲν γὰρ ἐκ μεταβολῆς γίνεται τῶν στερεῶν
σωμάτων, ὡς ἡ μυρμηκία καὶ ἡ λεύκη καὶ ὁ ἀλφὸς καὶ
ὁ σφάκελος, ἐλέφας τε καὶ ψώρα καὶ λέπρα, τινὰ δ᾽
οὐδ᾽ ὅλως ὄντα πρότερον ὕστερα γίνεται, καθάπερ καὶ
ἡ μελικηρὶς, ἀθερώματά τε καὶ στεατώματα· καὶ καθ᾽
ἕτερον τρόπον ἕλμινθες καὶ ἀσκαρίδες καὶ κηρία·
καλοῦσι γὰρ οὕτω μὲν μακρὰν καὶ πλατεῖαν ἕλμινθα·
1005K καὶ πάντα τὰ πρόσθεν | εἰρημένα, ἃ κατά τινα τῶν
ἀποστημάτων εὑρίσκεται, πώροις, ἢ λίθοις, ἢ ὀστοῖς,
ἢ θριξίν, ἤ τισιν ἑτέροις τῶν τοιούτων ἐοικότα. προσέχειν οὖν ἀεὶ χρὴ τὸν νοῦν ἐπὶ πάντων τῶν παρὰ
φύσιν ἀκριβῶς ἐπισκοπούμενον ἐκ τίνος γένους ἐστίν·
εἴπερ γε τὴν πρώτην ἔνδειξιν ὁρμητήριον ἐσομένην
ἁπασῶν τῶν ἐφεξῆς ὀρθῶς εἴπομεν ἐκ τοῦ γένους
λαμβάνεσθαι.

512

thicker than skin both to sight and touch, and furthermore, to those who judge on the grounds of it not growing hair. Therefore, when you have these things always to hand, you will be well supplied in terms of the treatments which you ought to carry out.

17. It is time now for me to proceed to other kinds of diseases associated with those previously spoken of. Thus, what is called herpes is from the same class as ulcerating erysipelas while sarcocele is from the same class as the scirrhosities (indurations). Ophiasis, alopecia and ptilosis are from that class of diseases in which something which accords with nature is destroyed, just as sessile warts are in the whole class contrary to nature. There is a threefold differentiation of all these. Some arise from a change of solid bodies, as, for example, sessile warts, leuke, alphos, sphacelus, elephas, psora and lepra. There are others that did not exist at all before and later arise, like meliceris, atheroma and steatoma and, in another way, helminths, ascarides and ceria (for so they call the long and flat helminths), and there are all those things mentioned before 1005K which are found in some of the abscesses, like gravel, stones, bones, hair or certain other such things. Therefore, you must always direct your attention, in the case of all the things contrary to nature, to considering accurately from which class they are, if in fact I am right in saying the primary indication will be taken to be the starting point of all the things to follow from the class.

τὸν γοῦν ἕρπητα χολώδης γεννᾷ χυμός· ὥστε κατά
γε τοῦτο ταὐτοῦ γένους ὑπάρχειν ἐρυσιπέλατι, καὶ
τοῦτό γ᾽ αὐτοῦ μᾶλλον ἔτι τὸ ἡλκωμένον. διαφέρει δὲ
τῇ λεπτότητι τοῦ χυμοῦ· πάνυ γάρ ἐστι λεπτὸς ὁ τὸν
ἕρπητα γεννῶν, ὡς μὴ μόνον διὰ πάντων διέρχεσθαι
τῶν ἔνδον μορίων, ὁπόσα σαρκώδη τὴν οὐσίαν ἐστίν,
ἀλλὰ καὶ δι᾽ αὐτοῦ τοῦ δέρματος ἄχρι τῆς ἐπιδερ-
μίδος, ἣν μόνην ἀναβιβρώσκει τε καὶ διεσθίει τῷ
στέγεσθαί τι πρὸς αὐτῆς· ὡς εἴ γε καὶ ταύτην διεξείη
τοῖς ἱδρῶσιν ὁμοίως, οὐκ ἂν ἕλκος εἰργάσατο. κοινὸν
γὰρ δὴ τοῦτο τοῖς γιγνομένοις ἐκ χυμοῦ δακνώδους
1006K ἕλκεσιν, ἅπερ αὐτόματα | ἕλκη προσαγορεύουσιν,
ἴσχεσθαί τε καὶ βραδύνειν ἐν τῇ διεξόδῳ τὸν ἐργαζόμενον αὐτὰ χυμόν. τῷ δ᾽ ἧττον καὶ μᾶλλον ἕτερον
ἑτέρου χυμὸν ἤτοι λεπτὸν ἢ παχὺν ὑπάρχειν αἱ κατὰ
τὸ βάθος ἐν τοῖς ἕλκεσιν γίνονται διαφοραί. τούτου
τοῦ γένους ἐστὶ καὶ ἡ φαγέδαινα καὶ οἱ ἡλκωμένοι τῶν
καρκίνων. ἐφ᾽ ὧν ἁπάντων ἡ μὲν κοινὴ θεραπεία
κωλύσαντα τὸν ἐπιρρέοντα χυμὸν ἰᾶσθαι τὸ ἕλκος· ἡ
δ᾽ ἰδία καθ᾽ ἕκαστον ἔκ τε τῆς τοῦ μορίου φύσεως
εὑρίσκεται καὶ τῆς ἰδέας τε καὶ ποσότητος τοῦ χυμοῦ.
λεπτότατος μὲν οὖν ἐν τοῖς τοιούτοις χυμοῖς ἐστιν ὁ
τὸν ἑλκούμενον ἕρπητα γεννῶν· παχύτατος δὲ ὁ τὸν
καρκίνον· ἐφεξῆς δὲ τούτων κατά γε τὸ πάχος ὁ τὰς
φαγεδαίνας ὀνομαζομένας. ὧν ἰδέαι τινές εἰσι τά τε
χειρώνια καὶ τηλέφια καλούμενα.

καὶ ἤδη τινὲς ἄλλαι τοιαίδε προσηγορίαι γεγόνασιν, ἄχρηστοί τε καὶ περίεργοι· πρὸς γάρ τοι τὴν

Anyway, a bilious humor generates herpes, so this is of the same class as erysipelas and even more than this, the ulcerated [form] of erysipelas. The difference lies in the thinness of the humor; for that which generates herpes is particularly thin so as not only to pass through all the parts within, however fleshy they are in substance, but also through the skin itself up to the epidermis, which alone it corrodes and eats through by being covered somewhat by it, since if it were also to go through this like sweat, it would not create an ulcer. For this is certainly common to those ulcers arising from a biting humor which they call "spontaneous ulcers," the humor which creates these being held back and delayed in its passage. By virtue of one humor being less or more than another, or thin or thick, the differentiae in relation to depth arise in the ulcers. Of this class also are the phagedaenae and the ulcerations of the cancers. In all these, the common treatment is to cure the ulcer by preventing the influx of humor. The specific treatment for each will be discovered in terms of the nature of the part and the form and quantity of the humor. Therefore, the thinnest among such humors is what generates the herpetic ulceration, and the thickest is what generates the cancerous. Next of these, at least in terms of thickness, is what generates the so-called "phagadaenae." There are certain forms of these called chironian and telephian.[49]

And now certain other such terms have come into existence but they are useless and superfluous, for what we

1006K

[49] See Book 2, note 5.

θεραπείαν ἐπίστασθαι χρὴ τό τε πλῆθος τοῦ χυμοῦ
καὶ τὴν δύναμιν καὶ τὴν σύστασιν, οἷον εὐθέως ἐπὶ
τῶν ἑρπήτων, ἐπειδὴ λεπτός ἐστιν ὁ χυμός, ἐκ τοῦ
γένους ὢν δηλονότι τῆς ξανθῆς χολῆς, ὅταν ἀναδείρῃ
1007K τὴν ἐπιδερμίδα | διαφορηθείς, ἐπιτρέπει συνουλωθῆ-
ναι τῷ ἕλκει. ἐὰν μὲν οὖν φθάσῃ τις ἐκκαθάρας τὸ
σύμπαν σῶμα, μετὰ τοῦ τοῖς ἀναστέλλουσί τε καὶ
ἀποκρουομένοις τοὺς ἐπιρρέοντας χυμοὺς χρήσασθαι
φαρμάκοις ἰάσατο τὸν ἕρπητα. μηδέτερον δὲ ἐργασά-
μενος τούτων, ἀλλὰ μόνοις ἀρκεσθεὶς τοῖς ἐπουλοῦσι,
τὴν ἡλκωμένην ἐπιδερμίδα ταύτην μὲν ἰάσατο, τὴν
συνεχῆ δ᾽ αὐτῇ παθεῖν οὐκ ἐκώλυσεν. εἶτ᾽ αὖθις πάλιν
ἐκείνης ἐπουλουμένης ἢ συνεχὴς ἀναδέρεται, καὶ τοῦτ᾽
ἐπὶ πλεῖστον γίγνεται, καθάπερ ἕρποντος τοῦ πάθους
ἄχρι περ ἂν ὁ ἐργαζόμενος αὐτὰ χυμὸς ἐκκενωθῇ.

γυνὴ γοῦν τις ἐν Ῥώμῃ τῶν ἐπιφανῶν ἕρπητα κατὰ
τὸ σφυρὸν ἔχουσα πρῶτον μὲν ἐχρήσατο τῷ διὰ
φύκους φαρμάκῳ, τάχιστα δὲ ἐπουλωθέντος αὐτοῦ τὸ
συνεχὲς εὐθὺς ἐπιπολῆς ἀνεδάρη δέρμα, καθάπερ ἐξ
ἀποσύρματος. ᾧ πάλιν ἐπιτιθέντος τοῦ φαρμάκου τὸ
συνεχὲς αὖθις ἡλκώθη. καὶ τοῦτ᾽ οὐκ ἐπαύετο γιγνόμενον, ἀλλ᾽ ἧκεν ὕστερον ἡ ἕλκωσις ἐπὶ τὸ γόνυ,
πάντα μᾶλλον αὐτῆς παθεῖν ἑτοίμης οὔσης ἢ καθαρθῆναι χολαγωγῷ φαρμάκῳ. καὶ τοίνυν ὅπερ εἴωθεν ἐν
1008K τοῖς τοιούτοις γίγνεσθαι, | διὰ τὸ τοὺς πλείστους αἰ-
τιᾶσθαι τὰ ἀναίτια, καταγνοῦσα τοῦ διὰ φύκους φαρμάκου, τῶν ἄλλων τι προσφέρειν ἐκέλευσεν. ἐχρώ-

must know with regard to treatment is the amount of the humor, and its potency and consistency. For example, in herpes, since the humor is thin and clearly from the class of those humors of yellow bile, whenever it is dispersed, it excoriates the epidermis and allows the ulcer to be complete cicatrized. Therefore, if someone carries out prior purging of the whole body along with the use of medications holding back and repelling the flowing humors, he will cure the herpes. On the other hand, someone who does neither of these two things but is satisfied with scarring over alone, will cure this epidermal ulceration but will not prevent the epidermis contiguous with it from being affected. Then again, contrariwise, when that is being scarred over or excoriated continuously, this occurs to a marked degree, just as when there is an herpetic affection, until such time as the humor creating it is evacuated.

At all events, a certain well-known woman in Rome, who had herpes on the ankle, first used the medication made from seaweed, and although the ulcer very quickly became scarred over, the contiguous skin of the surface was immediately excoriated as if from an abrasion. When the medication was again applied to it, what was contiguous again became ulcerated. And this did not stop occurring, so later the ulceration reached the knee, since she was more ready to suffer anything rather than be purged by a cholagogic medication. And accordingly, for this is what customarily happens in such instances because most people blame what is blameless, she gave a judgment against the medication made from seaweed, and directed me to apply one of the other medications. Therefore, I

μεθα οὖν ἐφεξῆς τῷ διὰ σάνδικος. ὡς δὲ καὶ τοῦτο τὸ
μὲν ἡλκωμένον ἐπούλου, τὸ δ' ἑλκούμενον οὐκ ἐκώ-
λυεν, ἀνελήλυθε δὲ τὸ πάθος ἐγγὺς ἤδη τοῦ βουβῶνος
ὑπὸ τῆς ἀνάγκης βιασθεῖσα γάλακτος ὀρρὸν ὑπέσχετο λήψεσθαι. παρεμβαλόντες οὖν ἡμεῖς αὐτῷ λάθρα
σκαμμωνίας ἐλάχιστον, ἄκουσαν αὐτὴν ἐκκαθάραντες
ἐθεραπεύσαμεν.

οὗπερ οὖν ἕνεκα ταῦτα λέγεται πάλιν ἀναμνήσω.
τὸν κοινὸν σκοπὸν ἐπὶ τῶν ὑπὸ ταὐτὸ γένος ἁπάντων
νοσημάτων ἐπειδὰν λάβῃς, οὐκ ἀναιρήσεις μὲν αὐτὸν
ἐν τοῖς κατὰ μέρος, εἰς διαφορὰν δ' ἄξεις ἕκαστον
πρέπουσαν ταῖς τε διαθέσεσι καὶ ταῖς ἐργαζομέναις
αὐτὰς αἰτίαις, ὥσπερ ἐπὶ τῶν αὐτομάτων ἑλκῶν ἐδείχθη. κενώσεις γὰρ δηλονότι τὸν πλεονάζοντα χυμὸν
ἐνίοτε μὲν τῷ τὴν ὠχρὰν χολὴν ἐκκαθαίροντι φαρ-
μάκῳ· πολλάκις δὲ τῷ τὴν μέλαιναν, ἔστι δ' ὅτε μικτῷ,
χολήν τε ἅμα καὶ φλέγμα κενοῦντι, ὥσπερ ἐπὶ θατέρου τῶν ἑρπήτων, ὃν ἀπὸ τῆς πρὸς τὰς κέγχρους
1009K ὁμοιότητος ὀνομάζουσι | κεγχρίαν. οὗτος γὰρ οὐκ εὐ-
θέως ἕλκος ἐργάζεται καθάπερ ὁ ἕτερος, ἀλλὰ μικρὰς
πάνυ φλυκταίνας ὥσπερ κέγχρους, αἱ καὶ αὐταὶ τοῦ
χρόνου προϊόντος εἰς ἕλκος τελευτῶσι.

καί τισιν οὐκ ἀλόγως ἔδοξεν ἐπιμεμίχθαι τῇ χολῇ
φλέγματος ἐν τῷ τοιούτῳ πάθει. γίγνεται δέ ποτε καὶ
χωρὶς τῆς τοῦ παντὸς σώματος ἰσχυρᾶς κακοχυμίας
ἐν μέρει ἕλκη, ἃ θεραπεύομεν ῥᾳδίως ὑπὸ φαρμάκων
μικτὴν ἐχόντων δύναμιν ἀποκρουστικήν τε καὶ δια-

next used the medication made from sandyx.[50] This did also effect cicatrization in what was ulcerated but it did not prevent the ulceration, and, as the affection had already advanced almost to the groin, she was compelled by necessity to promise to take the whey of milk. I then stealthily inserted the smallest amount of scammony into it and, having purged her despite her unwillingness, cured her.

I shall mention again the reason for saying these things. Whenever you take the common indicator in all diseases under the same class, you will not remove this from the individual diseases but will relate each to the differentiae which are appropriate to the conditions and the causes bringing these about, just as was shown in the case of spontaneous ulcers. For obviously you will evacuate the excess humor, sometimes by a medication purging yellow bile, but often by one purging black bile, and sometimes by one that is mixed, purging bile and phlegm at the same time, as in the case of the other herpetides which they call "miliary" from the resemblance to millet seeds. This does not immediately create an ulcer like the other, but very small pustules like millet seeds, and these themselves, over a period of time, progress and end up as ulcers.

1009K

To some it does not seem illogical that there is a mixture of phlegm with bile in such an affection. Sometimes, however, apart from there being a severe *kakochymia* of the whole body, ulcers also occur in a part which we treat easily with medications having a mixed potency, both repulsive

---

[50] Also called cerussa; see Dioscorides, V.103, where it is described as having "a cooling, pore-stopping, mollifying, filling, attenuating, and moreover a gently repressing of excrescencies, and a cicatrizing faculty" (Goodyer, p. 636).

φορητικήν. ἀποκρούεται μὲν οὖν τά τε στύφοντα καὶ τὰ χωρὶς τοῦ στύφειν ψύχοντα· διαφορεῖ δὲ τὰ θερμαίνοντα. καὶ δῆλον ὅτι κατὰ μὲν τὴν γένεσιν τῶν ἑλκῶν ἐπικρατεῖν χρὴ τὰ τὴν ἀποκρουστικὴν δύναμιν ἔχοντα· μηκέτι δ' ἐπιρρέοντος τοῦ χυμοῦ τοῦ μοχθηροῦ τῷ μορίῳ, τὴν διαφορητικήν. τὴν μὲν γὰρ ὀλίγην κακοχυμίαν κἂν ἀπώσηταί τις αὐτὴν ἐπί τε τὰ σπλάγχνα καὶ τὰς μεγάλας φλέβας, οὐδὲν ἐργάσεται κακὸν αἰσθητόν· εἰ δὲ ἀξιόλογος ᾖ, εἴς τι κύριον ἐνίοτε κατασκήπτει μόριον, ὅταν γε μὴ διὰ ῥώμην τῆς φύ-
1010K σεως ἐκκαθαιρούσης τὸ σῶμα κενωθῆναι | φθάσῃ διὰ τῶν διαχωρημάτων, ἢ τῶν οὔρων ἢ καὶ διὰ τοῦ περιέχοντος ὅλον τὸ σῶμα δέρματος. ἐπὶ μὲν οὖν τῆς ὠχρᾶς χολῆς αἱ κενώσεις ἑτοιμότεραι, τὸ δὲ φλέγμα, καὶ μάλισθ' ὅσον αὐτοῦ παχύτερόν τε καὶ γλίσχρον ἐστίν, ὡσαύτως δὲ καὶ ἡ μέλαινα χολὴ δυσκόλως ἐκκενοῦται· καὶ διὰ τοῦτο δεόμεθα καθαίροντος φαρμάκου. κατὰ δὲ τοὺς ἕρπητας, ἐπειδὴ λεπτός ἐστιν ὁ τὸ πάθος ἐργαζόμενος χυμός, ἀρκεῖ καὶ λαπάξαι γαστέρα διὰ τῶν ἐπιτυχόντων, ἢ οὖρα κινῆσαι διὰ τῶν μετρίως οὐρητικῶν.

ἀλλ' ἐπειδὴ περὶ φαρμάκων μεθόδου γέγραπταί τι κἂν τοῖς ἔμπροσθεν, οὐκ ὀλίγα δὲ καὶ κατὰ τὰς ἰδίας αὐτῶν εἴρηται πραγματείας, ἄμεινον ἂν εἴη μηκέτ' ἐκτείνειν τὸν λόγον· ἱκανὰ γὰρ καὶ ταῦτα τῷ γε προσέχοντι τὸν νοῦν, οὐδὲ γὰρ ἡμεῖς αὐτὰ παρὰ τῶν Μουσῶν ἐμάθομεν. ἀλλ' ἡ τῶν πραγμάτων φύσις ἀνδρὶ συνετῷ καὶ φιλοπόνῳ καὶ γεγυμνασμένῳ τὸν

and discutient. The astringents repel and are cooling, apart from being astringent, while those that disperse are heating. And it is clear that, in relation to the genesis of the ulcers, those [medications] which have the repulsive potency must prevail, but when there is no longer flow of the bad humor to the part, those having the discutient potency must prevail. Even if a minor *kakochymia* is repulsed to the viscera and the great veins, it will do no perceptible harm. If, however, the *kakochymia* is major, it sometimes attacks an important part, at least whenever the body has not been evacuated beforehand by means of the strength of the natural purging occurring through the feces or   1010K
urine, or also through the skin surrounding the whole body. In the case of yellow bile, the evacuations occur more readily, whereas with phlegm, and particularly phlegm that is thicker and viscid, and in like manner too, black bile, they are evacuated with difficulty and, because of this, we need a purging medication. In relation to the herpetides, since the humor which creates the affection is thin, it is sufficient to empty the belly through proven measures or to stimulate urine flow with moderate diuretics.

But since I have written something about a method of medications in what has gone before, and have also spoken at length of this in the specific treatises on them,[51] it would be better not to extend the discussion further, for these are adequate, at least for someone who pays attention to them, because we do not learn these things from the Muses. But the nature of the matters suggests to a man who is intelligent, diligent and trained in his mind, what must be done

[51] See the three major pharmacological treatises listed in Book 13, note 16.

νοῦν ὑπαγορεύει τὸ ποιητέον· ὅταν δὲ καὶ τὰς ὁδοὺς
τῆς εὑρέσεως τὶς ὑφ᾽ ἑτέρου διδαχθείσας ἔχῃ, ῥᾷστον
αὐτῷ προσέρχεσθαι κατ᾽ αὐτάς. ἱκανὸν δέ σοι μαρτύ-
1011K ριον ἔστω τὸ τοὺς τοιούτους ἄνδρας | ἔργοις μεγίστοις
κοσμῆσαι τὴν τέχνην· ὅσοι δ᾽ ἀσύνετοι, μηδὲν εὑρίσκεσθαι πλέον αὐτοῖς, εἰ καὶ δι᾽ ὅλου τοῦ βίου
μυρίων ἔργων ἰατρικῶν αὐτόπται γίγνοιντο.

πάμπολλα γοῦν ἐπινοεῖται μέχρι τήμερον, οὐδέπω
τοῖς ἔμπροσθεν εὑρημένα· καθάπερ νῦν ἐπὶ Ῥώμης
ἐπενόησέ τις ἰᾶσθαι διὰ στόματος ἀκροχορδόνας τε
καὶ μυρμηκίας. ἀλλ᾽ ἐπὶ μὲν τῶν ἀκροχορδόνων, ὡς ἂν
ἐξεχουσῶν τοῦ δέρματος, οὐδὲν θαυμαστόν· τὸ δὲ τῶν
μυρμηκιῶν καὶ μάλισθ᾽ ὅσαι τελέως εἰσὶν ἰσόπεδοι τῷ
δέρματι, θαυμαστὸν ἐδόκει. ἀλλ᾽ ὅμως καὶ ταύτας
πρῶτον μὲν τῇ θέσει τῶν χειλῶν ὥσπερ βδάλλων
ἐπεσπᾶτό τε κἀκ τῆς ῥίζης ἐμόχλευεν· εἶτα τοῖς προσ-
θίοις ὀδοῦσιν παραλαμβάνων ἀθρόως ἐξέσπα. καὶ
μὴν καὶ διὰ μυρσίνης σμίλης καὶ διὰ τοῦ καλουμένου
σκολοπομαχαιρίου γεγυμνασμένος ταῖς χερσὶ ῥᾳδίως
ἄν τις ἐκκόψειεν αὐτάς, ἰδίαν ἐχούσας περιγραφήν, ᾗ
χωρίζονται τοῦ πέριξ δέρματος. ὡσαύτως δὲ καὶ διά
τινος ἰσχυροῦ πτεροῦ, περιτιθεμένου κυκλοτερῶς τῇ
μυρμηκίᾳ, ποιούμεθα τὴν ἄρσιν αὐτῆς. χρὴ δὲ σύμμετρον ἔχειν δηλονότι τὴν ἑαυτοῦ κενὴν σύριγγα τὸ
1012K πτερὸν τῷ πάχει μυρμηκίας, | ἵνα τε πανταχόθεν
αὐτὴν ἀκριβῶς σφίγγῃ, κἄπειτα περιστρέψαι μετὰ
τοῦ κάτω βιάζεσθαι· τάχιστά τε γὰρ ἂν οὕτω καὶ σὺν
αὐτῇ τῇ ῥίζῃ τὴν μυρμηκίαν ὅλην ἐκβάλλοις. εὔδηλον

and, whenever someone follows the paths of discovery having been taught by another, it is very easy for him to come to these paths. Let this be enough evidence for you that men like this are an adornment to our craft by their very great works. However, as for those who are devoid of understanding, nothing more will be discovered by them, even if, through their whole life, they happen to see with their own eyes countless medical undertakings. 1011K

At all events, very many things are contrived even today which had not yet been discovered by our predecessors, just as now in Rome someone had the idea of curing thin-necked and sessile warts with his mouth. Truly, in the case of the thin-necked warts, as they project from the skin, this is not surprising whereas, in the case of sessile warts, and especially those that are completely level with the skin, it does seem surprising. Nevertheless, in respect of these too, after first placing their lips in position, they draw them as if by sucking, and begin to dislodge them from the root, and then, taking hold of them with their front teeth, they remove them altogether. But also someone practiced with a convex scalpel or what is called a *scolopomachairion* (pointed surgical knife) might easily cut these out with his hands as they have a distinct outline by which they are demarcated from the surrounding skin. In the same way too, with a strong quill placed around the sessile wart in circular fashion, we effect its removal. It is necessary, clearly, for the empty channel of the quill to be equal in diameter to the sessile wart so that it encloses it accurately everywhere, 1012K and then the quill is rotated with a downward force, for in this way you very quickly remove the whole sessile wart along with its actual root. It is also clear that the margin of

δὲ καὶ ὅτι τὸ πέρας τοῦ περιγλύφοντος αὐτὴν πτεροῦ
λεπτόν τε ἅμα καὶ ὀξὺ καὶ ἰσχυρὸν εἶναι χρή. διὸ τά
τε τῶν παλαιῶν ἀλεκτρυόνων πτερὰ εἰς τοῦτο χρήσιμα καὶ μᾶλλον ἔτι τὰ τῶν ἀετῶν. ἀποτέμνειν δ' αὐτῶν
χρὴ πρὸς τῇ ῥίζῃ τοσοῦτον, ὡς περιλαβεῖν ἱκανῶς τὴν
μυρμηκίαν. εὐθὺς δὲ ἀπὸ τῆς ἀποτομῆς, εἰ καλῶς
γίγνοιτο, καὶ τὴν ὀξύτητα παρέξεις αὐτῷ. καὶ τοῦτ'
οὖν ὁ λογισμὸς εὗρεν, οὐ περίπτωσις.

ὅτι δὲ διὰ τῶν ἑλκόντων σφοδρῶς φαρμάκων ἀνα-
σπασθήσεται καὶ ὅτι διὰ τῶν σηπόντων νεκρωθήσε-
ται, τῷ λογισμῷ μέν τις εὗρε· θαρρήσας δὲ χρῆσθαι,
πρὸς τῆς πείρας ἐμαρτυρήθη. τινὰ μὲν γὰρ ἄντικρύς
ἐστι καὶ πρὸ τῆς πείρας πιστά, καθάπερ εἰ τύχοι τὸν
ἀρτίως ἐμπεπαρμένον ἐξελεῖν σκόλοπα καὶ τὸ βέλος,
ὅσα τε τοῖς ὀφθαλμοῖς ἐμπίπτει ψαμμία. τινὰ δὲ
ἐπινοεῖται μὲν ὑπὸ τοῦ λογισμοῦ, βεβαιοῦται δὲ ὑπὸ
1013K τῆς πείρας. ἕνεκα δὲ τοῦ ῥᾷον εὑρίσκειν σε | καὶ κατὰ
σαυτὸν ὁδοὺς τοιαύτας εἰς εὐπορίαν ἰαμάτων οὐκ
ὀκνήσω προσθεῖναί τι παράδειγμα τῶν ἐκ τοῦ προκειμένου γένους. ὀνομάζομεν δὲ αὐτὸ κατὰ τὸν ἀριθμὸν
τῶν μορίων. ἐπειδὴ γὰρ ἔνια μὲν ἐλλείπει, καθάπερ
ὁδοὺς ἢ δάκτυλος ἢ ῥινὸς πτερύγιον, ἢ ὠτός τι μόριον,
ἢ αἰδοίου δέρμα, τινὰ δὲ πλεονάζει, καθάπερ ἕκτος
δάκτυλος, αἵ τ' ἐξοστώσεις καλούμεναι καὶ τῶν ὀδόντων οἱ παραφυόμενοι τοῖς κατὰ φύσιν· ἐξελεῖν μέντοι
τὸ περιττὸν οὐδὲν χαλεπόν, ἕτερον δὲ γεννῆσαι τῷ
μηκέτ' ὄντι παραπλήσιον ἐπὶ τινῶν μὲν ῥάδιον, ἐπὶ
τινῶν δὲ χαλεπόν, ἐπὶ τινῶν ἀδύνατον. ἐὰν μὲν δὴ

the quill, as you trim it, must be thin and, at the same time, both sharp and strong. For this reason, the quills from old roosters are useful for this purpose and still more so those from eagles. You must cut toward the root of these to such a degree that you encompass the sessile wart sufficiently. If the quill is well prepared, you will immediately establish its sharpness by the cut. This is discovered by reason and not by trial and error.

That they will be drawn out strongly by drawing medications and that they will be necrosed by putrefying medications, someone discovers by reason whereas to use these things confidently is established by experience. Some things are utterly reliable even prior to experience; for example, to remove a stake or dart, if it should happen to have just become embedded, and grains of sand that fall on the eyes. Some things are understood by reason and confirmed through experience. So that you can discover more easily, and also for yourself, such paths to a wealth of cures, 1013K I shall not hesitate to add some example of these from the class under consideration. We name this in respect to the number of the parts. For when some are lacking, such as a tooth, or finger, or an ala of the nose, or some part of the ear, or the skin of the penis, or when some are in excess, like a sixth digit, or a so-called exostosis, or teeth which grow up beside those that accord with nature, it is not difficult to take away what is superfluous, whereas to create something similar to what does not yet exist is in some cases easy, in some cases difficult, and in some cases impossible. Certainly, if something fleshy is missing, it should

σαρκῶδες ᾖ τὸ λεῖπον, οὐ χαλεπῶς ἄν τις αὐτὸ γεννή-
σειεν· εἴρηται δ' ἔμπροσθεν ἡ μέθοδος ἐπὶ τῶν κοίλων
ἑλκῶν. ἐὰν δὲ ὀστοῦν ᾖ, αὐτὸ μὲν ἀδύνατον, ἀντ' αὐτοῦ
δὲ ἕτερόν τι σκληρὸν οὐκ ἀδύνατον ἐργάσασθαι. λέλεκται δέ τι καὶ περὶ τῆς τῶν φλεβῶν γενέσεως
ἔμπροσθεν, ὡς ἐνίοτε μὲν αἰσθητῶς εἴδομεν ἑτέρας
νέας γεννηθείσας, ἐνίοτε δ' ἅπαντα μηχανωμένων οὐκ
ἔφυσαν. |

1014K 18. Εἰ δὲ δάκτυλος ὁ λείπων ἤ τι τοιοῦτον εἴη,
παντάπασιν ἀδύνατος ἡ γένεσις αὐτοῦ. λέλεκται δέ τι
καὶ περὶ ποσθῆς ἔμπροσθεν, ὅπως ἄν τις αὐτὴν ἐργάσηται. τὰ δ' ἐπὶ ῥινὸς ἢ ὠτὸς ἢ χείλους ἐλλείποντα
γεννῆσαι μὲν ἀδύνατον, εὐπρεπῆ δ' ἐργάσασθαι δυνατόν, ἐὰν ὑποδείρας τις ἑκατέρωθεν τὸ δέρμα, κἄπειτα
συναγαγὼν κολλῆσαι δυνηθῇ. καθ' ἕτερον δὲ τρόπον
ἐκ τούτου τοῦ γένους ἐστὶ τῶν νοσημάτων ἀθερώματά
τε καὶ στεατώματα καὶ μελικηρίδες, ἀσκαρίδες τε καὶ
κηρία καὶ ἕλμινθες, οἵ τε κατ' ἄρθρα καὶ πνεύμονα
πῶροι, καὶ οἱ κατὰ νεφροὺς καὶ κύστιν λίθοι· κοινὸν
γὰρ ἐπὶ πάντων αὐτῶν ἡ τῆς οὐσίας γένεσις οὐκ
οὔσης προτέρον. ἔμπαλιν δὲ κατὰ τὰς ἀλωπεκίας καὶ
τὰς ὀφιάσεις καὶ τὰς πτιλώσεις, ἔτι τε τὴν φαλάκρω-
σιν, ἀπώλεια μορίου τινός ἐστι χρησίμως γεγονότος.
ὥσπερ οὖν ἐπὶ πάντων τῶν ἄλλων, ἐν οἷς γεννηθῆναί
τι βουλόμεθα, τὰς κινήσεις τῆς φύσεως ἀκωλύτους
ἐργαζόμεθα, κατὰ τὸν αὐτὸν τρόπον ἐπὶ τῶν ἀπολω-
λυιῶν τριχῶν. ἔργον γὰρ τῆς φύσεώς ἐστιν, ὥσπερ ἡ

not be difficult for someone to generate this, the method being that spoken of before in the case of hollow wounds and ulcers. If, however, it is bony, then it is impossible, although it is not impossible to create something else hard. What has been said before about the creation of veins is that sometimes we see other, new ones being generated perceptibly, although sometimes all of the methods devised do not generate them.

18. If a finger or something like this is lacking, creation is altogether impossible. I have also said something about the foreskin before; that is, how someone might create this, while the creation of what is lacking in the nose, ears or lips is impossible. It is possible to create something acceptable if you strip off some skin from either side and are then able, having brought it together, to conglutinate it. Different from this class of diseases are atheroma, steatoma, meliceris, ascarides, ceria, helminthes, the chalk stones in joints and lungs, and the stones in kidneys and bladder, for what is common in them all is the creation of a substance not previously in existence. Contrariwise, in the alopecias, ophiases and ptiloses, and even more, in baldness, there is destruction of some part which had been useful. Therefore, just as in all the other instances in which we wish something to be generated, we carry out the actions as though they were those of Nature unimpeded, in the same way [do we do so] in the case of hairs that are destroyed. For this is an action of Nature like the generation

1014K

τῆς σαρκὸς γένεσις ἐν τοῖς κοίλοις ἕλκεσιν, οὕτω καὶ |
1015K ἡ τῶν τριχῶν ἐπί τε τῆς κεφαλῆς καὶ τῶν βλεφάρων.

ἀναμνησθεὶς οὖν ὧν ἔμαθες ἐν τοῖς φυσικοῖς λόγοις περὶ τριχῶν γενέσεως, ἐξ ἐκείνων εὑρήσεις τὰς
τῆς ἀπωλείας αὐτῶν αἰτίας. ἐδείχθη δ' ὅτι τῶν διαπνεομένων χυμῶν ὅσον ἰλυῶδές ἐστιν ἐξ ἀρχῆς τ'
εὐθέως τὰς τρίχας ἐγέννησε καὶ τοῦ λοιποῦ καθ'
ὑπόφυσιν αὐξάνει. τοῦτ' οὖν, ὅταν ἤτοι γ' ἀπόληται
παντάπασιν ἢ μοχθηρὸν γενηθῇ, φθείρεσθαι τὰς τρίχας ἀναγκαῖόν ἐστι. καὶ γὰρ οὖν καὶ τὰ φυτὰ κατὰ
διττὴν αἰτίαν ἀπόλλυται, ποτὲ μὲν ἀποροῦντα τῆς
τρεφούσης ὑγρότητος, ἔστι δ' ὅτε οὐκ οἰκείῳ χρώμενα.
τελέως μὲν ἀπολουμένου τοῦ τρέφοντος χυμοῦ τὰς
τρίχας ἡ φαλάκρωσις γίνεται, μοχθηροῦ δ' ἀποτελε-
σθέντος αἵ τ' ὀφιάσεις καλούμεναι καὶ αἱ ἀλωπεκίαι.

πρῶτον μὲν οὖν εὑρήσεις ἐκ τῆς κατὰ φύσιν αὐτῶν
διοικήσεως τὴν παρὰ φύσιν αἰτίαν. ἐφεξῆς δὲ τῆς
θεραπείας εὐπορήσεις ἐνδεικτικῶς ὑπὸ τῶν κοινῶν
ἀγομένης σκοπῶν ἅπασι τοῖς προειρημένοις, ἐφ' ὧν
ἐκ κακοχυμίας ἤτοι γ' ἄλλο τι πάθος ἢ ἕλκος γίνεται·
κωλῦσαι μὲν γὰρ χρὴ τὸ ἐπιρρέον, ἐκδαπανῆσαι δὲ
1016K καὶ διαφορῆσαι τὸ φθάσαν ἐν τῷ πεπονθότι | μορίῳ
περιέχεσθαι. καθάπερ οὖν ἐπ' ἐκείνων ἑλκῶν καθάρ-
σεις τοῦ λυποῦντος χυμοῦ πρῶτον ἐγίγνοντό σοι, τὸν
αὐτὸν τρόπον ἐπὶ τῶν τριχῶν αὐτῶν, ἀπ' αὐτῶν δ'
ἄρξῃ τῆς θεραπείας, ἐπισκεψάμενος ἀκριβῶς ὁποία
τις ἡ χρόα γέγονε τοῦ δέρματος, ἐξ οὗ τὰς τρίχας
ὁρᾷς ἀπολλυμένας. εἰ μὲν γὰρ λευκοτέρα τοῦ κατὰ
φύσιν, ἐπὶ τὴν τῶν φλεγματωδῶν χυμῶν ἀφικνοῦ

of flesh in the hollow wounds and ulcers, and, in like manner, that of the hair in the case of the head and eyelids. 1015K

Having called to mind those things you learned about
the genesis of hair in the physiological books,[52] you will
discover from those the causes of their destruction. It was
shown that the transpired humors which are muddy from
the beginning immediately generate hair and increase the
underlying attachments of what remains. This [humor],
whenever it is altogether destroyed or generated badly, is
necessarily destructive to the hair. In addition, the roots
are also destroyed by virtue of a twofold cause: sometimes
there is a lack of the nourishing fluid and sometimes it is
not used properly. Finally, when the humor nourishing the
hair is washed away, baldness arises, while when it has
been rendered bad, the so-called ophiases and alopecias
arise.

First, therefore, from the regulation of those things in
accord with nature, you will find the cause of what is contrary to nature. Next, you will find a way of treatment indicatively brought on by the common indicators for all
the things previously mentioned, in the case of which,
from a *kakochymia*, either some other affection or an ulcer
arises. It is necessary to prevent the flow and to consume
and disperse what had previously come to be encompassed
in the affected part. Therefore, just as in those ulcers evac- 1016K
uations of the harmful humor are primary for you, in the
same way, in the case of hair itself, you will begin the treatment from these [humors], observing accurately the color
of the skin of the part where you see the hair being destroyed. If it is whiter than normal, you move to the purg-

---

[52] See *De usu partium*, Book 11, M. T. May (1968), pp. 530–
36.

κάθαρσιν· εἰ δ' ὠχροτέρα πως, ἐπὶ τὴν τῆς τοιαύτης
χολῆς, ὥσπερ γε καὶ εἰ μελαντέρα, τὰ τῶν μελανῶν
ἀγωγὰ δώσεις φάρμακα. πρὸς δὲ τὴν ἀκριβεστέραν
τῆς κακοχυμίας διάγνωσιν οὐ μικρὰ κἀκ τῆς προηγησαμένης διαίτης ὠφεληθήσῃ, μεμαθηκώς γε τίνα μὲν
ἐδεσμάτων τὸν μελαγχολικὸν ἀθροίζει χυμόν, τίνα δὲ
τὸν τῆς ὠχρᾶς χολῆς καὶ τοῦ φλέγματος.

ὅταν οὖν ἤδη θαρρήσῃς ὡς ἐπὶ καθαρωτάτῳ
σώματι, τὸν ἐν τῷ πεπονθότι δέρματι περιεχόμενον
χυμὸν ἐκδαπανήσεις τοῖς διαφορητικοῖς φαρμάκοις,
φυλαττόμενος ἐν αὐτοῖς οὕτω θερμὰ καὶ δριμέα προσάγειν φάρμακα ὡς ἑλκωθῆναι τὸ δέρμα. καὶ μέντοι καὶ
τὰ ξηραίνοντα σφοδρῶς φυλάττεσθαι χρή, μή πως
ἅμα τῷ μοχθηρῷ χυμῷ συνεκδαπανήσῃς καὶ τὸν
ἐπιρρέοντα χρηστόν, ὥσπερ ἐπὶ τῇ φαλακρώσει |
1017K γίνεται. ταῦτ' οὖν ἐννοήσας ἐγὼ πρῶτον ἔμιξα τοῖς
τὰς ἀλωπεκίας θεραπεύουσι φαρμάκοις βραχύ τι θαψίας. εἶτα προσέχων καθ' ἑκάστην ἡμέραν, ὅπως ἡ
τοῦ κάμνοντος φύσις ὑπ' αὐτοῦ διατίθεται, κἀπειδὰν
ἤτοι γε οἰδισκόμενον ἐπὶ πλέον ἢ ἀναδερόμενόν πως
ἴδω, τοῦ φαρμάκου μὲν ἀφίσταμαι κατ' ἐκείνην τὴν
ἡμέραν, ἐπαλείφω δὲ τετηκότι στέατι τὸ μόριον ἤτοι
γ' ὄρνιθος ἢ χηνός, ἐπειδὴ λεπτομερέστερα ταυτά
ἐστι καὶ κατὰ βάθος εἰσδύεται. κἄπειτα κατὰ τὴν
ὑστεραίαν, εἰ μὲν ἐπιμένοι τι τῶν εἰρημένων, ὁμοίως
ἐπαλείφω· μὴ μενόντων δὲ τῷ φαρμάκῳ χρῶμαι· ὅπως
δ' εἰς βάθος δύοιτο, προσανατρίβω τὸ δέρμα τῇ σιν-
δόνι μέχρι τοῦ σαφῶς ἐρυθρὸν γενέσθαι. εἴ γε μὴν

ing of the phlegmatic humors. If, however, it is in some way more yellow, move to the purging of bile of the same [color] just as, if it is blacker, you will give the medications that draw away the black [bile]. In respect of the more accurate diagnosis of the *kakochymia*, you will be helped to no small extent by the preceding regimen, having learned, at least, what foods gather together the melancholic humor, what the humor of yellow bile, and what that of phlegm.

Therefore, when you are already confident that, as in the case of the very purified body, you will consume the humor contained in the affected skin with discutient medications, taking care to apply those that are hot and pungent in these instances in such a way that the skin is not ulcerated. However, you should also be on your guard against medications that are excessively drying lest you somehow completely consume the good humor that is flowing toward it along with the bad humor, as happens with baldness. So, after   1017K
taking note of this, I first mix a little thapsia with the medications for treating the alopecias. Then, paying attention each day to how the nature of the patient is affected by this, whenever I see him either swelling to a greater extent or being excoriated in some way, I withdraw the medication on that day and smear the part with melted fat, either from a rooster or a goose, since these [fats] are more fine-particled and penetrate deeply. Then, on the next day, if something remains of those things spoken of, I smear in like manner, whereas if nothing remains, I use the medication. So that it might penetrate into the depths, I rub the skin with a cloth beforehand until it becomes obviously red. If, after bath-

μετὰ τὸ λουτρὸν χρῆσθαι βούλοιο τῷ τοιούτῳ φαρμάκῳ παρὰ τοῦ βαλανείου, τοῦθ' ἕξεις γινόμενον ὃ πρόσθεν ὑπὸ τῆς ἀνατρίψεως. καὶ τοὺς ἀπολλύντας δὲ τὰς ἐκ τῶν βλεφάρων τρίχας, οὓς ὀνομάζουσι πτίλους, ὁμοίοις μὲν τῷ γένει θεραπεύσεις φαρμάκοις.

19. Ἐκλέξῃ δ' ὕλην ἐπιτήδειον τοῖς ὀφθαλμοῖς μετὰ τοῦ φροντίζειν δηλονότι καὶ τοῦ μὴ παραρρεῖν
1018K ἔσω τὸ | φάρμακον εἰς τοὺς χιτῶνας αὐτῶν. ὅθεν ἀμείνω τὰ ξηρά, περὶ ὧν ὥσπερ καὶ τῶν ἄλλων ἐν ταῖς ἡμετέραις τῶν φαρμάκων πραγματείαις λέλεκται. νυνὶ γάρ, ὅπερ ἔφην, ἀρκεῖ μόνα τὰ γένη διεξέρχεσθαι τῶν φαρμάκων, ἄνευ τῶν κατὰ μέρος ὑλῶν, ὅπως μὴ πολλάκις ἀναγκαζοίμην ὑπὲρ αὐτῶν λέγειν. ὅσα μὲν τοίνυν ἀλλότρια τῆς κατὰ φύσιν ἐστὶ διοικήσεως, ἐξαιρεῖν ὅλα αὐτὰ προσήκει· ὅσα δ' οἰκεῖα μέν, ἀλλὰ διέφθαρταί πως, ἀνασῴζειν αὐτά, καθόσον ἐνδέχεται. λέλεκται δ' ἔμπροσθεν ὡς ἐπαμφοτερίζει τινά. λέλεκται δὲ ὡς καὶ τῶν ἀλλοτρίων ἔνια ταῖς οὐσίαις ὅλαις ἐστὶν ἀλλότρια, λογικῆς ζητήσεως οὐκ ἐς τὴν θεραπείαν χρησίμης οὔσης κατὰ τοῦτον τὸν τόπον ἐπ' ἐνίων παθῶν. τὸ γάρ τοι πτερύγιον ὅτι μὲν ἀλλότριόν ἐστι τῆς ὑγιεινῆς καταστάσεως εὔδηλον εἶναι νομίζω πᾶσιν· οὐ μὴν ἀλλότριόν γε κατὰ τὴν οὐσίαν ἐστίν, ὥσπερ ἀθέρωμα καὶ μελικηρίς. ἴασις δὲ καὶ τούτου μικροῦ μὲν ὄντος ἔτι καὶ μαλακοῦ διὰ τῶν ῥυπτόντων φαρμάκων, οἷά πέρ ἐστι καὶ τὰ τραχωματικὰ καλού-
1019K μενα· μεγάλου | δὲ καὶ σκληροῦ γενομένου διὰ χειρουργίας. ὁμοίως δὲ καὶ τῶν ὑδατίδων τὰς μεγάλας

ing, you wish to use such a medication near the bathhouse, you will have happen what was previously happening due to the rubbing. And with medications that are similar in class, you will cure what people call "ptili" which destroy the eyelashes.

19. Choose a material suitable to the eyes, with consideration clearly also that the medication is not one to run off into their tunics. On which account, the dry medications, 1018K about which I have spoken (as I also have the others) in the treatises on medications, are better. As I was saying, for the present it is enough to go over the classes of medications alone without [going over] the materials individually, so that I am not frequently compelled to speak about them. Therefore, it is appropriate to exclude all those things that are alien to the orderly arrangement that accords with nature. However, it is appropriate to preserve those that are compatible [with this arrangement] but have in some way been destroyed, as far as is possible. I have said before that some are both. I have also said that some of those that are alien are alien in their whole substance and on theoretical grounds are not useful for treatment in relation to this place, in some affections. For, certainly, that the pterygium is alien to the healthy state is, I think, clear to everyone, but it is not, in fact, alien in terms of substance, as atheroma and meliceris are. And the cure of this (i.e. pterygium), when it is still small and soft, is by cleansing medications, like the so-called "trachomatics," but when it becomes large and hard, [the cure] is by surgery. You must 1019K also treat the large hydatids in similar fashion, although the

533

θεραπευτέον· τὰς μικρὰς δὲ τὰ ξηραντικὰ τῶν φαρμάκων ὀνίνησι. τὸ δὲ χαλάζιον, ἔστι γὰρ ἕν τι καὶ
τοῦτο τῶν ἐν ὀφθαλμοῖς γινομένων, ὅλῳ τῷ γένει
παρὰ φύσιν ὑπάρχον ἐκκόπτεσθαι δεῖται. οὕτω δὲ καὶ
τὸ πῦον ὑπὸ τῶν ὑποπύων ὀνομαζομένων ὀφθαλμῶν·
ἀλλὰ τοῦτο μὲν ὡς τὰ πολλὰ διαφορεῖται φαρμάκοις.
ὑπόχυμα δὲ ἀρχόμενον μὲν διαφορεῖται, σύστασιν δ᾽
ἱκανὴν λαβὸν οὐκέτι.

τῶν καθ᾽ ἡμᾶς δέ τις ὀφθαλμικῶν Ἰοῦστος ὄνομα
καὶ διὰ κατασείσεως τῆς κεφαλῆς πολλοὺς τῶν ὑποπύων ἐθεράπευσε, καθίζων μὲν αὐτοὺς ὀρθίους ἐπὶ
δίφρου, περιλαμβάνων δὲ τὴν κεφαλὴν ἑκατέρωθεν ἐκ
τῶν πλαγίων, εἶτα διασείων οὕτως ὥσθ᾽ ὁρᾶν ἡμᾶς
ἐναργῶς κάτω χωροῦν τὸ πῦον. ἔμενε δὲ κάτω καίτοι
τῶν ὑποχυμάτων μὴ μενόντων, εἰ μὴ πάνυ τις ἀκριβῶς αὐτὰ σφηνώσειε διὰ τὸ βαρὺ τῆς οὐσίας. κου-
φότερον γάρ ἐστιν, ὡς ἂν εἴποι τις εἰκάζων, ἢ νεφε-
λωδέστερον τὸ ὑπόχυμα τοῦ πύου· πλὴν ὅσα καὶ
τούτων αὐτῶν ἔνια, λέγω δὴ τῶν ὑποχυμάτων, ὀρρω-
1020K δεστέρας | ὑγρότητός ἐστιν. ἃ δὴ καὶ περικεντούντων
διαλύεται μὲν ἐν τῷ παραυτίκα, χρόνῳ δ᾽ ὕστερον οὐ
μακρῷ καθάπερ τις ἰλὺς ὑποχωρεῖ κάτω. τὸ δ᾽ ἐν τοῖς
ὀφθαλμοῖς πῦον ὅταν διαφορῆσαι βουλώμεθα, τοῖς
διὰ σμύρνης μάλιστα κολλυρίοις χρηστέον, ἃ δὴ καὶ
καλοῦσιν ἰδίως διάσμυρνα· τούτων δ᾽ ἧττον, ἄμεινον
δὲ τῶν ἄλλων ἐνεργεῖ τὰ διὰ λιβάνου, τὰ δ᾽ ἱκανῶς

drying medications help those that are small. The chalazion, for this is one of the things occurring in the eyes, needs to be cut out, being contrary to nature in its whole class. The same applies also [in the case of] the pus due to eyes that are termed "suppurating," but in many instances, this is dispersed by medications. Hypochyma (cataract) in its early stages is dispersed, but when it is sufficiently established, this is no longer so.

Of the ophthalmic doctors in our own time, a certain Justus by name treated many of those with hypopyon (suppuration of the eyes) by shaking of the head, sitting them down on a couch in an upright position, encompassing the head on either side from the sides, then shaking [it] violently in such a manner that we might clearly see the pus move downward.[53] And, indeed, it stayed down, even though the hypochyma did not remain, unless particularly somebody were to plug it up with precision, because of the weight of the substance. It is lighter, as someone might say in making a comparison, the hypochyma being more cloudlike than pus. Apart from those, there are also some of them (I speak of the hypochymas) that are more wheylike than watery. Also, those that have been pierced on all sides are dissolved straightaway, although a short time later it is as if some slime passes off below. Whenever we wish to disperse the pus in the eyes, we must use particularly the collyriums made with myrrh, which they also call, in a special sense, *diasmyrnon*.[54] Inferior to these but better than others are those that act through frankincense.

1020K

[53] Little is known about Justus the Ophthalmologist; see the Introduction, section 2.   [54] See Galen, *De simplicium medicamentorum temperamentis et facultatibus*, XII.257K, and *De compositione medicamentorum secundum locos*, XII.806K.

ξηραίνοντα παραχρῆμα μὲν ἱκανῶς ἐκκενοῖ, τὸ δ'
ὑπόλοιπον πήγνυσι δυσλύτως, ὡς ἐπὶ τῶν σκιρρουμένων ἔμπροσθεν εἴρηται. πολλάκις δὲ πῦον ἀθρόως
ἐκενώσαμεν διελόντες τὸν κερατοειδῆ μικρὸν ὑπεράνω
τοῦ χωρίου, καθ' ὃ συμφύονται πρὸς ἀλλήλους ἅπαντες οἱ χιτῶνες. ὀνομάζουσι δὲ ἔνιοι μὲν ἴριν, ἔνιοι δὲ
στεφάνην τὸ χωρίον. ὥστε καὶ τοῦτο τὸ πάθημα τοῖς
τρισὶν ὑποπίπτει τρόποις τῆς κενώσεως καὶ διὰ χειρουργίας ἀθρόως ἐκκενούσης καὶ διὰ φαρμάκου κατὰ
βραχὺ καὶ πρὸς ἀκυρώτερον ἀπαγόμενον τόπον, ὡς ἐν
ταῖς κατασείσεσι.

　　τοῦ γένους δ' εἰσίν, ὡς ἔφην, τῶν ὅλαις ταῖς οὐσίαις παρὰ φύσιν ἐχόντων καὶ αἱ ἀσκαρίδες, αἵ θ'
1021K ἕλμινθες, | εἴτ' οὖν στρογγύλαι τινές εἰσιν εἴτε καὶ
πλατεῖαι. διὸ καὶ τελέως αὐτὰς ἐξαιρεῖν χρὴ τοῦ
σώματος. ἐξαιρήσεις[6] δὲ ἀποκτείνας. ἀποκτενεῖς δὲ
τοῖς πικροῖς φαρμάκοις· ζῶσαι μὲν γὰρ ἀντέχονται
τῶν ἐντέρων, ἀποθανοῦσαι δὲ συνεκκενοῦνται τῇ κό-
πρῳ. συνεκκενοῦνται δὲ ζῶσαι μὲν ἔτι σκοτωθεῖσαι
καὶ ὡς ἂν εἴποι τις ἡμιθνῆτες γινόμεναι. τὰς μὲν οὖν
στρογγύλας ἕλμινθας ἱκανὸν ἀποκτεῖναι ἀψίνθιον. ἡ
πλατεῖα δὲ ἰσχυροτέρων δεῖται φαρμάκων, ὁποῖόν
ἐστι καὶ ἡ πτέρις, ἔτι δὲ καὶ ἡ καλουμένη ἀσκαρίς.
ἀλλὰ περὶ μὲν τῆς τῶν φαρμάκων εὐπορίας οὐ νῦν
πρόκειται λέγειν. ἐνταῦθα οὖν ἤδη τελευτάτω καὶ
οὗτος ὁ λόγος.

　　[6] K; ἐξερήσεις (-ε- pro -αι-) B

Those that dry excessively immediately evacuate strongly but what remains is fixed indissolubly, as was said before in the case of the indurations. Often I evacuated the pus all together, having divided the external coat of the eye just above the tunic at the place where all the tunics grow together with each other. Some call the place the "iris," others the "crown." As a result, this affection also falls under three kinds of evacuation: suddenly by surgery, gradually by medication, and by being led off to a less important place as in the shakings.

Of the class of those things that are contrary to nature in their whole substance, there are also, as I said, the ascarides and the helminthes, some of which are round and some flat. On which account, it is necessary to remove them completely from the body. You will remove them once you have killed them, and you will kill them with bitter medications, for they live clinging to the intestines and will die when they are evacuated with the feces. When they are evacuated, they live still stupefied and, as someone might say, are half-dead. Wormwood is sufficient to kill the round helminths; the flat ones need stronger medications, such as *pteris*,[55] as does the so-called ascaris. But now is not an appropriate time to speak about the provision of the medications. Therefore, let this discussion come to an end here and now.

1021K

[55] See Dioscorides, IV.186, 187.

INDEX OF NAMES

Capitalized Roman numerals represent volume numbers. Lower-case Roman numerals indicate page numbers in the introduction. References to the text and translation are given by the Kühn page number in Arabic numerals.

540

INDEX OF NAMES

GENERAL INDEX

Capitalized Roman numerals represent volume numbers. Lower-case Roman numerals indicate page numbers in the introduction. References to the text and translation are given by the Kühn page number in Arabic numerals. Where *et passim* appears at the end of a subject, it applies to all volumes.

547

GENERAL INDEX

Aloes (ἀλόη), I.cxxvi; II.320–22,
337, 374, 382, 515–16;
III.857, 927, 971

Alopecia (ἀλωπεκία), I.cxiii, 82;
III.1004, 1014–15; treatment
of, III.1017

Alphos (ἀλφός), I.cxiii; III.1004

Alum (στυπτερία), I.cxxvi, 199

Ambrosia, I.cxxvi; II.585;
III.986

Anaesthesia (ἀναισθησία)/
Dysaesthesia (δυσαισθη-
σία), I.cxiii, 60, 96

Analogists, I.76

Anastomosis, I.233; II.311;
III.999

Anatomy, I.xxxiv, xxxvi; II.319,
348–49, 409, 411, 422;
III.730, 928, 934, 968

Andron, troche of, I.cxxvi;
II.330, 354–55, 405; III.981

Aneurysm (ἀνεύρυσμα), I.cxiii;
II.320, 335

Anger (θυμός), II.535, 552–55;
III.666, 670, 679, 688, 692,
841

Anthrax (ἄνθραξ), I.cxiii, 82;
III.980, 982

Anus: anal/perianal warts,
III.988; putrefaction of
wounds, II.325; wounds and
ulcers, II.380–83

Anxiety (φροντίς), II.535, 552–
55; III.666, 679, 685, 692,
841

Apagma, alternative term for
fracture, II.424

Apepsia (ἀπεψία), I.cxiii, 60, 82,
98, 102; II.513, 564, 571, 576,

579–80, 584, 601, 624;
III.667, 777, 821

Aphrogala (ἀφρόγαλα), I.cxxvi;
II.468

Aphronitron (ἀφρόνιτρον),
I.cxxvi; II.547, 569; III.913,
954, 966

Apnoea (ἄπνοια), I.cxiii, 60, 82

Apomel (ἀπόμελι), I.cxxvi–
cxxvii; II.567; III.756

Apoplexy (ἀποπληξία), I.cxiii–
cxiv, 72; II.638; III.931

Apples, II.330, 342, 466, 469,
573; III.793, 905–6, 911

Aristolochia (ἀριστολοχία),
I.cxxvii, 163–65, 177–78;
II.446, 569; III.994, 996

Aromatic reed (κάλαμος
ἀρωματικός), I.cxxvii; II.466

Arteries, I.47, 101; arteriotomy,
III.941; function of, I.lxi–lxii;
et passim. *See also* Blood vessels

Arthritis (ἀρθρῖτις), I.cxiv, 82;
II.513; III.803, 956, 994–95

Artomel (ἀρτόμελι), I.cxxvii;
III.692, 781, 796, 910

Ascarides (ἀσκαρίδες), I.cxiv;
III.1004, 1014, 1020–21

Ascites (ἀσκίτες), III.987

Asia/Asia Minor, I.xiii, 6; II.404,
454, 483; III.791, 832–33,
835, 980

Asparagus (ἀσφάραγος),
I.cxxvii; II.469

Asphalt (ἄσφαλτος), I.cxxvii;
II.213, 342; III.757

Ass's milk (ὄνου γάλα), I.cxxvii;
II.365, 474; III.727

548

GENERAL INDEX

Copper (burnt) (χαλκὸς κεκαυμένος), I.cxxix, 199, 298

Copper (flake) (λεπὶς χαλκοῦ), I.cxxix, 199, 298; III.927

Copper (flower) (ἄνθος χαλκοῦ), I.cxxx; III.927

Copperas water, II.325

Coriander (κορίαννον), I.cxxx; III.952

Corpuscles, I.xxxvii, xxxix, 117, 268–70

Coryza (κόρυζα), I.cxv; II.513, 554

Cos, I.xxv, xxxiv, 5–6

Cranium, III.788–89, 934; sutures of, III.933–34

Cretan Alexanders. *See* Alexanders

Crocus. *See* Saffron

Cucumber (σίκυος), I.cxxx; II.382; III.959, 996

Cumin (κύμινον), I.cxxx; II.578; III.964

Cupping glass, I.260; II.331–32; III.785, 798, 869–70, 896, 925–26, 929, 964; use in hemorrhage, II.316

Cynanche (κυνάγχη), I.cxvi; III.691, 932

Cyperus (κύπειρον/κύπερος), I.cxxx; III.911

Cyrenian sap, II.393

Date palm (Βάλανος φοίνικος), I.cxxx; II.466, 575

Debility (ἀρρωστία), I.91

Definition and terminology, I.128–56

Delirium (παρακοπή/

παραφροσύνη), I.cxvi, 82, 149, 220, 290; II.457; III.699–700, 930

Demonstration, I.30, 32–39, 76, 107–9, 112–13

Dialectics, I.5, 9, 18, 28, 30, 71

Diapedesis (διαπήδησις), I.lxxxvii, 233; II.311, 332

Diarrhea, I.158; II.513, 571, 575–76

Diaspermaton, I.cxxx; II.345, 372, 374

Diatritarians, II.582, 584; III.673. *See also* Three-day period

Diet, II.561–63; in fever, II.588–89; Hippocrates on, II.543–45, 591; in relation to *krasis*, II.592; thinning, II.332; et passim

Differentia (διαφορά), I.xci, 21–27, 123, 227; II.645–46; of diseases, II.531; of diseases and symptoms, I.15, 20; of things contrary to nature, I.78–79; et passim

Dill (ἄνηθον), I.cxxx; II.381, 548; III.726, 822, 848

Diphryges (διφρυγής), I.cxxx; II.356; III.927

Disease (νόσος/νόσημα), I.lxxii, lxxix–lxxx, 79–81, 89–91; classification of, I.84–86, 117–18, 125–26; compound, I.233; definition of, I.40–63, 71–75, 116; duration of, I.72–74; indications in, I.157–59; terminology of, I.127–28, 155; et passim

GENERAL INDEX

GENERAL INDEX

412–16; ulcers of, I.297;
wounds of, II.419; et passim
Iris (ἶρις), I.cxxxii, 163–65,
177–78; II.356, 466, 554, 567;
III.793, 911
Isis, I.cxxxii; II.349, 454
Italy/Italian, I.6; II.363, 483,
633; III.791, 834
Ivy (κισσός), II.330, 365

Jejunum (νῆστις), II.528;
III.699, 804, 922; wounds of,
II.419
Joint stones, III.1014

Kachexia (καχεξία), I.lxxxviii,
68, 263; III.670, 840, 953
Kakochymia (κακοχυμία),
I.lxxiii, lxxxviii, 68, 166, 186,
190, 227, 239, 242, 248–49,
256, 276, 291, 294; II.373–74,
376, 387, 392, 514, 516, 521,
527–28, 535, 564–65, 571,
583–84; III.788, 837, 850,
857, 882–84, 891–92, 1009,
1015
Kakoethical (κακοήθης), I.lxxiii,
lxxxviii–lxxxix, 247, 254, 257–
64; fevers, II. 536; III.731;
ulcers, I.250, 275–76, 289–
92; II.333
Kidneys, I.300; II.525; III.699,
731, 794; inflammation of,
III.903; stones of, III.999–
1000, 1009; et passim
Krasis (κρᾶσις), I.lxi, lxxxix,
105, 121, 174, 178–80, 182,
185, 195, 209–19, 235, 243,
295; II.336, 399, 463, 552–57,

563–64, 571, 583–84, 588,
607, 622, 646; III.671, 679,
681, 684–85, 690, 697, 724,
735–43, 748, 776, 779, 838,
842, 993–95; of ambient air,
II.647–48; recognition/diag-
nosis of, II.653–56; et passim

Ladanum (λήδανον), I.cxxxii;
II.466
Laodicea, I.xxxix; III.791
Lead (μόλυβδος), I.cxxxii, 196,
298
Leek (πράσον), II.548
Lemnian earth; I.298; II.329,
337
Lentiscinum (σχίνινος), III.791
Lepra (λέπρα), I.cxix; II.368;
III.1004
Lesbos, III.832–35, 996
Lethargy (λήθαργος), I.cxix,
139, 142; III.929–32
Leuke (λεύκη), I.cxix, 82;
III.1004
Leukophlegmatic dropsy
(λευκοφλεγμασία), I.cxix–
cxx, 82
Libya, I.263
Ligaments and tendons, I.216,
221, 232; induration of,
III.958–59; injury of, II.409–
10; structure of, II.409
Ligatures, I.230–31, 239;
II.333, 403, 411; materials of,
III.942–43; use in hemor-
rhage, II.317–20
Lime (τίτανος), II.325
Linseed (λινόσπερμον),
I.cxxxii; III.796, 868

558